INDUSTRIAL ELECTRICITY

In the manufacture of this book, the publishers have observed the recommendations of the War Production Board and any variation from previous printings of the same book is the result of this effort to conserve paper and other critical materials as an aid to war effort.

INDUSTRIAL ELECTRICITY

A SIMPLE TREATISE OF FUNDAMENTALS OF ELECTRICITY, MACHINES, AND CONTROLLING EQUIPMENT

By

JOHN M. NADON

SUPERVISOR OF ELECTRICAL ENGINEERING INSTRUCTIONS; APPRENTICE SCHOOL, FORD MOTOR COMPANY

And

BERT J. GELMINE, M. S.

INSTRUCTOR OF INDUSTRIAL ELECTRICITY; APPRENTICE SCHOOL, FORD MOTOR COMPANY

ELEVENTH PRINTING

NEW YORK

D. VAN NOSTRAND COMPANY, INC.

250 FOURTH AVENUE

First Published May 1939
Reprinted March 1940—January 1941
September 1941—January 1942—April 1942
May 1942—August 1942—September 1942
December 1942—May 1943

PRINTED IN THE U.S.A.
GEORGE S. FERGUSON COMPANY, INC.
PHILADELPHIA, PA.

PREFACE

For many years the course in elementary electricity given at the Apprentice School of the Ford Motor Company has been taught from lesson sheets prepared by the authors of this book. These sheets have been frequently revised, both to keep the material abreast of modern practice and to insure presentation of the subject in the simplest and most understandable manner. This book is the product of a further revision of the material contained in these sheets with such added information as the authors have deemed necessary.

The book is designed not only for industrial schools but also for vocational and technical schools, and its contents reflect the requirements of such schools. Specifically the authors have kept in mind the needs of those who intend to make electrical work in industry their vocation.

It is sometimes said that those who intend to engage in electrical work should specialize in the fundamentals and principles of electricity rather than accumulate facts about special equipment. However, fundamentals alone are not enough; the prospective electrical worker must also learn their application to present-day equipment. For this reason, this book covers not only the essentials of electricity but also devotes considerable attention to methods of connecting, to operating characteristics, and to the industrial applications of electrical machines and controls. Controllers are an important part of the book because an electrician in industry devotes most of his time to working with them.

Since numerous electronic devices are being used in industry and their uses are steadily increasing, electronic tubes and their electrical applications are discussed. One vital application of the theory of electronic tubes has to do with electric welding systems. In late years, electric welding has to a great extent superseded riveting and screw fastening and has brought about the use of a variety of welding equipment.

The book is intended to be elementary and yet to cover the field with thoroughness. The only mathematics prerequisite is a knowledge of arithmetic; such additional mathematics as is required for a comprehensive understanding of this book is treated in two special chapters.

The authors desire to thank the various manufacturers who have supplied information and diagrams and wish to express their appreciation for assistance received from Theodore Killian, Eugene F. McAuliffe, and C. Lee Dryden, instructors in the Electrical Department of the Apprentice School of the Ford Motor Company, for suggestions, for making drawings, and for proof reading. As this edition represents its first publication in book form, errors that have escaped the eyes of the authors are doubtless present in the text. The authors will be deeply grateful to those who report such errors for correction in future printings.

J.M.N.
B.J.G.

January 1939
Dearborn, Michigan

CONTENTS

CHAPTER 1 PAGE

ELECTRICAL ALGEBRA 1

Positive and Negative Numbers—Addition—Subtraction—Multiplication—Continued Products—Factors—Algebraic Expressions—Addition and Subtraction of Like Terms—Powers—Law of Exponents in Multiplication—Division—Roots—Subscripts—Equations—Parentheses—Factoring—Reading and Plotting of Curves.

CHAPTER 2

FUNDAMENTALS OF ELECTRICITY 21

Structure of Matter—Electrons—Kinds of Atoms—Direction of Electron Flow in a Conductor—Current—Voltage—Voltage and Current Measurement—Resistance—Ohm's Law—Series Circuit—Parallel Circuit—Series-Parallel Circuits—Electroplating—Cells—Grouping of Cells—Lead Storage Cell—Batteries — Internal Resistance — Charging Rate — Care of a Battery—Sulphation—Freezing—Work—Power—Efficiency—Measurement of Heat—Heating Effect of Electric Current—Torque—Transmission of Power.

CHAPTER 3

WIRING 61

Circular Mils—Mil-foot Wire—Square Mils—Resistivity—Wire Gages—Current Carrying Capacity of Conductors—Wiring Systems—Magnet Wire—Temperature Scales—Change in Resistance—Wiring Methods—Rigid Conduit—Benders—Right Angle and End Bends—Reaming—Offsets—Equipment Used in Conduit Systems—Cabinets—Boxes—Fittings—Multiple Conduit Runs—Wire Fishing—Size of Conduit—Electric Metallic Tubing—Flexible Conduit—Armored Cable—Non-Metallic Sheathed Cable—Surface Metal Raceway—Insulation—Flexible Cords—Underwriter's Knot—Switches.

CHAPTER 4 PAGE

MAGNETISM 100

Magnets—Direction of Magnetic Field—Properties of Magnetic Lines — Weber's Theory — Magnetic Screens — Compass — Earth's Magnetic Field—Thumb Rule—Magnetic Field about Two Parallel Conductors—Ohm's Law for Magnetic Circuits—Reluctance — Permeability — Flux Density — Series Magnetic Circuits—Parallel Magnetic Circuits—Number of Turns and Size of Wire — Hysteresis — Lifting Magnets — Magnetic Chucks, Clutches, Brakes, and Separator Pulley — Magnetic Pull.

CHAPTER 5

ARMATURE WINDING 130

Fleming's Right-hand Rule—Voltage in a Rotating Coil—Lap Winding—Pitch—Double Layer Winding—Multiple Coils—Position of Brushes—Duplex Lap Winding—Armature with Two Commutators—Equalizer Connections—Wave Winding—Dummy Coils — Multiplex Wave Winding — Paths — Frogleg Winding—Selection of Type of Winding—Generated Voltage—Winding Small Armatures—Loop, Split, Fractional Pitch, and V-Windings—Armature Testing—Growler Test.

CHAPTER 6

GENERATORS 170

Construction of Generators—Field Excitation—Armature Reaction—Compensating Armature Reaction—Commutation—Commutating Poles—Saturation Curve—Separately Excited Generator—Self-Excited Generator—Building Up of a Generator—Compound Generator—Generators in Parallel—Connecting a Generator to a Bus Line.

CHAPTER 7

MOTORS 197

Principle of the Motor—Fleming's Left-hand Rule—Torque of a Motor—Power Required to Drive a Generator—Counter Electromotive Force—Armature Reaction—Commutating Poles—Shunt, Series, and Compound Motors—Speed Adjustment.

CHAPTER 8 PAGE

CONTROLLERS 216

Types of Controllers—Magnetic Blowout—Magnetic Switch—Low Voltage and Overload Protection—Dynamic Braking—Counter E M F and Current Limit Starters—Reversing—Time Element Starters—Limit Switches—Time Current Relay—Hoisting—Lowering—Ward-Leonard System.

CHAPTER 9

GEOMETRY AND TRIGONOMETRY 260

Angles—Definitions—Trigonometry Tables—Solution of Right Triangles—Oblique Triangles—Law of Sines—Law of Cosines—Vectors—Resultant of Several Vectors.

CHAPTER 10

ALTERNATING CURRENT 283

Hydraulic Analogy—Advantages of Alternating Current—Generation of a Sine Wave—Effective Values—Vector Representation—Self-Inductance—Computation of Inductive Reactance—Condensers—Condensers in Series and in Parallel—Condensers in A C Circuits—Computation of Inductive Reactance—Calculation of Condenser Capacity—Dielectric Strength—Power in A C Circuits.

CHAPTER 11

A C CIRCUITS 318

Resistance and Inductance in Series—Resistance and Capacity in Series—Resistance, Inductance, and Capacity in Series—Impedance—Resistance, Inductance, and Capacity in Parallel—Series Parallel Circuits—Power Factor Correction—Cost of Power.

CHAPTER 12

TRANSFORMERS 337

Efficiency of Transmission—Transformer—Principle—Construction of Transformers—General Transformer Equation—Efficiency—Methods of Cooling Transformers—Tap Changer—Auto-Transformers—Current and Potential Transformers.

CHAPTER 13 PAGE

POLYPHASE SYSTEMS 355

Two-Phase and Three-Phase Alternators—Star and Delta Connections—Power in a Three-Phase System—Measurement of Power—Transformer Polarity—Transformers in Three-Phase Systems.

CHAPTER 14

ALTERNATORS 375

Revolving Field Alternators—Frequency—Forms of Alternator Windings—Alternator Characteristics—Paralleling Alternators—Synchroscope—Alternator Panel—Division of Load—Effect of Varying Field Excitation—Hunting.

CHAPTER 15

INDUCTION MOTORS 397

Rotating Magnetic Field—Speed—Torque—Starting Condition—Efficiency — Loading a Squirrel Cage Motor — Effect of a Higher Rotor Resistance — Wound Rotor Motors — Double Squirrel Cage Rotor—Adjustable Speed Motors—High Frequency Motors—Frequency Converter—Synchronous Motors—Starting Synchronous Motors—Varying Load and Field Excitation of Synchronous Motors—Uses of Synchronous Motors—Single-Phase Motors—Single-Phase Stator Winding.

CHAPTER 16

STATOR WINDING 435

Single Line Diagrams—Long and Short Jumpers—Three-Phase Winding—Parallel Connections—Double Voltage Winding—Unequal Coil Grouping — Change in Voltage, Speed, and Frequency.

CHAPTER 17

MULTI-SPEED INDUCTION MOTORS 453

Constant Horsepower, Constant Torque, and Variable Torque Multi-Speed Induction Motors — Multi-Speed Synchronous Motors.

CHAPTER 18

PAGE

A C CONTROLLERS 465

Across-the-Line Starters — Reversing Controllers — Controllers for High Torque Starting—Synchronous Motor Controller—Friction Brake — Dynamic Braking — Plugging — Star-Delta Starting—Definite Time Relays—Resistance Starter—Controllers for Wound Rotor Induction Motors—Compensators—Oil Switches—Starting Several Motors from a Single Compensator—Controllers for Two-Speed, Three-Speed, Four-Speed Motors—Reversing Multi-Speed Motors.

CHAPTER 19

ELECTRICAL INSTRUMENTS 504

Permanent Magnet and Electrodynamometer Type Instruments—Wattmeter — Polyphase Wattmeter — Iron Vane and Inclined Coil Instruments—Induction Type Instrument—Hot Wire and Thermocouple Instrument—Wheatstone and Slide Wire Bridges —Megger—Direct Current and Alternating Current Watthour Meters—Three-Phase Watthour Meter—Power Factor Meter —Synchroscope — Frequency Meter — Voltage and Power Relays—Power Directional Relays—Phase Balance Current Relay —Differential Protection—Alternator and Motor Differential Relays—Transformer Differential Relay.

CHAPTER 20

ELECTRON TUBES 541

Vacuum Tubes — Tube Rectifiers — Copper Oxide Rectifiers — Three-Electrode Vacuum Tubes—Plate Resistance—Plate Current—Hot Cathode, Cold Cathode, and Mercury Pool Cathode Gas Filled Tubes—Phototube—Industrial Uses of Phototubes—Temperature Control—Photocells.

CHAPTER 21

WELDING SYSTEMS 560

Arc Welding — Methods of Reducing Current Fluctuations — Welding Generators — Automatic Arc Welding — A C Arc Welding—Spot and Butt Welding—Control for Portable Spot

CHAPTER 21 (*Continued*) PAGE

WELDING SYSTEMS (*Continued*)

Welders—Automatic Spot Welder—Constant Heat Welding—Seam Welding—Mercury Pool Electron Tube Control.

CHAPTER 22

ELECTRICAL DRAWING 587

Symbols—Main and Simplified Wiring Diagrams—Control Sequence Table — Wiring Plans — Conduit Chart — Single Line Diagrams.

CHAPTER 1

ELECTRICAL ALGEBRA

IT MUST be realized that a thorough understanding of equations is essential if one intends to master the fundamentals of electricity. They are used extensively in electrical work. Equations may be completely understood, however, only when one is familiar with certain fundamentals of algebra. These will be considered first.

Positive and Negative Numbers—In arithmetic the + and − signs are merely symbols of operation, which denote addition and subtraction. In algebra their meaning is extended.

The scale of the thermometer extends in both directions from zero, indicating numbers of opposite nature (see fig. 1-1). The thermometer may indicate 7° above zero or 7° below zero. To distinguish between the two readings it is convenient to call the readings above zero, positive, and those below zero, negative. Then 7 degrees above zero is written +7, and 7 degrees below zero is written −7. Thus we have a new meaning for the + and − signs. They not only indicate addition and subtraction, but also the direction the quantity is from zero. If there is no sign preceding the number, the + sign is understood.

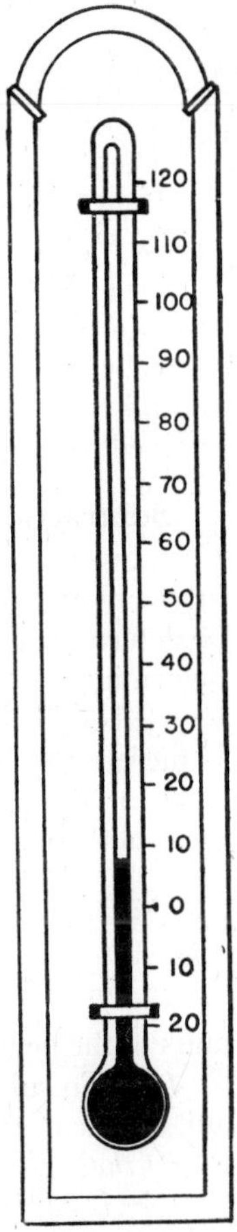

FIG. 1-1

Addition—Algebraic addition is the process of combining positive and negative numbers. This is easily done by means of a number scale, which is shown in fig. 1-2.

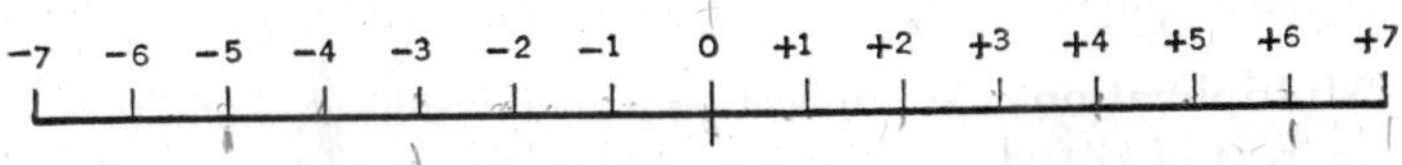

FIG. 1-2

(1) To add +5 and +6, begin at +5 and count 6 units in the positive direction, arriving at +11. Then $(+5) + (+6) = +11$.

(2) To add −5 and −6, begin at −5 and count 6 units in the negative direction, arriving at −11. Then $(-5) + (-6) = -11$.

(3) To add −5 and +6, begin at −5 and count 6 units in the positive direction, arriving at +1. Then $(-5) + (+6) = (+1)$.

(4) To add +5 and −6, begin at +5 and count 6 units in the negative direction, arriving at −1. Then $(+5) + (-6) = (-1)$.

These illustrative examples suggest the following rules:

Rule 1. *To add two numbers of like signs, add the numbers as in arithmetic and prefix the common sign to the result.*

Rule 2. *To add two numbers of unlike signs, subtract the smaller from the larger and prefix the sign of the larger number to the result.*

When more than two numbers are added, the best procedure is to add all the positive numbers, then all the negative numbers, and then proceed as explained in Rule 2.

Subtraction—In the illustrative addition examples, two numbers were given and their sum computed. In subtraction the sum of two numbers and one of the numbers are given. The other number is to be determined. If it is desired to subtract, for instance +2 from +6, determine what number added to +2 gives +6. This procedure may be used with both positive and negative numbers. Thus, subtraction is merely the inverse of addition.

(1) To subtract +2 from +6, begin at +2 and count to +6. Four units in the positive direction are counted. Then $(+6) - (+2) = +4$.

(2) To subtract −2 from +6, begin at −2 and count to +6. Eight units in the positive direction are counted. Then $(+6) - (-2) = +8$.

(3) To subtract +6 from −2, begin at +6 and count to −2. Eight units in the negative direction are counted. Then $(-2) - (+6) = -8$.

(4) To subtract −2 from −6, begin at −2 and count to −6. Four units in the negative direction are counted. Then $(-6) - (-2) = -4$.

In these illustrations note that subtracting any positive or negative number is equivalent to adding the same number with its sign changed.

Rule 3. *To subtract one number from another, change the sign of the number to be subtracted and proceed as in addition.*

Multiplication—Multiplication is merely repeated addition. It is the process of taking one number as many times as there are units in the other. Four possible cases arise.

1. $(+3) \times (+5)$ means add $+5$ three times, and this gives $(+5) + (+5) + (+5) = +15$. Then $(+3) \times (+5) = +15$.

2. $(+3) \times (-5)$ means add -5 three times, and this gives $(-5) + (-5) + (-5) = -15$. Then $(+3) \times (-5) = -15$.

3. $(-3) \times (+5)$ means subtract $+5$ three times, but subtracting $+5$ is the same as adding -5 (see Rule 3). Therefore, $-(+5) - (+5) - (+5) = (-5) + (-5) + (-5) = -15$. Then $(-3) \times (+5) = -15$.

4. $(-3) \times (-5)$ means subtract -5 three times, but subtracting -5 is the same as adding $+5$. Therefore, $-(-5) - (-5) - (-5) = (+5) + (+5) + (+5) = +15$.

In these illustrations when both numbers are positive or both negative, their product is positive; when one number is positive and the other is negative, their product is negative.

Rule 4. *The product of two numbers with like signs is a positive number, and the product of two numbers with unlike signs is a negative number.*

Continued Products—To find the product of three or more numbers, find the product of the first two, and then multiply the result by the third number. Continue this procedure until all the numbers have been used. In an example such as $(+2) \times (-3) \times (-5)$, $+2$ is multiplied by -3, giving -6. Their result is then multiplied by -5, giving $+30$. Changing the order of the multiplication does not change the result.

Factors—A factor of a product is any one of the numbers which multiplied together give the product. In $2 \times 3 \times 6 = 36$, the numbers 2, 3, and 6 are the factors.

Exercises

Perform the indicated addition:

1. $(+8) + (-4)$
2. $(-9) + (+6)$
3. $(-8) + (-2)$
4. $(+5) + (-12)$
5. $(-4) + (+7)$
6. $(-16) + (+7)$
7. $(+8) + (-2) + (-3)$
8. $(-4) + (+3) + (-4)$

Determine the missing values in the following:

9. $(+8) + (?) = +4$
10. $(-10) + (?) = -17$
11. $(?) + (+12) = -3$
12. $(-8) + (?) = -13$

Perform the indicated subtraction:

13. $(+6)-(+3)$
14. $(-8)-(+4)$
15. $(-11)-(-17)$
16. $(+12)-(-6)$
17. $(-9)-(+12)$
18. $(-13)-(+6)$

Find the following products:

19. $(+3)(-4)$
20. $(-5)(+8)$
21. $(8)(-1)$
22. $(-3)(-2)$
23. $(+9)(-7)$
24. $(-3)(-6)$
25. $(-3)(8)(-2)$
26. $(-5)(+2)(+4)$
27. $(-3)(-4)(-6)$

In algebra letters are frequently used instead of numbers. By means of letters it is possible to express, in general, the relation between certain quantities regardless of their numerical values. This is best illustrated by means of a few examples.

EXAMPLE—The cross section of a bus-bar may be found by multiplying its width by its thickness. Letter A may be used to represent the area, W the width, and T the thickness. This rule may be expressed as

$$A = W \times T$$

If W is ¼ and T is 6, $A = \frac{1}{4} \times 6$ or $1\frac{1}{2}$.

EXAMPLE—The circumference of a circle is found by multiplying 3.1416 by the diameter. This may be expressed as

$$C = 3.1416 \times D$$

where C = the circumference and D = the diameter.

If D is 5, $C = 3.1416 \times 5 = 15.708$.

If the sign of operation is omitted, multiplication is understood. For instance, the above two rules are generally written $A = WT$ and $C = 3.1416D$.

Algebraic Expressions—*Any group of numbers and letters joined by signs of operation, such as $E - 2R - 10$, is called an expression. The parts of an expression connected by plus or minus signs or both are called terms.* In the expression $E - 2R - 10$, the terms are E, $-2R$, and -10. Terms with like literal factors, such as 51R, 21R, and 31R, are called like terms.

Addition and Subtraction of Like Terms—Adding 5 two's, 3 two's, and 10 two's gives 18 two's. This may be expressed:

$$\begin{array}{r} 5 \times 2 \\ 3 \times 2 \\ 10 \times 2 \\ \hline 18 \times 2 \end{array}$$

Using letters instead of the common numbers, we have

$$\begin{array}{r} 5A \\ 3A \\ 10A \\ \hline 18A \end{array}$$

The common factor A is brought down and the coefficients 5, 3, and 10 are added and the sum written as the coefficient in the answer.

In arithmetic we learned that unlike things, such as 8 batteries and 2 motors, cannot be added. In the same way, unlike terms, such as 8b and 2m, cannot be added; they can only be indicated, as 8b + 2m. Observe the following:

Examples in Addition			Examples in Subtraction		
4ab	12ab	−16IR			
−6ab	4c	3IR	4IR	14ab	16IR
16ab	6ab	5I	−6IR	−4c	3I
14ab	18ab + 4c	−13IR + 5I	10IR	14ab + 4c	16IR − 3I

Rule 5. *Terms with the same literal expression can be added or subtracted by adding or subtracting the numerical coefficient. If the literal expressions are not alike, the operation can only be indicated.*

Powers—If a number is taken several times as a factor, the number is said to be raised to a power. Thus $2 \times 2 \times 2$ may be expressed a shorter way by writing 2^3, which means that 2 is taken as a factor three times and equals 8. 8 is the third power or cube of 2. 2^3 is read 2 to the third power, or 2 cubed. 2^2 means 2×2 and is read 2 to the second power or 2 squared.

A number written to the right and a little above another number to indicate the number of times the latter is taken as a factor is called an exponent. The exponent 1 is never written, since 2^1 is the same as 2.

Exponents may be affixed to letters as well as numbers. Thus aaaa may be expressed a^4. In a term such as $2a^2$, the exponent applies only to the letter. The exponent may be made to apply to the number also by enclosing in parentheses the letter and the number, as $(2a)^2$.

Law of Exponents in Multiplication—Since a^2 means aa, and a^3 means aaa, then $a^2 \times a^3 = (aa)\ (aaa) = a^5$. That is, $a^2 \times a^3 = a^5$. From this and all similar problems the following rule can be stated.

Rule 6. *The product of two or more powers of the same base is equal to the base with an exponent equal to the sum of the exponents of the powers.*

When terms with different bases are to be multiplied, their product can only be indicated. Thus $a^3 \times y^2 = a^3y^2$. When several different bases occur in each factor, the powers with the same base can be combined. For instance, $2a^2b^3c \times 3a^3b^4d = 2 \times 3a^2a^3b^3b^4cd = 6a^5b^7cd$.

Division—In the division of positive and negative numbers, the laws of signs and of exponents may be developed from a study of the laws that apply to signs and exponents in multiplication. Division is the inverse of multiplication.

Since $(+4) \times (+3) = +12$, then $(+12) \div (+3) = +4$.

Similarly, since $(-4) \times (+3) = -12$, then $(-12) \div (+3) = -4$.

Also, $(-12) \div (-4) = +3$.

Again, since $(-4) \times (-3) = +12$, then $(+12) \div (-3) = -4$.

From these examples, the law of signs for division may be stated as follows:

Rule 7. *The quotient of two numbers with like signs is a positive number, and the quotient of two numbers with unlike signs is a negative number.*

Law of Exponents in Division—Since in multiplication, $a^3a^4 = a^7$, then $a^7 \div a^4 = a^3$, and $a^7 \div a^3 = a^4$.

The division of powers may also be performed by cancellation. Since $a^5 = aaaaa$, and $a^3 = aaa$, then

$$\frac{a^5}{a^3} = \frac{\cancel{a}\cancel{a}\cancel{a}aa}{\cancel{a}\cancel{a}\cancel{a}} = a^2$$

Rule 8. *In the division of powers with the same base, the exponent in the quotient equals the exponent in the numerator minus the exponent in the denominator.*

When several different bases occur in each factor, the division may be performed by cancellation. Since $8a^3b^2$ means 8aaabb, and $2a^2b$ means 2aab, then

$$\frac{8a^3b^2}{2a^2b} = \frac{\overset{4}{\cancel{8}}\cancel{a}\cancel{a}a\cancel{b}b}{\cancel{2}\cancel{a}\cancel{a}\cancel{b}} = 4ab$$

Since this process is too long for rapid work, the quotient should be found by subtracting the exponents. Thus,

$$\frac{8a^3b^2}{2a^2b} = 4a^{3-2}b^{2-1} = 4ab$$

Root—If a quantity can be divided into two equal factors, one of the factors is called the square root of the quantity. Since $25 = 5 \times 5$, then 5 is the square root of 25. Square root, which is indicated by the sign $\sqrt{}$, is the inverse of the square. For instance, $5^2 = 25$ and $\sqrt{25} = 5$. The expression $\sqrt[3]{}$ indicates that the cube root is to be taken, that is, one of the three equal factors, as $\sqrt[3]{27} = \sqrt[3]{3 \times 3 \times 3} = 3$. The roots of a few quantities are given below:

$$\sqrt{a^2} = a;\ \sqrt{a^2b^2} = ab;\ \sqrt{9a^2} = 3a;\ \sqrt[3]{64} = 4;\ \sqrt[3]{a^3b^3} = ab;\ \sqrt[3]{8a^3} = 2a$$

Subscripts—Different quantities of the same thing are often represented by the same letter with a number or letter to the right and a little below it. For instance, when an equation involves two different diameters, one may be represented by D_1 and the other by D_2. These are read D sub 1 and D sub 2. Other examples are X_L, R_1^2, and $\sqrt{Z_R}$. These are read X sub L, R sub 1 squared and the square root of Z sub R. Subscripts, unlike exponents, possess no numerical significance. They merely indicate that the numerical values of the quantities are different. Thus, R_1 and R_2 may be as different in values as A and B.

Exercises

Perform the indicated operation:

1. $(+35I) + (-30I)$
2. $(-16L) - (+2L)$
3. $(+3R) + (-R) + (-7R)$
4. $5a + 7 - 3a$
5. $12xy - 7xy + 3x$
6. $14ac - 9xy + ac - 2xy$
7. If $R = 3$ and $I = 5$, determine the value of R^3, I^2, $5R^2$, and 2^3I.

Multiply:

8. a^7a^3
9. $(2a^2b)(-4abc)$
10. $(xy)(-6xy)$
11. $(4x^2y^2z)(2xy)^3z$
12. $(IR)^2(I^2R^2)R$
13. $a - 3$ by a
14. $E + IR$ by I
15. $m^2 - 2mn + 4n^2$ by $-2mn$
16. $-4a^2 - 10ab + 40a^2$ by $2ab^2$

Perform the indicated division:

17. $(+32) \div (-4)$
18. $(-25) \div (+5)$
19. $(-36) \div (-4)$
20. $(+12a^4b) \div (-3a)$
21. $(-2x^2y) \div (-4xy^2)$
22. $x^5 \div x$
23. $\dfrac{R + Rat}{R}$
24. $\dfrac{16np^2 - 36n^2p^2}{2np^2}$

25. Determine: $\sqrt{9}$; $\sqrt{E^2}$; $\sqrt[3]{27}$; $\sqrt{4a^2}$; $\sqrt{a^2b^2}$; $\sqrt{\dfrac{a^4}{4}}$; $\sqrt[3]{27a^3}$

Simplify the following by combining terms (in problems 28 and 29, first find the L. C. D. and then combine terms):

26. $\dfrac{2x^2y + 3xy^2 - 4x^2y}{2xy}$
27. $\dfrac{x - 5x^3 + 15x^5}{-5x^2}$
28. $\dfrac{3}{a} - \dfrac{5}{2a}$
29. $\dfrac{\dfrac{b}{2c} - \dfrac{4bc}{c^2}}{-\dfrac{b}{c^2}}$

Equations—The circumference of a circle may be expressed as $C = 3.1416D$; similarly, the area of a circle may be expressed as $A = .7854D^2$. Rules stated in this way are said to be expressed by an equation. *The expressions on the left side of the equal sign are called left members and those on the right side are called right members.*

Equations are used extensively in the study of electricity, because they convey in condensed form information that would otherwise be imparted only by means of long and cumbersome rules. Because an equation occupies little space, it is possible for the eye to catch at a glance the rule given. Moreover, it is much easier to remember an equation than a rule expressed in words. Then, too, the equation makes clearer the relation between any quantities in it. *An equation much used in electricity will be considered to illustrate the last statement.*

EXAMPLE—The twisting effect of a motor (torque) is found by multiplying the developed horsepower by 5252 and dividing by the speed. Expressed by an equation,

$$T = \frac{5252HP}{N}$$

where T = torque in pound-feet, HP = horsepower, and N = speed in revolutions per minute. In this equation, if N is kept constant and HP, the quantity in the numerator, is increased, T becomes greater in the same proportion. If HP is kept constant and N, the quantity in the denominator, is increased, T decreases in the same proportion. Then T varies directly as HP and inversely as N.

Rule 9. *A quantity in the numerator of one member varies directly as the quantity in the numerator of the other and inversely as the quantity in the denominator of the other.*

It is often necessary to express an equation, such as those given in the electrical course, in different ways in order to fit different needs. For instance, suppose it is desired to determine HP in the above equation when T and N are known. To do this, the equation must be expressed as

$$HP = \frac{TN}{5252}$$

In this case the equation is said to be solved for HP in terms of the other letters.

Solving for a letter in an equation is often necessary, and the rule used is founded upon the following evident principle: *If exactly the same operation is performed on the two equal members of an equation, the resulting members will be equal.*

This important principle may be expressed otherwise by the following axioms (an axiom is a statement so evident that it requires no proof):

1. The same quantity may be added to both members of an equation without destroying their equality.
2. The same quantity may be subtracted from both members of an equation without destroying their equality.
3. Both members of an equation may be multiplied by the same quantity without destroying their equality.
4. Both members of an equation may be divided by the same quantity without destroying their equality.

5. Both members of an equation may be raised to the same power without destroying their equality.
6. The same root may be extracted of both members of an equation without destroying their equality.

By means of these principles, it is possible to get the letter being solved for on one side of the equal sign and all the other letters on the other side.

EXAMPLE—Solve for E_1 in the equation $E = E_1 + E_2$.

To obtain E_1 alone on one side, E_2 must be subtracted from both members. This gives $E - E_2 = E_1 + E_2 - E_2$. Since $+E_2 - E_2 = O$, then the equation becomes $E - E_2 = E_1$. The same result is obtained if E_2 is transposed to the opposite member and its sign changed.

Hence, any term may be transposed from one member to the other if its sign is changed.

EXAMPLE—Solve for R in the equation $Z = \sqrt{R^2 + X^2}$

Squaring both sides to eliminate the square root sign gives $Z^2 = R^2 + X^2$. Transposing X^2 to the left member gives $Z^2 - X^2 = R^2$, and taking the square root of both sides gives $\sqrt{Z^2 - X^2} = R$.

EXAMPLE—Solve for B in the equation $F = \frac{8.85BIL}{10^8}$.

Multiplying both members by 10^8 to remove 10^8 from the right member gives $10^8F = 8.85BIL$. Dividing by 8.85IL to remove all quantities except B from the right member,

$$\frac{10^8F}{8.85IL} = B$$

The same result is obtained if 10^8 in the denominator is transferred to the numerator in the opposite member, and if 8.85IL in the numerator is transferred to the denominator in the opposite member.

Hence, in an equation with only one term in each member, any quantity may be transferred from one member to the other if it is transferred from the denominator on one side to the numerator on the other.

If a member has more than one term, a value in its numerator and a like value in its denominator cannot be canceled. For instance, in solving for E_1 in the equation $E = E_1 + E_2$, one is apt to proceed as follows:

Dividing both members by E_2, $\frac{E}{E_2} = \frac{E_1 + E_2}{E_2}$

Canceling the E_2's in the right member is incorrect because only one of the two terms in that member is divided by E_2.

Exercises

1. In the equation $H = .000948I^2Rt$ explain how t varies with the other letters.

Solve for:

2. y_b when $y_f = y_b - 2m$
3. I when $E_g - IR = E_t$
4. y_s when $y_b = y_sC_s + 1$
5. E_c when $I = \frac{E_a - E_c}{R}$
6. R when $E = IR$
7. L when $X_L = 2\pi fL$
8. N when $f = \frac{PN}{2 \times 60}$
9. C when $X_c = \frac{1}{2\pi fC}$
10. I when $H = .000948\ I^2Rt$
11. X when $Z = \sqrt{R^2 + X^2}$
12. I_p when $I_pE_p = I_sE_s$
13. T_p when $\frac{E_p}{T_p} = \frac{E_s}{T_s}$
14. m when $y_c = \frac{C - 2m}{P}$
15. E_2 when $\frac{T_1}{T_2} = \frac{E_1^2}{E_2^2}$
16. N_r when $S = \frac{N_s - N_r}{N_s}$

In using an equation to solve a problem, the values of all the letters must be known except one. The steps involved in determining the value of this unknown letter are: (1) Removing of fractions by multiplying both members by the least common denominator; (2) Transposing to get all unknown quantities in one member and all known quantities in the other; (3) Collecting like terms; (4) Dividing by the coefficient of the unknown.

EXAMPLE—Determine the value of X if $\frac{1}{6}X - \frac{1}{2} = \frac{3}{4}$.

Multiplying by least common denominator (12), $2X - 6 = 9$

Transposing, $2X = 9 + 6 = 15$

Dividing by the coefficient, $X = \frac{15}{2}$ or $7\frac{1}{2}$

The answer should be tested to be certain that the problem is solved correctly. This is accomplished by substituting the value obtained for the unknown in the original equation. If the two members become identical, then the solution is correct.

Substituting in the above example, $\left(\frac{1}{6}\right)\left(\frac{15}{2}\right) - \frac{1}{2} = \frac{3}{4}$

Collecting terms in the left member, $\frac{3}{4} = \frac{3}{4}$

Example—Find the value of R if $\frac{3R}{2} + 40 = 4R - 7 + \frac{6}{7}R.$

Multiplying by the L C D, which is 14, $21R + 560 = 56R - 98 + 12R$

Transposing, $21R - 56R - 12R = -98 - 560$

Collecting terms, $-47R = -658$

Dividing by the coefficient, $R = 14$

Checking, $\frac{3 \times 14}{2} + 40 = 4 \times 14 - 7 + \frac{6}{7} \times 14$

Collecting terms in each member, $61 = 61$

Example—Solve for L if $\frac{7}{2L} = \frac{2}{3} + \frac{4}{3L}.$

Multiplying by the L C D, which is 6L, $21 = 4L + 8$

Transposing, $21 - 8 = 4L$

Collecting terms in the left member, $13 = 4L$ or $4L = 13$

Dividing by the coefficient, $L = \frac{13}{4}$ or $3\frac{1}{4}$

Checking, $\frac{7}{2\left(\frac{13}{4}\right)} = \frac{2}{3} + \frac{4}{3\left(\frac{13}{4}\right)}$

Collecting terms, $\frac{14}{13} = \frac{14}{13}$

Exercises

Determine the value of the unknowns in the following equations:

1. $\frac{3}{5}X = 15$
2. $m - 11m = 48$
3. $2E - 7 + 3E = 28$
4. $15 - I = 5 - 3I$
5. $\frac{4}{3}R - \frac{2}{5}R = \frac{14}{75}$
6. $\frac{S}{2} - 15 = \frac{S}{3}$
7. $3 + .5R = 4.25$
8. $X - 16 + 3X = 0$
9. $\frac{2}{5} = \frac{1}{T}$
10. $10 = \frac{90}{Z^2}$
11. $\frac{27}{2a} = \frac{3}{4}$
12. $\frac{1}{a} + \frac{3}{a} = \frac{4}{5}$
13. $\frac{8 + m}{5} = m$
14. $\frac{13}{2k} + \frac{5}{3k} = \frac{7}{6}$
15. $\frac{2}{2 + m} = \frac{6}{m}$

Determine the missing values underneath the equations:

16. $R = R_1 + R_2 + R_3$

R	R_1	R_2	R_3
14	4	?	2
16	?	7	5

17. $I = \frac{E}{R}$

I	E	R
?	110	16
5	220	?
4	?	30

18. $R = \frac{K1}{cm}$

R	K	1	cm
6	10.4	1000	?
15	10.4	?	404
?	670	15	102

19. $H = .000948I^2Rt$

H	I	R	t
?	5	150	60
250,000	?	200	20
350,000	25	260	?

20. $S = \frac{N_s - N_r}{N_s}$

S	N_s	N_r
?	1200	1140
.3	3600	?
5%	?	570

21. $W = IE$ and $I = \frac{E}{R}$

W	I	E	R
450	?	110	?
?	5	?	55
300	?	?	200

22. $R_f = R_o + R_o at$

R_f	R_o	a	t
?	200	.004	55
50	20	.004	?
200	?	.004	30

23. $\frac{1}{R} = \frac{1}{R_1} + \frac{1}{R_2}$

R	R_1	R_2
?	8	3
6	?	12
1	4	?

24. $T = Fr$ and $HP = \frac{NT}{5252}$

HP	N	T	F	r
?	1200	20	?	1
5	1725	?	?	1.5
?	850	?	15	1.5

Parentheses—The sign most commonly used for grouping is the parentheses, written ( ). It signifies that the terms, or parts of terms, enclosed are to be treated as a single quantity. The parts enclosed then become a single term or part of a term. Thus, $2(5+3)$ means that the quantity $5+3$ is to be multiplied by 2, which gives 16. $A+(B+C)$ means that the sum of B and C is to be added to A, and $A-(B+C)$ means that the sum of B and C is to be subtracted from A.

In solving certain equations it is necessary to insert or remove parentheses.

EXAMPLE—Remove the parentheses from the expression $X^2+(-X+3)$. This means that $-X+3$ is to be added to X^2.

$$\text{Adding,}\quad \begin{array}{r} X^2 \\ -X+3 \\ \hline X^2-X+3 \end{array}$$

Therefore, $X^2+(-X+3)=X^2-X+3$. Note that in the result the sign of X is negative and the sign of 3 is positive, just as they were within the parentheses.

Rule 10. *Parentheses preceded by a plus sign may be removed without making any other change.*

EXAMPLE—Remove the parentheses from the expression $X^2-(-X+3)$. This means that the quantity $-X+3$ is to be subtracted from X^2.

$$\text{Subtracting,}\quad \begin{array}{r} X^2 \\ -X+3 \\ \hline X^2+X-3 \end{array}$$

Therefore, $X^2-(-X+3)=X^2+X-3$. Note that in the result the signs of X and 3 have changed.

Rule 11. *Parentheses preceded by a minus sign may be removed if the sign of every term within the parentheses is changed.*

EXAMPLE—Remove the parentheses and collect like terms in the expression $2(2a-6b)+(4a-b)-(5a-4b)$.

When the parentheses enclosing $2a-6b$ are removed, each term must be multiplied by 2. When the parentheses enclosing $4a-b$ are removed, the sign of each of these terms remain as they were, because the expression is preceded by a plus sign. When the parentheses enclosing $5a-4b$ are removed, the sign of each of these terms must be changed, because the expression is preceded by a minus sign.

Therefore, $2(2a-6b)+(4a-b)-(5a-4b)=4a-12b+4a-b-5a+4b=3a-9b$.

If terms are placed within parentheses preceded by a plus sign, their signs must not be changed; if placed within parentheses preceded by a minus sign, their signs must be changed.

Then $X^2-X-Y+3=X^2+(-X-Y+3)$
or $X^2-(+X+Y-3)$.

Factoring—When an expression consists of several terms all of which contain a common factor, the expression may be written with two factors. On dividing by the common factor the other is obtained.

EXAMPLE—Determine the two factors in the expression $2ab+4ab^2+8a^2b$.

Dividing by the common factor, which is $2ab$, gives $1+2b+4a$. Hence, $2ab+4ab^2+8a^2b=2ab(1+2b+4a)$.

In some equations, the letter to be solved for occurs more than once. In this case it is necessary to factor out this letter from the terms in which it appears before solving for it.

EXAMPLE—Solve for I in the equation $E=\sqrt{I^2R^2+I^2X^2}$.

Factoring out I^2 in the right member gives $E=\sqrt{I^2(R^2+X^2)}=I\sqrt{R^2+X^2}$.

$$\text{Then}\quad \frac{E}{\sqrt{R^2+X^2}}=I$$

Exercises

Remove the parentheses and collect like terms:

1. $3a+(5a-2)$
2. $3a+(-a-4)$
3. $6R-(2+3R)$
4. $9K-(-3K-1)$
5. $4L-3(5-4L)$
6. $-(x-y)-(2y-x)$

Enclose the last two terms of the following expressions in parentheses preceded by (a) a plus sign; (b) a minus sign.

7. $a-b+c$
8. X^2+X-10
9. $R-5S-S^2$

Write as two factors:

10. $.5I+.5$
11. R_o+R_oat
12. $4xy+8x$
13. $10R^2+6R$
14. $\frac{2\pi R_1-2\pi R_2}{2}$
15. $\pi R_1{}^2-\pi R_2{}^2$
16. $3x^2y-6xy^2$
17. $\sqrt{I^2R^2+I^2X^2}$
18. $\sqrt{(IR)^2+(IX_L-IX_c)^2}$

In the following equations, which nearly all pertain to electricity, solve for:

19. t when $R_f = R_o + R_o at$

20. F when $C = \frac{5}{9}(F - 32)$

21. N_s when $S = \frac{N_s - N_r}{N_s}$

22. R_i when $R_f = \frac{R_i(1 + at_f)}{1 + at_i}$

23. R_o when $R_f = R_o + R_o at$

24. I when $E = \sqrt{(IR)^2 + (IX)^2}$

25. n when $C = \frac{2248AK(n-1)}{10^{10}d}$

26. X when $\frac{R_2}{R_1} = \frac{X}{L - X}$

Another Important Use of Equations—One common type of problem is solved as follows: (1) Represent the unknown quantity by a letter. (2) If there are more unknowns than one, express the others in terms of the first. (3) Write an equation from the information given in the problem. (4) Solve for the unknown letter. (5) Solve for the other unknown quantities with the expression written for case 2.

EXAMPLE—A bus-bar must be 24 times as wide as it is thick. What must be its dimensions to be 1½ square inches in cross section?

Let $T =$ its thickness

Then $24T =$ its width

And $(T)(24T) = \frac{3}{2}$

Or $24T^2 = \frac{3}{2}$

$T^2 = \frac{1}{16}$ or $T = \frac{1}{4}$ inch in thickness

$24T = 24 \times \frac{1}{4} = 6$ inches in width

EXAMPLE—A room is 6 feet longer than it is wide. If its perimeter (distance around) is 52 feet, determine its dimensions.

Let $W =$ its width

Then $W + 6 =$ its length

And $2W + 2(W + 6) =$ its perimeter

Therefore $2W + 2(W + 6) = 52$ feet

Or $4W = 40$, or $W = 10$ feet in width

$W + 6 = 10 + 6 = 16$ feet in length

Exercises

1. A switch and its cover together cost 76 cents. If the switch cost 24 cents more than the cover, how much did each cost?
2. A generator and switch board together cost \$8050. If the switchboard cost $2\frac{1}{2}$ times as much as the generator, determine the cost of each.
3. A bus-bar must be 12 times as wide as it is thick. What must be its dimensions to be $2\frac{1}{4}$ square inches in cross section?
4. The length of a rectangle is five times the width, and the perimeter (distance around) is 30 feet. Find the length and width.
5. The average grade made by a student in three subjects is 88. What grade must he make in the fourth subject so that his average is 90?
6. The weight of a motor is $2\frac{1}{2}$ times that of its starter and the weight of the rheostat is one-third the weight of the starter. If the three together weigh 69 lbs., determine the weight of each.
7. A pipe 10 feet long is to be cut so that the first piece is three times as long as the second and the second half as long as the third. Determine the length of each.
8. An electrician can install a switch board in three days and his helper in four days. In how many days can both together install it?

Curve Reading—Curves are used to show clearly and in little space what would otherwise require numerous words to state. The

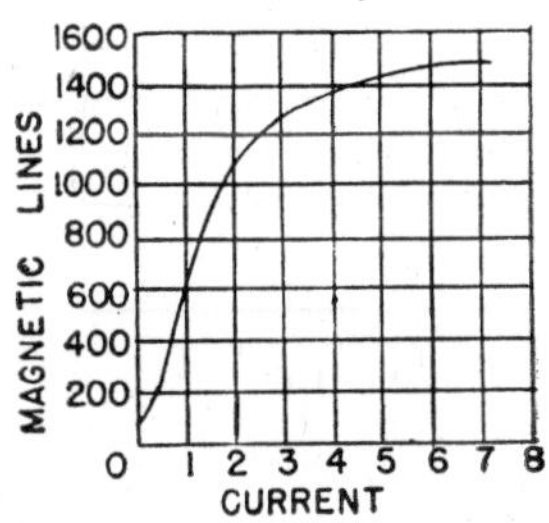

Fig. 1-3

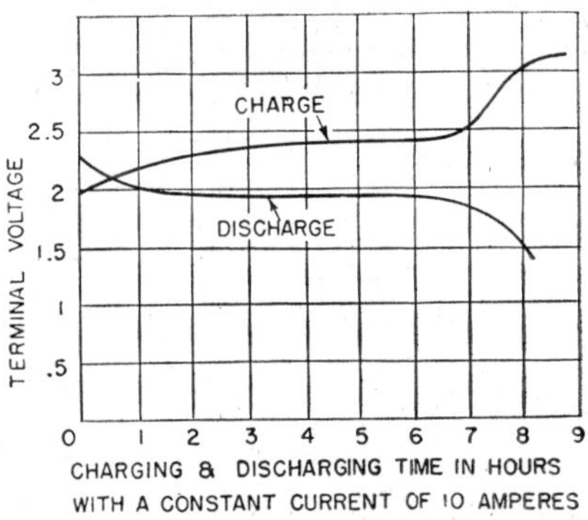

Fig. 1-4

characteristics of electric machinery are frequently shown in this way. In Figure 1-3 is illustrated one application of a curve, which shows how the amount of magnetism produced in an iron core coil varies with the current through it. This curve indicates that with a constant increase in current, the magnetism increases slightly at first, soon rises rapidly until considerable magnetism is produced, and then again increases at a slow rate.

Figure 1-4 shows that the terminal voltage of one cell of a fully charged battery is 2.3 volts. As the battery is discharged, the voltage drops very slowly until it reaches about 1.9 volts, and then drops rapidly.

The upper curve shows how the terminal voltage of a battery varies while it is being charged. The voltage rises rapidly at first for a short period, then rises slowly until the battery is fully charged. After this the rise is rapid until a voltage of about 3.2 is reached; then the voltage remains practically constant.

Plotting Curves—Suppose that it is desired to plot a discharge curve for a particular battery, such as the one shown in fig. 1-4. To obtain the data for plotting the curve, an adjustable resistance is connected across a cell to maintain the rated electric current, and

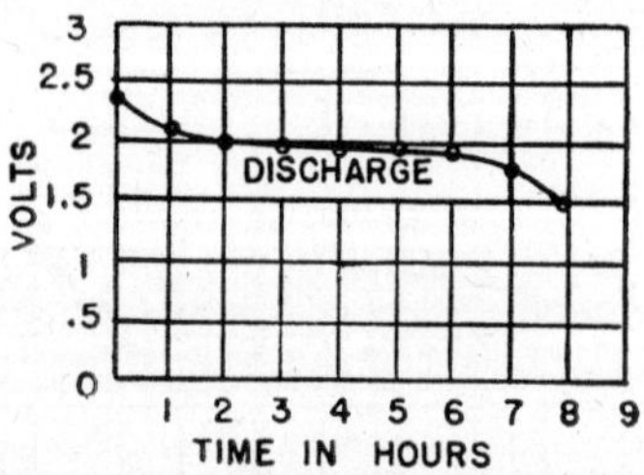

FIG. 1-5

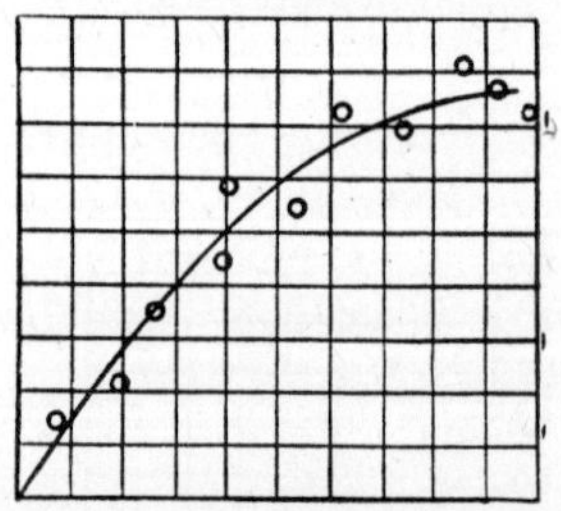
FIG. 1-6

voltage measurements are made every hour. The data obtained is recorded as indicated in Table 1-1. On the graph of fig. 1-5 a point is indicated for each set of readings. Each point is placed on a vertical line corresponding to the time the reading was taken and on a horizontal line corresponding to the voltage. After all the points are marked, a smooth line is drawn through them.

If readings obtained during an experiment are not accurate, a smooth curve cannot be drawn through all the points. In this case the curve should be drawn to give an average of the values. This is illustrated in fig. 1-6.

It is sometimes necessary to plot negative as well as positive numbers. As an illustration, suppose two quantities are related according to the equation $y = x + 2$. By giving x a series of values, a corresponding series of values can be obtained for y. Letting $x = 8$, then $y = 8 + 2$ or 10; letting $x = 6$, then $y = 8$, etc. Table 1-2 gives values of y for different values of x.

TIME	VOLTS
0	2.3
1	2.1
2	2.
3	1.95
4	1.94
5	1.93
6	1.92
7	1.8
8	1.5

TABLE 1-1

x	y
8	10
6	8
4	6
2	4
0	2
–2	0
–4	–2
–6	–4
–8	–6

TABLE 1-2

The type of graph required for plotting negative and positive numbers is shown in fig. 1-7. Line x is called the x axis, line y is called the y axis, and the point of intersection O of these lines is called the origin. Values to the right of O are positive and those to the left of O are negative; those upward are positive and those downward are negative. To locate the first point given in Table 1-2, which is (8, 10), count on the x axis 8 units to the right of the origin and from this point count upward 10 units. This point is shown at P in fig. 1-7; the second point is shown at Q; etc.

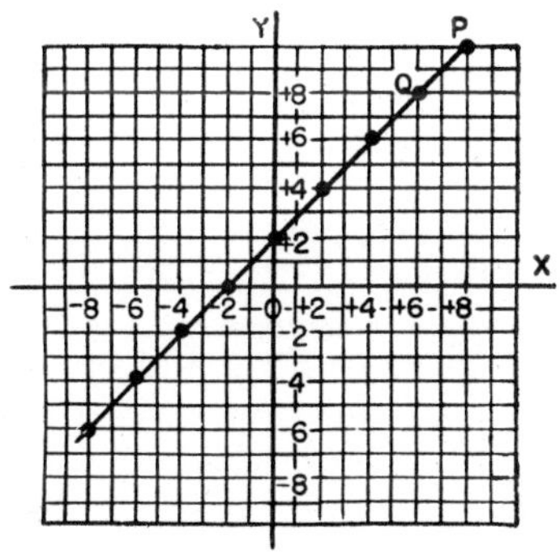

FIG. 1-7

Exercises

In the following problems use ¼ inch as the unit of measure.
Plot curves for the values given in the following:

1.

I	e
0	0
.8	20
1.6	40
3	60
4	70
5.5	80
8	82
12	80

2.

x	y
−6	15
−4	11
−2	7
0	3
2	−1
4	−5
6	−9

3.

I	e
−6	5
−4	0
−2	−3
0	−4
2	−3
4	0
6	5

4. Relatively to two axes at right angles to each other, indicate the following points: (5, 2); (0, 4); (3, 0); (−2, 6); (3, −4); and (−2, −4).

Plot the following equations:

5. $y = 2x^2 - 4$ 6. $2x + y = 4$ 7. $xy = 4$

CHAPTER 2

FUNDAMENTALS OF ELECTRICITY

It has been known for many years that if sticks of sealing wax are rubbed with wool or if glass rods are rubbed with silk, these materials become electrified (or charged). If two charged sticks of sealing wax or two charged glass rods are brought near each other, repulsion occurs. Attraction occurs, however, if a charged stick of sealing wax is brought near a charged glass rod. This illustrates that the charge on the sealing wax differs from the charge on the glass. Such a test on other materials reveals either a charge like that found on sealing wax or one like that found on glass. To distinguish between these two charges, the one on the sealing wax is called negative and the one on the glass rod is called positive.

The reason that materials become electrified on being rubbed is now known, and this will be explained after the structure of matter has been considered.

Structure of Matter—Matter is anything that has weight. Any visible quantity of matter contains an inconceivable number of particles called molecules. A molecule is the smallest particle into which a substance can be divided without changing its identity. If the division is carried further, which can be done chemically, each molecule can be divided into two or more particles called atoms. Suppose that a certain quantity of water is divided into the smallest parts in which it can exist as water. These parts are called molecules. Each molecule can be further broken up into three atoms: two atoms of hydrogen and one atom of oxygen. Both hydrogen and oxygen are gases, and have entirely different properties from water.

Electrons—Each atom, in turn, consists of an equal number of protons and electrons. The electrons possess the same kind of charge as the sealing wax and are therefore considered negative. The protons possess the same kind of charge as the glass rod and are therefore positive. All the protons and about half the electrons are concentrated in the central portion of the atom, which is called the nucleus. The remaining electrons (which of course are negative) revolve about the nucleus in definite paths, as the planets revolve about the sun. A model of a copper atom is shown in fig. 2-1.

Electrons are all alike; each weights $317 \div 10^{31}$ of an ounce and has a diameter of $157 \div 10^{15}$ of an inch. The protons (which are

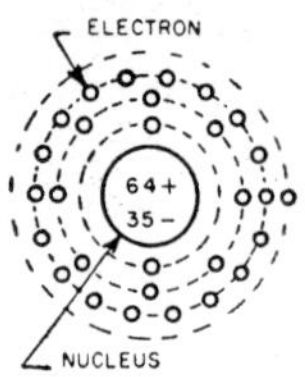

FIG. 2-1

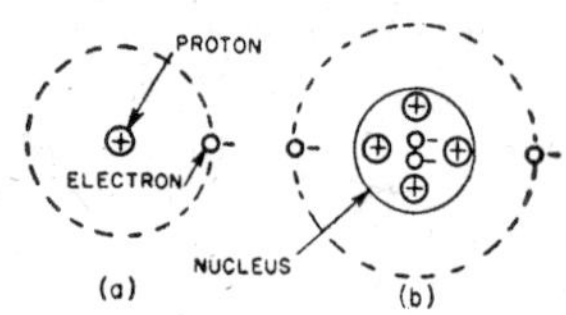

FIG. 2-2

also all alike) each weigh about two thousand times as much as an electron.

Kinds of Atoms—There are actually only 92 different kinds of atoms. The number of external electrons in them varies by one, and the charge on each nucleus varies in proportion. Thus the heavier atoms contain many more electrons and protons than the lighter ones. Hydrogen has the simplest atomic structure, consisting of one proton and one electron (fig. 2-2a). Next in simplicity is the helium atom, having four protons and two electrons in the nucleus and two external electrons (fig. 2-2b). Note that in this atom, as well as in all others, the negative and positive charges are just equal. All the elements may be analyzed in this way, finishing with uranium which has 92 external electrons.

Suppose that some of the external electrons shown in fig. 2-1 are removed. Then the positive and negative charges are no longer equal. The excess of protons in the nucleus gives the atom a positive charge. If the removed electrons are allowed to return to the nucleus, a neutral atom again results. Thus a positively charged body is one which is deprived of electrons. Apparently, a negatively charged body is one which has an excess of electrons.

The sealing wax became negative when it was rubbed because electrons were deposited upon it after they left the fur. If the fur is tested it will be found to have a positive charge or a lack of electrons. The glass rod rubbed with silk became positive because the electrons left it and were deposited upon the silk. *Therefore, whenever a*

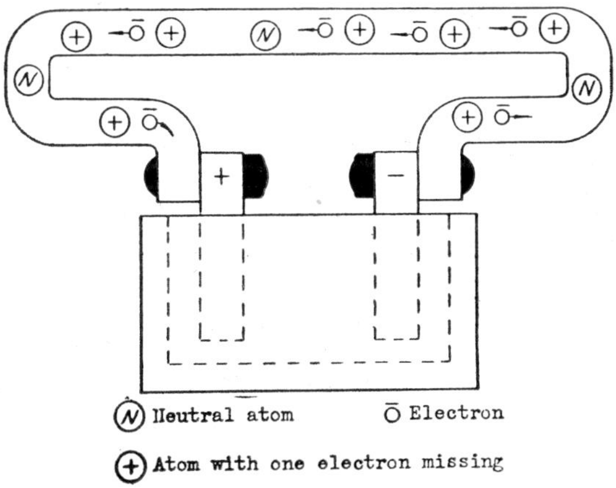

FIG. 2-3

material has an excess of electrons it is negatively charged, and when it has a lack of electrons it is positively charged.

Direction of Electrons in a Conductor—In non-metallic materials, such as rubber and glass, the electrons are held rather rigidly to their nuclei. But in metals, such as copper and iron, some of the electrons are held loosely to their nuclei and can be made to flow. These are called free electrons.

Suppose that a heavy copper wire or some other material that has free electrons is connected to the terminals of a battery (fig. 2-3), making one end of the wire positive and the other negative. The free electrons will be attracted toward the positive terminal and will be repelled from the negative terminal. This will cause the electrons to drift toward the positive. They next pass through the battery, then again through the wire, and so on. This flow of electrons constitutes an electric current. The current through a small lamp is about 10^{18} electrons per second. *In a wire the electrons move from negative to positive and this flow constitutes an electric current.* The early experimenters, not knowing the actual direction of flow, arbitrarily assumed it to be from positive to negative. *Due to this unfortunate circumstance, it is necessary to accustom ourselves to thinking that the current flow is in the opposite direction to the flow of electrons.*

Ampere—To specify the rate at which water is flowing through a pipe, the unit of measure called the gallon is used and the rate is specified in gallons per second. Similarly, to specify the rate at which an electric current flows through a conductor, a unit of measure called the coulomb is used and the rate is specified in coulombs per second, which is the flow of 628×10^{16} electrons per second. The expression "coulomb per second" is seldom used in practice. Instead, a shorter term, "ampere," is generally used and means the same thing. It is unnecessary to say "per second" because ampere means coulombs per second. The relation between the coulomb and the ampere may be expressed as

$$Q = It \qquad (2\text{-}1)$$

where Q = quantity of coulombs, I = current in amperes, and t = time in seconds.

EXAMPLE—A battery sends a current of 10 amperes through a circuit for one hour. (a) How many coulombs will flow through the circuit? (b) How many electrons per second?

(a) $Q = 10 \times 3600 = 36{,}000$ coulombs.

(b) $10 \times 628 \times 10^{16} = 628 \times 10^{17}$ electrons per second.

Current Measurement—To measure the flow of water through a pipe, a flowmeter must be inserted into the pipe through which the water flows, as shown in fig. 2-4a, so that all the water flowing through the pipe also flows through the meter.

In like manner, to measure the flow of current through a wire, a current meter (called an ammeter) must be inserted into the circuit so that all the current to be measured flows through the instrument (fig. 2-4b). The ammeter has a low resistance so as not to hinder the flow of current. Its principle of operation, and that of other indicating instruments, will be studied in chapter 19.

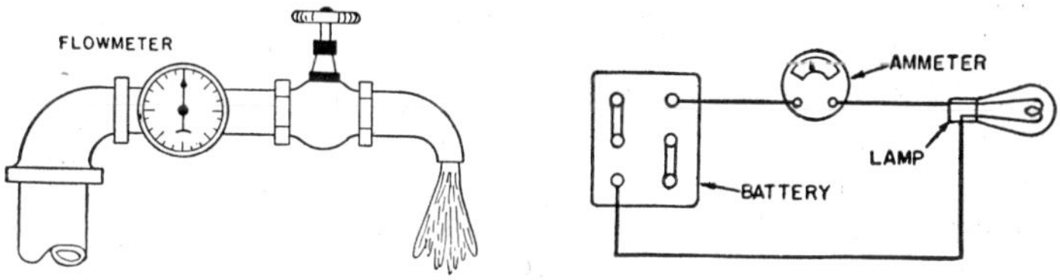

(a) FIG. 2-4 (b)

Voltage—In fig. 2-5a the action of the blades causes the water to enter the pump at D at a low pressure and leave at A at a high pressure. Section AB has little friction, and the pressure at B is therefore only slightly less than at A, say 5 pounds per square inch. Section BC has considerable friction, and the pressure at C is therefore considerably less than at B, say 100 pounds per square inch. CD has the same friction as AB and therefore the same drop in pressure. The drop

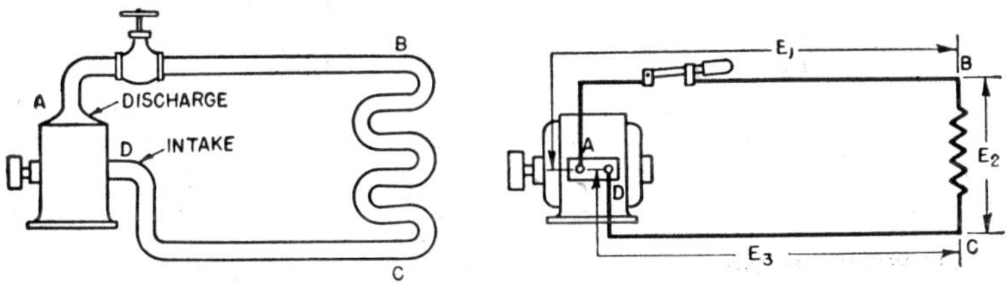

(a) FIG. 2-5 (b)

in pressure from A to D along the pipe line is 5 + 100 + 5 or 110 pounds per square inch, which equals the pressure difference supplied by the pump.

A similar condition exists in an electric circuit, where the pressure drops are expressed in "volts." The unit "volt" is used in an electric circuit just as the unit "pounds per square inch" is used in the hydraulic circuit. Referring to fig. 2-5b, the section of the wire from A to B has a low resistance and the pressure drop is low, say 5 volts.

BC has a high resistance and the drop is therefore high, say 100 volts. CD has the same resistance as AB and therefore has the same pressure drop. The drop in pressure along the conductor is 5 + 100 + 5 or 110 volts, which equals the pressure difference supplied by the generator.

In an electric circuit, the term "electromotive force" (denoted by e m f) is used to indicate the voltage generated at the source, such as that developed by a battery or a generator. The term "voltage drop" is used to denote the drop in pressure in a particular part of a circuit. In fig. 2-5b or in any other series circuit, if E equals the supplied voltage and E_1, E_2, and E_3 equal the voltage drops in the sections, then

$$E = E_1 + E_2 + E_3 \qquad (2\text{-}2)$$

Voltage Measurement—The amount of water flowing through a pipe (say pipe XY in fig. 2-6a) depends upon the pressure difference

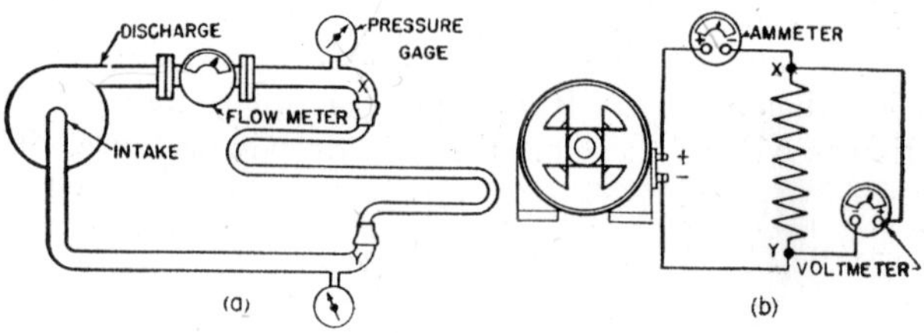

FIG. 2-6

between the two ends. To measure this difference it is necessary to tap a pressure gage at each end of the pipe as shown. The difference in the two readings will be the resultant pressure effective in forcing water through the pipe.

Similarly, the amount of current flowing through a conductor (say conductor XY in fig. 2-6b) depends upon the pressure difference (voltage) between the two ends. To measure this difference, it is necessary to tap the voltmeter leads to the ends (X and Y) of the wire. *The voltmeter indicates the pressure difference between its ends.* It will therefore indicate the voltage across wire XY.

Resistance—As previously stated, some materials have an abundance of free electrons, which require a low pressure to move them from atom to atom, and establish a high current. Such materials are known as good conductors. Other materials have few free electrons. In these the same electric pressure can move only a few electrons from atom to atom, establishing a low current. These are considered poor conductors. The progressive motion of free electrons is hindered in all materials, because they collide with the atoms of the substance used. The opposition to the flow of electrons (due to the bonds between protons and electrons, as well as to collisions) is called electrical resistance. The methods of computing resistance will be considered later.

There are two common methods of avoiding the use of very large or very small numbers. The first is to group a definite number of small units and call the group by a new name. For example, 12 inches are grouped and known as one foot, three feet become one yard, 5280 feet equal one mile, etc. The second method is to attach a prefix, denoting the quantity, to the name of the unit. This is best illustrated by the following list.

Quantity	Prefix	Example
1000	Kilo	Kilowatt
1,000,000	Mega	Megavolt
.000001	Micro	Microampere
.001	Milli	Milliampere

OHM'S LAW AND SERIES CIRCUITS

Ohm's Law—The amount of current flowing through a circuit is directly proportional to the electromotive force or voltage and inversely proportional to the resistance of the circuit. This relationship is called Ohm's law and is expressed as

$$I = \frac{E}{R} \tag{2-3}$$

where I = current in amperes, E = electromotive force in volts, and R = resistance in ohms.

EXAMPLE—How much current does a 30 ohm heater take if the applied e m f is 120 volts?

$$I = \frac{120}{30} = 4 \text{ amperes}$$

EXAMPLE—Find the resistance of the electric iron shown in fig. 2-7 if the current flowing is 3 amperes.

$$R = \frac{E}{I} = \frac{120}{3} = 40 \text{ ohms}$$

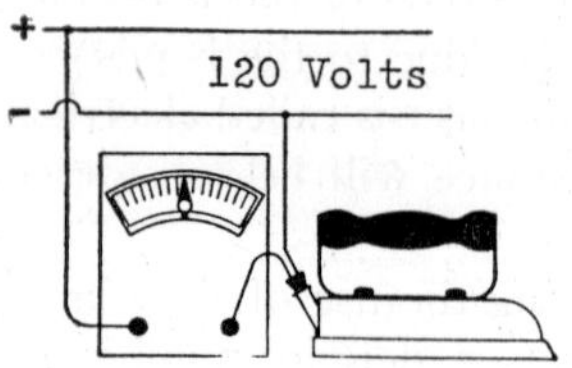

FIG. 2-7

Ohm's law may be applied to a circuit as a whole or to any of its parts. This may be stated as follows:

(a) *The total current flowing in a circuit equals the total voltage applied to the circuit divided by the total resistance of the circuit.*

(b) *The current flowing in a part of a circuit equals the voltage applied to that part divided by the resistance of that part.*

Series Circuit—If several electrical appliances are connected end to end to form one path for the flow of an electric current (fig. 2-8a), they are said to be in series. The laws of series circuits can be illustrated by the hydraulic analogy of fig. 2-8b, which shows pipes of different diameters connected together. The same number of gallons per second must flow through each pipe, regardless of its size, because the water that flows through one pipe must also flow through the others.

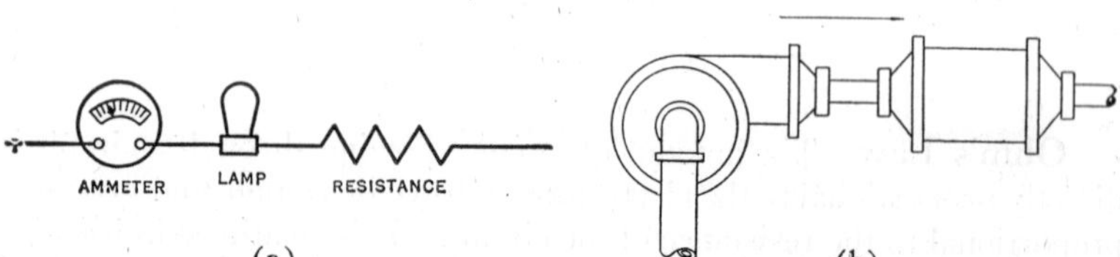

(a) (b)

FIG. 2-8

Similarly, in the series circuit shown in fig. 2-8a, no greater current can flow through the ammeter than flows through the lamp and the resistor, even though the resistance of the ammeter is much less. Then the current is the same in all the parts.

Laws of Series Circuits—In series circuits, the following laws apply:

(1) *The current is the same in all parts of the circuit.*

(2) *The total voltage equals the sum of the voltages across the different parts of the circuit.* This was expressed by equation 2-2, which is as follows:

$$E = E_1 + E_2 + E_3 \qquad (2\text{-}4)$$

(3) Apply Ohm's law to the whole circuit shown in fig. 2-9,

$$E = IR \qquad (2\text{-}5)$$

Expressing the voltage drop across the various parts,

$$E_1 = IR_1, \quad E_2 = IR_2, \quad E_3 = IR_3 \qquad (2\text{-}6)$$

Substituting 2-5 and 2-6 into 2-4,

$$IR = IR_1 + IR_2 + IR_3$$

Dividing both members of the equation by I,

$$R = R_1 + R_2 + R_3 \qquad (2\text{-}7)$$

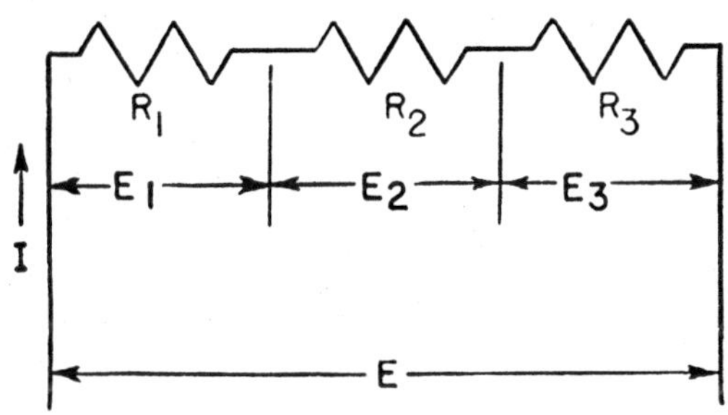

Fig. 2-9

Thus, in a series circuit the total resistance equals the sum of the resistances of the separate parts.

Example—What voltage must be supplied by a generator to force 2 amperes through a series circuit consisting of a 100 ohm lamp, a 20 ohm resistor, and a 40 ohm soldering iron.

$$R = 100 + 20 + 40 = 160 \text{ ohms}$$

$$E = 2 \times 160 = 320 \text{ volts}$$

Example—Determine the line voltage (E) in fig. 2-10 if the voltmeter indicates 20 volts.

The current through the 10 ohm resistor is $20 \div 10$ or 2 amperes. The two lamps are in series with the conductor and therefore 2 amperes must also pass through them. Then

$$E_2 = 2 \times 15 = 30 \text{ volts}$$

$$E_3 = 2 \times 30 = 60 \text{ volts}$$

$$E = 20 + 30 + 60 = 110 \text{ volts}$$

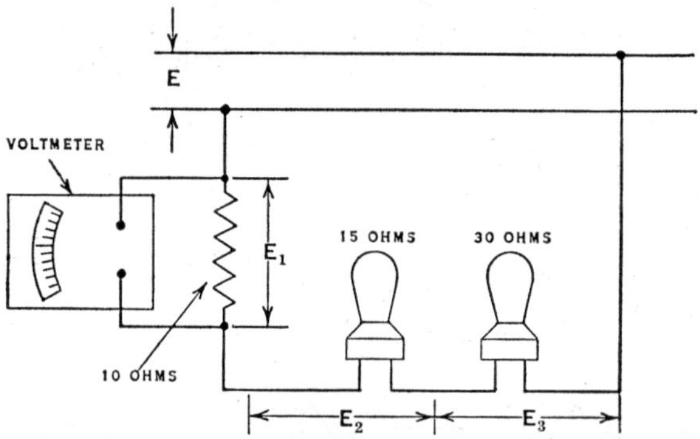

FIG. 2-10

The most common method of regulating the amount of current supplied to an appliance is to vary the amount of resistance in the circuit. This may be accomplished, without opening the circuit, by means of a rheostat. One common type of rheostat is shown in fig. 2-11. It

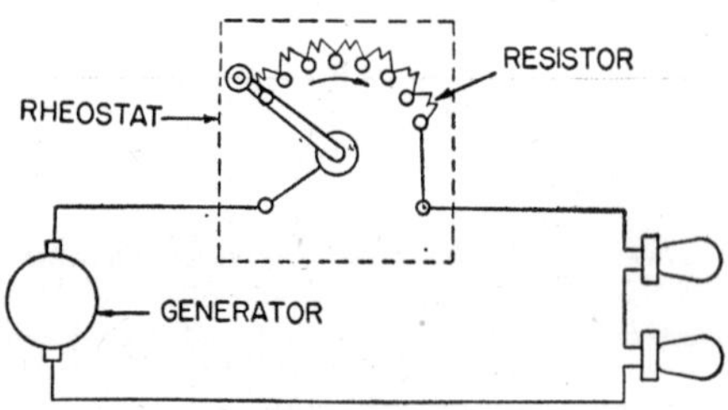

FIG. 2-11

consists of series resistors connected to segments over which a contact arm is moved. When the contact arm is in the position shown, all the resistors are in series with the load and the current is low. When the arm is moved to the right, resistors are cut out of the circuit and more current flows. Thus the current may be regulated by shifting the arm.

PARALLEL CIRCUITS

Current in a Parallel Circuit—A parallel circuit is a circuit that has two or more paths through which the current flows. Fig. 2-12a shows a hydraulic parallel circuit. The water flows out of the pump to X and there divides, flowing through pipes a, b, and c. Flowing out at Y, it returns to the pump, etc. Of course, the number of gallons per second flowing in at X must equal the sum of the gallons per second through pipes a, b, and c.

The electric circuit shown in fig. 2-12b may be analyzed in the same manner. The electric current flows from the generator to X and there divides, flowing through conductors a, b, and c. It then flows out at Y and returns to the generator, etc. As in the hydraulic circuit, the current flowing in at X must equal the sum of the currents flowing through conductors a, b, and c. This may be expressed as follows:

$$I = I_1 + I_2 + I_3 \tag{2-8}$$

Voltage in a Parallel Circuit—The pressure drop across the three parallel pipes in fig. 2-12a is found by subtracting the gage reading at Y from the gage reading at X. This gives the drop in pressure across the three pipes in parallel, or across any one of them. If the pressure at one end of pipe a is 120 pounds per square inch and at the other end is 10 pounds per square inch, the drop in pressure across this branch is 110 pounds per square inch. The drop in pressure across the other two branches must be the same as it is across pipe a, because they are connected between the same two points.

Similarly, if the voltage drop across conductor a in fig. 2-12b is 110 volts, the voltage drop across the other two paths must be the same as the drop across a.

Laws of Parallel Circuits—In a parallel circuit, the following laws apply:

(1) *The total current supplied to the system equals the sum of the currents through the several paths.* This was expressed by equation (2-8).

(2) *The voltage across a parallel combination is the same as the voltage across each branch.*

(3) Applying Ohm's law to the circuit of fig. 2-13, as a whole,

$$I = \frac{E}{R} \tag{2-9}$$

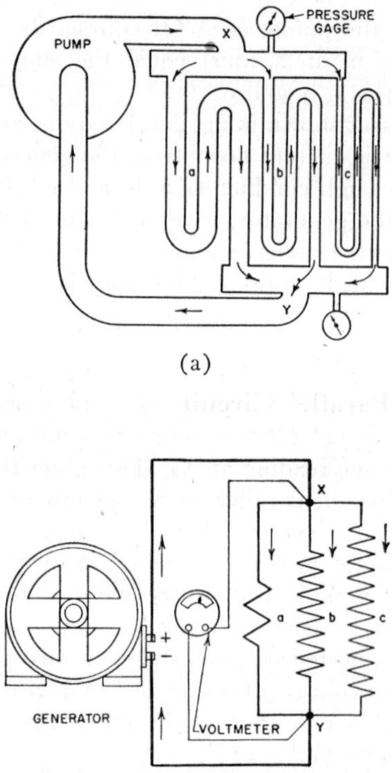

(a)

(b)

FIG. 2-12

Applying Ohm's law to the individual paths,

$$I_1 = \frac{E}{R_1}, \quad I_2 = \frac{E}{R_2}, \quad I_3 = \frac{E}{R_3} \tag{2-10}$$

Substituting 2-9 and 2-10 into 2-8,

$$\frac{E}{R} = \frac{E}{R_1} + \frac{E}{R_2} + \frac{E}{R_3}$$

Dividing both members of the equation by E,

$$\frac{1}{R} = \frac{1}{R_1} + \frac{1}{R_2} + \frac{1}{R_3} \qquad (2\text{-}11)$$

Thus, in a parallel circuit, the reciprocal of the total resistance equals the sum of the reciprocals of the resistances of the individual branches.

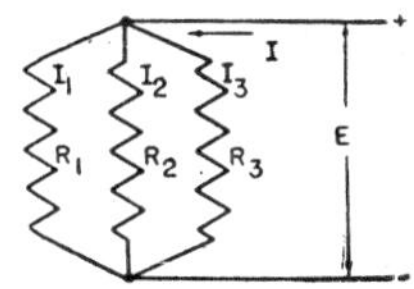

FIG. 2-13

EXAMPLE—Find (a) the combined resistance and (b) the total current taken by the circuit shown in fig. 2-14.

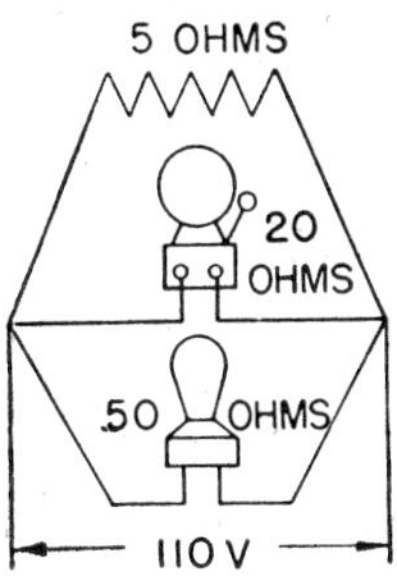

FIG. 2-14

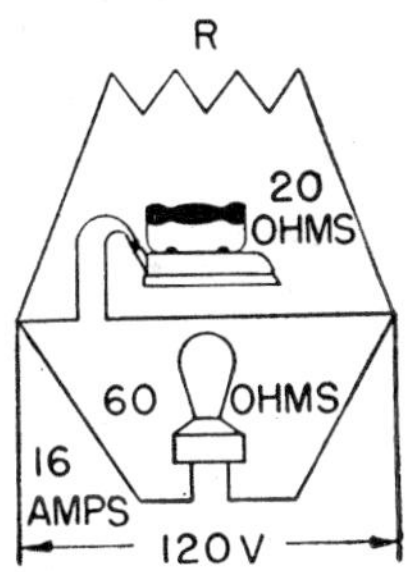

FIG. 2-15

(a) $\frac{1}{R} = \frac{1}{5} + \frac{1}{20} + \frac{1}{50} = \frac{27}{100}$

$$R = \frac{100}{27} = 3.7 \text{ ohms}$$

(b) $I = \frac{110}{3.7} = 29.7 \text{ amperes}$

EXAMPLE—Find the value of R in fig. 2-15.

The voltage across R is 120. Before the resistance of R can be determined, the current through it must be found, and that can be done only after the current through the other two branches is known. The current through the 20 ohm branch is 120 ÷ 20 or 6 amperes, and the current through the 60 ohm branch is 120 ÷ 60 or 2 amperes. Then the current through R is 16 − 6 − 2 or 8 amperes. Then R = 120 ÷ 8 = 15 ohms.

If the resistances of the branches in a parallel circuit are equal, the total resistance may be found by dividing the resistance of one branch by the number of branches.

EXAMPLE—Five lamps are connected in parallel across a 120 volt line. If each lamp has a resistance of 200 ohms, determine their combined resistance:

$$R = \frac{200}{5} = 40 \text{ ohms}$$

Series-Parallel Circuits—Only simple series and simple parallel circuits have been considered thus far. A combination of series and parallel circuits often exists, however, and a particular arrangement is sometimes complicated. In solving such problems, it should be remembered that any number of resistances in series can be replaced by one resistance having a value equal to the sum of the individual resistances. Also, any number of resistances in parallel can be replaced by a resistance having a value equal to the reciprocal of the sum of the reciprocals of the individual resistances.

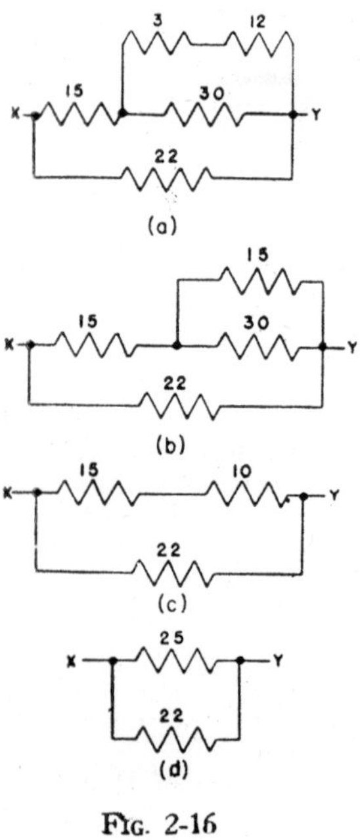

FIG. 2-16

EXAMPLE—Determine the combined resistance of the circuit shown in fig. 2-16a.

The 3 ohm resistance and the 12 ohm resistance are in series and may be replaced by a 15 ohm resistance (fig. 2-16b). Now a 15 ohm resistance and a 30 ohm resistance are in parallel, and these may be replaced by a 10 ohm resistance (fig. 2-16c). The circuit then has a 10 ohm resistance and a 15 ohm resistance in series, which may be replaced by a 25 ohm resistance (fig. 2-16d). Ultimately, a 25 ohm resistance and a 22 ohm resistance are in parallel, which are combined to give 11.7 ohms.

Incandescent lamps, designed for indoor lighting, are usually connected in parallel. The resistance of any two lamps, even of the same make, is seldom the same; neither is the voltage across each lamp connected to the line the same. However, for convenience in calculating the voltage drop in the line wires, it is the practice to assume that lamps of the same rating take the same amount of current. The error introduced in making this assumption is negligible.

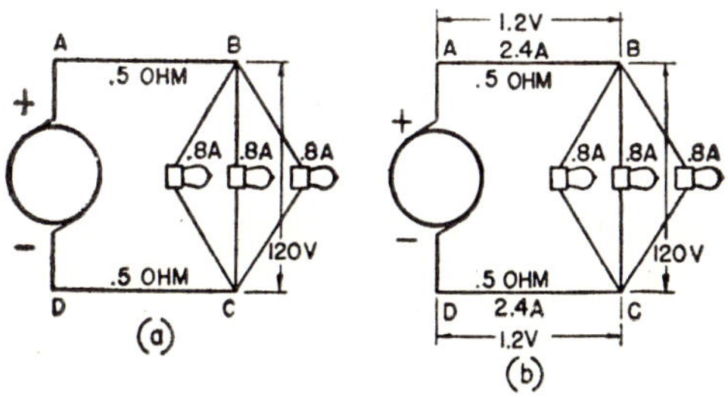

FIG. 2-17

In practice, the simple series circuit and simple parallel circuit seldom occur, but such connections are found in various parts of more complicated systems. By considering each of these parts separately, it is usually not difficult to find the current, voltage, and resistance of each part in a system. The method of doing this will be made clear by several examples.

EXAMPLE—In the lighting system shown in fig. 2-17a, each lamp takes .8 ampere at 120 volts. Find the voltage supplied by the generator.

It is apparent that the three lamps are in parallel and that this combination is in series with the two line wires AB and CD. Then the current through this parallel combination is 3 × .8 or 2.4 amperes. Since the two line wires AB and CD are in series with the parallel combination, the 2.4 amperes also flow through these wires. This current distribution is shown in fig. 2-17b. The drop in voltage (IR drop) in each of the two sections AB and CD is 2.4 × .5 = 1.2 volts.

The voltage across the three parts in series is now known. The voltage across the entire series system equals the sum of the voltages across the separate parts, which is $1.2 + 120 + 1.2 = 122.4$ volts. Thus the generator must supply 122.4 volts to force the current through the line wires and lamps.

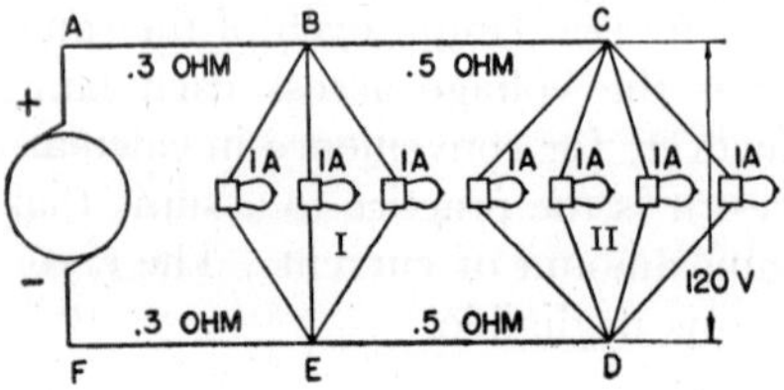

Fig. 2-18

When solving problems of this nature, proceed as follows: First, find the current through each section of the circuit; second, find the voltage across each section; and third, combine the voltages according to the rule for a series circuit.

Example—If each lamp in fig. 2-18 takes 1 ampere, determine what the voltage across the generator must be to produce 120 volts across the group of lamps located more distant from the generator.

To determine the current through the various sections of a complex system, it is usually more convenient to begin with the lamps farthest from the generator. Thus, in fig. 2-18, group II requires 4 amperes. Therefore sections BC and DE must also carry 4 amperes. Group I requires 3 amperes. Then the section AB must carry sufficient current to supply 4 amperes to group II and 3 amperes to group I, which is a total of $4 + 3$ or 7 amperes. The 3 amperes leaving group I and the 4 amperes leaving group II pass through section EF. This section, then, must also carry 7 amperes. The current distribution is shown in fig. 2-19.

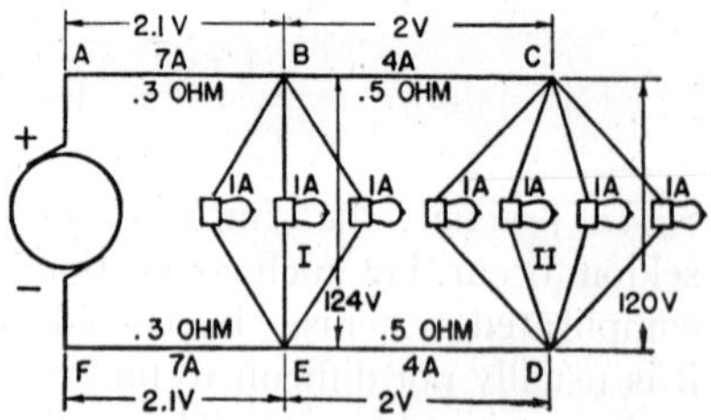

Fig. 2-19

The IR drop in each section must now be determined. Since BC and DE each have a resistance of .5 ohm, $4 \times .5$ or 2 volts are required to force the 4 amperes through each section. Since there are 120 volts across group II, then there must be $120 + 2 + 2 = 124$ volts across group I. The voltage required to force 7 amperes through each of the .3 ohm sections, AB and EF, is $7 \times .3 = 2.1$ volts. The e m f at the terminals of the generator must be 2.1 + 124 + 2.1 = 128.2 volts.

EXAMPLE. Fig. 2-20 shows a trolley line fed by two 550 volt generators, one at each end. Two cars are on the line at the points indicated. Car I takes 250 amperes and car II takes 200 amperes. The resistance of the trolley and track are as marked. Find the voltage across each car.

Assume that current flows through the various sections as shown. In the solution, if a current has a negative value the assumed direction is wrong and the arrow should be reversed. However, the numerical value of the current is correct.

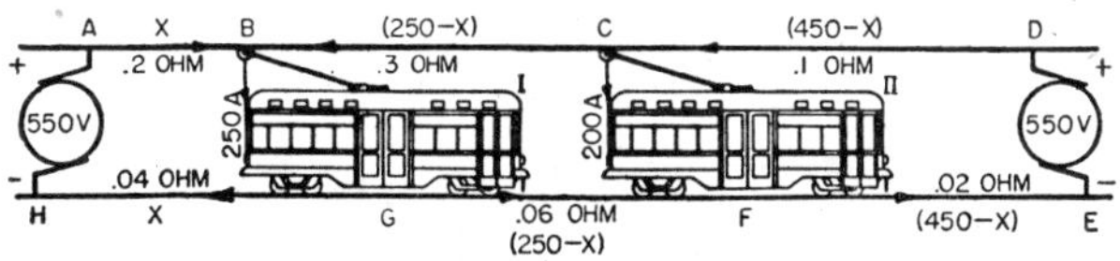

FIG. 2-20

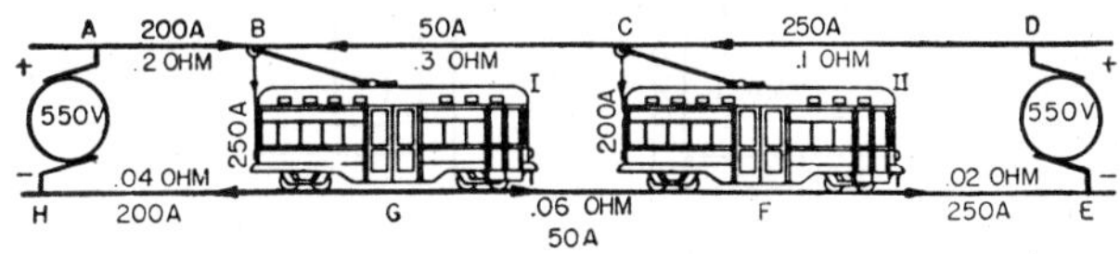

FIG. 2-21

Let X represent the current through section AB. Since the same current flows through GH, this current may also be represented by X. The current in sections BC and FG is (250 − X) amperes and the current in sections CD and EF is (450 − X) amperes.

The voltage across car I is 550 − .24X. This voltage is also equal to 550 − .12(450 − X) − .36(250 − X). Then 550 − .24X = 550 − .12(450 − X) − .36(250 − X).

Solving for the unknown, X = 200 amperes (current through sections AB and GH). It is now an easy matter to find the current in the remaining sections (these are indicated in fig. 2-21).

The voltage across car I is 550 − (.24 × 200) = 502 volts. The voltage across car II is 502 + (.36 × 50) = 520 volts.

A large variety of mesh-work problems may be solved by using these principles.

ELECTROPLATING AND BATTERIES

Certain liquids, such as distilled water, alcohol, and kerosene, do not conduct electricity. Other liquids, such as acid and salt solutions, readily conduct electricity. These latter are known as electrolytes. When current passes through an electrolyte, chemical decomposition takes place. This is known as electrolysis.

Electroplating—A very important industrial application of electrolysis is electroplating, which is the process of coating materials with metals by means of an electric current. Suppose it is desired to copper plate a carbon brush. The equipment required is shown in fig. 2-22. The

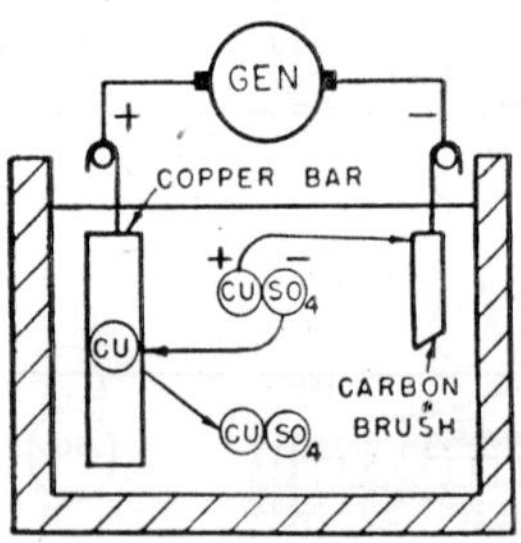

FIG. 2-22

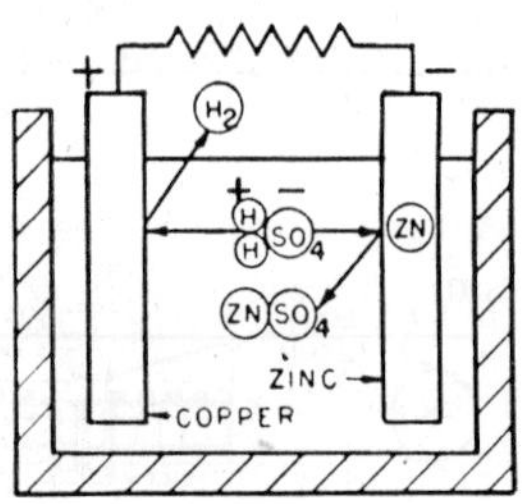

FIG. 2-23

tank contains a solution of copper sulphate. One molecule of copper sulphate consists of one atom of copper (denoted by Cu), one atom of sulphur (denoted by S), and four atoms of oxygen (denoted by O_4), and is represented by the formula $CuSO_4$. In a solution of copper sulphate, each molecule divides into two parts as shown in fig. 2-22. Each part is called an ion and has an electric charge. Each of the SO_4 ions has two extra electrons, making each negatively charged. The two extra electrons on each SO_4 ion come from a Cu ion. Then each Cu ion must have two electrons less than normal and is therefore positively charged. Positive Cu ions migrate to the negative electrode because unlike charges attract, each takes on two electrons to neutralize its charge, and is firmly attached to the brush. The negative SO_4 ions migrate to the positive electrode, each gives off two electrons to neutralize its charge, and combines with one atom of copper to form one molecule of copper sulphate. Then as much copper returns to the solution as is taken from it, and the concentration of the solution does not change. Thus it is possible to continue plating until the copper bar is entirely consumed.

This same method may be used for nickel plating, in which case a solution containing nickel must be used (such as nickel sulphate, $NiSO_4$). The bar suspended from the positive electrode into the solution should also be nickel.

Cells—Chemical energy may be converted into electrical energy by means of cells, each of which generally consists of two plates immersed in an electrolyte. Cells are divided into two kinds—primary cells and secondary cells.

After a primary cell has become run down, its solution must be removed and a new solution supplied. One electrode goes into solution, and must be replaced. A secondary cell differs from a primary cell in that it may be recuperated, after it has run down, by sending a current through it in the opposite direction to that during discharge.

Primary Cell—Although various combinations of metals and solutions may be used in making a cell, a simple cell may be made by immersing a copper plate and a zinc plate in a diluted solution of sulphuric acid, as shown in fig. 2-23 (copper is denoted by Cu and zinc by Zn). Each molecule of sulphuric acid consists of two atoms of hydrogen, one atom of sulphur, and four atoms of oxygen, and is represented by the formula H_2SO_4. Each molecule of sulphuric acid breaks up into a negatively charged ion (SO_4) and two positively charged ions ($2H_2$). The SO_4 ions migrate to the Zn electrode, because Zn is more active than H_2, and each unites with one atom of Zn to form $ZnSO_4$, which passes into solution. The zinc plate gains electrons from the SO_4 ions during this action and thus becomes negative. The H_2 ions migrate to the copper, each takes on one electron, and passes off into the air. This action causes one electrode to become negative and the other positive, and therefore a difference of potential is established between them (about 1.1 volts). When a load is attached to the cell, the chemical action maintains a voltage, which causes electrons to pass through the external circuit from the Zn to the Cu. This action takes place until all the Zn has been converted to $ZnSO_4$ or the sulphuric acid has been consumed.

Dry Battery—The most common type of primary cell is the so-called dry cell, which is shown in fig. 2-24. The negative electrode is a zinc cup, which is lined with blotting paper. The positive electrode is a carbon rod centrally located. The space between the blotting paper and the carbon rod contains a paste of sal ammoniac, zinc chloride, manganese dioxide, and ground carbon. This paste is not dry, and if it becomes dry the cell ceases to develop an e m f. The sal ammoniac is the electrolyte, the zinc chloride is added to improve the action, and the manganese dioxide is used to remove the hydrogen bubbles which otherwise would collect on the carbon electrode and increase the internal resistance of the cell.

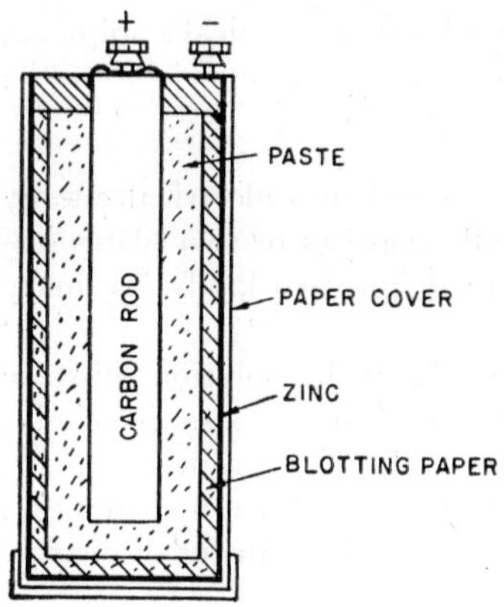

FIG. 2-24

The e m f of a new dry cell is about 1.5 volts and its internal resistance is about .1 ohm. A six inch cell has a capacity of about 32 ampere-hours when discharged through a 20 ohm resistance until the voltage has reduced to .5. Even if a dry cell is not used, its deterioration is such that at room temperature it becomes useless in about one year. If a cell is kept in a cool place, it has a much longer life.

Grouping of Cells—The number of cells used and the method of grouping them depends upon the voltage and current desired. If cells of equal rating are joined in series (fig. 2-25a), the total voltage

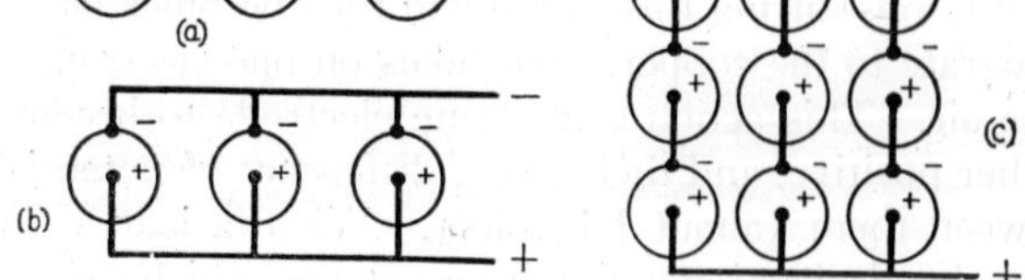

FIG. 2-25

equals the sum of the separate voltages, and the current capacity is the same as that of one cell. If cells of equal rating are connected in parallel (fig. 2-25b), the total e m f is the same as that of one cell, and the current capacity equals that of one cell times the number of cells. Cells are sometimes arranged as shown in fig. 2-25c. This is known as a series-parallel arrangement. In this case the total e m f is the same as that of a series group. The current capacity is the same as that of one cell times the number of series groups in parallel.

Lead Storage Cell—In the most common type of storage cell, the negative electrode consists of plates made of spongy lead (denoted by Pb), and the positive electrode consists of plates made of lead peroxide (denoted by PbO_2, which means that each molecule contains one atom of lead and two atoms of oxygen). The electrolyte consists of diluted sulphuric acid. Thin sheets of wood or perforated hard rubber are inserted as insulation between the plates (see fig. 2-26). If no insulation were used, a short circuit would exist between the plates, causing the battery to discharge internally.

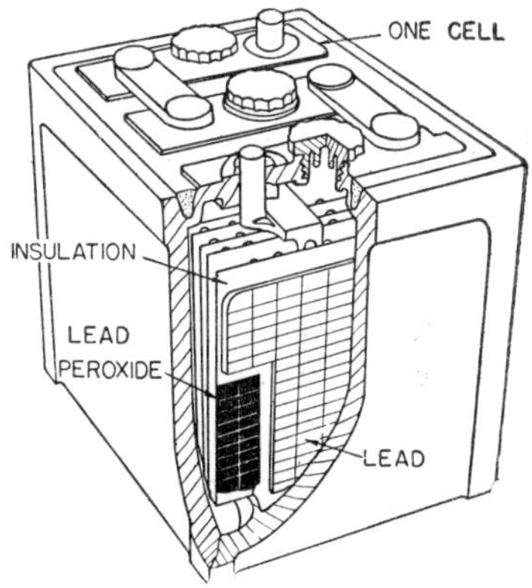

FIG. 2-26

The spongy lead and lead peroxide are relatively poor conductors, and are too soft to be formed into plates. These substances are molded into frames made of some hard material that is a good conductor and does not act as a third electrode, in which case acid would produce action between the frames and the substances molded into them. A frame is generally a casting of an alloy of lead and antimony.

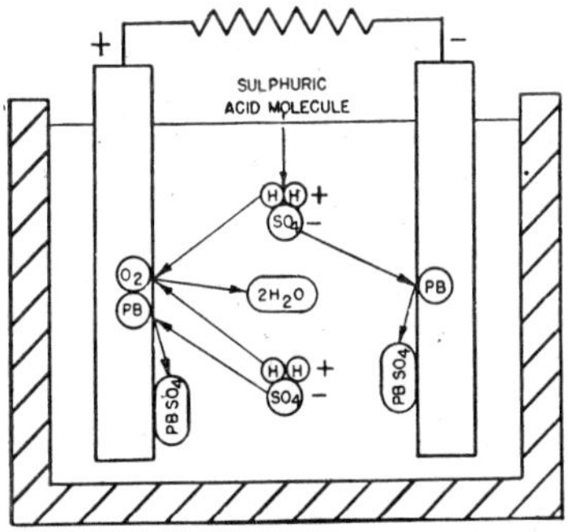

FIG. 2-27

Chemical Action of the Lead Storage Cell During Discharge—Each molecule of sulphuric acid breaks up into two positively charged hydrogen ions (2H) and a negatively charged sulphate ion (SO_4), as indicated in fig. 2-27.

Each SO_4 ion near the lead plate combines with one atom in the lead plate to form lead sulphate ($PbSO_4$), and gives up two electrons. Each pair of associated hydrogen ions are attracted to the lead peroxide plate, takes on two electrons, and then unites with one atom of oxygen to form a molecule of water (H_2O). This action adds electrons to the lead, making it negative, and removes electrons from the lead peroxide, making it positive. This establishes a difference of potential or voltage.

Each SO_4 ion near the lead peroxide plate combines with a Pb atom in a PbO_2 molecule to form a $PbSO_4$ molecule. Each pair of hydrogen ions unite with one atom of oxygen in the PbO_2 molecule to form a molecule of water.

These chemical actions occur simultaneously and may be expressed by the following equation:

$$PbO_2 + Pb + 2H_2SO_4 \rightarrow 2PbSO_4 + 2H_2O$$

This shows that lead sulphate is being formed during discharge. This increases the internal resistance of the cell and the acid becomes weaker. After considerable $PbSO_4$ is formed, the cell becomes too weak for practical use.

Chemical Action of a Lead Storage Cell During Charge— The cell may be restored to its former state by sending a current through it in the opposite direction to that during discharge (fig. 2-28).

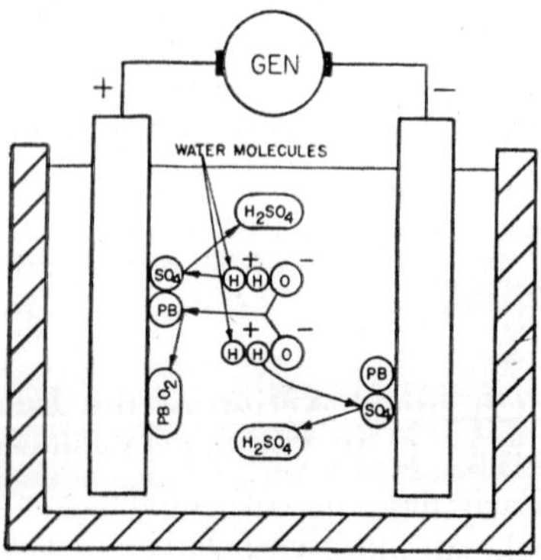

Fig. 2-28

In this case each molecule of water breaks up. Each pair of hydrogen ions near the negative plate unites with an SO_4 ion on the negative plate to form a molecule of sulphuric acid, and leaves an atom of spongy lead. Each pair of hydrogen ions near the positive plate unite with an SO_4 ion on the positive plate to form a molecule of sulphuric acid. Pairs of liberated oxygen ions combine with each Pb atom in the positive plate to form PbO_2.

Thus, during charge, the water breaks up and combines with the lead sulphate, forming spongy lead at the negative plate and lead peroxide at the positive plate. This chemical action may be expressed as follows:

$$2PbSO_4 + 2H_2O \rightarrow PbO_2 + Pb + 2H_2SO_4$$

Note, that this chemical action is just the opposite to that which occurs during discharge.

Fig. 2-29 shows typical charge and discharge curves of a storage cell, in which it will be noted that discharge occurs at nearly a constant e m f of 2 volts.

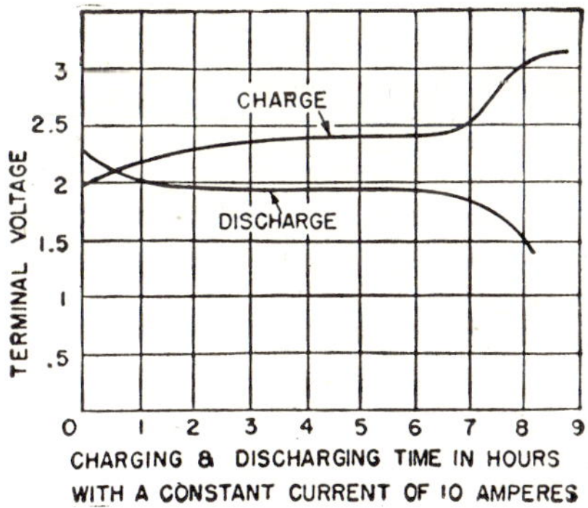

Fig. 2-29

Batteries—A battery usually consists of three cells placed in separate compartments of a single container. These cells are connected in series to obtain maximum voltage (see fig. 2-26).

Rating of a Battery—A battery is usually rated according to the maximum current it can supply during an eight hour period, which is known as the normal rate of discharge. A battery that can deliver a maximum of 10 amperes for eight hours is considered an 80 ampere-hour battery.

The ampere-hour capacity of a cell can be estimated by considering that one ampere can be supplied for eight hours for each 25 square inches of surface on the positive plates. In determining the number of square inches in the surfaces, both sides of the positive plates are to be considered.

Internal Resistance—The internal resistance of a storage battery increases during discharge, due to the formation of lead sulphate, which has a higher resistance than lead or lead peroxide. When charged, a battery has a very low internal resistance (about 0.0025 ohm). Hence, it may discharge at a very high rate.

Charging Rate—A discharged storage battery can be charged with a high current without overheating. Charging may start with a current equal numerically to the ampere-hour rating of the battery. For instance, in the case of a discharged 80 ampere-hour battery, charging may be started at 80 amperes. As the battery becomes charged, the rate should be correspondingly decreased. The rate should never be such as to allow violent gassing to take place, because in such a case a considerable amount of the energy supplied to the battery is used in breaking the water up into hydrogen and oxygen and is not effective in charging. Moreover, during gassing the battery becomes heated. This is detrimental since heat causes some of the material to come loose from the plates, thus reducing the amount of active material, damaging the separators, and causing buckling of the plates.

Most chargers used in small installations cannot supply this high current. These chargers supply a maximum current of about 10 amperes, which is supplied until gassing begins. The current is then reduced to about half value, and is supplied until gassing again begins. The current is decreased repeatedly in this way until charging is completed.

Care of a Battery—Water continually vaporizes from a battery; thus distilled water must be added occasionally to keep the plates

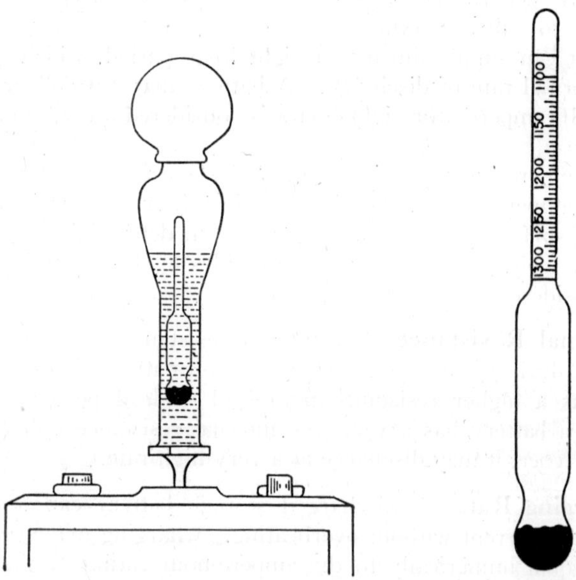

(a) Fig. 2-30 (b)

submerged. Undistilled water should not be used, because it may contain impurities which will react with the plates. The acid does not vaporize and none should be added unless some is spilled.

A battery should never be excessively overcharged, especially at a high rate. This causes severe gassing, and hence excessive heat.

The condition of the electrolyte can be determined by a hydrometer (see fig. 2-30), which consists of a small glass float with a calibrated stem placed in a syringe. This float indicates 1000 when immersed in pure water, 1850 when immersed in pure sulphuric acid, and 1300 when immersed in a battery fully charged.

Whether or not a battery is completely charged cannot always be reliably determined by taking a hydrometer reading, unless the reading at the time of complete discharge is known. If, during the charging period, the hydrometer reading increases 100 above the discharged reading, the battery has undoubtedly become charged.

Another method of determining if a battery is charged is to measure the terminal voltage of each cell, with each discharging at normal rate. If the voltage is two or above, each cell is charged.

The top of a storage battery should be kept free from moisture and dirt, which permit current to leak between electrodes, causing the lugs and supports to corrode. This action is reduced by applying a light coat of grease where corrosion is apt to occur.

Sulphation—It has been explained that, during discharge, both plates change to lead sulphate, which is a very porous material that readily changes back to lead and lead peroxide during charge. If a battery gets too hot, is left discharged for a long time, or has too much sulphuric acid, a deposit of hard sulphate forms on the plates. This sulphate, if excessive, is insoluble and cannot be reconverted to lead or lead peroxide. It increases the internal resistance and isolates the active material, decreasing the capacity of the battery.

Hydrometer Reading	Freezing Point (°F)
1050	27 above 0
1100	19 " 0
1150	6 " 0
1200	16 below 0
1250	62 " 0

Freezing—The temperature at which an electrolyte freezes depends upon the hydrometer reading (see table above). This table shows that the freezing point is considerably reduced with an increasing charge on the battery, so that if the battery is well charged there is no danger of it freezing. When the electrolyte in a cell freezes, the plates are usually injured and the case cracks.

Work—The purpose of every machine is to do a certain amount of work. Some common examples of work-performing machines are: a punch press which stamps holes in metal, a crane that lifts a load of iron, a conveyor that moves stock from one place to another, etc. In each of these illustrations it should be noted that a force acts through space. A floor upon which a machine rests does no work because the weight supported is not moved. The amount of work done depends upon the magnitude of the acting force and the distance through which the force acts. Thus

$$\text{Work} = \text{force} \times \text{space} \qquad (2\text{-}12)$$

A common unit of work is known as the foot-pound. A foot-pound of work is done when a force of one pound acts through a distance of one foot.

EXAMPLE—A man lifts 500 iron rods, each weighing 30 pounds, from the floor to a conveyor 3 feet high. How much work is done?

To place one rod on the conveyor, the work done is 30×3 or 90 foot-pounds. The total work done is then 90×500 or 45,000 foot-pounds.

Power—In the above example observe that the amount of time required to do the work is not considered. The same amount of work is done, regardless of the time used. A weaker man, B, for instance, might do the same amount of work as A, provided he had sufficient time. However, A is said to possess greater power because he can do the work in less time. In a given time, a crane can do much more work than either man; consequently, the crane has the greatest power. Thus power differs from work in that time is involved. *Power means the rate at which work is done.* This may be expressed:

$$\text{Power} = \frac{\text{work}}{\text{time}} \qquad 2\text{-}13$$

Common units of power are horsepower (H P), watts (W), and kilowatts (kw).

In English-speaking countries the unit "horsepower" is widely used to express the mechanical output of machines. The origin of this unit is briefly as follows:

In early times water was pumped from mines by horses. Later, horses were replaced by steam engines, which were rated as having the power of many horses. The rating of these engines was made possible by James Watt, who determined, by experiment, the number of foot-pounds the average horse used for this purpose could do. He found that for a short time the average horse could work at the rate of 33,000 foot-pounds per minute or 550 foot-pounds per second. This rate was standardized as the "horsepower." Thus,

$$HP = \frac{\text{foot-pounds per minute}}{33{,}000} \qquad 2\text{-}14$$

EXAMPLE—Suppose that in the problem considered above, the man did 45,000 foot-pounds of work in 20 minutes. Find the average horsepower expended.

The work done per minute was 45,000 ÷ 20 or 2250 foot-pounds. The total power was 2250 ÷ 33,000 or .068 horsepower.

Watts and Kilowatts—It was shown that the flow of electricity through a conductor is similar to the flow of water through a pipe. A flow of water is usually expressed in gallons or pounds per minute; a flow of electricity is usually expressed in coulombs per second, called amperes. The power required to keep water flowing against a certain head is numerically equal to the flow in pounds per minute times the head or pressure in feet. Thus,

$$\text{Power} = \text{pounds per minute} \times \text{feet} \qquad (2\text{-}15)$$

Then to determine the horsepower required to force water against a given head, equation 2-14 must be expressed as follows:

$$HP = \frac{\text{pounds per minute} \times \text{feet}}{33{,}000}$$

Similarly, the power required to keep electricity flowing against a certain resistance is numerically equal to the flow in coulombs per second (amperes) times the pressure in volts. This product gives the power in watts. Thus,

$$\text{Watts} = \text{amperes} \times \text{volts}$$

or expressed in algebraic form,

$$W = IE \qquad (2\text{-}16)$$

Thus to determine the power in an appliance, multiply the voltage across the appliance by the current through it.

EXAMPLE—A heating element takes 20 amperes at 110 volts pressure. What power does it consume?

The power is:

$$W = 20 \times 110 = 2200 \text{ watts}$$

Since $E = IR$ (Ohm's law), then equation 2-16 may be expressed as

$$W = I \times IR = I^2R \qquad (2\text{-}17)$$

In equation 2-16, if we substitute for I its value $\left(I = \frac{E}{R}\right)$, we get

$$W = \frac{E}{R} \times E = \frac{E^2}{R} \qquad (2\text{-}18)$$

Equation 2-17 should be used when the current and resistance are known, while equation 2-18 should be used when the voltage and resistance are known.

When a large amount of power is considered, a larger unit of power, known as the kilowatt, is used.

$$1 \text{ kilowatt} = 1000 \text{ watts}$$

We have now learned that mechanical power is usually expressed in horsepower, and that electrical power is expressed in watts. Since it is often necessary to convert power from one unit to the other, a knowledge of the relation of the two is therefore essential. By experiment it was found that

$$1 \text{ H P} = 746 \text{ watts or } .746 \text{ kilowatt}$$

Efficiency—Since any electrical appliance has some resistance, heat is produced when a current is sent through it. This heat is wasted unless it is used for heating purposes. Thus no electrical machine converts all the electrical power it receives into mechanical power, or vice versa. The power that a machine receives is called the input, and the power it gives out is called the output. The fraction of the input that is delivered is usually expressed in percentage and is called the efficiency. This may be expressed by an equation, as follows:

$$\text{Input} \times \text{efficiency} = \text{output, or efficiency} = \frac{\text{output}}{\text{input}} \qquad (2\text{-}19)$$

In using this equation it should be realized that both the output and input must be expressed in the same unit, that is, either in horsepower, watts, or kilowatts.

EXAMPLE—A 5 H P motor takes 4.5 k w at full load. Find its efficiency.

$$\text{Output} - 5 \times 746 = 3730 \text{ watts}$$

$$\text{Input} = 4.5 \times 1000 = 4500 \text{ watts}$$

$$\text{Efficiency} = \frac{3730}{4500} = .829 \text{ or } 82.9\%$$

Electrical Units of Work—Equation 2-13 may be expressed as

$$\text{Work} = \text{power} \times \text{time} \qquad (2\text{-}20)$$

This means that the total work done is equal to the rate of doing work (power) times the time. Suppose that a turbine is working at the average rate of 500 horsepower for 24 hours. Then the turbine does 500 × 24 or 12,000 horsepower-hours of work. Similarly, if a generator is supplying electric power at the average rate of 500 kilowatts for 24 hours, the generator supplies 500 × 24 or 12,000 kilowatt-hours of work.

EXAMPLE—A 200 k w generator runs at full load for 16 hours. How much work is done?

$$200 \times 16 = 3200 \text{ k w h}$$

$$3200 \text{ k w h} = 3200 \div .746 \text{ or } 4290 \text{ h p h}$$

EXAMPLE—For a two-month period a small consumer is charged 10 cents a k w h for the first 18 k w h, 4 cents a k w h for the next 100 k w h, and 2½ cents a k w h for the remaining consumption. If 150 k w h are used, determine the cost.

For the first 18 k w h the cost is 18 × .10 or $1.80.

For the next 100 k w h the cost is 100 × .04 or $4.00.

For the remaining k w h (150 − 100 − 18 = 32) it will cost 32 × .025 = $.80.

The total cost is then $1.80 + $4.00 + $.80 = $6.60.

Measurement of Heat—Before we can understand how the heating effect of electricity is determined, we must become familiar with the unit in which heat is measured. This unit is called the British Thermal Unit (abbreviated to B T U), and its size has been established as follows:

A British Thermal Unit is the amount of heat necessary to raise one pound of water one degree Fahrenheit.

When it is desired to find the amount of heat necessary to raise a quantity of water to a certain temperature from an initial temperature, it is only necessary to multiply the weight of the water by the temperature rise in degrees Fahrenheit. Expressing this by an equation,

$$H = Qt \qquad (2\text{-}21)$$

where H = the amount of heat in B T U, Q = quantity of water in pounds, and t = the change in temperature in degrees Fahrenheit.

EXAMPLE—How many B T U of heat are required to raise one gallon of water ($8\frac{1}{3}$ lbs.) from 75° F to the boiling point (212° F)?

Substituting in the above equation,

$$H = 8\tfrac{1}{3}(212 - 75) = 1141.66 \text{ B T U}$$

Heating Effect of Electric Current—A convenient method of measuring the amount of heat produced by an electric current passing through a resistance is by means of a calorimeter (fig. 2-31). This is simply a vessel containing water, in which a thermometer and a one-ohm resistance are placed. The vessel is usually surrounded by a material which is a poor conductor of heat, so that the heat generated will not be carried away by the surrounding air. The number of B T U developed in the wire may be determined by Equation 2-21. It has been found that when one ampere flows through a one-ohm resistance for one second, .000948 B T U of heat will be generated. Hence, one watt produces .000948 B T U per second.

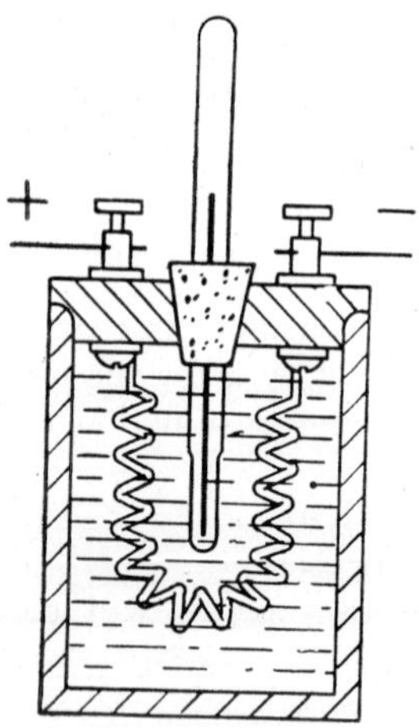

FIG. 2-31

From this we conclude that the amount of heat produced per second by a current passing through a resistance R is $.000948I^2R$. The total heat produced for any length of time is $.000948I^2R$ multiplied by the time in seconds (t). This may be expressed as

$$H = .000948I^2Rt \qquad (2\text{-}22)$$

EXAMPLE—How many B T U are generated in one hour by a heating element which carries 20 amperes and has 2 ohms resistance?

$$H = .000948 \times 20^2 \times 2 \times 3600 = 2730.24 \text{ B T U}$$

Electric motors are used so extensively for supplying mechanical power because they are easily controlled and have a relatively high efficiency. The efficiency of a motor is determined by dividing the output by the input. The input in watts is found by multiplying the line voltage by the supplied current, the voltage and current being measured by instruments.

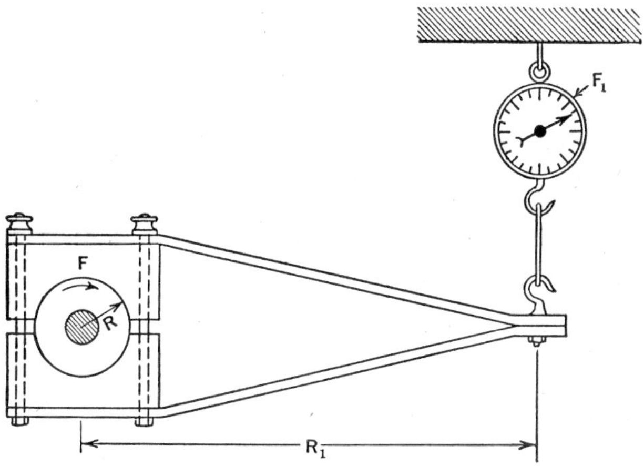

FIG. 2-32

The output of small and medium size motors may be determined by means of a prony brake, a common form of which is shown in fig. 2-32. It consists of two blocks of wood shaped to fit round the pulley of the motor. The pressure of these blocks against the pulley may be varied by means of the two hand wheels at the top. The two bars fastened at the top and bottom of the brake form an arm which is attached to a scale. When the armature turns, the friction between pulley and wooden blocks converts the power supplied by the motor into heat.

In developing an equation for the horsepower converted into heat, let R represent the radius of the pulley in feet, and F the force supplied at the surface of the pulley. Then when the pulley completes one revolution, the force F moves a distance of $2\pi R$ feet, and therefore $F(2\pi R)$ foot-pounds of work is done. If the pulley rotates at the rate of N revolutions per minute, $2\pi FRN$ foot-pounds of work are done. Since one horsepower is developed for every 33,000 foot-pounds of work done per minute, then to convert foot-pounds per minute into horsepower, we must divide by 33,000. Hence,

$$H\,P = \frac{2\pi FRN}{33000}$$

$$\text{or}\quad H\,P = \frac{FRN}{5252} \qquad (2\text{-}23)$$

Torque—The twisting effect produced by a motor does not depend only upon the magnitude of the force F, but also upon the radial distance this force acts. The twisting effect is expressed as the product of the force F and the radius R through which the force acts, and is called torque. Torque is represented by T and is expressed in pound-feet, as

$$T = FR \qquad (2\text{-}24)$$

Then Equation 2-23 may be expressed as

$$H\,P = \frac{TN}{5252} \qquad (2\text{-}25)$$

EXAMPLE—A motor rated at 12 horsepower runs at 850 r p m. What twisting effect (torque) will the motor supply at its rated load?

From Equation 2-25,

$$T = \frac{5252\,H\,P}{N} = \frac{5252 \times 12}{850} = 74.146 \text{ pound-feet}$$

In the case of the brake shown in fig. 2-32, it should be realized that the product of scale reading F_1 and the length of arm R_1 gives a torque which opposes the torque produced by the motor and is equal to it. Since, in practice, the scale reading F_1 and the arm length R_1 are easily obtained, the torque delivered by the motor may be determined by taking the product of these values. That is,

$$T = F_1R_1 \qquad (2\text{-}26)$$

The horsepower output is then found by substituting in Equation 2-25.

POWER, TORQUE, AND WORK

EXAMPLE—A prony brake attached to a motor as shown in fig. 2-32 has an arm 10.5 inches long. The scale indicates 15 pounds when the pulley revolves at 1800 r p m. Determine the output of the motor at this particular load.

From equation 2-26, $T = 15 \times \frac{10.5}{12} = 13.125$ pound-feet

From equation 2-25, $H\,P = \frac{13.125 \times 1800}{5252} = 4.498$ or 4.5 horse power

EXAMPLE—Find the efficiency of the motor in the above example if it takes 18 amperes at a pressure of 240 volts.

The output of the motor is 4.5 horsepower or 3357 watts. The input is 18 × 240 or 4320 watts. The efficiency is then

$$\frac{3357}{4320} = .777 \text{ or } 77.7\%$$

Transmission of Power—Machines may be divided into two classes—the driving machine, which usually consists of some type of electric motor, and the driven machine, such as a press, a lathe, a planer, or a grinder.

The usual connecting links between the driving and the driven machine are belts on pulleys (or chains on gears), direct gear drives, or shafts.

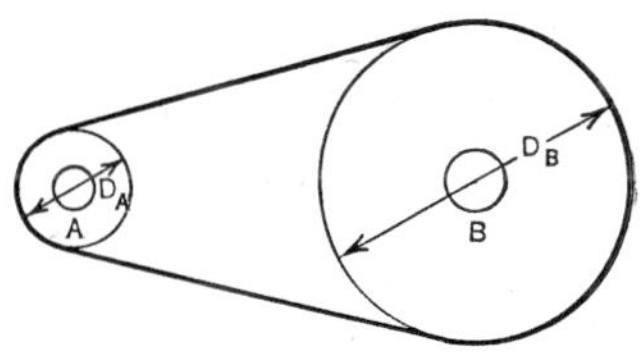

FIG. 2-33

When two pulleys are connected by a belt (fig. 2-33), their surface speeds are the same. Letting D represent the diameter and N represent the revolutions per minute, the surface speed of A is $\pi D_A N_A$, and the surface speed of B is $\pi D_B N_B$. It follows that $\pi D_A N_A = \pi D_B N_B$, or

$$\frac{N_A}{N_B} = \frac{D_B}{D_A} \tag{2-27}$$

This equation shows that the speed of these pulleys is inversely proportional to their diameters. For instance, if B has twice the diameter of A, it will run one-half as fast.

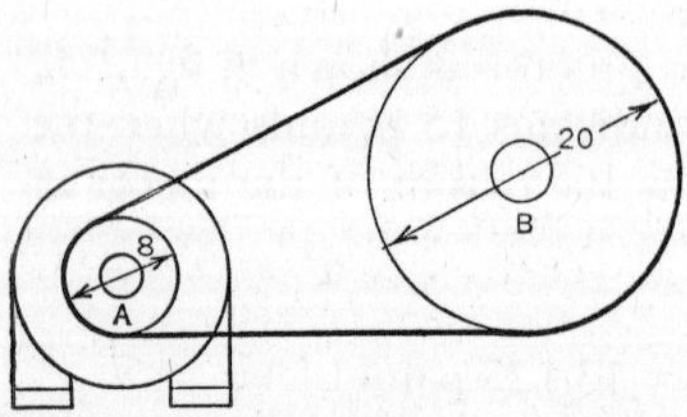

FIG. 2-34

EXAMPLE—Fig. 2-34 shows a 10 H P motor supplying power to a machine by means of a belt drive. If the motor runs at 900 r p m, determine the speed, torque, and horsepower that the driven pulley supplies to the load.

The speed of B is

$$N_B = \frac{N_A D_A}{D_B} = \frac{900 \times 8}{20} = 360 \text{ r p m}$$

The surface force on the two pulleys is the same. Hence for A (or B),

$$F = \frac{5252 \text{ H P}}{RN} = \frac{5252 \times 10}{300} = 175.06 \text{ pounds}$$

Then the torque supplied by B is

$$\frac{10}{12} \times 175.06 = 145.88 \text{ pound-feet.}$$

If friction is neglected, the horsepower delivered by B is the same as that delivered by A, because the product RN in equation 2-23 will be the same in either case.

Equation 2-27, which gives the relative speeds of two pulleys connected by a belt, also applies to two gears in mesh (fig. 2-35). However, in this case we realize that the teeth ratio is proportional to the diameter ratio. Letting T represent the number of teeth, then

$$\frac{N_A}{N_B} = \frac{T_B}{T_A} \qquad (2\text{-}28)$$

Shafts—When two pulleys, two gears, or a pulley and a gear are connected by means of a shaft, the twisting effects of the two are always equal. The following example is given to show how a problem involving at least one of these conditions is solved.

EXAMPLE—Fig. 2-36 shows a 5 H P motor running at 1200 r p m. On its shaft is a 4 inch gear in mesh with a 24 inch gear. The second gear is attached to shaft C, which is 6 inches in diameter. What maximum weight W can a rope winding on this shaft raise?

From the equation 2-23 the force on the surface of A is

$$F = \frac{5252 \text{ H P}}{\text{RN}} = \frac{5252 \times 5}{\frac{2}{12} \times 1200} = 131.3 \text{ lbs.}$$

Since the force acting on B is also 131.3 lbs., the twisting effect of B is

$$T_B = \frac{12}{12} \times 131.3 = 131.3 \text{ pound-feet}$$

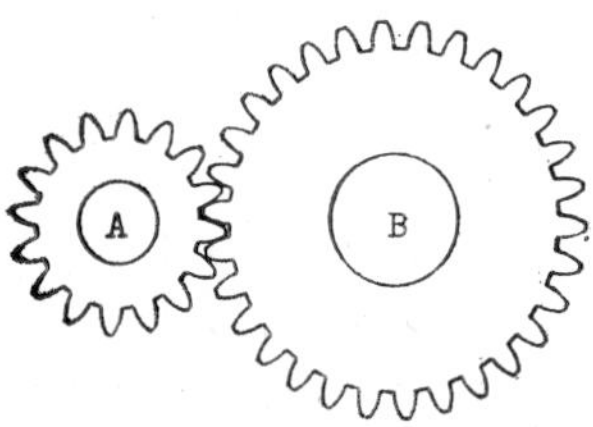

Fig. 2-35

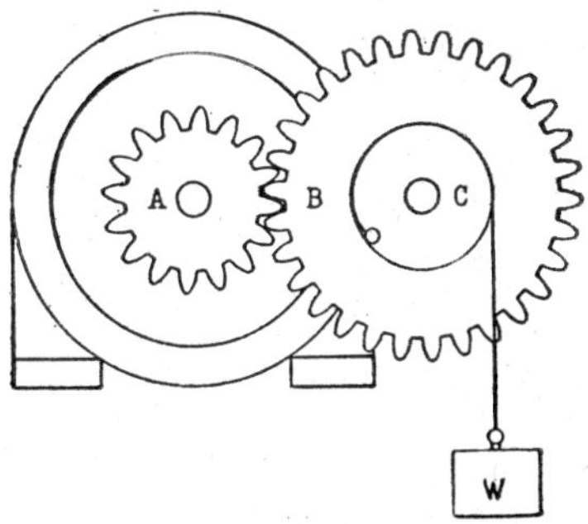

Fig. 2-36

The torque due to the weight W at the end of the rope will be equal and opposite that of pulley B, hence the force F on the surface of shaft C becomes

$$131.3 \div \frac{3}{12} = 525.2 \text{ lbs.}$$

Thus with this combination the 5 H P motor will raise 525.2 pounds at its rated load.

Exercises

1. Define the following: (a) molecule, (b) atom, (c) electron, (d) free electron.
2. What is the difference between atoms of different substances?
3. What is meant by resistance? What causes it?
4. What is an ohm? A megohm? A microhm?
5. What is meant by an ampere? A milliampere? A microampere?
6. What is meant by a volt? A millivolt? A microvolt?
7. Explain voltage, current, and resistance by a water analogy.
8. Find the number of coulombs passing through a 20 ampere motor running for 10 minutes.
9. If a storage battery supplies 900 coulombs in five minutes to a constant load, how much current is it delivering?

10. It takes 360,000 coulombs to charge a certain storage battery. If the charging rate is 15 amperes, how long will it take to fully charge the battery?
11. Why is an electric current assumed to flow in a direction opposite to that in which it actually flows?
12. Why is copper such a good conductor of electricity and mica such a poor one?
13. Write Ohm's law three ways.
14. A voltmeter, which has a resistance of 10,000 ohms, indicates 120 volts How much current passes through it?
15. A soldering iron connected to a 120 volt line takes 2.5 amperes. What is its resistance?
16. What is the voltage drop in a rheostat if its resistance is 100 ohms and .2 of an ampere flows through it?
17. How much current flows through a 100 ohm lamp and a 160 ohm lamp connected in series to a 120 volt line?
18. A 30 ohm resistor, a 100 ohm lamp, and a 150 ohm lamp are connected in series. If there is a 60 volt drop in the resistor, determine the supplied voltage.
19. A 13,000 ohm voltmeter and a 17,000 ohm voltmeter are connected in series across a 550 volt line. What will each indicate?
20. How much resistance must be connected in series with a 50 ohm lamp, rated at 2 amperes, if it is to be used on a 120 volt line?
21. An arc lamp takes 12 amperes at 120 volts. How much resistance must be connected in series with it to reduce the current to 5 amperes?
22. A 4 ohm, a 6 ohm, and a 7 ohm cast iron grid resistor are connected in series and take 10 amperes. How much additional resistance must be connected in the circuit to reduce the current to 8 amperes, if the supplied voltage remains unchanged?
23. A 23 ohm, a 30 ohm, and an 80 ohm appliance are connected in series to a 250 volt generator. If the current supplied by the generator is 1.8 amperes, determine the resistance of the connecting wires.
24. Five lamps of equal resistances are connected in parallel across a 120 volt line. If they take 3 amperes, what is the resistance of each?
25. Three heating elements having resistances of 30, 60, and 80 ohms are connected in parallel. If a current of 8 amperes flows through the 30 ohm branch, determine the current through each of the other two branches.
26. An unknown resistance, a resistance of 15 ohms, and a resistance of 20 ohms are connected in parallel across a 120 volt generator. If the generator supplies 22 amperes, determine the value of the unknown resistance.
27. What is the combined resistance of a .3 ohm and a .75 ohm cast iron grid resistor connected in parallel?
28. The combined resistance of two lamps in parallel is 80 ohms. If the resistance of one is 200 ohms, what is the resistance of the other?

29. What voltage will be required to force a current of 10 amperes through a 16 ohm resistor and a 20 ohm resistor connected in parallel?
30. The current through a 4 ohm coil and an 8 ohm coil connected in parallel is 10 amperes. Determine (a) the voltage across the coils and (b) the current through each coil.
31. Three lamps having resistances of 60, 80, and 100 ohms are connected in parallel. If the current through them is 8 amperes, determine the current through each.
32. Three equal resistors connected in series have a combined resistance of 25 ohms. What would be their combined resistance if they were connected in parallel?

FIND THE TOTAL RESISTANCE IN EACH OF THE FOLLOWING:

(33)

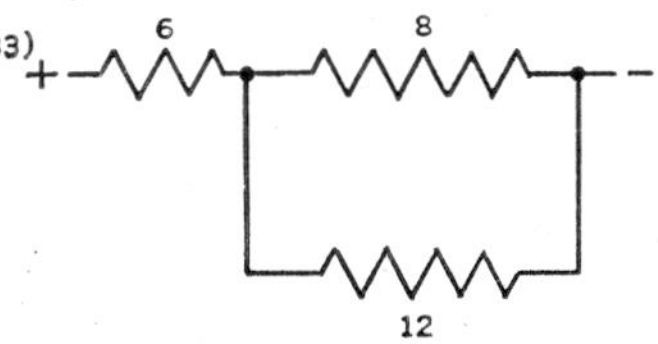

(34)

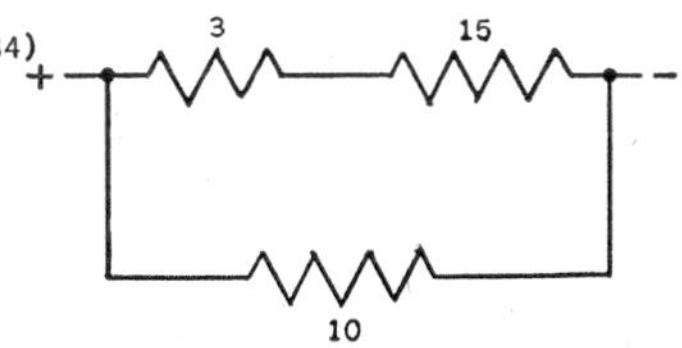

(35)

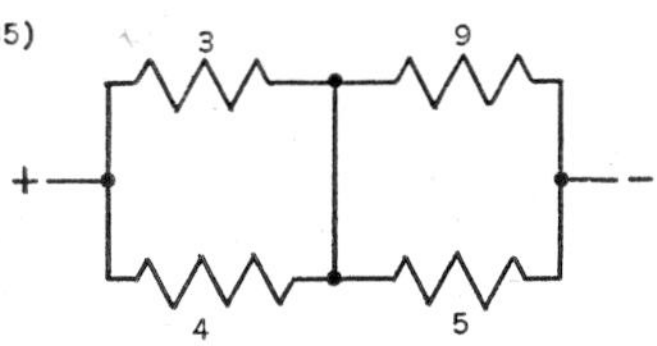

(36)

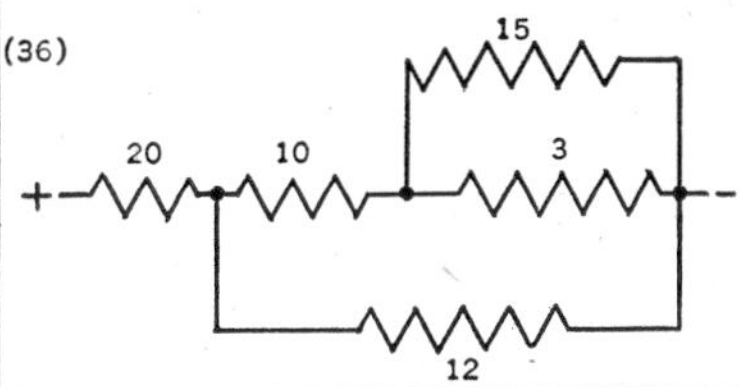

FIND THE VALUE OF R IN EACH OF THE FOLLOWING:

(37)

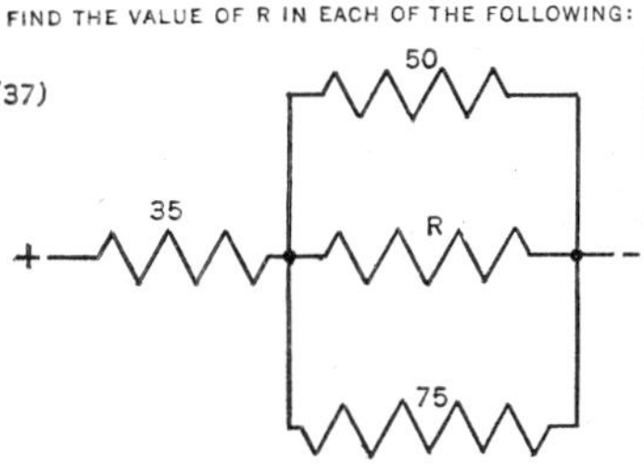

TOTAL RESISTANCE = 50 OHMS

(38)

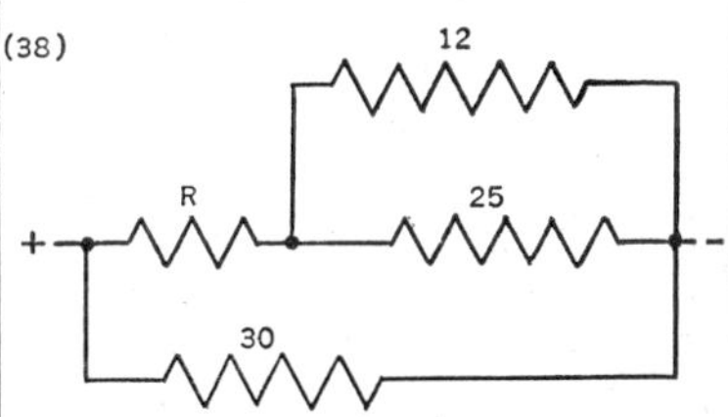

TOTAL RESISTANCE = 16 OHMS

(39)

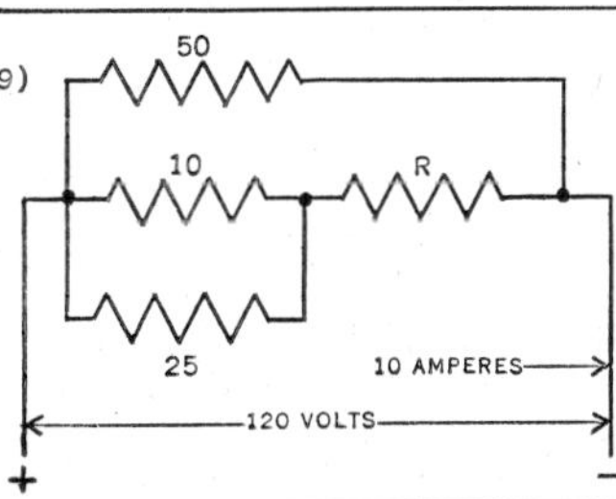

(40)

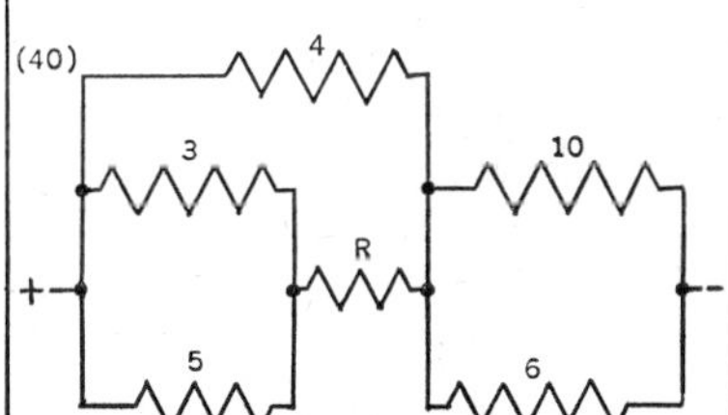

TOTAL RESISTANCE = 6 OHMS

41. In the circuit of fig. 2-37 each lamp takes 2 amperes. Compute (a) the voltage drop in each line wire, and (b) the voltage across each lamp.
42. If the pressure across the generator of fig. 2-38 is 120 volts, (a) what will be the voltage across lamp L_1, and (b) the voltage across lamp L_2?
43. If a current of .8 ampere flows through each lamp in the circuit of fig. 2-39 (a) what current flows through each line wire; (b) what is the voltage across each lamp; (c) what is the voltage across the generator?

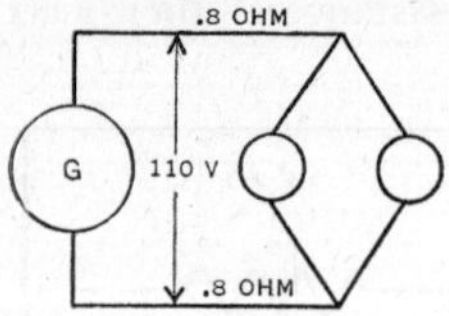

FIG. 2-37

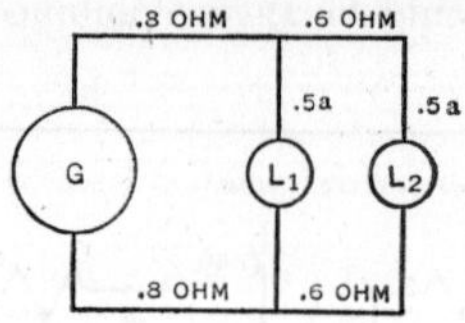

FIG. 2-38

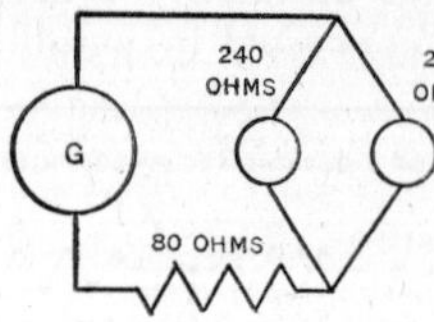

FIG. 2-39

44. A resistance of 6 ohms, a parallel combination of two resistances of 6 and 8 ohms, and a parallel combination of three resistances of 10, 16, and 18 ohms are connected in series. What is the resistance of the combination?
45. Three incandescent lamps having resistances of 94, 80, and 76 ohms are connected in parallel. If the total current flowing through them is 8 amperes, determine (a) the voltage across the combination, and (b) the current that flows through each branch.
46. A 550 volt generator supplies 30 amperes to each of two motors. The resistance of each line wire between the generator and the first motor is .4 ohm, and the resistance of each line wire between the first and second motor is .5 ohm. Determine the voltage across each motor.

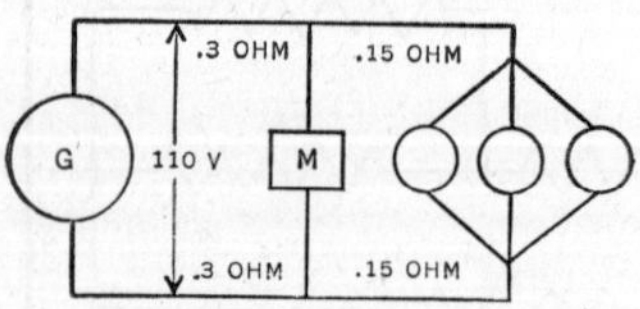

FIG. 2-40

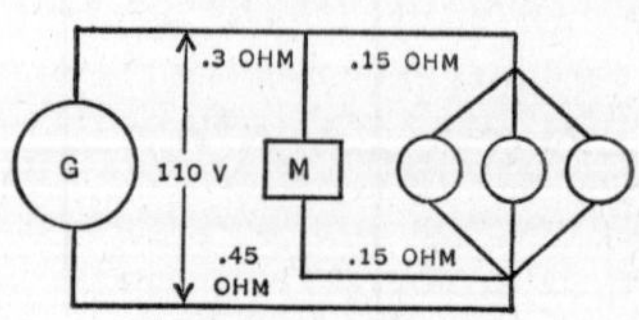

FIG. 2-41

47. What must be the voltage at the generator in problem 46 if the voltage across the second motor is to be 550 volts?
48. The motor in the circuit of fig. 2-40 takes 20 amperes and each lamp takes 2 amperes. The terminal voltage of the generator is 110 volts. Find: (a) the voltage across the motor; (b) across the lamps.

49. If the lamps and motor of problem 48 are connected as shown in fig. 2-41, what is the voltage (a) across the motor, and (b) across the lamps?
50. What is the difference between a primary cell and a secondary cell?
51. What change in the electrolyte occurs during charge and discharge?
52. Can you determine the charge of a battery by taking a hydrometer reading? Explain.
53. Why should the current be decreased during the charging period?
54. Why is pronounced gassing objectionable in a battery?
55. What is the effect of overcharging?

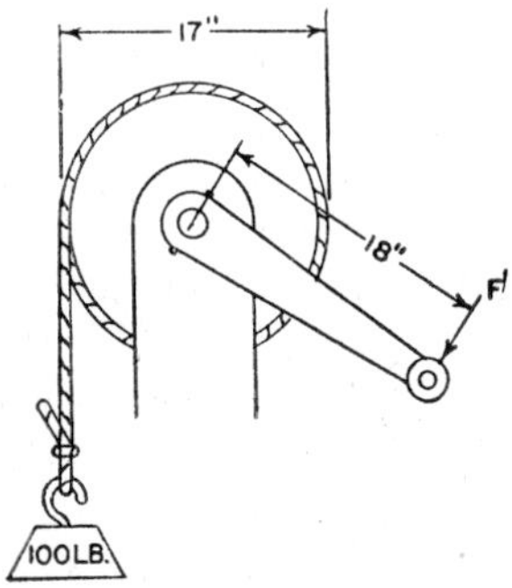

FIG. 2-42

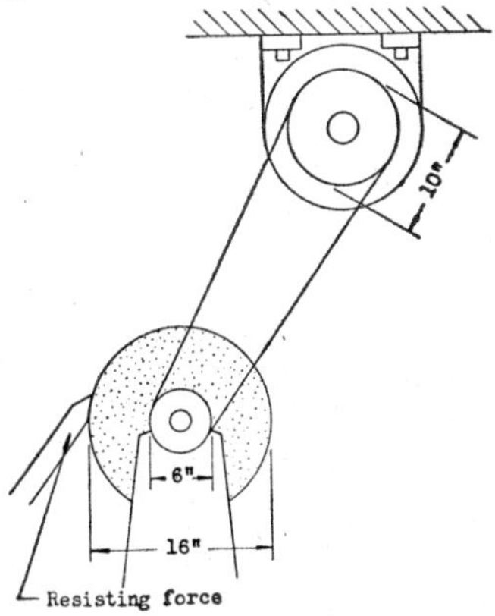

FIG. 2-43

56. In two minutes a man raises a 50 pound weight to a distance of 6 feet. Determine (a) the amount of work done, and (b) horsepower of the man while raising the weight.
57. Compute the number of watts supplied to a 20 ohm electric heater when 8 amperes are flowing through it.
58. What full load current will a 220 volt, 15 horsepower generator deliver?
59. Express 1200 watts in kilowatts. In horsepower.
60. At its rated load a motor is 80 per cent efficient and takes 50 amperes at 110 volts. What is its horsepower output?
61. Determine the number of B T U necessary to raise the temperature of 10 pounds of water 50° F.
62. Find the number of B T U generated by a current of 20 amperes flowing through a 50 ohm resistance for five minutes.
63. Find the force F that must be exerted on the handle of the windlass in Fig. 2-42 in order to raise the weight.

64. A certain motor at full load supplies 12 horsepower to a prony brake and takes 46 amperes at a pressure of 240 volts. Determine its efficiency at full load.
65. Suppose that the prony brake shown in fig. 2-32 is attached to a motor to be tested and that the arm of the brake is 18 inches long. If the motor runs at 1200 r p m and the scale indicates 30 pounds, what horsepower is delivered?
66. What torque does the motor in problem 65 supply?
67. In fig. 2-43 determine the horsepower and speed of the motor driving the grinding wheel, if this wheel rotates at 2000 r p m and has a surface resisting force of 18 pounds.
68. Pumps in the power house of the Ford Motor Company raise 8,300,000,000 pounds of water 2 feet in 24 hours. Compute the horsepower of the electric motors used to drive these pumps if they are 65 per cent efficient.
69. A 12 inch pulley on a motor shaft is belted to a 20 inch pulley. If the torque supplied to the 12 inch pulley is 50 pound-feet, what is the torque delivered by the 20 inch pulley?
70. A 14 inch pulley on the shaft of a 10 horsepower motor drives by belt a 24 inch pulley at 480 r p m. Determine the torque supplied by the motor at full load.
71. A 40 tooth gear on a 5 horsepower motor drives an 18 tooth gear at 2000 r p m. At full load what torque is supplied by the motor?
72. A 36 inch pulley on the shaft of a 12 horsepower motor drives by belt a 30 inch pulley on a second shaft at 1800 r p m. What horsepower motor running at 1200 r p m with an attached pulley must be used to raise the speed of the second shaft to 2400 r p m and supply the same torque to the load?

CHAPTER 3

WIRING

Circular Mils—Due to the fact that a wire generally has a very small diameter, it is convenient to express its thickness in thousandths of an inch. One thousandth of an inch is therefore the unit of measure and is called a mil. Instead of saying that a certain wire has a diameter of .032 inch, it is said to have a diameter of 32 mils. The area of a circular conductor is expressed in terms of a unit of area called the circular mil (fig. 3-1a), which is the area of a circle that has a diameter of one mil. The diameter in mils is denoted by "m" and the area in circular mils by "cm."

The area of a circular conductor is expressed in circular mils because there is a very simple relation between mils and circular mils. This relation may be developed in the following way.

Let A equal the area of a circle expressed in square inches and D equal the diameter in inches. Then

$$A = .7854D^2 \qquad (3\text{-}1)$$

The area of one circular mil is therefore $.7854 \times .001^2 = .0000007854$ square inch. There are as many circular mils in a conductor as the area of one circular mil is contained in the area (A) of the conductor. Therefore,

$$cm = \frac{A}{.0000007854} = \frac{.7854D^2}{.0000007854} \qquad (3\text{-}2)$$

$$\text{But } D = \frac{m}{1000}, \text{ and therefore } D^2 = \frac{m^2}{1{,}000{,}000} \qquad (3\text{-}3)$$

Substituting equation (3-3) into equation (3-2)

$$cm = \frac{.7854 \times \dfrac{m^2}{1{,}000{,}000}}{.0000007854} = m^2 \qquad (3\text{-}4)$$

Therefore,

$$cm = m^2 \qquad (3\text{-}5)$$

Equation (3-5) may be expressed as follows: *To find the number of circular mils in a conductor, square the diameter expressed in mils.*

EXAMPLE—Find the number of circular mils in a conductor .025 inch in diameter.

$$.025 \text{ inch} = 25 \text{ mils}$$

$$\text{Then, cm} = 25^2 = 625 \text{ circular mils}$$

Instead of determining the area of a conductor in square inches, it is easier to determine it in circular mils because the value .7854 is not involved. Also, the answer is a whole number.

EXAMPLE—What is the diameter of a wire in inches if it has an area of 5184 circular mils?

$$m = \sqrt{\text{cm}} = \sqrt{5184} = 72 \text{ mils or } .072 \text{ inch}$$

Mil-foot Wire—A wire one mil in diameter and one foot long is considered a mil-foot wire (fig. 3-2). One foot of any wire may

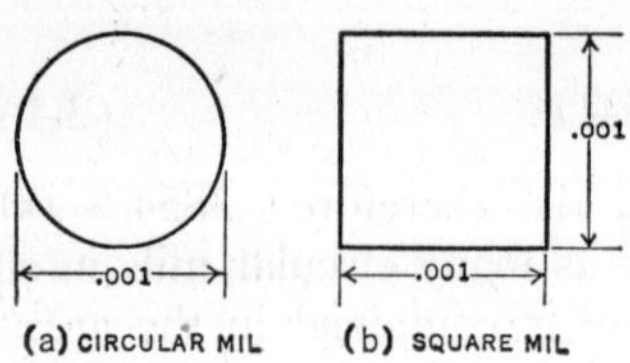

FIG. 3-1

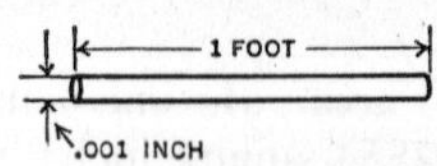

FIG. 3-2

be considered to be composed of a number of mil-foot wires in parallel. The resistance of any number of similar wires in parallel is equal to the resistance of one of them divided by the total number. It has been found experimentally that the resistance of a hard drawn copper wire one mil in diameter and one foot long is 10.4 ohms. The resistance of a wire of the same material and length but 10 circular mils in cross section is 10.4 ÷ 10 or 1.04 ohms. The resistance of a wire varies directly with its length. If the 10 circular mil copper wire is 1000 feet long, its resistance is 1.04 × 1000 or 1040 ohms.

To find the resistance (R) of any wire, then, it is merely necessary to divide the resistance of one mil-foot of the wire (K) by the cross section in circular mils (cm) and multiply by the length (l). The same idea is conveyed by the following equation.

$$R = \frac{Kl}{cm} \qquad (3\text{-}6)$$

EXAMPLE—Find the resistance of a copper wire .15 inch in diameter and one mile long.

A copper wire .15 inch in diameter contains 22,500 circular mils. Then

$$R = \frac{10.4 \times 5280}{22{,}500} = 2.44 \text{ ohms}$$

Square Mils—Areas of rectangular conductors, such as bus bars, are more conveniently expressed in terms of a unit of area called the square mil. One square mil is the area of a square with sides one mil in length (fig. 3-1b), and is denoted by "sq. m." The number of square mils in a rectangular conductor may be determined by multiplying together the thickness and width expressed in mils.

EXAMPLE—Find the area in square mils of a bus bar ¼ inch thick and 2 inches wide.

¼ inch equals 250 mils and 2 inches equals 2000 mils. Then the area is 250 × 2000 or 500,000 square mils.

The area of a rectangular conductor in square mils may also be found by multiplying its area (A) in square inches by 1,000,000. That is,

$$\text{sq. m} = 1{,}000{,}000A \qquad (3\text{-}7)$$

EXAMPLE—How many square mils are contained in a conductor .75 square inch in cross section?

$$\text{sq. m} = 1{,}000{,}000 \times .75 = 750{,}000 \text{ square mils}$$

There are as many circular mils in a rectangular conductor as the area of one circular mil is contained in the area (A) of the conductor. Then the number of circular mils is

$$cm = \frac{A}{.0000007854} \qquad (3\text{-}8)$$

According to equation 3-7, however, $A = \dfrac{\text{sq. m.}}{1{,}000{,}000}$

Substituting $\frac{\text{sq. m}}{1{,}000{,}000}$ for A in equation 3-8, then, we have

$$cm = \frac{\frac{\text{sq. m}}{1{,}000{,}000}}{.0000007854} = \frac{\text{sq. m}}{.7854}$$

$$\text{Then, } cm = \frac{\text{sq. m}}{.7854} \qquad (3\text{-}9)$$

Therefore, to find the number of circular mils in a rectangular conductor, divide the area expressed in square mils by .7854.

EXAMPLE—Find the number of circular mils in a conductor ⅛ by 1 inch.

⅛ inch = 125 mils and 1 inch = 1000 mils. Then the area is 125 × 1000 or 125,000 square mils. Substituting in equation (3-9)

$$cm = \frac{125{,}000}{.7854} = 159{,}154 \text{ circular mils}$$

The resistance of a rectangular conductor may be computed by first finding its area in square mils, reducing this to circular mils, and then substituting in equation 3-6.

TABLE 3-I

Materials	Resistivity (K) at 20° C	Temperature Coefficient (a) per Degree C*
Aluminum	17.7	0.0043
Carbon	–	–0.0005
Constantan	296.	–
Copper	10.4	0.0043
German Silver Wire	200	0.0004
Iron Wire	60	0.006
Iron (cast)	500	–
Manganin	266	0.00002
Nichrome	610	0.0004
Nickel	60	0.006
Silver	9.5	0.004
Steel (soft)	90	0.0044
Steel (hard)	275	0.0016
Tungsten (annealed)	26	0.005
Tungsten (hard drawn)	33	0.005

* Average of values between 0° and 100° centigrade.

Resistivity—Table 3-I gives the resistance per mil-foot, called the resistivity, of the materials most commonly used as conductors. It will be seen that silver has the lowest resistivity. While this metal has excellent properties, it is not used to any great extent because of its high cost.

Nearly all electric conductors are made of copper because this material is nearly as good a conductor as silver, will bend easily without breaking, is strong, is inexpensive, and can be soldered easily.

The resistivity of aluminum is about 1.6 times as great as copper, but due to the fact that aluminum is only 30 per cent as heavy, it has less resistance per pound than copper, for a given length. For equal resistance per foot, aluminum has a greater cross section than copper, and this limits its use with an insulating cover and inclosure. Aluminum has less strength and melts at a lower temperature than copper, and is not readily soldered. It is used to some extent for bus bars, because the large cross section which is required provides more radiating surface than copper would give.

All other materials given in Table 3-I have a much higher resistivity than copper. These materials are generally used in circuits in which resistance is desired to restrict the flow of current or produce heat.

Wire Gages—Wires are manufactured in sizes numbered according to what was formerly known as the Brown & Sharpe (B & S) gage but is now known as the American wire gage (A W G). The numbering of these gages is given in Table 3-II. It will be seen that the numbers are in reverse order, that is, the large numbers denote a small size of wire. For every third gage of smaller size, the wire halves in cross section and therefore doubles in resistance. Therefore, the ratio of cross sections between any two sizes must be $\sqrt[3]{2}$ or 1.26 to 1. Furthermore, a No. 10 wire is practically 1/10 of an inch in diameter, has a cross section of 10,000 circular mils, and has a resistance of about one ohm per 1000 feet. With these facts in mind, it is possible to estimate quickly the area and resistance of any size of wire and any length, without using the table.

For example, to estimate the cross section and resistance for 1000 feet of No. 17 wire, count from No. 10 to No. 16 by threes. This requires two steps, indicating an area one-fourth that of No. 10, or 2500 circular mils, and a resistance of four times that of No. 10, or 4 ohms. Since the ratio of one size to the next is 1.26, the area of

TABLE 3-II

AWG No.	Diameter in Mils	Area in Circular Mils	Ohms per 1000 ft. at 20° C	AWG No.	Diameter in Mils	Area in Circular Mils	Ohms per 1000 ft. at 20° C
000	409.6	167772	.062	20	32.0	1024	10.15
00	364.8	133079	.078	21	28.5	812	12.80
0	324.9	105560	.098	22	25.3	640	16.14
1	289.3	83694	.124	23	22.6	511	20.36
2	257.6	66358	.156	24	20.1	404	25.67
3	229.4	52624	.197	25	17.9	320	32.37
4	204.3	41738	.248	26	15.9	253	40.81
5	181.9	33088	.313	27	14.2	202	51.47
6	162.0	26244	.395	28	12.6	159	64.91
7	144.3	20822	.498	29	11.3	128	81.80
8	128.5	16512	.628	30	10.0	100	103.2
9	114.4	13087	.792	31	8.9	79	131.0
10	101.9	10384	.999	32	7.9	62	164.8
11	90.7	8226	1.260	33	7.1	50	207.0
12	80.8	6529	1.588	34	6.3	40	260.0
13	71.9	5170	2.003	35	5.6	31	329.0
14	64.1	4109	2.521	36	5.0	25	414.8
15	57.1	3260	3.184	37	4.5	20	522.3
16	50.8	2581	4.016	38	4.0	16	659.0
17	45.3	2052	5.064	39	3.5	12	835.0
18	40.3	1624	6.385	40	3.1	10	1040.0
19	35.9	1289	8.051	..	..	..	..

No. 17 is $2500 \div 1.26 = 1984$ circular mils, and its resistance is 4×1.26 or 5.04 ohms.

Current Carrying Capacity of Conductors—When current passes through a conductor, heat is generated and its temperature rises. This rise continues until a temperature is reached at which the air carries away the heat as fast as it is produced (the rate at which heat is carried away is proportional to the temperature difference between the wire and the surrounding air). If the current is high, the temperature reached may be high enough to damage the insulation. For power wiring, where the current may flow for long periods, it should not exceed the values given in Table 3-III, which were taken from the National Electric Code. For aluminum wire, the current should not exceed 84 per cent of the values given.

TABLE 3-III

Size of Conductor	Current Carrying Capacity in Amperes: Rubber Insulation	Varnished Cambric Insulation	Other Insulation
18	3		6
16	6		10
14	15	18	20
12	20	25	30
10	25	30	35
8	35	40	50
6	50	60	70
5	55	65	80
4	70	85	90
3	80	95	100
2	90	110	125
1	100	120	150
0	125	150	200
00	150	180	225
000	175	210	275
0000	225	270	325

In selecting wire sizes, it is essential that the wire be large enough to carry the current without overheating and also large enough to keep a low voltage drop in the line. The requirement of the National Electric Code for conductors supplying current to a motor is that they shall have a capacity not less than 125% of the motor full load current. For feeders supplying power to lamps, the voltage drop should not exceed 3 per cent of that impressed on the input end, because even a low voltage drop causes an appreciable reduction in the efficiency of the lamps. For example, a voltage drop of 10 per cent causes the lamps to supply 30 per cent less light, while the power consumed is reduced only 19 per cent.

If the load consists of motors or heating units, a drop of 10 per cent is not objectionable.

EXAMPLE—What size wire is required for transmitting 10 amperes from a 120 volt generator to a bank of lamps 875 feet away if the line drop is not to exceed 3 per cent of the generated voltage?

The permissible line drop is $.03 \times 120$ or 3.6 volts, and the line resistance must be $3.6 \div 10$ or .36 ohm.

The length of the line is 2×875 or 1750 feet.

The resistance per 1000 feet of wire, then, is $.36 \div 1.75$ or .206 ohm. The wire size nearest this resistance is No. 3 (see Table 3-II). According to Table 3-III, this gage is large enough to carry the current without overheating.

Wiring Systems—Lamps of 120 volt rating are the most common because they are durable and economical. Lamps with a higher rating have a longer and thinner filament, which is more apt to break or burn out, and lamps with a lower rating require a lower voltage supply line. Such a line will have greater copper losses unless larger line wires are used (a more costly installation). A common method of reducing line losses and the amount of copper required for transmission will now be considered.

EXAMPLE—Determine the line loss when eight 120 volt, 60 watt lamps are connected at the end of a 4 ohm line which supplies 120 volts.

Each lamp takes $60 \div 120$ or .5 amp, and the line current is $.5 \times 8$ or 4 amps (see fig. 3-3). The loss in the line wires is therefore $4^2 \times 4$ or 64 watts.

EXAMPLE—Solve the above problem if 240 volts are supplied to the lamps paired off in series as shown in fig. 3-4.

In this case the line current is 2 amperes and the line loss is $2^2 \times 4$

or 16 watts, one-fourth as much loss as in the above problem. This is due to the fact that the line loss varies as the square of the current. For a given power line, then, when the voltage is doubled, the line loss becomes one-fourth as much, or for the same line loss the line resistance can be four times as much, which is obtained by using wires of one fourth the cross section. This suggests the following rule: *When the voltage of a power line is doubled, one-fourth as much copper is required for the same line loss.*

The decrease in power loss, obtained merely by doubling the voltage, suggests connecting 120 volt lamps to a 240 volt line as shown in fig. 3-4. With this system, however, two lamps must be used when only one is needed, four when three are needed, etc. To eliminate this undesirable feature and still supply power to 120 volt lamps at 240 volts, a third wire called the neutral is added to a 240 volt line such as the one shown in fig. 3-4, forming what is known

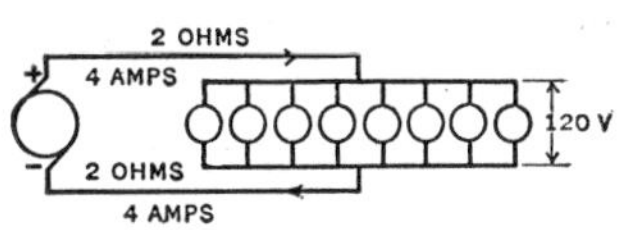

FIG. 3-3

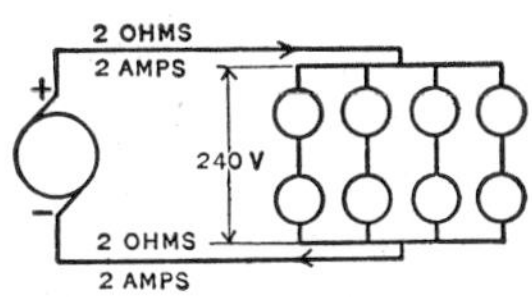

FIG. 3-4

as a three wire system. The National Electric Code requests that this neutral wire be of the same size as each of the other two, which is necessary when the lights on only one side of the system are turned on.

It was shown above that a 240 volt, two wire system requires one-fourth the copper of a 120 volt system. By employing a neutral wire the same size as each of the others (see fig. 3-5), the amount of copper required is $\frac{1}{4} + \frac{1}{8}$ or $\frac{3}{8}$ of that required in a 120 volt system.

In a three wire system, a small unit called a balancer set is generally used, to which the neutral is attached. In dealing with lighting loads, however, it is more convenient to consider the power as being supplied from two generators in series, as shown in fig. 3-6.

Three Wire Systems—When the lamps on each side of the neutral take the same amount of current, as in fig. 3-5, the neutral carries no current and the system is said to be balanced. When the lamps on one side take a different amount of current to those on the other side, the system is said to be unbalanced. In fig. 3-6, the lamps on one side of the neutral take 3 amps and those on the other side take 1 amp. The neutral must carry the difference of the currents in the two outer wires, or 2 amps. In practice, an effort is made to keep the system balanced because in such a system the current and voltage distributions are the same as in the 240 volt, two wire system with lamps connected series parallel. A balanced condition seldom exists, however, because more lamps are usually turned on on one side of the neutral than on the other. In an unbalanced system, the voltage

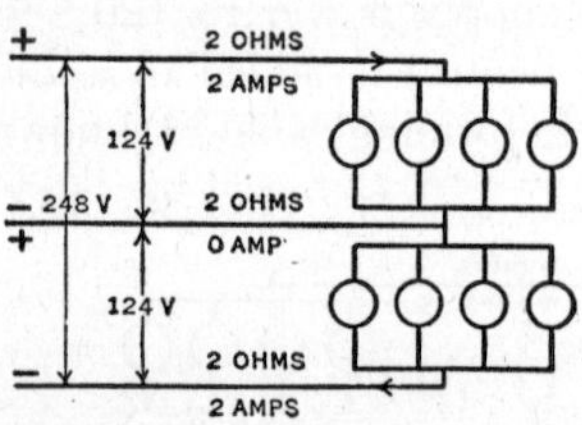

FIG. 3-5

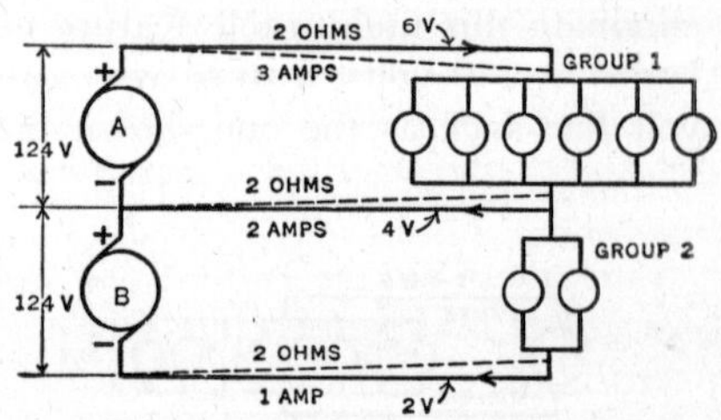

FIG. 3-6

at the load, between the neutral and the + wire, is not the same as that between the neutral and the − wire.

EXAMPLE—Determine the voltages across group 1 and group 2 in fig. 3-6.

The voltage drop for each line wire is indicated graphically by dotted lines. The fall of potential, which, in a line wire, always occurs in the direction in which the current flows, is indicated by a downward slope. In determining the voltage across group 1, the dotted lines indicate that the voltage drop in the positive line and in the neutral line must be subtracted from the voltage of generator A, which gives $124 - 6 - 4 = 114$ volts. In determining the voltage across group 2, the dotted lines indicate that the voltage drop in the neutral must be added to the voltage of generator B, while the voltage drop in the negative wire must be subtracted, giving $124 + 4 - 2 = 126$ volts. Since the lamps are all rated at 120 volts, the voltage across group 1 is too low and the voltage across group 2 is too high.

Open Neutral in a Three Wire System—If the neutral in an unbalanced system becomes open, current and voltage distributions make severe changes.

EXAMPLE—Determine the current through and the voltage across each lamp of group 1 and group 2 in fig. 3-6 if the neutral wire is opened.

With an open neutral wire, groups 1 and 2 form a series circuit across 248 volts. Since each lamp has a rating of 60 watts and 120 volts, each has a resistance of 240 ohms. The resistance of group 1 is one-sixth of 240 or 40 ohms and the resistance of group 2 is one-half of 240 or 120 ohms. The total resistance, then, is $40 + 120 + 4$ or 164 ohms, and the line current is $248 \div 164$ or 1.512 amps. The current through each lamp of group 1 is $1.512 \div 6$ or .252 amp, and the voltage across these lamps is $.252 \times 240$ or 60.48 volts. The current through each lamp of group 2 is $1.512 \div 2$ or .756 amp, and the voltage across these lamps is $.756 \times 240$ or 181.44 volts. Thus the voltage across the lamps in group 1 is below their rated value, causing them to barely glow, and the voltage across the lamps in group 2 is above their rated value, causing them to burn out. In practice, the neutral is never fused, so that such trouble cannot occur.

Magnet Wire—Coils are wound with wire insulated by enamel, single or double layers of silk or cotton over bare wire, or single or double layers of silk or cotton over enameled wire. When the space is limited, enamel or single silk covered wire is generally used. Enameled wire costs less. Its insulation has a breakdown temperature of about 550° F and will withstand an electric pressure of 500 volts per mil in thickness. However, this wire must be handled carefully to prevent scratching or cracking the insulation. Where this happens, shorts or grounds may occur. Silk and cotton are injured by charring at a temperature of about 240° F. These materials insulate wire without enamel by spacing it. Wire with either of these insulations will withstand 125 volts per mil. The insulating ability of silk and cotton is considerably improved by baking the coil to remove moisture, dipping it in an insulating varnish while it is hot, and then baking it for several hours. When space permits, double cotton covered wire should be used, as its cost is low and its insulation is very good after it has been impregnated with varnish.

The number of wires that may be placed in one square inch of cross section is given in Table 3-IV. This data is used extensively in figuring coils.

TABLE 3-IV

AWG No.	Area in Circular Mils	Approximate turns per square inch for different types of insulation				AWG No.	Area in Circular Mils	Approximate turns per square inch for different types of insulation					
		Single Cotton	Double Cotton	Enamel	Enamel Single Cotton			Single Cotton	Double Cotton	Enamel	Enamel Single Silk	Single Silk	Enamel Single Cotton
000	167772	..	5.4	..	..	20	1024	775	625	912	812	861	724
00	133079	..	6.8	..	..	21	812	940	750	1150	1000	1065	895
0	105560	..	9	..	..	22	640	1149	910	1429	1225	1330	1072
1	83694	..	11	..	..	23	511	1400	1075	1779	1500	1649	1305
2	66358	..	14	..	..	24	404	1700	1260	2240	1865	2051	1575
3	52624	..	18	..	..	25	320	2059	1505	2820	2300	2540	1905
4	41738	23	22	..	..	26	253	2500	1750	3560	2850	3100	2310
5	33088	29	27	..	..	27	202	3030	2020	4419	3465	3790	2775
6	26244	36	34	..	..	28	159	3669	2305	5579	4220	4710	3345
7	20822	45	42	..	..	29	128	4300	2700	6900	5100	5605	3900
8	16512	55	51	..	..	30	100	5039	3015	8700	6165	6930	4660
9	13087	69	63	..	..	31	79	5919	3460	10700	7345	8472	5275
10	10384	87	79	93	84	32	62	7059	3910	13500	8885	10125	6245
11	8226	108	98	115	104	33	50	8119	4370	17000	10700	12150	7355
12	6529	134	120	145	130	34	40	9600	4870	21100	12800	14350	8310
13	5170	169	149	184	162	35	31	10900	5390	26300	15000	17397	8700
14	4109	210	183	232	197	36	25	12200	6920	32000	17300	20402	10705
15	3260	261	223	292	251	37	20	..	..	39750	20500	23710	..
16	2581	320	271	365	305	38	16	..	..	49350	23500	27495	..
17	2052	397	330	459	374	39	12	..	..	61200	..	..	..
18	1624	492	400	571	453	40	10	..	..	76100	..	..	..
19	1289	592	479	718	552	..	..	..	..	..	..	..	..

EXAMPLE—Determine the cross section of a 200 turn coil wound with No. 18 enamel, single cotton covered wire.

According to Table 3-IV, 453 turns of this wire occupy one square inch. Then 200 turns occupy 200/453 of one square inch or .44 square inch.

Temperature Scales—Temperature is expressed in the centigrade scale in some cases and in the Fahrenheit scale in others. It is therefore necessary to be familiar with both and to be able to determine the temperatures on one scale that correspond to those on the other. The two scales are shown in fig. 3-7. Zero and 100 degrees are taken as the freezing and the boiling points respectively on the centigrade scale, while 32 and 212 are taken on the Fahrenheit scale. Between these two points, then, the centigrade scale has 100 divisions and the Fahrenheit scale has 180 divisions. Consequently, a Fahrenheit degree is 5/9 as large as a centigrade degree. To determine a centigrade temperature (C) corresponding to a Fahrenheit temperature (F), subtract 32 and multiply by 5/9. That is,

$$C = \frac{5}{9}(F - 32)$$

To determine a Fahrenheit temperature corresponding to a centigrade temperature, multiply by 9/5 and add 32. That is,

$$F = \frac{9}{5}C + 32$$

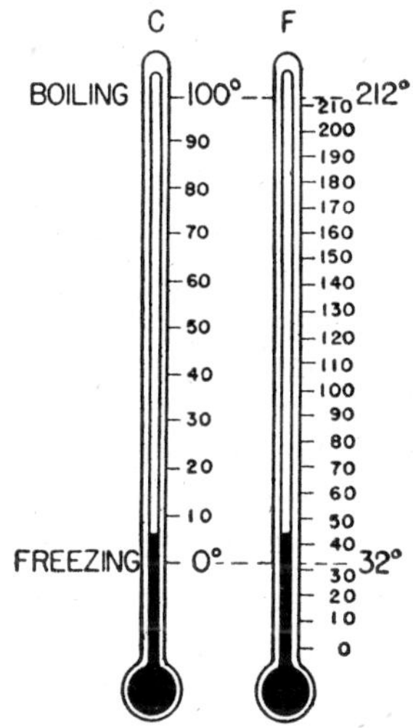

FIG. 3-7

EXAMPLE—What temperature in degrees centigrade corresponds to 68 degrees Fahrenheit?

$$C = \frac{5}{9}(68 - 32) = 20°$$

Change in Resistance Due to a Change in Temperature—The resistance of pure metals, such as silver, copper, aluminum, etc., increases as the temperature becomes greater. Some alloys, such as constantan and manganin, change very little in resistance, and these are used in measuring instruments, where it is desired to have the resistance of coils remain constant so that the instruments indicate accurately. The resistance of carbon and electrolytes decreases with a rise in temperature.

The amount by which the resistance of a substance changes per degree change in temperature for each ohm of resistance at 0° C is called the temperature coefficient of resistance. For pure metals, this coefficient (denoted by the letter "a") has a value of approximately .004 (see Table 3-I).

Letting R_o represent the resistance at 0° C, t_f the final temperature, then R_oa is the change in resistance for one degree rise in temperature, while R_oat_f is the change in resistance for t_f degrees rise in temperature.

Then the final resistance (R_f) will equal the resistance at 0° C, which is R_o, plus the change in resistance (R_oat_f). That is,

$$R_f = R_o + R_oat_f = R_o(1 + at_f) \qquad (3\text{-}10)$$

Example—The resistance of a nichrome heating element is 50 ohms at 0° C. What is its resistance at 1000° C?

According to Table 3-I, the value of "a" for nichrome is .0004. Then

$$R_f = 50(1 + .0004 \times 1000) = 70 \text{ ohms}$$

In some problems it is necessary to determine the final resistance when a change in temperature occurs from a value other than 0° C. This may be done without first finding the resistance at 0° C, as would be required by using equation (3-10).

Letting R_i equal the initial resistance and t_i the initial temperature, then

$$R_i = R_o(1 + at_i) \qquad (3\text{-}11)$$

Solving for R_o in equation 3-11 and substituting its value in equation 3-10, we have

$$R_f = \frac{R_i(1 + at_f)}{(1 + at_i)} \qquad (3\text{-}12)$$

Example—The resistance of a coil wound with iron wire is 200 ohms at 20° C. What is its resistance at 80° C?

$$R_f = \frac{200(1 + .006 \times 80)}{(1 + .006 \times 20)} = 264.3 \text{ ohms}$$

It is sometimes necessary to determine the temperature rise that a machine undergoes when started and run at full load for a long period of time. This may be done by measuring its resistance before starting the motor, at a room temperature of 20° C, and again after the run. Then, substitute in equation (3-12), expressed as follows:

$$t_f - t_i = \left(\frac{1}{a} + t_i\right)\frac{R_f - R_i}{R_i} \qquad (3\text{-}13)$$

EXAMPLE—The winding of a motor has a resistance of 10 ohms when the machine is started and has a resistance of 12 ohms after it has run several hours at full load. If the room temperature is 20° C, what is the temperature rise of the winding?

$$t_f - t_i = \left(\frac{1}{.0043} + 20\right)\left(\frac{12 - 10}{10}\right) = 50.5 \text{ degrees}$$

Wiring Methods—While a large number of wiring methods are permitted by the National Electric Code, many of these are seldom employed in practice because of poor appearance, lack of durability, high cost of installation, etc. In the following pages only the wiring methods in common use will be considered, such as wiring with rigid conduit, electric metallic tubing, flexible conduit, armored cable, leaded armored cable, non-metallic sheated cable, and surface metal raceway.

Rigid Conduit—A rigid conduit wiring system is one in which the conductors are installed in iron or steel pipes and attached equip-

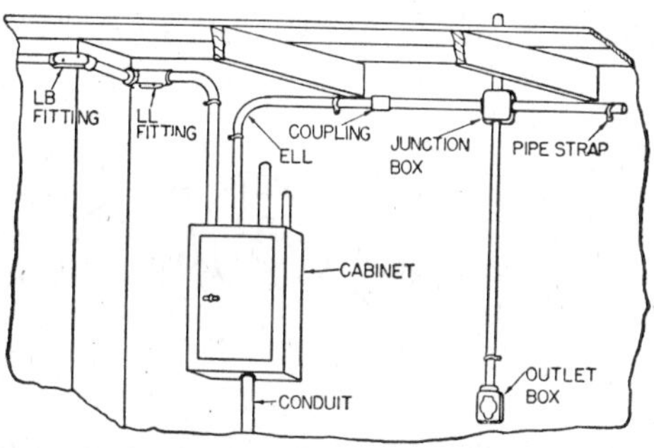

FIG. 3-8

ment(see fig. 3-8). These pipes are called conduit. A good grade has both the interior and the exterior coated with zinc and enamel.

This grade should be used in damp places and in concrete. A less expensive grade has its interior and exterior surfaces coated with either enamel or zinc, or has its interior surface coated with zinc and its exterior surface coated with enamel. Rigid conduit has the same dimensions and thread size as gas pipe, but is softer and therefore can be bent more easily. The Code requests that the radius of the inner edge of any bend be not less than six times the internal diameter of the conduit. If the radius of the bend is made smaller than this, the conduit is apt to be injured or its internal diameter reduced. The actual diameter of conduit is slightly greater than the so-called size. This is shown in Table 3-V.

TABLE 3-V

Size of Conduit	Internal Diameter in Inches	Internal Area in Square Inches
1/2	.622	.3
3/4	.824	.53
1	1.049	.86
1-1/4	1.380	1.5
1-1/2	1.610	2.04
2	2.067	3.36
2-1/2	2.469	4.79
3	3.068	7.38

Benders—Various devices for bending conduit are manufactured. Some of the most common types are shown in fig. 3-9. The bender shown in fig. 3-9a is commonly called a hickey and is used for bending

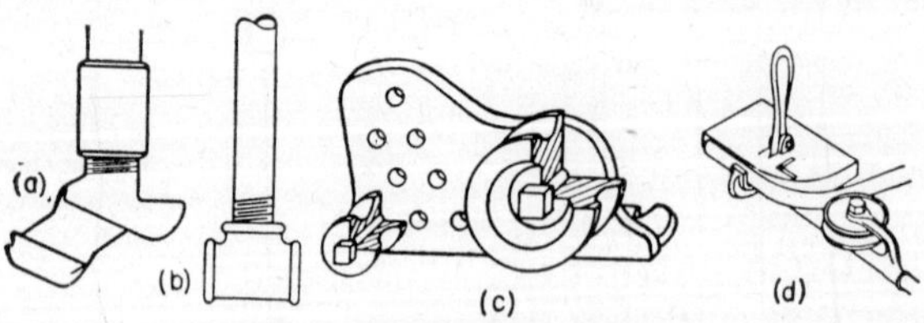

FIG. 3-9

small size conduit. A hickey may be made by screwing a one-inch pipe, about 3½ feet long, into a tee (fig. 3-9b). This hickey is less suitable than the one shown in fig. 3-9a, because the head cannot be put around the conduit from the side, but must be put in place from the end. Then, too, it produces kinks unless it is shifted frequently while making a bend. For bending large size conduit, the benders shown in figs. 3-9c and 3-9d are generally used.

Making a Right Angle Bend—To make a right angle bend with the hickey shown in fig. 3-9a, proceed as follows: Measure from

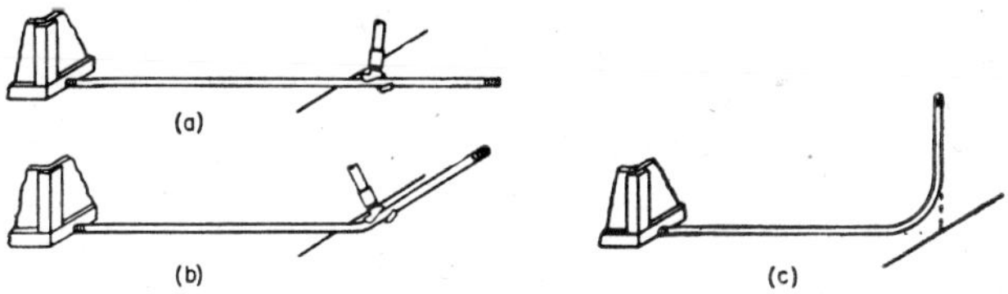

FIG. 3-10

the point where the conduit is to start to the outside of the bend to be formed, and then lay off this distance from a wall. This is indicated by the line in fig. 3-10a. Place one end of the conduit against the wall and slip the hickey over the other end to a point about two inches beyond the line. Stand on the conduit with one foot where the bend is to start and pull on the bender until the end beyond the bend forms a small angle (fig. 3-10b). Now slide the head of the hickey slightly

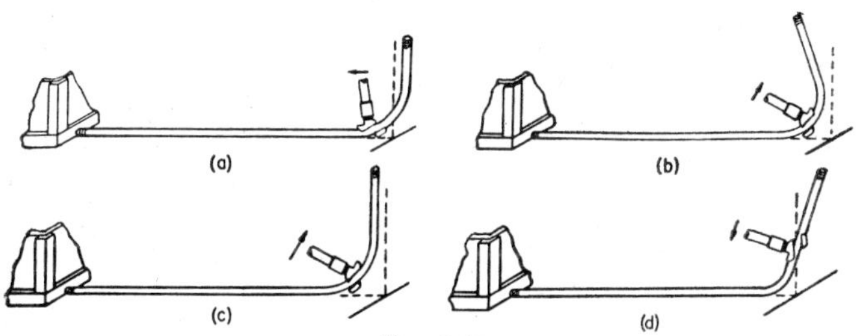

FIG. 3-11

nearer the wall and bend further. Continue this procedure until the outer edge of the raised portion aligns vertically with the line drawn on the floor (fig. 3-10c).

If the vertical portion of the conduit lies beyond the line, slip the hickey from its last position so that it is about ¾ of an inch nearer the wall, and bend the conduit further (fig. 3-11a). Next turn the head around and move it near the center of the bend (fig. 3-11b), and then pull up on the handle until the raised section is vertical.

If the raised end falls short of the line, place the hickey at about the center of the bend (fig. 3-11c) and pull up on the handle. Then turn the head of the hickey around, move it up the pipe, and pull down until the raised section is vertical.

End Bend—It is sometimes necessary to have a small radius bend at the end of a conduit run, threaded as near the bend as possible (fig. 3-12a). To make such a bend, a coupling is first screwed on to prevent injury to the threads. Then a small radius bend is made near the end and the coupling used in protecting the threads is removed. Next the die in the stock is reversed (fig. 3-12b) and the conduit is threaded to the bend. The die is then removed and the surplus length cut off.

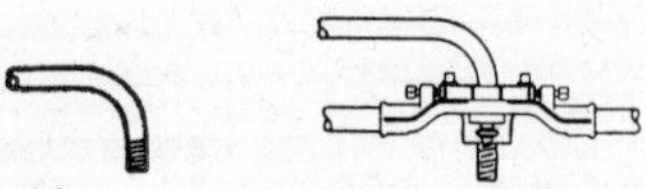
(a) (b)
FIG. 3-12

Reaming—After a piece of conduit has been cut, its ends should be reamed to remove the burr formed by the cutting (fig. 3-13). If this is not removed, it will probably injure the insulation on the

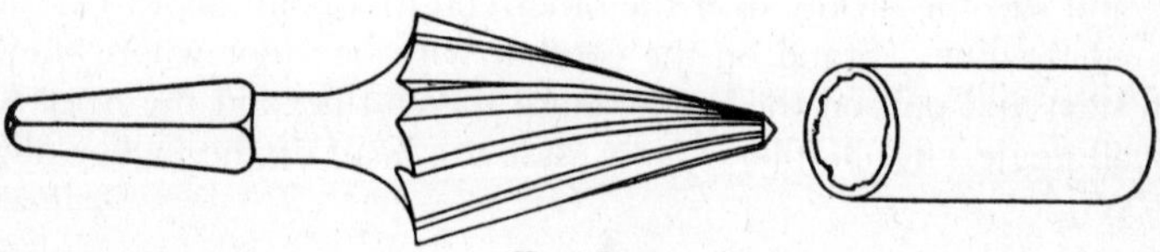
FIG. 3-13

conductors when they are drawn into the conduit. The burr is usually removed from small conduit by a reamer (fig. 3-13), which is turned by a brace, and from large conduit by means of a half-round file.

Making an Offset—Fig. 3-14 shows a properly made offset. The distance within which the bends fall, which is indicated by B, should be as short as possible if the conduit is exposed, but the radius of each bend should not be less than that permitted by the Code. A convenient way of making the offset shown in fig. 3-14a is to make the measurements A, B, and C. Then mark these measurements on the floor and draw the additional lines shown in fig. 3-14b. The run must lead up to 1, extend diagonally to 2, and then extend parallel to the straight section to 1 as shown in fig. 3-14b. To make the first bend, mark off the distance A from one end of the conduit, slip on the hickey so that it will line up with this mark, place it in the position shown in fig. 3-14c, and then press down on the unmeasured end of the conduit until the bend is found to pass through the center of the rectangle in fig. 3-14b. Next slip the conduit to the point where the second bend begins, turn the unmeasured end upward, and pull down on this end until the offset is complete (see fig. 3-14 d.)

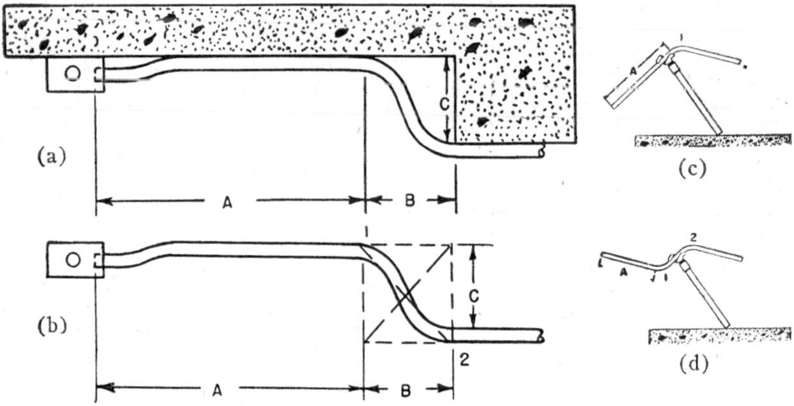

FIG. 3-14

Equipment Used in Conduit Systems—Conduit must always be fastened rigidly in place, and its ends must be connected to cabinets, boxes, or fittings.

Cabinets—A cabinet is made of sheet steel and encloses a panel board on which branch circuits are connected to a feeder. Surround-

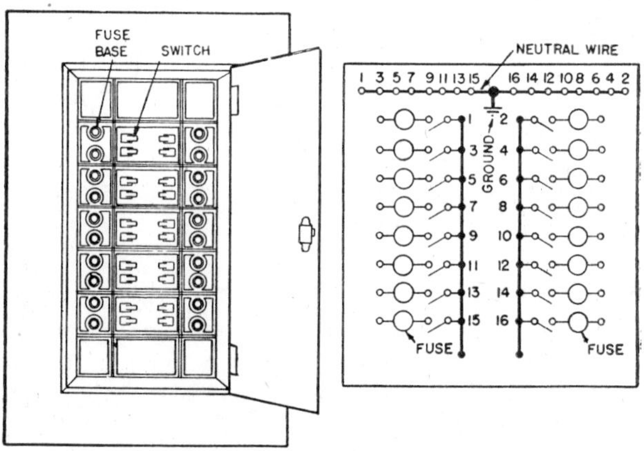

(a) FIG. 3-15 (b)

ing the cabinet is a wiring gutter to house the section of the conductors extending from the conduit to the fuses. Fig. 3-15a shows a three

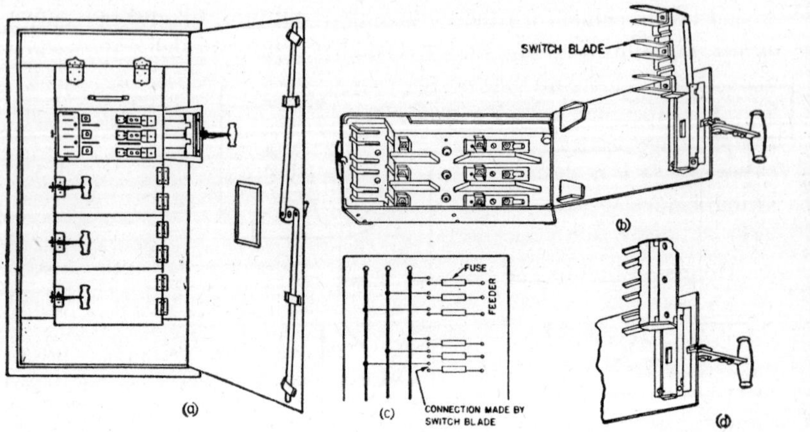

Fig. 3-16

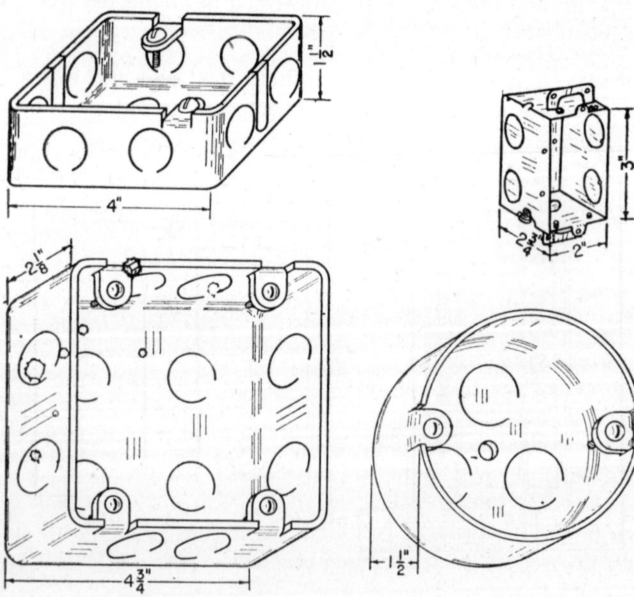

(a) Fig. 3-17 (b)

wire lighting cabinet. It has a grounded unfused neutral, as shown in fig. 3-15b. Fig. 3-16a shows a power cabinet, in which the switch blades are mounted on the doors. When a door is opened (fig. 3-16b), a circuit, together with its fuses, is entirely disconnected from the feeders (fig. 3-16c). With this construction, then, no live parts are exposed. If it is desired to keep the branch circuit disconnected, the switch blades are changed to the "off" position, as shown in fig. 3-16d.

Boxes—Conduit systems contain both junction and outlet boxes. A junction box is placed where several conduit runs intersect (see fig. 3-8). It provides means for pulling in the wire and serves as an inclosure for the connections that must be made. Junction boxes metallically connect conduit runs, making it possible to ground the entire system with one ground connection. They are also used as outlet boxes. For half-inch conduit, these boxes are generally made in two sizes, as shown in fig. 3-17a.

An outlet box (fig. 3-17b) may be used to inclose a switch which controls the current in a circuit, or can be fitted with a special cover through which light and power loads can be connected.

For concealed work, all boxes should be mounted so that their outer edges are flush with the wall surface (fig. 3-18).

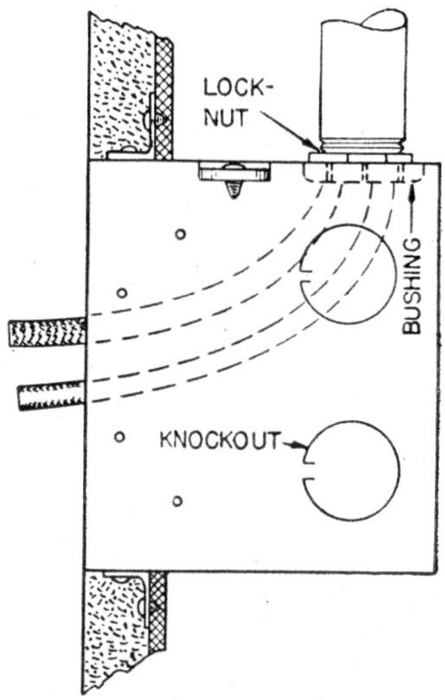

Fig. 3-18

Fittings—Some common types of conduit fittings are shown in fig. 3-19. They are called by various names, such as condulets, unilets, and pipe taplets. These fittings are made with two types of projections. One type has a female pipe thread so that it can be screwed directly to a threaded piece of conduit. The other has a projection which is fastened to a threadless piece of conduit (fig. 3-20). The use of fittings with such projections is increasing because of a considerable saving in time. Both types are shown in fig. 3-19.

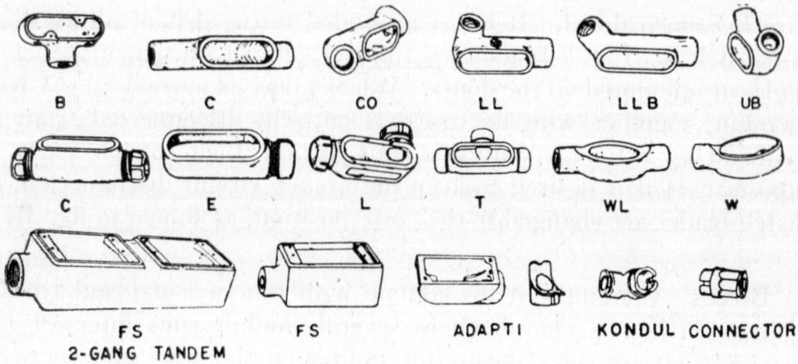

FIG. 3-19

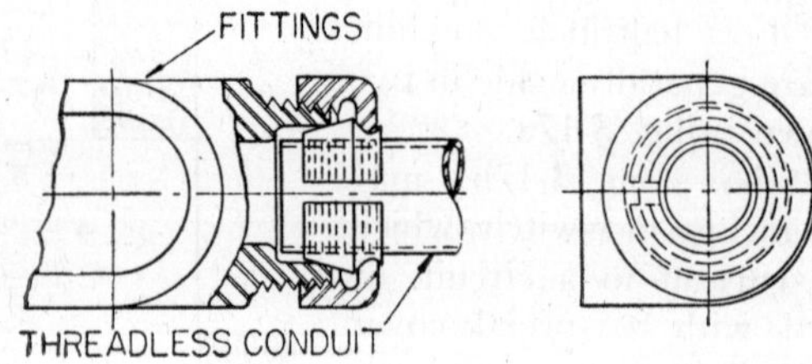

FIG. 3-20

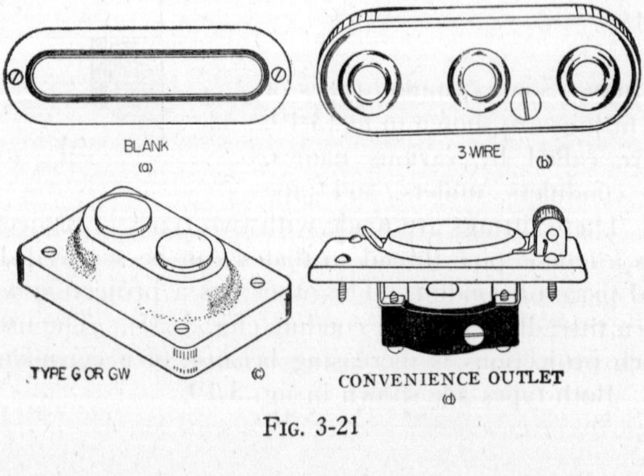

FIG. 3-21

The principal difference between a box and a fitting is a difference in construction that makes it necessary to use different methods of attaching conduits to them. In many cases the two serve the same purpose. Both are used for exposed work but fittings are preferable because they are more nearly watertight, occupy less space, and are more attractive.

If fittings merely inclose splices made in conductors, a blank cover is used (fig. 3-21a). If the conductors are to protrude from a fitting, a porcelain cover is used (fig. 3-21b). Switches and plugs are also attached to fittings (figs. 3-21c and 3-21d).

Multiple Conduit Run—In fig. 3-22a a multiple run made with standard elbows is shown. Such a run will have a much neater appearance if the conduit is parallel, not only in the straight portions but also at the bends (fig. 3-22b). It is therefore necessary for the bends to be handmade. In making them, it is advisable to lay out the contour of each bend on the floor and then form the bends to conform with the lines. This is done in the following way.

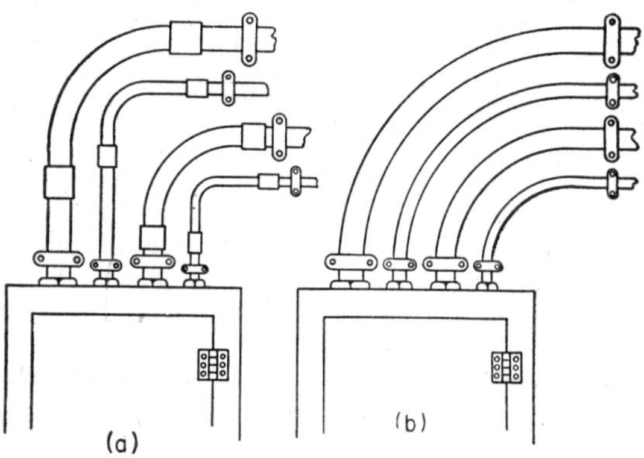

FIG. 3-22

Draw two lines at right angles to each other. With a compass formed by tying a cord to a piece of chalk, and the intersection of the two lines as a center, draw arcs for each conduit of the multiple run, as shown in fig. 3-23a. When a bend is being made, it should be laid alongside the chalk line drawn for it, from time to time, to make sure that the contour is correct (fig. 3-23b).

A conduit installation that does not look well is permissible if it is concealed, but should not be tolerated if exposed. Bends in conduit should not have kinks (fig. 3-24a). Conduit should enter at right angles to openings (fig. 3-24b). If it does not, the locknuts and bushings cannot be screwed up tight and will therefore become loose. The straight sections of an offset should fit against the wall (fig. 3-24c). When a run is to extend around a corner, a fitting should be used and its opening should lie in the direction the wire is to be pulled. In extending a conduit run around a beam, either method shown in fig. 3-25 may be employed. If fittings are used, they should be installed so as to present a symmetrical appearance, as shown in fig. 3-25a. This method is not always used, however, because it is expensive. If the bends are handmade, the conduit should fit flush against the beam, as shown in fig. 3-25b. The bends should be made with as small a radius as possible.

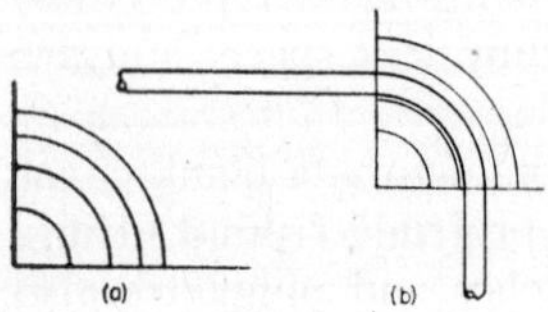

Fig. 3-23

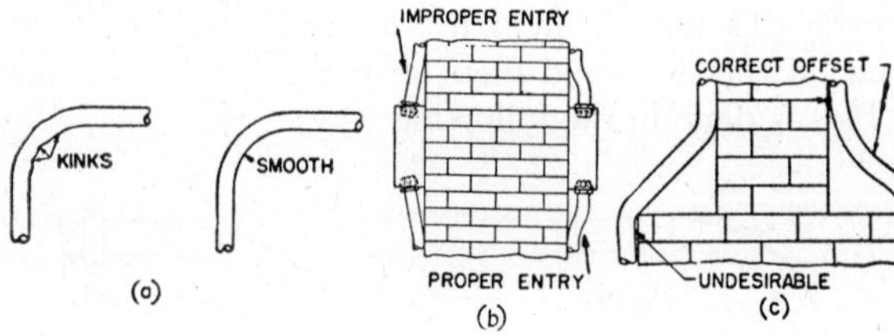

Fig. 3-24

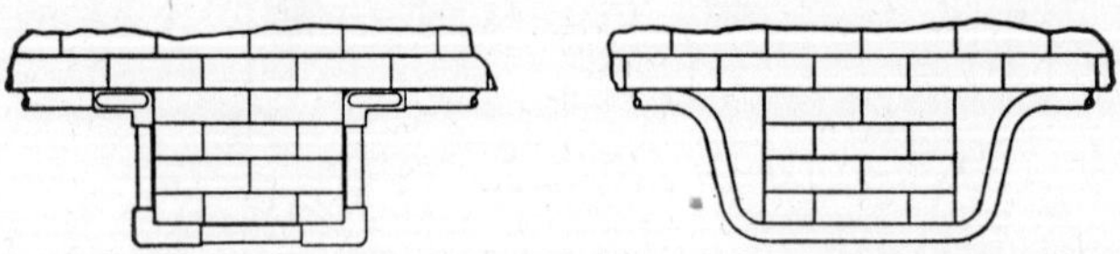

(a) Fig. 3-25 (b)

In general, the conduit system of wiring is considered the best, but it is also the most expensive. It may be concealed or exposed. If the conduit is in a dry location, the conductors need only be rubber covered, but if it is in a moist location they must be lead covered as well. Rigid conduit is generally used only in damp places or in locations where the wire is apt to be injured.

Method of Pulling in Conductors—Conductors must be inserted in conduit runs only after the conduit system has been completely installed. If the runs are plastered over, the conductors must be inserted only after the plaster has dried and all of the other work is finished. By wiring in this manner, much time is saved and better work is obtained. Also, the system is laid out so that the wire may be withdrawn and reinserted at any time. This must be done if the wire becomes faulty.

Methods of Fishing—The most common method of inserting conductors that cannot be pushed through a conduit run is as follows: First, push through a tempered steel tape (known as a fish tape)

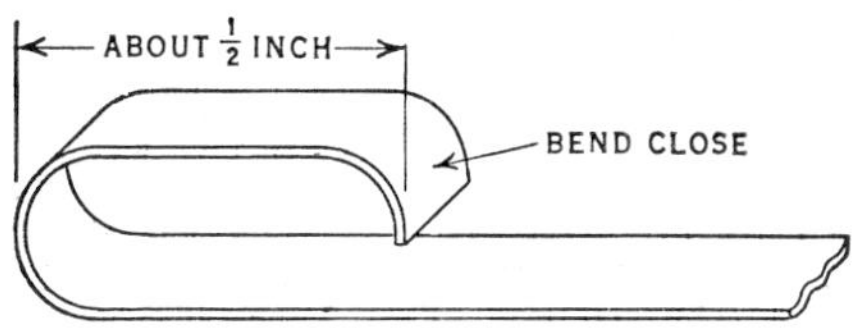

FIG. 3-26

which has a hook bent at its end, as shown in fig. 3-26 (this hook usually sliding easily through couplings and elbows). Next, fasten

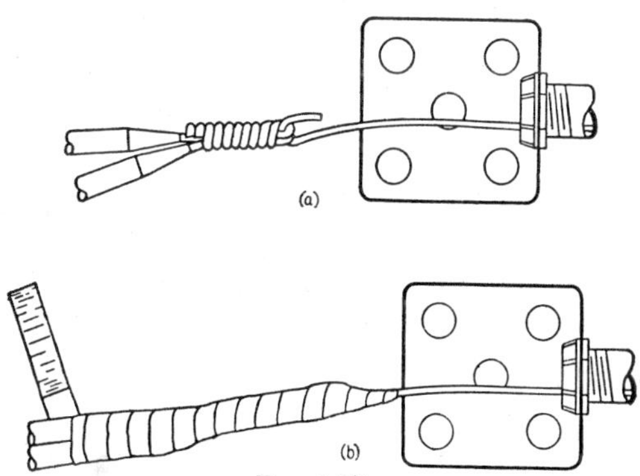

FIG. 3-27

the conductors to the hook, as shown in fig. 3-27a, and tape the joint (fig. 3-27b). The wire is then pulled into the conduit.

If an attempt is made to push a tape through a conduit run that contains a large number of bends, the tape frequently buckles at one of the bends as shown in fig. 3-28. Even if the tape is pushed through,

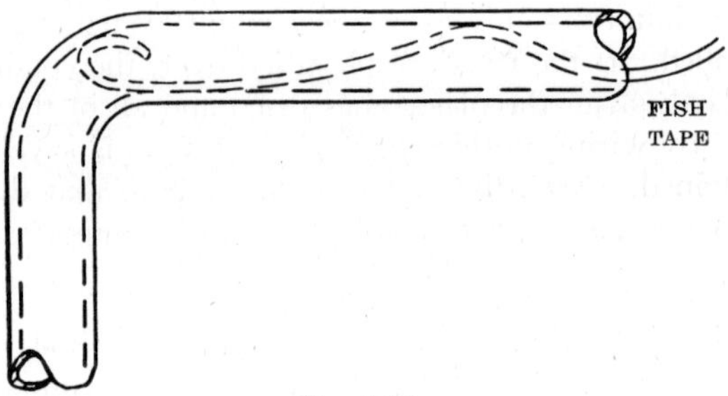

FIG. 3-28

it becomes difficult to pull in the wire because each bend increases the friction between the wire and the conduit. For this reason, the Code specifies that a run between boxes or fittings must not have more than the equivalent of four right angle bends.

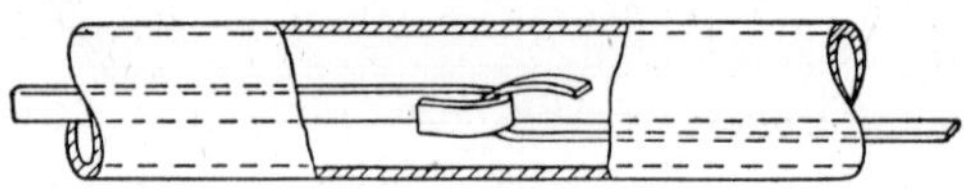

FIG. 3-29

If a fish tape cannot be pushed through from one end, a second tape may be pushed in from the other end until it comes in contact with the first and then manipulated until the two hook ends of the tapes engage (fig. 3-29). After this, the first tape is pulled through by the second.

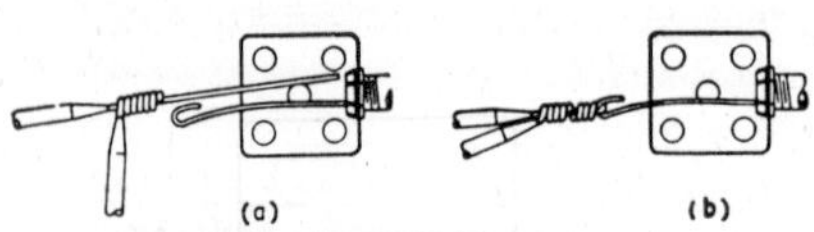

FIG. 3-30

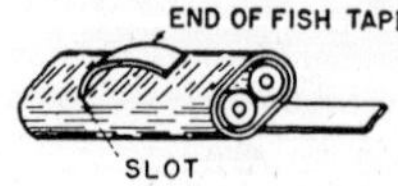

FIG. 3-31

The connection between the conductors and the fish tape must not give way when stress is applied and must offer a negligible amount of friction when the conductors are drawn into the conduit. An excellent method of attaching two conductors to a fish tape so as to satisfy these requirements is illustrated in fig. 3-30. The second conductor is attached to the first (fig. 3-30a), the first conductor is attached to the fish tape (fig. 3-30b), and the joint is taped.

If twin conductors are to be pulled through a somewhat straight and short run, the conductors may be attached to the tape as shown in fig. 3-31. The connection is made by cutting a slit in the braid and hooking the fish tape in it. This method of connecting the conductors should be used when possible, as it saves time.

Pulling Conductors into the Conduit—Conductors are usually drawn into the conduit by one person pulling the fish tape out at one

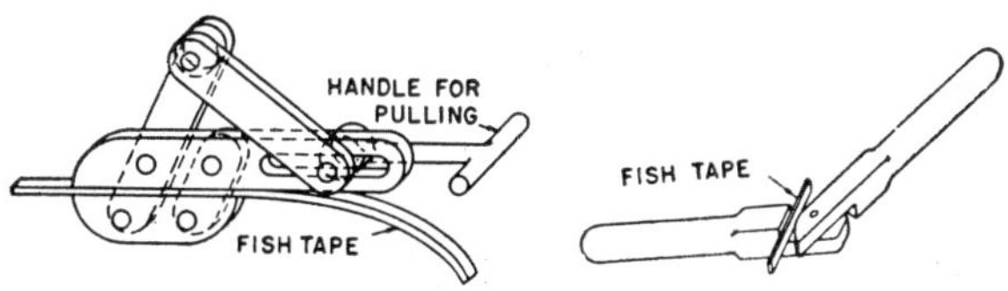

Fig. 3-32

end and another forcing the conductors in at the other end. If the pull is hard, the tape is pulled out by means of a pair of pliers or, preferably, by means of a fish tape puller (fig. 3-32).

Method of Determining Size of Conduit—According to the Code, the size of rigid conduit required for a given number of conductors of the same size is given in Table 3-VI.

If the conductors are of different sizes, the sum of their total areas, expressed as a percentage of the internal area of the conduit, should not exceed the values given in Table 3-VII. The area over the braid of rubber covered conductors is given in Table 3-VIII.

Example—Determine the size of the conduit needed for a run using four No. 6 gage wires, three No. 8 gage wires, and two No. 12 gage wires.

According to Table 3-VIII, the cross-sectional area of the conductors for this conduit is $(4 \times .13) + (3 \times .071) + (2 \times .038)$ or .809 square inch.

By Table 3-VII, the internal area of the conduit must not be less than $.809 \div .40$ or 2.02 square inches. A 1½ inch conduit has an internal area of 2.04 square inches (see Table 3-V), and is therefore large enough for the conductors.

TABLE 3-VI

Size of Conductor	Minimum Size in Inches of Conduit for 1 to 9 Conductors								
	1	2	3	4	5	6	7	8	9
No. 14	½	½	½	½	¾	¾	¾	1	1
12	½	½	½	¾	¾	1	1	1	1¼
10	½	¾	¾	¾	1	1	1¼	1¼	1¼
8	½	¾	1	1	1¼	1¼	1¼	1¼	1½
6	½	1	1¼	1¼	1½	1½	2	2	2
5	¾	1¼	1¼	1¼	1½	2	2	2	2
4	¾	1¼	1¼	1½	2	2	2	2	2½
3	¾	1¼	1¼	1½	2	2	2	2½	2½
2	¾	1¼	1½	1½	2	2	2½	2½	2½
1	¾	1½	1½	2	2	2½	2½	3	3
0	1	1½	2	2	2½	2½	3	3	3
00	1	2	2	2½	2½	3	3	3	3½
000	1	2	2	2½	3	3	3	3½	3½
0000	1¼	2	2½	2½	3	3	3½	3½	4

TABLE 3-VII

Number of Wires	Per cent of Total Area in Conduit That May Be Occupied by Conductors
1	53
2	31
3	43
4 or over	40

TABLE 3-VIII

Size of Conductors	Approximate Area of Braid in Square Inches
14	.031
12	.038
10	.045
8	.071
6	.13
4	.16
2	.21
1	.27
0	.31
00	.35
000	.41
0000	.48

Electric Metallic Tubing—Electric metallic tubing resembles rigid conduit. It has the same internal diameter, but has a thinner wall, making the outer diameter smaller. This tubing will therefore carry the same number of wires as rigid conduit. The wall thickness for metallic tubing is too thin to be threaded and the connections are made by threadless fittings similar to those used with rigid conduit. Metallic tubing may be installed wherever rigid conduit is permitted, except where it may be subject to severe mechanical injury.

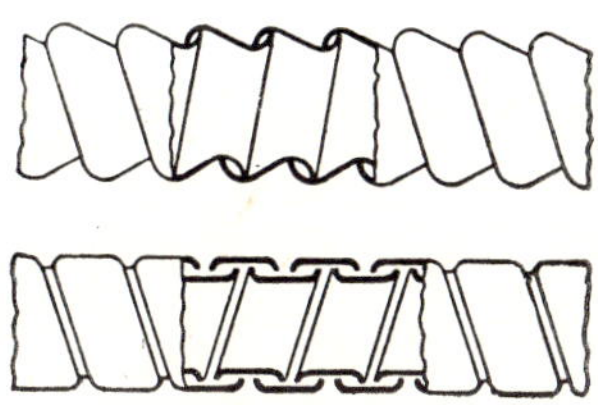

FIG. 3-33

Flexible Conduit—As shown in fig. 3-33, flexible conduit consists of one or two steel strips wound spirally, with interlocked edges, to form a steel tube that can be bent easily. This conduit is much easier to install than rigid conduit because the desired bends are easily

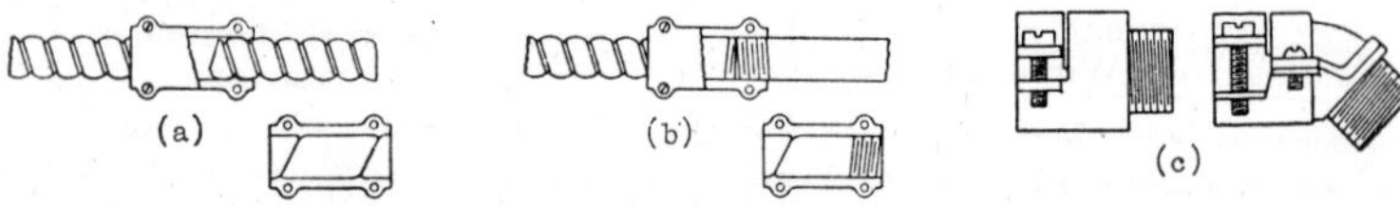

Fig. 3-34

formed by hand and couplings are seldom required, since it is manufactured in lengths ranging from 50 to 250 feet. Some common types of fittings for flexible conduit are shown in fig. 3-34. Flexible conduit, electric metallic tubing, and non-metallic sheathed cable may be connected to rigid conduit boxes by means of connectors. A connector for flexible conduit is shown in fig. 3-35.

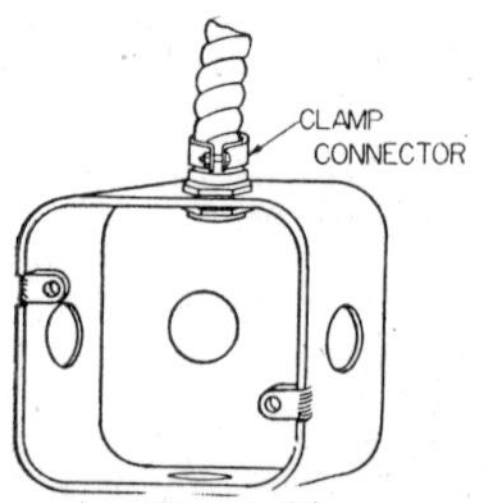

Fig. 3-35

Flexible conduit may be used in both concealed and exposed work, but must not be used in damp places, because it is not waterproof. When the run is in the same direction as the joists, the conduit should be fastened to the sides of the joists with straps. When the run is at right angles to the joists, the conduit

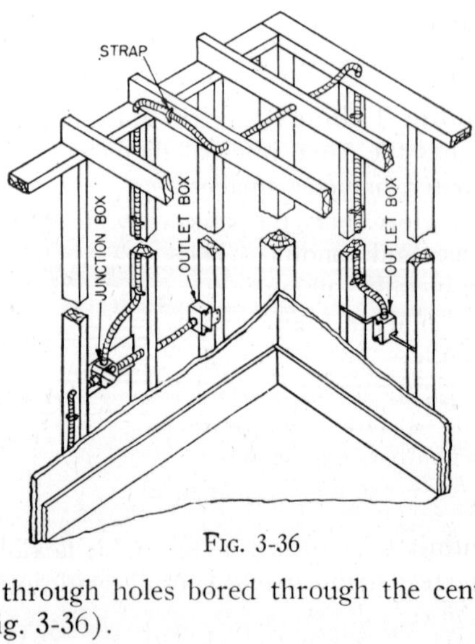

Fig. 3-36

should be run through holes bored through the centers of the joists (as shown in fig. 3-36).

Flexible conduit is used chiefly for wiring and making extensions in finished buildings. It is also used for extensions to portable and movable equipment and short lengths are used between stationary motors and rigid conduit. Its use in the latter case makes it possible to shift a motor slightly to tighten a belt or chain, or to mesh gears accurately.

Armored Cable—Armored cable, generally known as BX, consists of rubber insulated wire covered with an armor that has the same appearance as flexible conduit. This armor protects the wire from injury and to a slight extent from moisture. It is installed in the same manner as flexible conduit, except that about 8 inches of the armor must be stripped from the end before connecting to a box. The procedure is as follows: Using a fine tooth hacksaw, the armor is cut as shown in fig. 3-37a, about 8 inches from its end, being careful

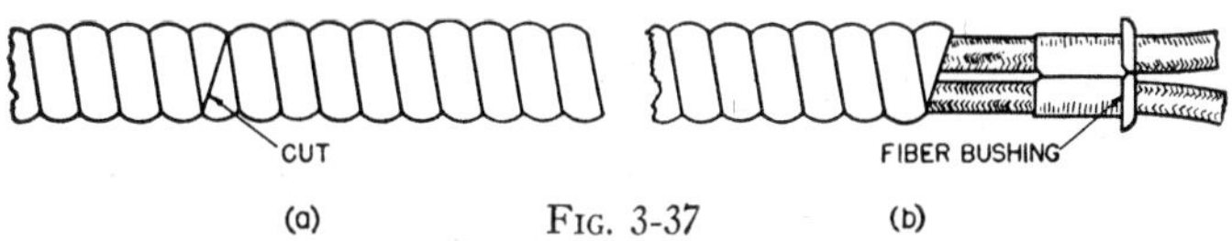

FIG. 3-37

not to injure the insulation on the wire. The 8 inch end is then broken off by bending it back and forth. The sharp edges are filed smooth and the wire insulation is protected by means of a fiber bushing (fig. 3-37b). The cable is then attached to the box in the same manner as flexible conduit. It must not be used in places subject to excessive humidity.

Armored cable is also made with a lead sheet between the rubber insulated conductors and the armor. Such cable is commonly referred to as BXL and may be used in places that are permanently damp, as in masonry, in the fill of buildings, and in underground runs. Both BX and BXL must contain no joints and splices between outlets.

Non-Metallic Sheathed Cable—A common type of non-metallic sheated cable is shown in fig. 3-38a. It is manufactured with either two or three wires. In one construction, one wire is not insulated and is used for grounding outlet boxes. The outer braids are impregnated with a moistureproof and fire-resisting compound.

Non-metallic sheathed cable is installed in the same way as armored cable and similar box connectors may be used with it (fig. 3-38b).

Boxes with attached connectors are generally used, however, as shown in fig. 3-38c.

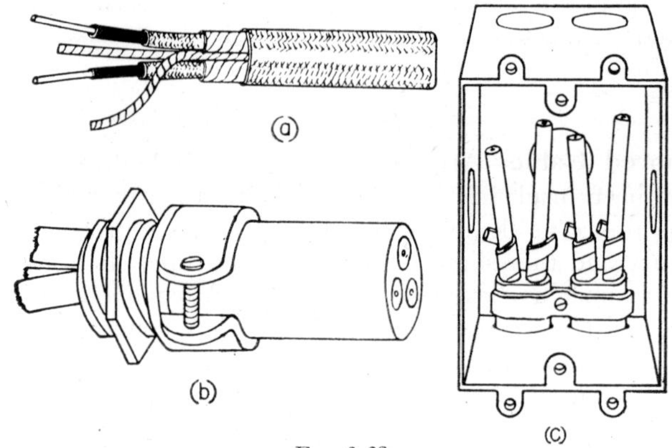

Fig. 3-38

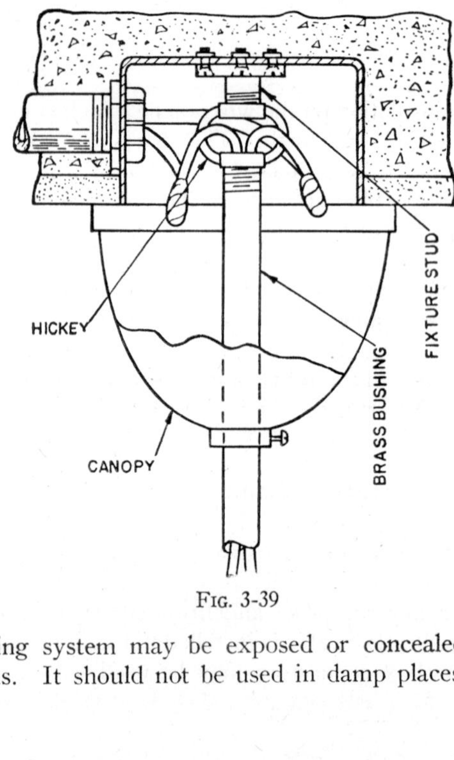

Fig. 3-39

This wiring system may be exposed or concealed in air spaces between walls. It should not be used in damp places nor be buried in plaster or concrete. This wiring method is widely used for house wiring because of its low cost.

With any of the wiring methods considered, ceiling fixtures are installed as shown in fig. 3-39.

Surface Metal Raceway—Surface metal raceway, also called metal molding, consists of a metal strip with edges that bend around a slightly curved base (fig. 3-40). This raceway requires the use of special boxes and fittings, some of which are shown in fig. 3-41. Fig. 3-42 illustrates how easily metal molding can be installed. One end of a piece of raceway is slipped on a projection on the box base and the other end is slipped on one end of an elbow. Another piece of raceway is

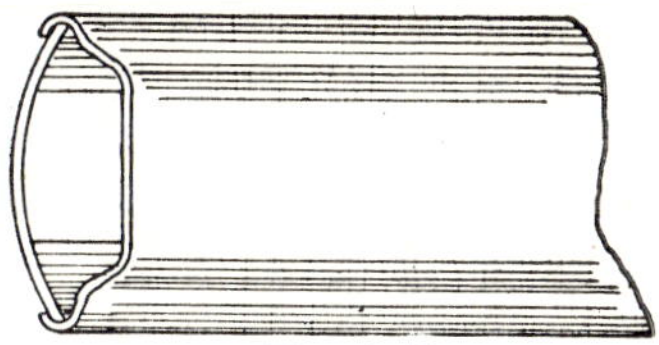

FIG. 3-40

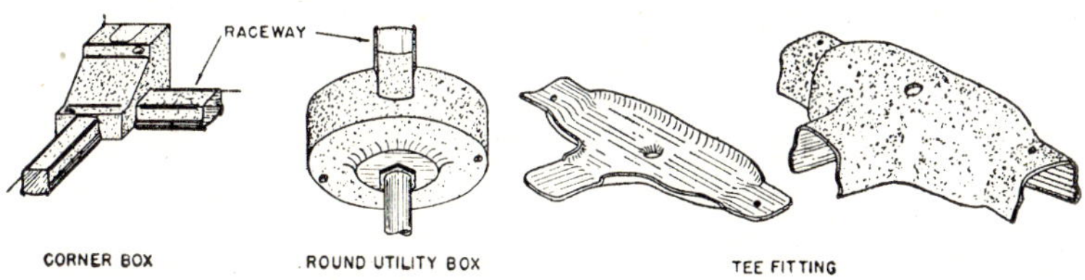

FIG. 3-41

slipped on the other end. Then the box and elbow are fastened with screws. The wire is pulled into the raceway, and after the desired

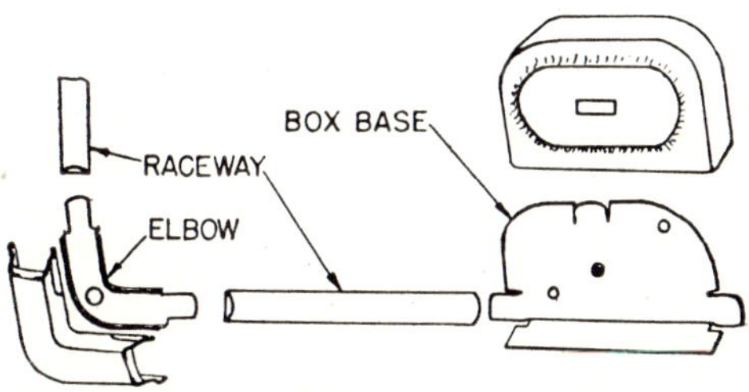

FIG. 3-42

connections are made, the box cover is fastened with screws and the elbow cover is snapped on.

Metal raceway may be used only in dry places. It may extend through walls, partitions, and dry floors, providing the molding is not broken in passing it through. It is used principally in wiring old buildings and in making extensions. For exposed work, surface metal raceway is preferable to other materials, because it has a much better appearance.

Insulation—Rubber is in more general use as an insulating material than any other substance, because it is a good insulator, is very tough, is waterproof, and is cheap. Rubber covered wire (type R) may be used in all installations except those where the temperature exceeds 120° F. Rubber deteriorates very slightly below this temperature but very rapidly above it, becoming brittle and breaking easily. For temperatures above 120°F, varnished cambric or asbestos covered wire should be used.

Varnished cambric insulation (type VC) consists of a cotton tape impregnated with insulating varnish, which is wrapped several times over the conductor. This may be used where the temperature does not exceed 167° F. Because of the higher permissible temperature, the current carrying capacity of varnished cambric wire is about 20 per cent higher than the capacity of rubber covered wire (see Table 3-III). Varnished cambric is not as easily injured by oil as is rubber. It is an excellent type of insulation and is frequently used on wires subject to high voltages.

Conductors insulated with a braid of asbestos should be used in hot dry places where other types of insulation would be damaged. Wire with this insulation (type A) is used for leads on heating elements and mogul type lamp sockets, and on the back of large switchboards. It should never be used for exterior wiring or in a place where moisture is present.

Slow burning wire (type SB) has three braids of cotton impregnated with a material that has both insulating and fire-resisting properties. This insulation will not withstand the temperatures that asbestos will withstand, but it is used for about the same purposes.

Flexible Cords—Some of the most common types of cords used in wiring pendants and portable equipment are lamp cord, reinforced cord, hard service cord, and heater cord. The conductors in these types consist of stranded wire wrapped with thread to keep the broken strands from punching through the insulation.

Lamp cord consists of two conductors covered with rubber and a cotton braid, which are twisted together (fig. 3-43a). This cord (type C) may be used in dry places, where it is not subject to hard usage.

Lamp cord is not highly recommended for any use, reinforced cord being preferred instead.

Reinforced cord (type P) is the same as lamp cord, but has, in addition, a rubber jacket applied over the two conductors (the jacket being covered with a cotton or silk braid (fig. 3-43b). This cord is also recommended for use in dry places where it is not subject to severe treatment, as in residences and office buildings.

Hard service cords (known as type SJ and type S) are recommended for hard usage. Both of these consist of rubber insulated conductors which have been twisted together and covered with a special high grade rubber (fig. 3-43c). Type SJ is not approved for

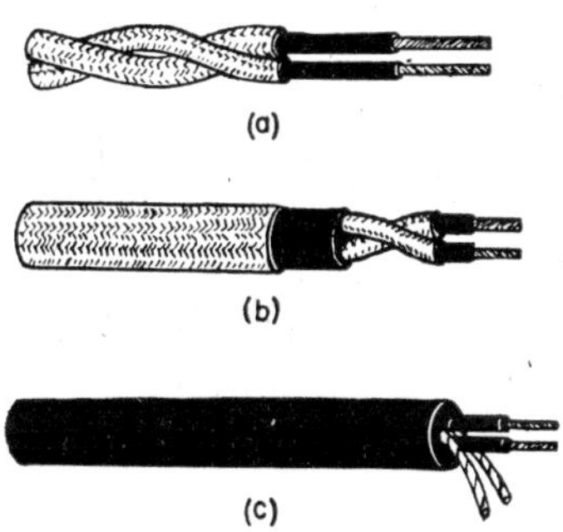

FIG. 3-43

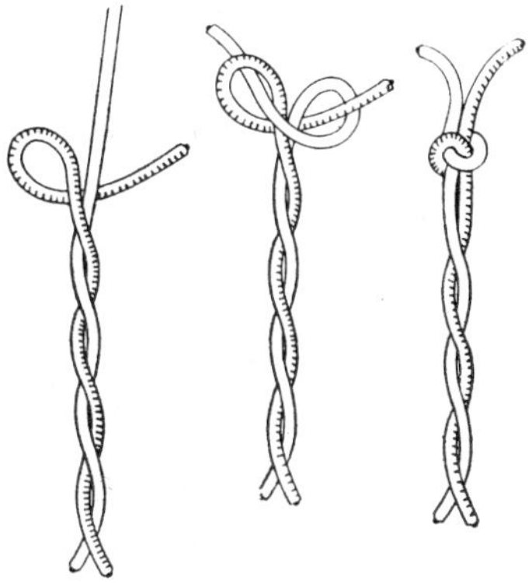

FIG. 3-44

factories, garages, or outside use. Type S differs from type SJ only in that it has a heavier outer cover. Type S cord may be used anywhere as long as the temperature does not exceed 120° F.

For portable heating equipment, heater cord must be used. Each conductor of this cord is covered with a 1/64 inch thickness of rubber, followed by a covering of asbestos. This may or may not be covered with a cotton braid.

Underwriter's Knot—The portion of a cord extending within a socket, plug, or fitting should be installed so that any stress imposed on it will not be transmitted to joints or binding screws. This condition is fulfilled by tying an underwriter's knot at the end of the cord, as shown in fig. 3-44. Such a knot can stand considerable stress without injury to the insulation.

Switches—Manually operated switches used in lighting and low power circuits are of three types: knife, snap, and flush. Knife switches may be single, double, or three pole, and may be single or double throw (fig. 3-45). A single pole switch disconnects one line wire, a two pole

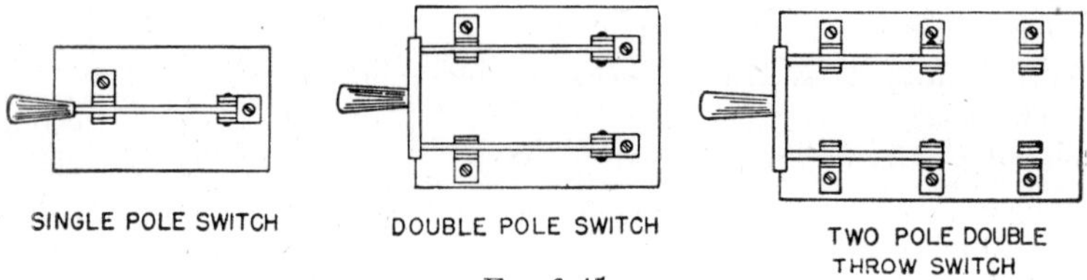

FIG. 3-45

switch disconnects two, etc. Snap and flush switches (fig. 3-46) may be single pole, double pole, three way, or four way.

A three-way switch has three terminals and one or two blades, which make the connections shown in fig. 3-47a. One or more lamps

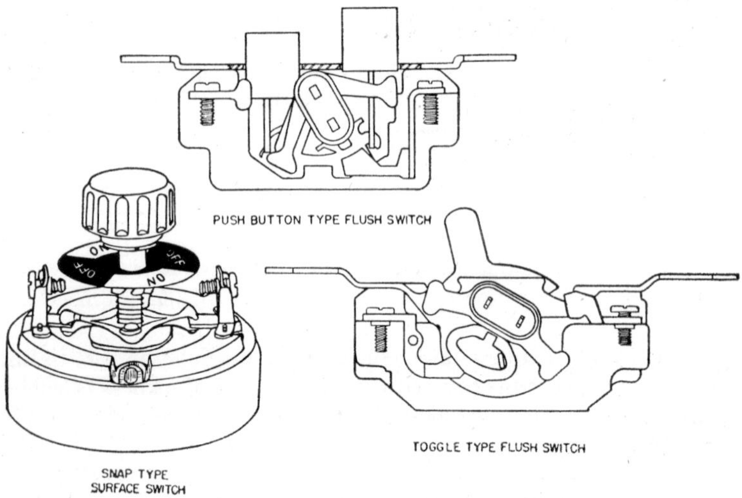

FIG. 3-46

may be controlled from two locations by means of two three-way switches (fig. 3-47b). This scheme of wiring is frequently used for stairway lighting, as it makes it possible to control the lights from either the bottom or the top of the stairway.

If it is desired to control lights from more than two points, four-way switches must be connected between three-way switches. Four-way switches have four terminals and two blades, which make the connections shown in fig. 3-47c. The method of connecting switches to control lamps from four points is shown in fig. 3-47d.

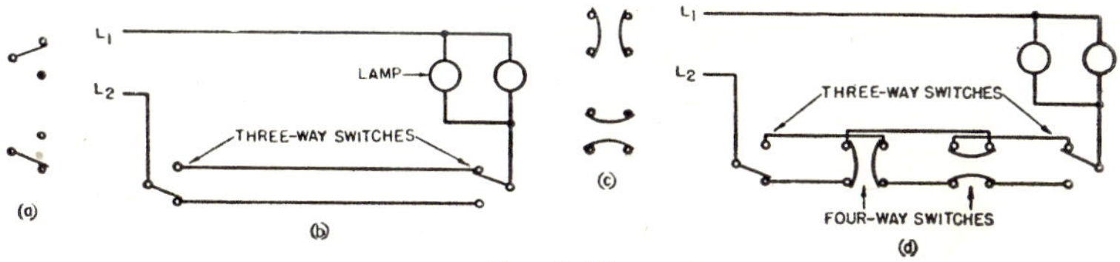

Fig. 3-47

Exercises

1. How many mils are there in .339 of an inch?
2. How many circular mils are contained in a wire 461 mils in diameter?
3. Find the number of circular mils in a conductor .182 of an inch in diameter.
4. What is the diameter of a conductor in inches that has a cross-sectional area of 243,049 circular mils?
5. Find the number of square mils in a rectangular conductor that is .675 of an inch thick and one inch wide.
6. How many circular mils are contained in a bus-bar 3/8 by 1.75 inches?
7. Calculate, in inches, the diameter of a circular conductor that will have the same area as a bus-bar 1/4 by 3 inches.
8. If a rectangular conductor is to be used instead of a cable containing 105,530 circular mils, what will be its cross-sectional area in square inches?
9. A cable is composed of 16 strands of wire and each strand is .144 of an inch in diameter. Find the area of the cable in circular mils.
10. Determine the resistance of one mile of aluminum wire .25 inch in diameter.
11. What length of nichrome wire, 1000 circular mils in cross section, is required to give a resistance of 5 ohms?
12. What size iron wire will have the same resistance as a No. 18 copper wire of the same length?
13. What size line wires are required for a motor taking 20 amperes if it is 1000 feet from a 240 volt generator (allow a line drop of 10 per cent)?
14. In problem 13 what smallest size line wires will carry the current without overheating?

15. Twelve lamps, connected as shown in fig. 3-4, are 400 feet from a 240 volt generator. If each takes 1.5 amperes, what size line wires are required (allow a line drop of 3 per cent)?
16. In fig. 3-48, find the voltage across each load if each lamp takes 2.5 amperes.

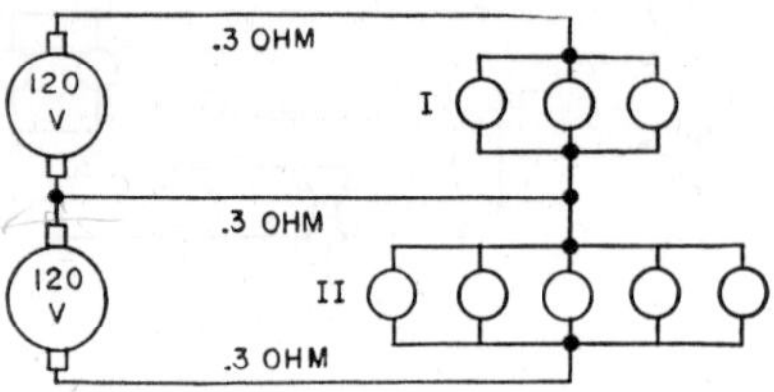

FIG. 3-48

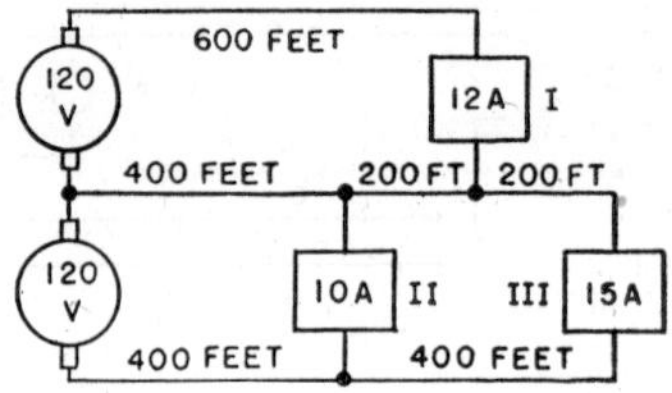

FIG. 3-49

17. Solve problem 16, considering the neutral wire broken.
18. Find the voltage across each load in fig. 3-49 if the line wires are No. 8 gage.
19. Determine the height of a coil .6 inch thick if it is wound with 700 turns of No. 16, double cotton covered wire.
20. A coil wound with No. 28, enameled, single silk covered wire has a cross section of .3 square inch. How many turns does it contain?
21. Express the following temperatures in degree centigrade: (a) 72° F, (b) 100° F, (c) 50° F.
22. Express the following temperatures in degrees Fahrenheit: (a) 20° C, (b) 68° C, (c) 12° C.
23. The resistance of a copper wire is 3 ohms at 0° C. What is its resistance at 20° C?
24. The resistance of a coil of nichrome wire is 100 ohms at 20° C. What is its resistance at 200° C?
25. The resistance of a transformer winding is .4 ohm at room temperature (20° C). After the transformer reaches its maximum temperature, its resistance is .46 ohm. Determine the temperature rise.
26. A winding in a motor has a resistance of 100 ohms at a temperature of 80° F. If the resistance of this winding increases to 120 ohms, what is its temperature rise?
27. When cold, an electric furnace with a nichrome heating element takes 100 amperes at 240 volts. After it is hot, the current drops to 90 amperes. What is the temperature of the furnace if the room temperature is 20° C?
28. (a) What is the internal diameter of a ½ inch conduit? (b) Of a 1¼ inch conduit?

29. Give the least radius permitted by the Code for the inner edge of a bend made in (a) a ½ inch conduit. (b) A 1 inch conduit. (c) A 2 inch conduit.
30. Give the steps that should be taken in making a short radius end bend.
31. Where should the conduit system of wiring be used?
32. For what purpose are fittings preferable to boxes?
33. Explain (a) one way fishing; (b) two way fishing.
34. When is two way fishing used?
35. What is the maximum number of right angle bends permitted by the Code between two outlets in a conduit run?
36. How many No. 12 wires may be pulled into conduits of the following sizes: (a) ½ inch; (b) ¾ inch; (c) 1 inch?
37. What size conduit is required for (a) five No. 4 wires; (b) eight No. 14 wires; (c) six No. 8 wires?
38. (a) What size conduit is required for two No. 4 and five No. 14 wires? (b) For three No. 6, four No. 10, two No. 12, and eight No. 14 wires?
39. What size line wires and conduit are required for a motor that takes 50 amperes at full load?
40. Determine the size of the line wires and conduit required by a 230 volt, 10 horsepower motor which has an efficiency of 85 per cent.
41. (a) What is the internal diameter of half inch electric metallic tubing? (b) One inch electric metallic tubing?
42. For what use is electric metallic tubing prohibited?
43. What are the chief uses of flexible conduit?
44. What is the difference between BX and BXL?
45. Give the industrial applications of armored cable and non-metallic sheathed cable.
46. How are armored cable and non-metallic sheathed cable installed (a) in the direction of the joists, and (b) at right angles to the joists?
47. Describe two methods of connecting non-metallic sheathed cable to boxes.
48. (a) For what use is non-metallic sheathed cable prohibited? (b) For what use is surface metal raceway prohibited?
49. Why is surface metal raceway more suitable for exposed work than other materials?
50. Give the uses of (a) rubber covered wire, (b) varnished cambric covered wire, and (c) asbestos covered wire.
51. Give the uses of flexible cord.
52. What is the difference between type SJ and type S cord?
53. By means of a diagram, show how a lamp is controlled from three switches.

CHAPTER 4

MAGNETISM

THE importance of a thorough understanding of magnetism cannot be over-estimated, for it is involved in the operation of practically all kinds of electrical apparatus, such as dynamos, transformers, indicating instruments, and lifting magnets.

Magnets may be divided into three classes—natural, artificial, and electro-magnets.

Natural Magnets—Long before the Christian era it was discovered that certain iron ore possessed the peculiar property of being able to attract small pieces of iron, steel, and a few other metals. It was later discovered that an elongated piece of this ore would assume a nearly north and south position if freely suspended. For that reason it was used as a compass and called lodestone, which means leading stone.

Artificial Magnets—The natural magnets described above no longer have any practical value, the various types of magnets now in commercial use being made artificially from iron or steel. A piece of iron or steel may be magnetized by rubbing it with a magnet, but a better method is to magnetize it with an electric current. A practical way to magnetize a steel bar, for instance, is to insert the bar in a coil of insulated wire and then pass a heavy direct current through the coil. If the bar is made of hardened steel, it will be difficult to magnetize, but when once magnetized, it will retain indefinitely a large percentage of its magnetic strength. If the bar is made of soft steel or iron, it will be easy to magnetize, but when the magnetizing force is removed, most of the magnetism will soon disappear. The magnetism that remains after the magnetizing force has been removed is called residual magnetism.

Field about a Bar Magnet—If a piece of cardboard is placed over a bar magnet and iron filings sprinkled over the cardboard, the filings will arrange themselves as shown in fig. 4-1. The arrangement of the filings will be more pronounced if the cardboard is tapped gently as the filings are sprinkled.

Direction of Magnetic Field—Fig. 4-1 shows that magnetism manifests itself as if it existed in lines, and for that reason we speak of lines of magnetism or magnetic lines. The space through which these lines pass is called the magnetic field. If a small compass is placed in various parts of a magnetic field, it will be found that the magnetic lines have a definite direction. The lines seem to emanate

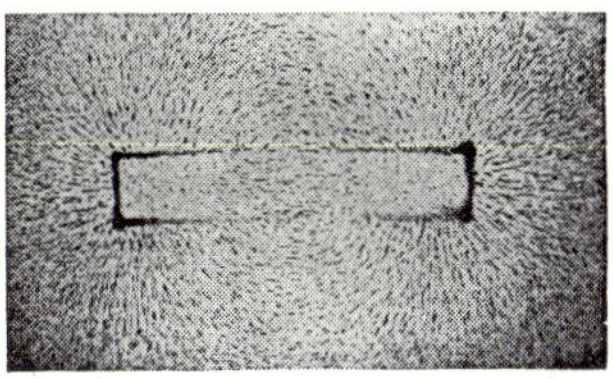

FIG. 4-1

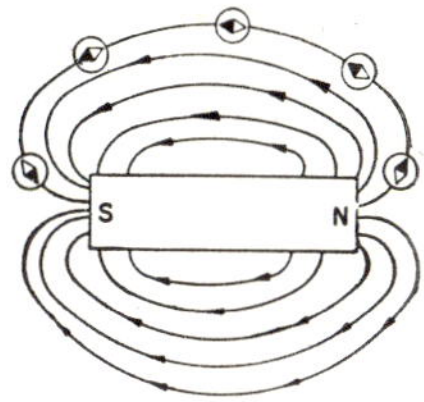

FIG. 4-2

from one end of the magnet, flow through the air to the opposite end, and return through the magnet itself, as shown in fig. 4-2. The two ends of the magnet are called the poles. These poles are distinguished by the position they take when they are suspended freely. The end which points toward the north pole is called the north-seeking pole or simply the north pole, and the other is called the south-seeking or south pole. In practice it is desirable to assume that the magnetic lines leave the magnet at the north pole and reenter at the south pole. Within the magnet the lines continue from the south pole to the north pole. Thus each line forms a closed loop.

Breaking a Magnet—As just explained, a magnetic bar has two poles. If a magnetized bar is broken into a number of pieces, as

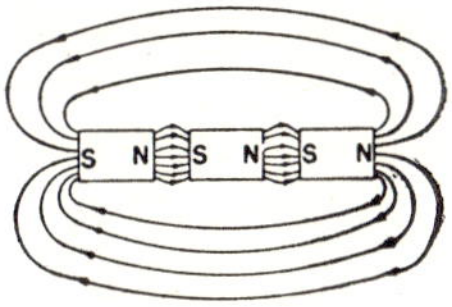

FIG. 4-3

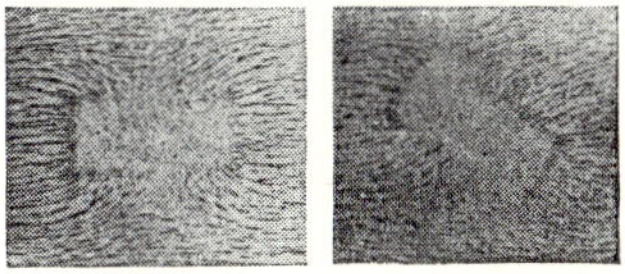

FIG. 4-4

shown in fig. 4-3, each piece will constitute a bar magnet with its north and south poles in the same respective directions as those of the original magnet. The reason for this will be evident if it is observed that the magnetic lines still continue to pass from one piece to the next, and in so doing, constitute a north pole where the magnetism leaves a piece and a south pole where the magnetism enters a piece.

Properties of Magnetic Lines—Magnetic lines have certain properties that should be clearly understood in order to understand what comes later.

1. There is no insulator for magnetic lines; they pass through all materials.
2. Magnetic lines pass easily through materials that can be magnetized. As shown in fig. 4-4, when a piece of iron is placed in a magnetic field, the lines crowd into the iron instead of passing through the air.
3. Magnetic lines tend to shorten as though they were stretched rubber bands. For this reason they are sometimes called lines of force.
4. Two magnetic lines that extend in the same direction tend to push each other apart.
5. Magnetic lines never cross.
6. Each magnetic line always forms a closed loop.

Attraction and Repulsion—If the north pole of a magnet is placed near the south pole of another magnet, there will be a mutual

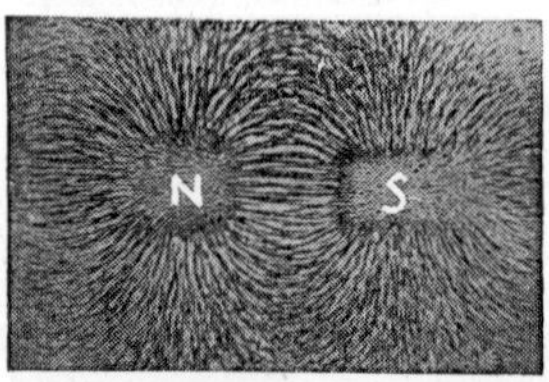

FIG. 4-5

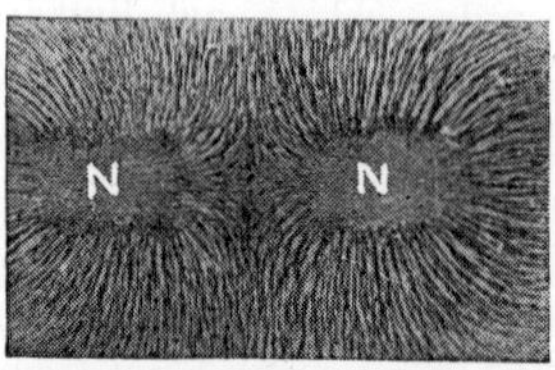

FIG. 4-6

attraction between the poles. The cause for this attraction is evident from fig. 4-5. The magnetic lines leave the N pole and enter the S pole. Since these lines are under tension, like rubber bands, they tend to shorten and to pull the two poles together.

When two like poles are placed near each other (fig. 4-6) the lines from one pole push the lines from the other pole aside causing a lateral crowding action which pushes the poles apart.

The rule, then, for magnetic attraction and repulsion is as follows: *Unlike magnetic poles mutually attract each other, while like magnetic poles mutually repel each other.*

Iron Attracted by a Magnet—Since magnetism passes more readily through iron or steel than through air, these materials are attracted by a magnet. For instance, when a piece of soft iron is placed near a magnet (fig. 4-7), the field is distorted by the lines

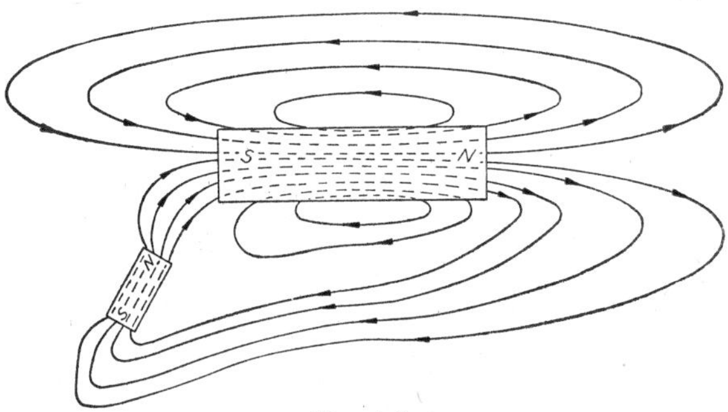

FIG. 4-7

tending to pass through the iron instead of through the air, because the iron offers an easier path. The iron now has a north and a south pole of its own, since magnetic lines enter one end and leave the other. As a result there will be two magnets with their unlike poles near each other; therefore attraction occurs.

Weber's Theory—The above phenomena regarding magnetism are explained by Weber's theory as follows: Matter is made up of small particles called molecules. There seems to be good evidence that each molecule is a magnet with a north and a south pole. In iron and steel these molecular magnets are comparatively strong, while in most other materials they are very weak.

In an unmagnetized bar they are arranged in a haphazard manner, as shown in fig. 4-8a, so that the various north and south poles all neutralize one another and no external magnetic effect is produced. Upon bringing a magnetizing force, such as a permanent magnet near one end of the unmagnetized bar, the little magnets tend to line up with their axes parallel (fig. 4-8b). This is due to the laws of magnetic attraction and repulsion. The north poles will all point in one direction, and the south poles will all point in the opposite direction. The bar will have become magnetized and will produce an external magnetic field. This also explains why breaking a magnet into several pieces (fig. 4-8c) leaves perfect magnets, each having a north and south pole.

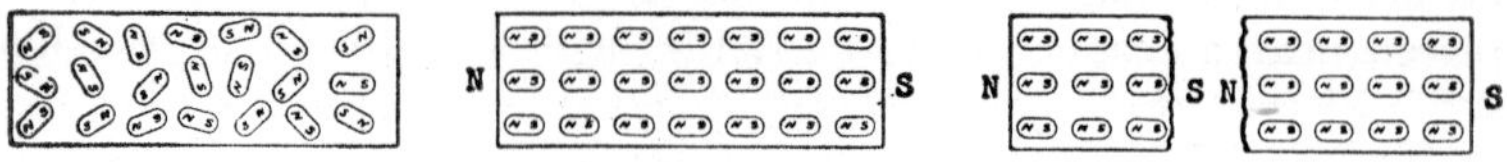

(a) (b) (c)

Fig. 4-8

The reason that steel is hard to magnetize is that there is a high frictional resistance between the molecules, making it difficult to swing them around so that their axes are parallel. However, once parallel, they are equally difficult to get out of line; therefore, steel retains its magnetism well.

Soft iron is easily magnetized because there is a low frictional resistance between the molecules, making it easy to swing them into line. However, wher the magnetizing force is removed, the molecules tend to lose their alignment; therefore, soft iron does not retain its magnetism very well. Hence, steel is always used where permanent magnets are desired.

Intermediate Poles—Usually a magnet has only two poles, but it is possible to magnetize a bar so that it will have more than two poles. Fig. 4-9 shows one method of securing such a magnet. The bar is placed, so that each end rests on a north pole, and is stroked at the middle with the south pole of third magnet. The bar will then be found to be magnetized with a south pole at each end and a north pole in the middle. It is evident that in this case we have in reality two magnets with the two north poles at the middle. Such an intermediate pole is frequently produced on iron and steel bars that have been in contact with lifting magnets.

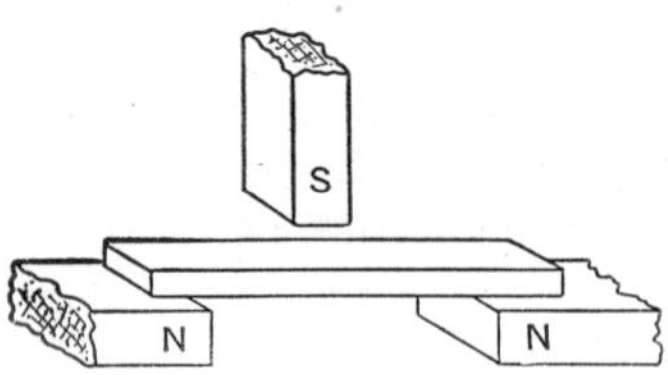

Fig. 4-9

Magnetic Screens—There is no known substance that will insulate against magnetic flux. A magnetic field readily passes through any such substances as glass, copper, wood, etc. Because of this fact, it is sometimes difficult to eliminate the effect which stray magnetic fields have upon electric measuring instruments. For instance, a conductor carrying a large current produces a strong magnetic field, which may affect meters in the vicinity. The meters are therefore shielded by placing them in a cast iron case (fig. 4-10) so that the stray lines of force take the path through the case and do not influence the meters.

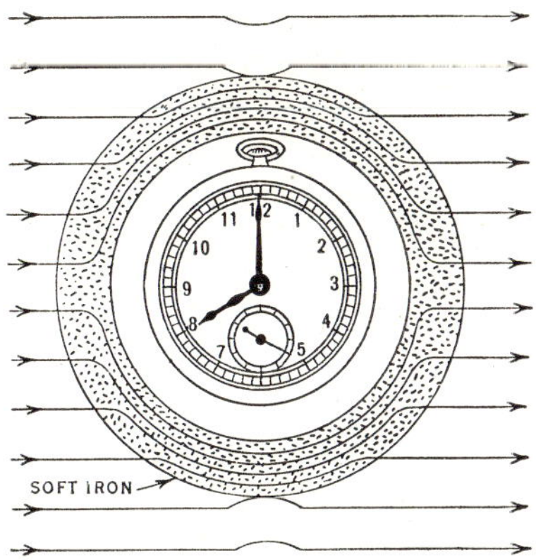

Fig. 4-10

Expensive watches carried in electric power plants should be kept in soft iron cases to protect the steel parts from becoming magnetized, for a magnetized watch either loses or gains time.

The Compass—A compass is merely a thin bar magnet, accurately balanced and suspended on a pivot so that it may rotate freely.

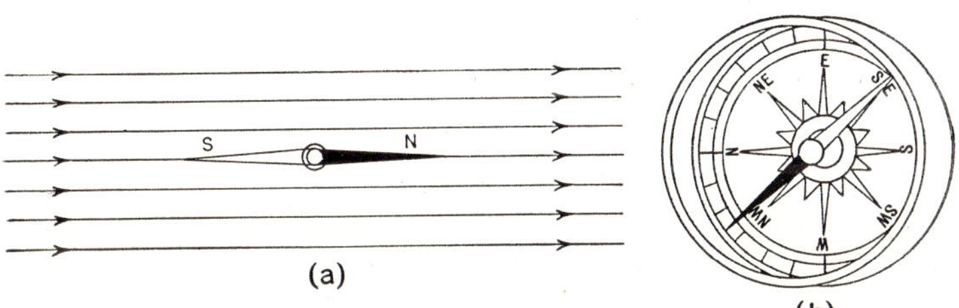

Fig. 4-11

This thin bar, which is usually called a needle, always tends to set itself in the direction of the magnetic field in which it is placed, the north end, which is usually colored blue, pointing in the direction of the magnetic lines. This is illustrated in fig. 4-11a.

A compass of the type shown in fig. 4-11b is much used in practice. It enables one to determine whether or not a material is magnetized and to determine the polarity of poles in a dynamo, which in turn indicates whether or not the field coils have been connected correctly. It also enables one to find faults in armatures of D.C. and A.C. machines.

The Earth's Magnetic Field—It has been known for a long time that the earth is a huge magnet, whose two poles coincide approximately with the geographical poles. Magnetic lines of force emanate from the magnetic north pole, proceed to the magnetic south pole, and return to the magnetic north pole through the earth. Thus, the space in which we live is permeated with magnetic lines, which ordinarily occupy roughly a north and south position. It is this magnetic field that causes a compass at the earth's surface to point toward the north.

In 1819, Oersted, a Danish physicist, made a discovery which aroused a great deal of interest because it showed the first evidence of a connection between magnetism and electricity. He found that an electric current is always accompanied by certain magnetic effects and that these effects obey certain definite laws. These important laws may be demonstrated by sending a large current through a vertical wire which passes through a horizontal piece of cardboard. A magnetic effect will be produced, which may be observed by sprinkling iron filings on the cardboard while it is being gently tapped. It will be seen that the filings arrange themselves in concentric rings about the wire. An examination of the filings will show that each magnetic line forms a complete circle by itself. By placing small compasses at various positions on the cardboard (fig. 4-12), one may see that the needles always point in a direction parallel to the circular magnetic lines. When the current flows through the wire as indicated, the needles will point in a counter-clockwise direction. If the current is reversed, the needles will point in a clockwise direction. Evidently the direction of the lines depends upon the direction of the current through the conductor.

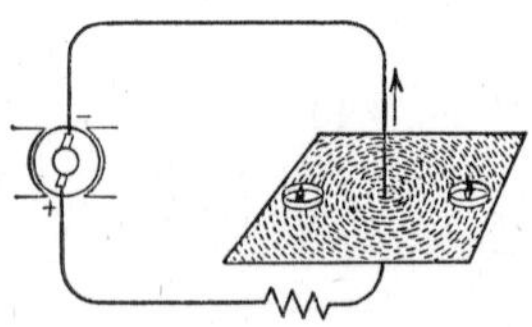

FIG. 4-12

Fig. 4-13a shows a section of a current-carrying wire with its magnetic field, while figs. 4-13b and 4-13c show cross sections of this wire. In fig. 4-13b note that the direction of the current is away from the observer, and that the magnetic lines are clockwise. In fig. 4-13c the field about the same wire is shown with the current reversed. Note that the direction of the current is toward the reader and that the lines extend counter-clockwise. A convenient rule for remembering

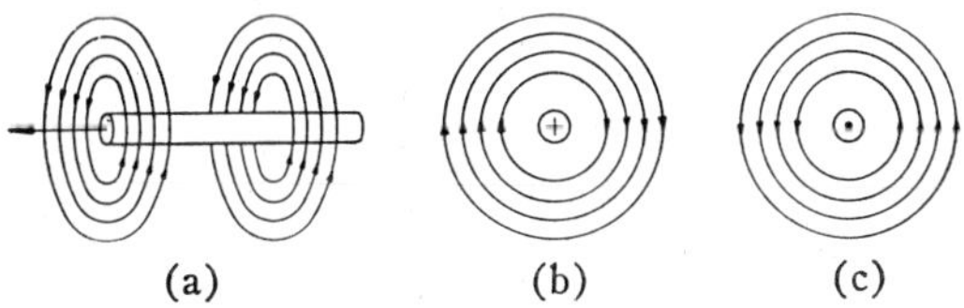

FIG. 4-13

the relation that exists between the magnetic lines around a conductor and the direction of current through the conductor is as follows:

Thumb Rule—*If the conductor is grasped in the right hand with the thumb pointing in the direction of the current, the fingers will point in the direction of the magnetic field.*

Thus, if the direction of current is known, the direction of the lines can be readily determined; or conversely, if the direction of the lines is known, the direction of the current can be readily obtained. The thumb rule is illustrated in fig. 4-14.

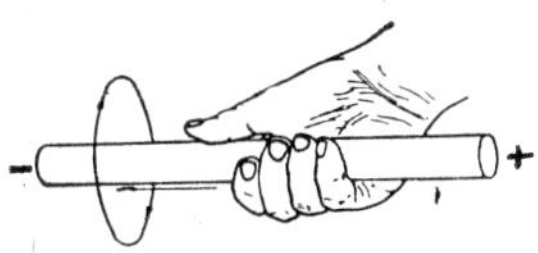

FIG. 4-14

Magnetic Field about Two Parallel Conductors—When two parallel conductors carry current in the same direction, the magnetic fields tend to draw them together (see fig. 4-15a, which shows

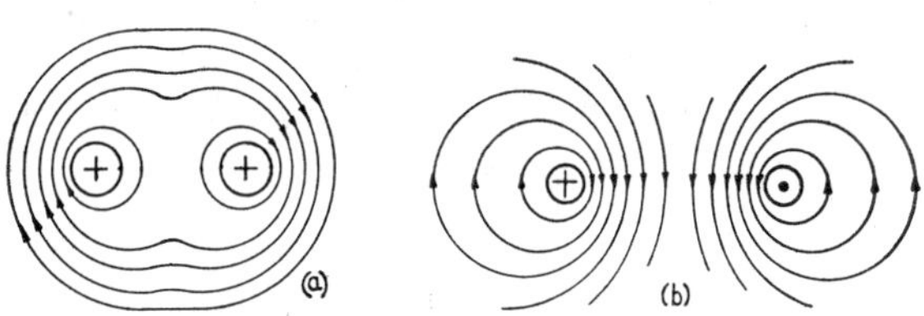

FIG. 4-15

the cross section of two parallel wires with the currents flowing away from the observer). The magnetic lines between the conductors neutralize, producing the field shown in fig. 4-15a. Since magnetic lines tend to contract, the conductors will be forced toward each other.

Fig. 4-15b shows parallel conductors with currents in opposite directions. Since the lines around one conductor are opposite in direction to those around the other, the two magnetic fields will not join as in fig. 4-15a, nor cross, but there will be a lateral crowding effect that tends to push the conductors apart. Under short-circuit conditions, bus bars have been torn loose from their supports by the powerful magnetic action of parallel currents.

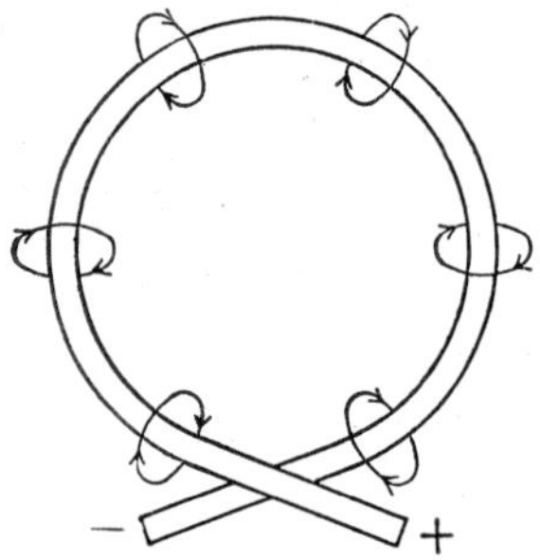

Fig. 4-16

If we apply the thumb rule to a wire bent into a loop as shown in fig. 4-16, we find that the lines of force which encircle the conductor will enter the same face of the loop and come out at the other face.

If several loops are placed together in a loose coil as shown in fig. 4-17, most of the lines will thread the whole coil and return on the outside of the coil to the other end. The reason that practically no magnetic lines encircle the loops of a closely wound coil, but thread the entire coil, can be seen by studying fig. 4-18, which is an enlarged cross section of fig. 4-17. The current enters at A, B, and C and comes out at D, E, and F. If A and B are sufficiently near each other, the field on the right side of A will neutralize the field on the left side of B, since the fluxes are in opposite directions. The

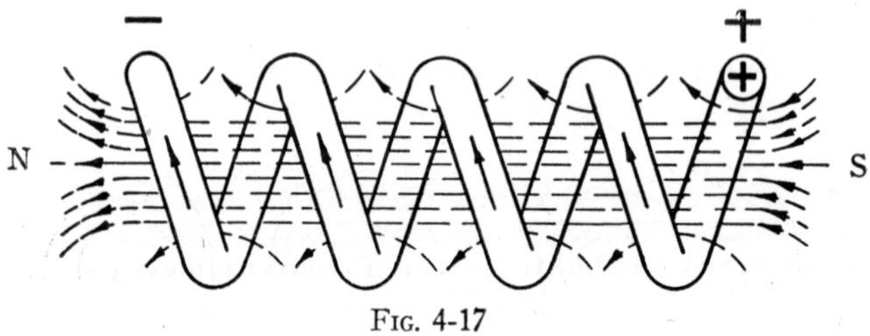

Fig. 4-17

same can be said about the fluxes between any two adjacent turns. Thus the magnetic lines are compelled to continue through the whole length of the coil and cannot pass between the turns. This produces the same shape of field that is found about a bar magnet, the end from which the lines come out being a north pole and the end in which the lines enter being a south pole.

Thumb Rule for Coil—The relation between the direction of current and direction of magnetic flux may be conveniently expressed by the thumb rule as follows:

Grasp the coil with the right hand so that the fingers point in the direction of the current in the coil. Then the thumb extended longitudinally will point in the direction of the flux, or toward the north pole (fig. 4-19).

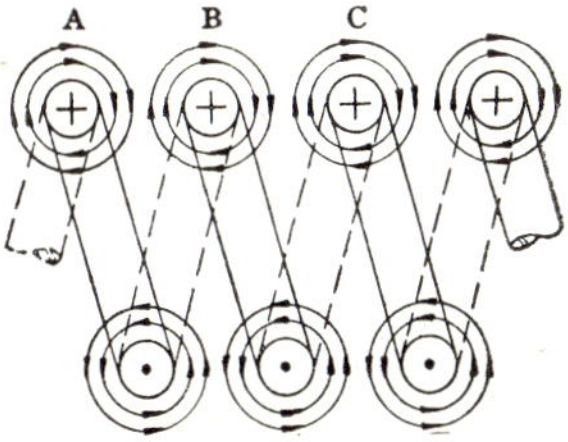

FIG. 4-18

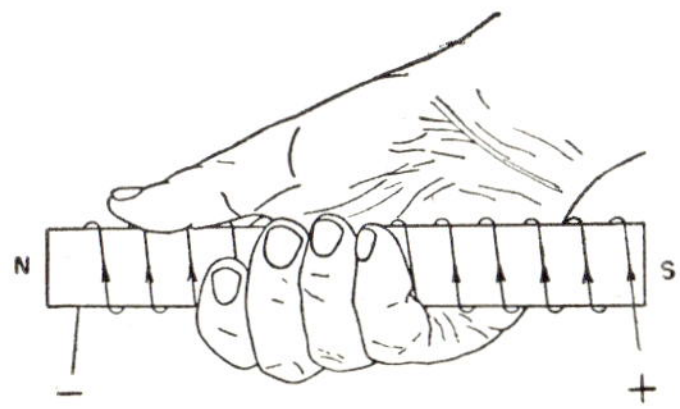

FIG. 4-19

Thus the polarity may be determined if the direction of the current is known, and conversely, the direction of the current may be determined if the polarity is known.

A clear understanding of the operating characteristics of electrical machinery and auxiliary equipment requires a knowledge of the basic laws of magnetism as well as a knowledge of electricity. Fortunately, the laws of magnetic circuits are similar to those of electric circuits. Hence, instead of having to become familiar with new fundamentals, it is only necessary to become familiar with the fundamentals and laws of the electric circuit restated for magnetic circuits. To show how this is done, a simple magnetic circuit will be considered (fig. 4-20). This consists of a coil with a circular air core. When a current is sent through this coil as indicated, a magnetic flux is set up in a clockwise direction. The amount of flux is expressed in magnetic lines, just as the amount of electric current is expressed in amperes.

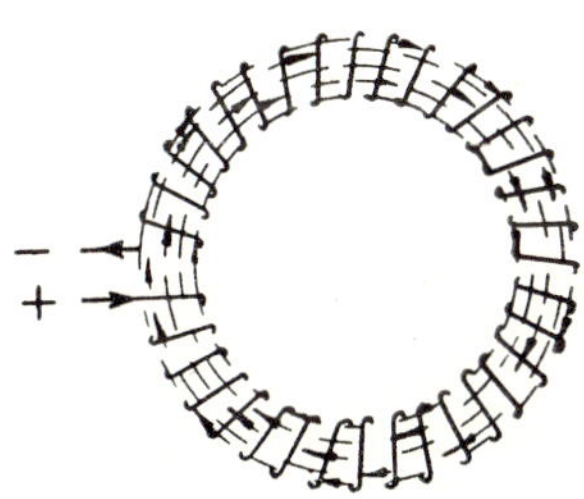

FIG. 4-20

This flux is set up by a magnetomotive force (abbreviated m m f) produced by the coil carrying current, just as the current in an electric circuit is set up by an electromotive force produced by a battery. The number of magnetic lines actually set up is proportional to only two factors—the current flowing through the coil and the number of turns in the coil. Thus the m m f is expressed in a unit called the ampere-turn.

The number of ampere-turns (IT) is found by multiplying the number of turns by the current through them. For instance, a 300 turn coil carrying 2 amperes produces a m m f of 300 × 2 or 600 ampere-turns. The same m m f could be produced by a 100 turn coil carrying 6 amperes.

All magnetic circuits, including the one shown in fig. 4-20, offer resistance to the flow of magnetic lines. This magnetic resistance, called reluctance, is represented by the scrip letter $\mathcal{R}$ and is measured in a unit that is often called the rel (no name has as yet been universally adapted for this unit).

Ohm's Law for Magnetic Circuits—The above shows that magnetic circuits are similar to electric circuits. There is a m m f which sets up a flux in the magnetic circuit, and a magnetic resistance, called reluctance, which opposes the flow of these lines. Then we may write a law for magnetic circuits which resembles Ohm's law for electric circuits.

Ohm's law for electric circuits is expressed as follows:

$$\text{Current} = \frac{\text{e m f}}{\text{Resistance}}, \quad \text{or Amperes} = \frac{\text{Volts}}{\text{Ohms}}, \quad \text{or in symbols}$$

$$I = \frac{E}{R} \qquad 4\text{-}1$$

Then for a magnetic circuit, Ohm's law may be expressed as follows:

$$\text{Flux} = \frac{\text{m m f}}{\text{Reluctance}}, \quad \text{or magnetic lines} = \frac{\text{Ampere-turns}}{\text{Rels}},$$

$$\text{or in symbols } \Phi = \frac{IT}{\mathcal{R}} \qquad 4\text{-}2$$

Equation 4-2 has as broad an application in magnetic circuits as equation 4-1 has in electric circuits.

EXAMPLE—In fig. 4-20 the reluctance of the magnetic circuit is 2 rels, and the coil consists of 1000 turns of wire which carry 10 amperes. Determine the number of magnetic lines set up.

$$\text{From equation 4-2, } \Phi = \frac{10 \times 1000}{2} = 5000 \text{ lines}$$

Reluctance—The reluctance of a magnetic circuit is determined in a manner similar to that used in determining the resistance of a conductor. A conductor's resistance is determined by substituting in the equation

$$R = \frac{Kl}{cm}$$

where K is the resistance of a unit piece (see equation 3-6).

Similarly, the reluctance of a magnetic circuit may be determined if the reluctance of a unit piece is known. Since magnetic circuits are usually rectangular in cross section, it is convenient to use a one inch cube as a unit piece. The reluctance of an inch cube of a material is called the reluctivity, and is represented by the Greek letter ν (pronounced nu). Then the reluctance of a magnetic circuit may be found by multiplying the reluctivity ν by the length l, and dividing by the cross-sectional area A. Thus

$$\mathcal{R} = \frac{\nu l}{A} \qquad (4\text{-}3)$$

The value of K in an electric circuit does not change, regardless of how much current is flowing through the conductor, provided the temperature is kept constant. (For instance, the value of K for drawn copper is 10.4.) Likewise, the value of ν in non-magnetic materials does not change, regardless of how many magnetic lines there are flowing through the circuit. (Its value for air and all non-magnetic materials is .313.) Then the reluctance of non-magnetic materials may be found by the equation

$$\mathcal{R} = \frac{.313\ l}{A} \qquad (4\text{-}4)$$

EXAMPLE—Compute the reluctance of a wooden ring 4 square inches in cross section, with an average diameter of 6 inches.

The length of the ring (πD) = 3.1416 × 6 = 18.8496 inches. Substituting in equation 4-4

$$\mathcal{R} = \frac{.313 \times 18.8496}{4} = 1.475 \text{ rels}$$

Permeability—The reluctivity of magnetic materials ranges from .007 to .00009, which is much less than the reluctivity of non-magnetic materials. To avoid dealing with small decimals, the term permeability is used, which is merely the reciprocal of reluctivity. Permeability is denoted by μ, a Greek letter pronounced mu. Thus

$$\mu = \frac{1}{\nu}$$

Since the reluctivity for non-magnetic materials is .313, their permeability is 1/.313 or 3.19.

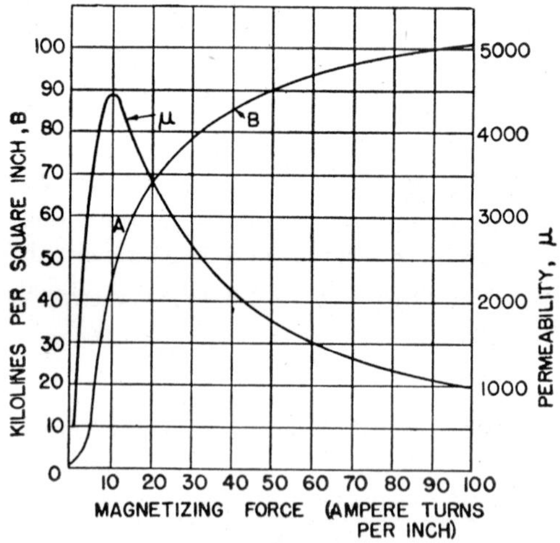

FIG. 4-21

The permeability or reluctivity of a magnetic material cannot be stated, even if its composition and heat treatment are known, because the flux does not change in direct proportion to the magnetizing force (ampere-turns). To make this clear, suppose we consider a steel ring surrounded by a coil. As the magnetizing force is gradually increased from zero to large values, the flux density increases as indicated by the magnetization curve B in fig. 4-21. It should be noted that the flux density increases slowly at first, but soon rises rapidly until at high values of magnetizing force it again increases only gradually. From this illustration we must conclude that the permeability of iron and steel depends upon the number of magnetic lines flowing in the material. It is not constant, varying with the flux density in the manner indicated by curve μ.

Note that near point A the m m f is very effective in producing flux, but beyond this point it must be increased tremendously to cause the flux density to increase slightly. In magnetic circuits it is not practical to use a flux density above point A, because too many ampere turns are required to produce the added flux. Point A, commonly called the saturation point, is not a point at which the magnetic material contains so many lines that no more can be added, but the point beyond which it is uneconomical to add more lines.

Thus a constant cannot be used for the permeability of magnetic materials, as in non-magnetic materials. However, curves may be used to show the permeability corresponding to different flux densities.

Flux Density—The number of lines per square inch is called the flux density and is denoted by B. The flux density B may be found by dividing the total flux by the cross-sectional area of the magnetic circuit A. Thus

$$B = \frac{\Phi}{A} \qquad (4\text{-}5)$$

The magnetizing force required per inch of a magnetic circuit is denoted by H, and depends only upon flux density and permeability. This may be proved as follows.

$$\Phi = \frac{IT}{\mathcal{R}} \qquad (4\text{-}6)$$

$$\text{But } \Phi = BA, \quad IT = Hl, \quad \text{and} \quad \mathcal{R} = \frac{1}{\mu A} \qquad (4\text{-}7)$$

Then substituting for Φ, IT, and $\mathcal{R}$ in equation 4-6 gives

$$BA = \frac{Hl}{\frac{1}{\mu A}} \text{ or } \mu = \frac{B}{H} \qquad (4\text{-}8)$$

Thus to determine the ampere-turns per inch of magnetic circuit, it is only necessary to know the flux density and permeability. Then instead of using curves which give the permeability corresponding to different flux densities, as suggested above, equation 4-8 makes it possible to plot curves giving the direct relation between flux density B and ampere-turns required per inch H. Fig. 4-22 shows such curves, which save considerable calculation, because instead of determining the ampere-turns by finding the reluctance and multiplying it by the flux, we find the ampere-turns per inch directly from the curves, and multiply by the length of the magnetic circuit.

In problems arising in practice, it is usually necessary to compute the number of ampere-turns required to produce a certain flux. To show how this is accomplished with the aid of the curves of fig. 4-22, the following example will be considered.

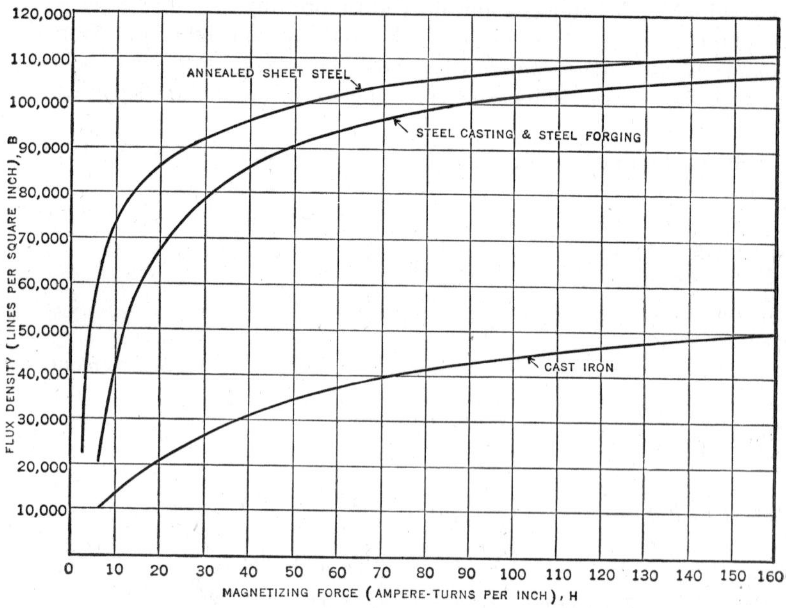

FIG. 4-22

EXAMPLE—The magnetic circuit of fig. 4-23 is made of cast iron and the coil contains 300 turns. Determine the current required to set up 100,000 magnetic lines.

$$\text{The flux density B} = \frac{100{,}000}{4} = 25{,}000 \text{ lines per square inch.}$$

For this flux density, the curve for cast iron shows that 28 ampere-turns per inch are needed. Then the required number of ampere-turns is 28×16 or 448.

Since the coil has 300 turns,

$$I = \frac{448}{300} = 1.49 \text{ amps}$$

Occasionally it is necessary to determine the permeability of certain materials at some specific flux density. The method of doing this is illustrated by an example.

EXAMPLE—Determine the permeability of annealed sheet steel at a flux density of 80,000 magnetic lines.

At 80,000 lines sheet steel requires 14 ampere-turns per inch (see fig. 4-22).

Substituting in equation 4-8,

$$\mu = \frac{80{,}000}{14} = 5714.3$$

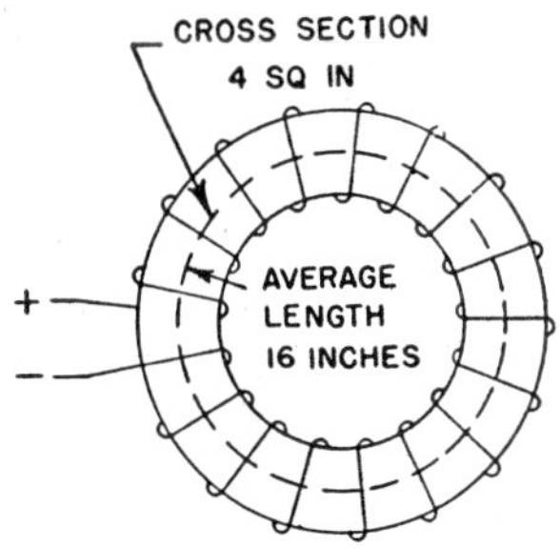

FIG. 4-23

Magnetic circuits may be joined in series or in parallel. Problems involving either of these cases may be solved by methods that are similar to those used in corresponding electric circuits.

Series Magnetic Circuits—In a series circuit the number of magnetic lines through all sections is the same, though the flux density may be different. If a series circuit is made up of different kinds of iron, contains air gaps, or is made up of parts that have different flux densities, the total m m f required is found by adding the ampere-turns of the separate parts.

The m m f required for an air gap or a piece of non-magnetic material may be determined by a simple equation, developed as follows:

$$\Phi = \frac{IT}{\mathcal{R}}$$

Substituting BA for Φ and $\frac{.313\ l}{A}$ for $\mathcal{R}$, we have the following equation:

$$BA = \frac{IT}{\frac{.313\ l}{A}}$$

Solving for IT,

$$IT = .313\ Bl \qquad (4\text{-}9)$$

The flux density in an air gap is very difficult to compute accurately, because the magnetic lines spread out on passing through the air, as indicated in fig. 4-24a. If an air gap is exceedingly short (not more than 3/16 of an inch), we may assume that it has the same flux density as the iron. Otherwise, it is a good approximation to assume that the flux fringes outward on each side for a distance equal to the length of the air gap. With this assumption the cross-sectional

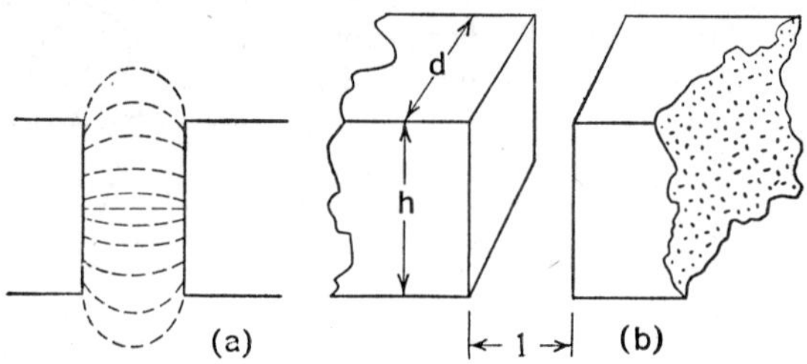

FIG. 4-24

area of a rectangular air gap shown in fig. 4-24b may be determined by the following equation:

$$A = (h + 1)(d + 1) \tag{4-10}$$

If the air gap is circular in cross section,

$$A = \frac{\pi(D + 1)^2}{4}$$

EXAMPLE—Determine the number of ampere-turns required to set up 210,000 magnetic lines through the circuit shown in fig. 4-25.

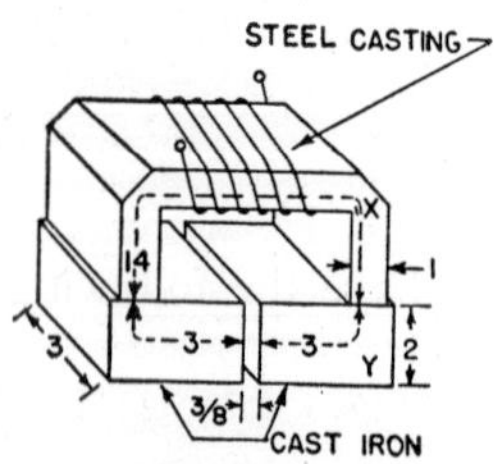

FIG. 4-25

In part X the flux density $B_X = \frac{210{,}000}{3}$ = 70,000 magnetic lines per square inch.

The ampere-turns per inch required for a density of 70,000 lines per square inch in steel casting is 21.5. For this part, 21.5 × 14 or 301 ampere-turns are required.

In part Y the flux density $B_Y = \frac{210{,}000}{6}$ = 35,000 magnetic lines per square inch.

The m m f per inch required for a density of 35,000 lines in cast iron is 53 ampere-turns. For the two cast iron parts, 53 × 6 or 318 ampere-turns are required.

According to equation 4-10, the air gap area $A = \left(2 + \frac{3}{8}\right)\left(3 + \frac{3}{8}\right)$ or 8 square inches.

The flux density is $B_a = \frac{210{,}000}{8}$ or 26,250 lines per square inch.

By equation 4-9, $IT_a = .313 \times 26{,}250 \times .375 = 3081$ ampere-turns.

The total number of ampere-turns required is 301 + 318 + 3081 = 3700.

More than 80 per cent of the total m m f is required to force the flux through the air gap. This is because the reluctance of the air is much greater than that of the iron. To use the smallest number of ampere-turns on a magnetic circuit, the air gap should be made as small as possible.

Parallel Magnetic Circuits—In practice magnetic circuits are frequently in parallel. The flow of magnetism through parallel circuits is governed by the same laws as the flow of current through electric circuits.

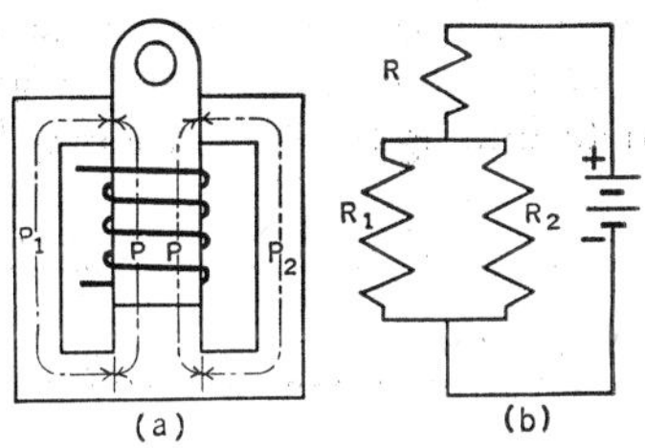

FIG. 4-26

When studying electric circuits, we learned that the e m f that forces current through one path also forces the currents through all the other parallel paths. A similar law also applies to parallel magnetic circuits; that is, the m m f that forces flux through one path also forces the flux through all the other parallel paths.

Fig. 4-26a shows the magnetic circuit of a plunger type of electromagnet. The magnetic lines on leaving the middle portion P divide equally and flow through the two portions P_1 and P_2, which have the same reluctance. Such a circuit is similar to the electric circuit of fig. 4-26b. In this circuit, resistance R carries the total current and corresponds to the portion P of the magnetic circuit. Resistances R_1 and R_2 are equal and in parallel; thus each carries half the current through R. Then R_1 and R_2 correspond to P_1 and P_2 of the magnetic circuit. The e m f necessary to force a current I through the electric circuit equals the e m f necessary to force I amperes through R plus the e m f necessary to force I/2 amperes through either R_1 or R_2.

Similarly, the m m f necessary to force Φ lines through the magnetic circuit equals the m m f necessary to force Φ lines through P plus the m m f necessary to force $\Phi/2$ lines through either P_1 or P_2.

Example—In fig. 4-26a the portion P is 4 square inches in cross section and 6 inches in length. The two portions P_1 and P_2 are each 2 square inches in cross section and 8.5 inches in length. If the entire magnetic circuit is a steel forging, determine the number of ampere-turns required to force 360,000 magnetic lines through P.

The flux density throughout the magnet $B = \frac{360{,}000}{4} = 90{,}000$ lines per square inch. For this density in a steel forging, 49 ampere-turns per inch are required. Then part P requires 49×6 or 294 ampere-turns, and part P_1 or P_2 requires 49×8.5 or 416.5 ampere-turns. The total m m f required is then $294 + 416.5$ or 710.5 ampere-turns.

Number of Turns and Size of Wire—After the number of ampere-turns for a magnetizing coil has been computed, it is at times necessary to determine the size of the wire and the number of turns that may be used to suit a given voltage. A suitable equation for determining the wire size may be developed as follows:

The resistance of the coil is $R = \frac{Kl}{cm}$ (4-11)

If l_t represents the average length per turn in inches, then $l = \frac{l_t T}{12}$

Substituting for l in equation 4-11, $R = \frac{K l_t T}{12\ cm}$

According to Ohm's law, $R = \frac{E}{I}$

Then $\frac{E}{I} = \frac{K l_t T}{12\ cm}$

Solving for cm, we have $cm = \frac{K l_t I T}{12 E}$ (4-12)

Equation 4-12 is a suitable equation for determining the wire size. The current through a coil is fixed by the current-carrying capacity of the wire, which for air cooled coils is 1200 circular mils per ampere.

EXAMPLE—In fig. 4-25, 3700 ampere-turns are required to set up the desired flux. If the magnetizing coil is connected to 12 volts and is one inch thick, determine the size of wire and the number of turns required.

Substituting in equation 4-12, $cm = \frac{10.4 \times 12 \times 3700}{12 \times 12}$

$= 3206.7$ circular mils.

(A number 15 wire will be used, which has a circular mil area of 3260.) Since 1200 circular mils are required per ampere, then

$$I = \frac{3260}{1200} = 2.7 \text{ amperes}$$

$$T = \frac{IT}{I} = \frac{3700}{2.7} = 1370.4 \text{ turns}$$

Hysteresis—If a coil on an iron core is energized, a flux will be set up which will increase in the manner suggested by curve ab in fig. 4-27. A similar curve for a steel casting was shown in fig. 4-22, and called a saturation or magnetization curve.

When the current in the coil of fig. 4-27 is decreased, the flux density will not decrease along line ab, but will decrease less rapidly as indicated by line bc. The magnetizing force is zero when c is reached, but the flux density is 50,000 lines per square inch. This flux, which remains after the magnetizing force has been removed, is called residual magnetism. The reason it remains may be explained as follows.

The molecules of a magnetic material are small magnets haphazardly arranged. When a magnetizing force is applied, the little magnets tend to line up with their axes parallel. Their motion is resisted by molecular friction, however, which produces heat. When the magnetizing force is removed, forces of attraction and repulsion tend to turn the little magnets back to their normal position. As before, friction tends to resist this motion and additional heat is produced. This molecular friction also prevents the molecules from completely returning to their normal position. Thus the material is magnetized to some degree even after the magnetizing force is completely removed.

At d (fig. 4-27) the residual magnetism has been reduced to zero by reversing the coil current and producing a m m f of 20 ampere-turns per inch.

As this reversed coil current is increased further, the flux density will build up as indicated by curve de. For equal magnetization, the flux density at e is the same as at b.

If the coil current is reduced to zero, the flux density will decrease along the curve ef and will again reach 50,000 magnetic lines at zero magnetization.

To reduce the flux to zero, it is necessary to supply a magnetizing force ag equal to ad.

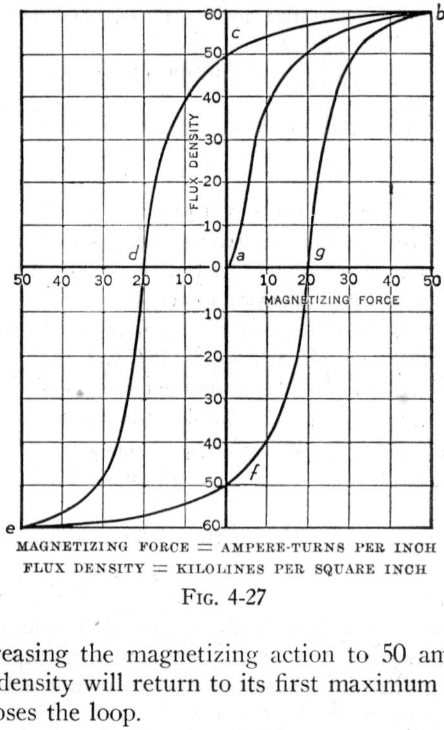

MAGNETIZING FORCE = AMPERE-TURNS PER INCH
FLUX DENSITY = KILOLINES PER SQUARE INCH

FIG. 4-27

While increasing the magnetizing action to 50 ampere-turns per inch, the flux density will return to its first maximum value, which is at b. This closes the loop.

It is thus seen that the flux lags behind the magnetizing force, and this lagging effect is called hysteresis. In DC and AC machines the flux is continually changing in this way, and as a result some power is converted into heat.

Industrial Applications—A group of industrial applications of magnetism which may be discussed conveniently at this time are lifting magnets, magnetic chucks, magnetic clutches, magnetic brakes, and magnetic separator pulleys.

Lifting Magnets—Lifting magnets are often used to transfer large amounts of iron and steel from one place to another. These magnets save considerable time and labor because they require no slings or chains for holding a load.

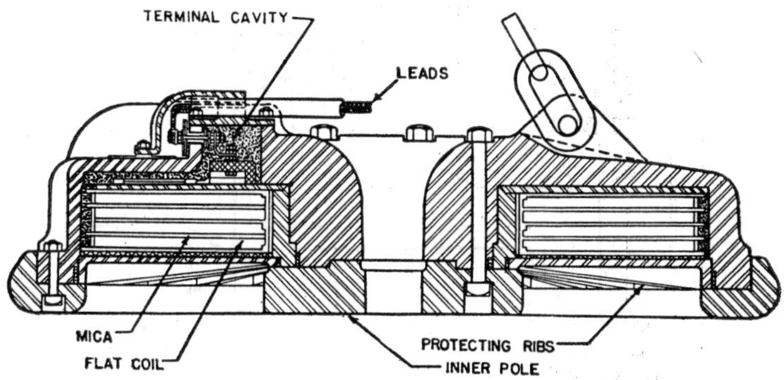

Fig. 4-28

Fig. 4-28 shows a typical magnet, consisting of a four-layer coil surrounded by a steel frame. The frame protects the winding and forms the magnetic circuit between the two poles, located at the center of the coil and at the outer ring. Between these two poles is a non-magnetic manganese steel casting with projecting ribs, which protects the coil and enables the magnet to withstand extremely severe service. The magnetic path between the faces of the two poles is completed through the material being handled.

A suitable controller must be provided to rapidly magnetize and demagnetize the lifting magnet. The construction and wiring diagram of a typical controller is shown in fig. 4-29. When it is desired to lift a load, the master switch S is moved to the "lift" position. This energizes coil C_1, which closes breakers 1 and connects the magnet across the line. The master switch will remain in this position until again moved by the operator.

When the operator wishes to drop the load, the master switch is moved to the "drop" position and left there. This allows breakers 1 to drop out and causes breaker 2 to close, thus connecting coil C_2 across the terminals of the magnet, with R_1 in series and R_2 in parallel with this coil. On dying out, the flux on the lifting magnet sets up a voltage which energizes coil C_2 for a few seconds (the time can be

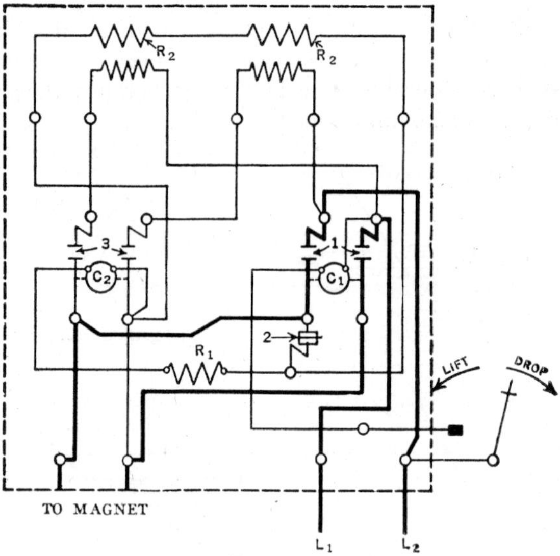

FIG. 4-29

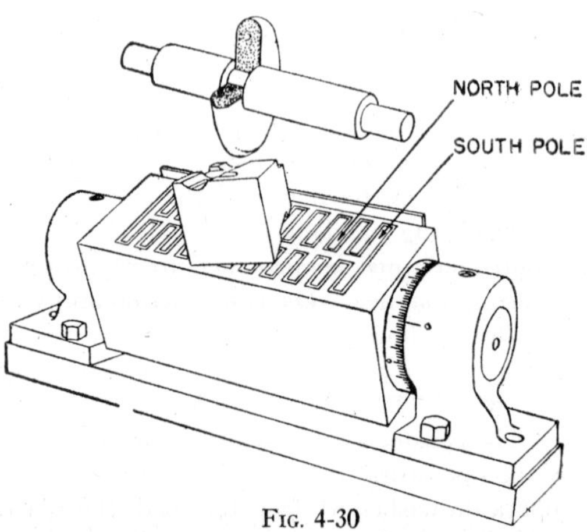

FIG. 4-30

regulated within certain limits by varying R_1 and R_2). Breakers 3 are closed while C_2 is energized, thus placing a reverse voltage across the magnet terminals. As a result, a reversed current flows through the coil of the magnet, setting up a flux that neutralizes the residual magnetism. This allows the load to drop off within 1 to 3 seconds, whereas about 15 seconds would be required without a reverse voltage.

Magnetic Chucks—Magnetic chucks are sometimes used on planers and grinders to avoid the use of clamps. One type, which consists of many coils surrounded by a steel housing, is shown in fig. 4-30. The upper portion of the housing is divided into a number of sections. These are magnetized by the coils to form poles of alternate polarity. Placing the work on these poles closes the magnetic circuit, thus holding it firmly in place.

Magnetic Clutches—Clutches are used to connect and disconnect a load from its driving source. They consist of two elements—a fric-

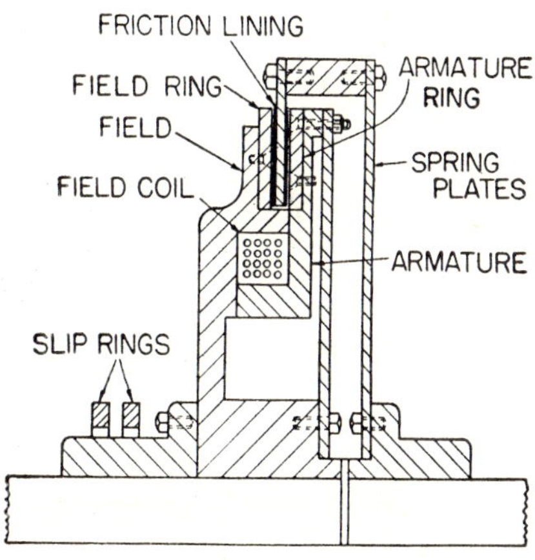

FIG. 4-31

tion element and a field member for exerting force on the friction element. Fig. 4-31 shows the construction of one type. One member carries a double friction lining, which lies between two rings attached to the field and the armature. When the field coil is energized, the field ring and the armature ring (see fig. 4-31) are drawn together, thus clamping the friction lining. This produces a driving force.

Magnetic Brakes—A type of brake frequently found on direct current cranes is shown in fig. 4-32. The field winding of this brake is connected in series with the armature circuit. When the power is removed from the motor, or when it is no longer supplied, due to some other reason, such as a blown-out fuse, the coil becomes de-energized and the heavy spring S applies the friction brake immediately. When power is supplied to the motor, the magnet becomes energized and pulls arm A downward. This action raises the shoes from the brake wheel, thus allowing the armature to rotate freely.

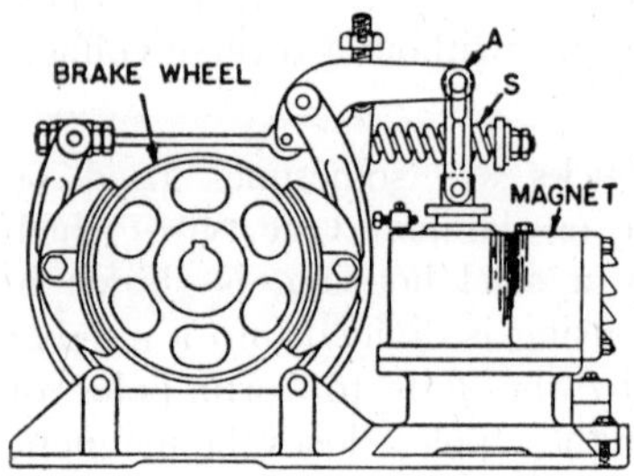

FIG. 4-32

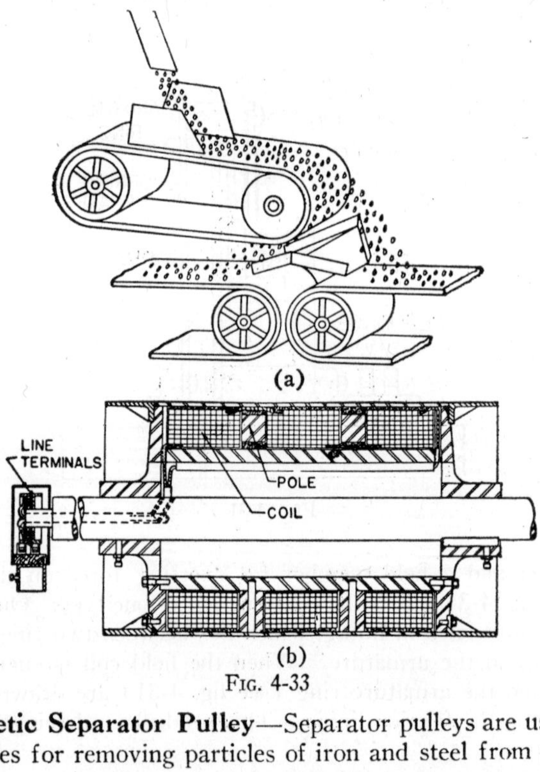

FIG. 4-33

Magnetic Separator Pulley—Separator pulleys are used mostly in foundries for removing particles of iron and steel from pulverized materials. The material is delivered onto a moving belt (fig. 4-33a), which passes over the magnetic pulley. The magnetic material sticks to the belt and is therefore carried further around than the non-magnetic material with which it is mixed. Thus the two drop on different conveyors. The magnetic pulley consists simply of a series of alternate north and south poles, each extending entirely around the pulley (see fig. 4-33b).

Magnetic Pull—It was shown that magnetic lines passing between iron or steel poles tend to pull them together. This pull may be calculated by the equation

$$F = \frac{B^2A}{7213 \times 10^4} \qquad (4\text{-}13)$$

where F = force in pounds, B = flux density in each pole, and A = cross-sectional area of each pole.

EXAMPLE—The magnetic circuit shown in fig. 4-34 has a cross section of 2 square inches. If the magnetizing action is produced by 280 ampere-turns, compute the attractive force between A and B.

The ampere-turns per inch is $H = 280 \div 14 = 20$. Fig. 4-22 shows that 20 ampere-turns per inch set up 68,000 magnetic lines per square inch. Substituting in equation 4-13

$$F = \frac{68{,}000^2 \times 2}{7213 \times 10^4} = 128.2 \text{ lbs. per pair of poles}$$

$$\text{Total force} = 128.2 \times 2 = 256.4 \text{ lbs.}$$

In the design of a plunger electromagnet, a low flux density should be selected for the poles (about 50,000 lines per square inch), so that the m m f (ampere-turns) required for the air-gap, and therefore the amount of copper, will not be excessive.

EXAMPLE—The electromagnet shown in fig. 4-35 is connected to a 50 volt source, has a .5 inch travel, and exerts a 20 pound pull. Determine the size of the wire, the number of turns, and the lettered dimensions.

The cross section of the plunger is

$$A = \frac{7213 \times 10^4 F}{B^2} = \frac{7213 \times 10^4 \times 20}{50{,}000^2}$$

$$A = .577 \text{ square inches}$$

Making the plunger square, then $M = \sqrt{.577} = .76$ inch.

Determining the cross section of the flux in the air-gap by substituting in equation 4-10 $A = (.76 + .5)\ (.76 + .5) = 1.59$ square inches. The total flux in the plunger, and also in the air-gap, is $50{,}000 \times .577$ or 28,850 lines, and the flux density in the air-gap is $B_a = 28{,}850 \div 1.59$ or 18,145 lines per square inch.

The number of ampere-turns required to force the flux through the steel is very small compared to the number required to force the flux through the air-gap, and consequently, only the ampere-turns required for the air-gap will be considered. Substituting in equation 4-9, $IT_a = .313 \times 18{,}145 \times .5 = 2839.69$ ampere-turns.

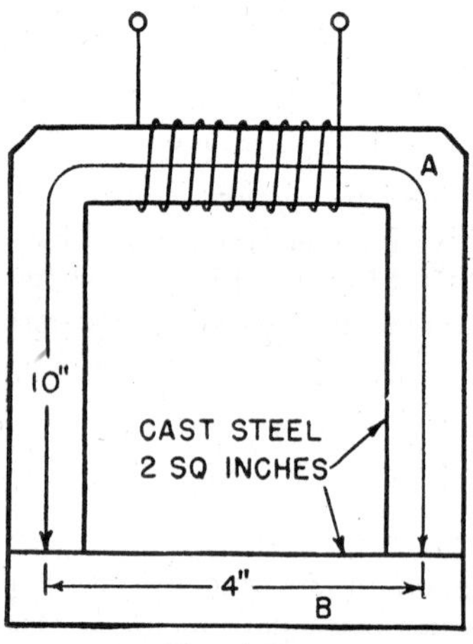

FIG. 4-34

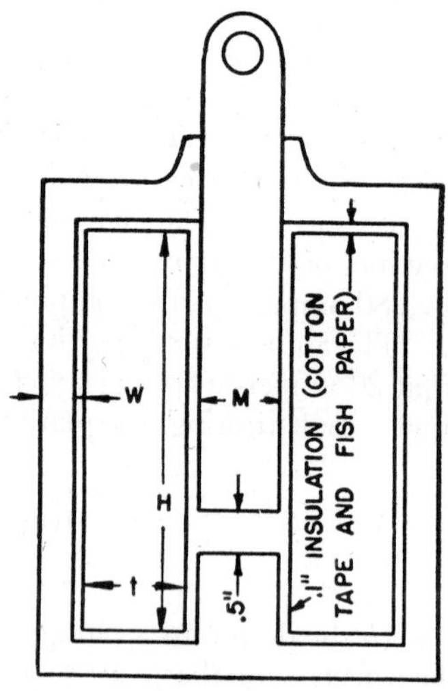

FIG. 4-35

Selecting a coil thickness (t) of 1 inch and allowing .2 inch for insulation thickness, the average length per turn is $l_t = 4(.76 + 1 + .2) = 7.84$ inches. Substituting in equation 4-12, then,

$$cm = \frac{10.4 \times 7.84 \times 2839.69}{12 \times 50} = 386$$

No. 24 gage wire, which has an area of 404 circular mils is the nearest size to this calculated value.

An air cooled coil will not overheat if the wire has a cross section of 1200 circular mils per ampere. Figuring the coil current on this basis, $I = 404 \div 1200 = .337$ ampere.

The number of turns in the coil is T = IT ÷ I = 2839.69 ÷ .337 = 8426.38. Single cotton covered wire will be selected, since it has sufficient insulation for 50 volts. According to Table 3-IV, 1700 turns of No. 24 gage, single cotton covered wire occupies one square inch. Then the cross section of the coil will be 8426.38 ÷ 1700 or 4.96 square inches. Since the coil is one inch thick, its height H must be 4.96 inches.

The section of the core surrounding the coil should have a flux density that does not exceed the saturation point, otherwise a high m m f will be required for this part of the magnetic circuit, which has been neglected. The amount of flux in this portion of the magnetic circuit is 28,850 ÷ 2 = 14,425 lines. Considering that this circuit is made of cast steel, for which a suitable flux density is 60,000 lines per square inch, the cross section of the surrounding portion must be 14,425 ÷ 60,000 or .24 square inch. It is desirable to make this cross section the same depth as the plunger, which is .76 inch. Then W will be .24 ÷ .76 = .318 inch.

Exercises

1. Determine the m m f produced by a coil of 2000 turns carrying .5 ampere.
2. How much current must flow through a coil of 1500 turns to set up 1,000,000 lines in a magnetic circuit having a reluctance of .007 rel?
3. What is the reluctance of a magnetic circuit in which a magnetomotive force of 45,000 ampere-turns sets up 200,000 lines?
4. A magnetic circuit similar to that of fig. 4-20 is 10 square inches in cross section and has an average length of 30 inches. Calculate the total flux produced if the coil has 500 turns and carries 10 amperes.
5. An 1800 turn coil on a .9 rel magnetic circuit has 55 ohms resistance and is connected to a 120 volt supply. Find the amount of flux produced.
6. How many turns of wire carrying 10 amperes must be wound on a brass ring 4 square inches in cross section and 24 inches long to set up 50,000 magnetic lines?
7. A magnetizing force of 20 ampere-turns per inch produces what flux density in cast iron? In cast steel? Determine the permeability of these materials with this magnetizing force.
8. A cast steel circuit 12 square inches in cross section is to carry 800,000 magnetic lines. Determine the magnetizing force required per inch.
9. How many ampere-turns are necessary to force 180,000 magnetic lines through a cast steel circuit 3 square inches in cross section and 9 inches long?

10. Determine the number of ampere-turns required to set up 240,000 magnetic lines in the magnetic circuit of fig. 4-25 if parts X and Y are both annealed sheet steel.
11. A ring of steel casting has an average circumference, not including a .15 inch air gap, of 18 inches, and has a circular cross section of 4 square inches. Calculate the number of ampere-turns required to set up 60,000 lines per square inch.
12. The electromagnet E in fig. 4-36 is .25 sq. in. in cross section and the armature A is .25 sq. in. in cross section. Neglecting flux fringing at the air gap, determine the number of ampere-turns required to set up 20,000 lines through the circuit if both E and A are steel castings.
13. In fig. 4-37 the electromagnet is of annealed sheet steel, portion A is 12 sq. in. in cross section, and portions B are each 8 sq. in. in cross section. Portion C is of cast iron 14 sq. in. in cross section. If the coil has 80 turns, how much current must flow through it to set up 960,000 lines through A?

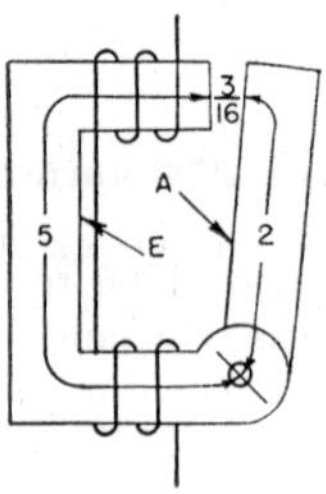

Fig. 4-36

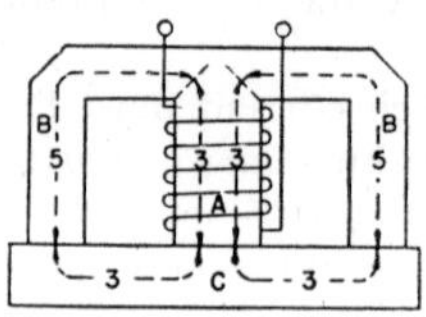

Fig. 4-37

14. A magnetic circuit of a two pole generator is shown in fig. 4-38. The yoke is a steel casting 14 sq. in. in cross section and the armature and field poles are of annealed sheet steel. One path through the armature is 12 sq. in. in cross section and the field poles are each 26 sq. in. in cross section. The air gap is 28 sq. in. in cross section. Determine the number of ampere-turns necessary on each field pole to set up 2,000,000 lines in the air gap.
15. Suppose that 4500 ampere-turns are required to set up the desired flux in a magnetic circuit that is 1 by 3 inches in cross section. If the coil on this magnetic circuit is 1 inch thick and is connected to a 12 volt supply, determine: (a) the size of wire required; (b) the current through the coil.
16. In fig. 4-34, with what force is armature B attracted if 2 amperes flow through the coil and it has 200 turns (both A and B are cast steel)?

17. With what force is part B in fig. 4-34 attracted if its poles are each ¼ inch from those of part A (the m m f required to force the flux through parts A and B is to be neglected)?
18. To ring the bell shown in fig. 4-39, the armature must be attracted with a force of 8 pounds. Part E is circular in cross section. (Since the bell is used only for short intervals, consider the wire to have a cross

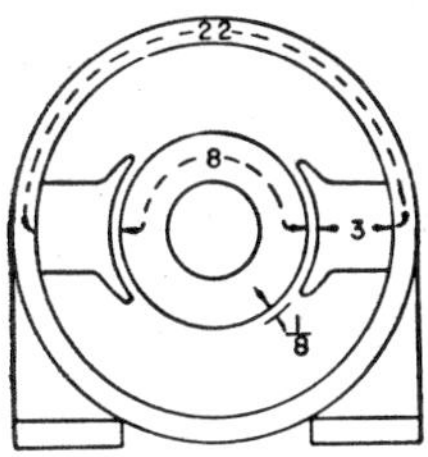

FIG. 4-38

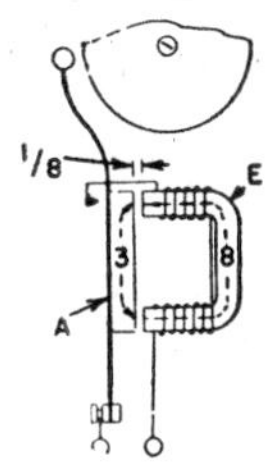

FIG 4-39

section of 500 circular mils per ampere.) The coil is .5 inch thick and is connected to a 6 volt line. Determine (a) the diameter of E, (b) the size of the wire in the coil, and (c) the number of turns in the coil.
19. A cast steel electromagnet of the type shown in fig. 4-35 has a ¾ inch travel, exerts a pull of 10 pounds, and is connected to a 40 volt line. Consider the coil to be one inch thick and determine (a) the size of the wire, (b) the number of turns in the coil, and (c) the dimensions shown in fig. 4-35.

CHAPTER 5

ARMATURE WINDING

A GENERATOR is a machine that converts mechanical energy into electrical energy. It consists fundamentally of a large number of conductors on an iron core, arranged to rotate in a magnetic field. The generator action depends upon the principle that an e m f is induced in a conductor cutting magnetic lines. Fig. 5-1a shows a conductor in a magnetic field. If it is moved either up or down, it cuts flux and an e m f is induced in it. If the conductor is moved either to the left or the right, however, it merely passes alongside the magnetic lines without cutting them, and therefore no e m f is induced.

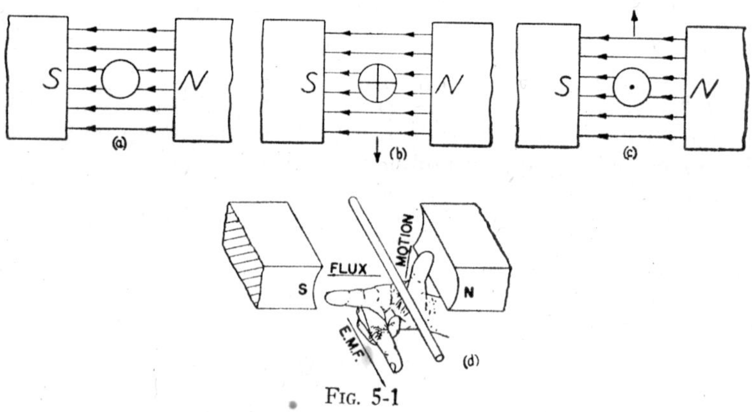

FIG. 5-1

Fleming's Right-hand Rule—There is a definite relation between the direction of flux, the direction of motion of the conductor, and the direction of the induced e m f. In fig. 5-1b the e m f is induced "in" when a conductor is moved downward. If the direction of motion is reversed, as indicated in fig. 5-1c, the direction of the e m f reverses. The direction of the e m f induced in a conductor can be found as follows: *Extend the thumb, the forefinger, and the middle finger of the right hand at right angles to one another (fig. 5-1d). Point the forefinger in the direction of the flux and the thumb in the direction in which the conductor is moving. Then the middle finger will point in the direction of the induced e m f.*

Voltage in a Rotating Coil—Suppose that a single turn coil (fig. 5-2a) rotates in a uniform magnetic field at a constant speed. As the coil changes position, the induced e m f changes, but the e m f induced in one side always adds to that induced in the other. When

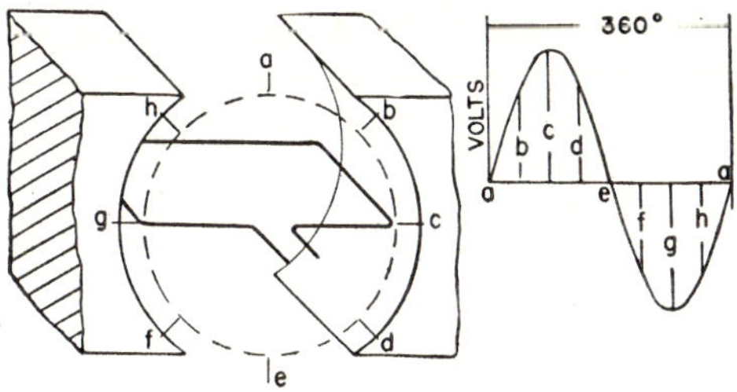

(a) (b)

Fig. 5-2

the coil is in position a, the e m f generated is zero, because the coil sides are not cutting magnetism, but are moving parallel to it. When the coil is in position b, its sides are moving at an angle to the flux, and therefore a low e m f is generated (indicated by b in fig. 5-2b). When the coil is in position c, the coil sides are moving at right angles to the flux, and are therefore cutting the flux at the maximum rate. Hence, at this instant the generated e m f is maximum. At position d, the generated e m f is less because the coil sides are again cutting at an angle, as at b. At e no lines are being cut, and as at a, no e m f is generated. At f, each coil side moves under a pole of opposite polarity, and the direction of the generated e m f is reversed. The e m f is maximum in this direction when the coil is at g and zero when it is again at a. This cycle repeats each revolution.

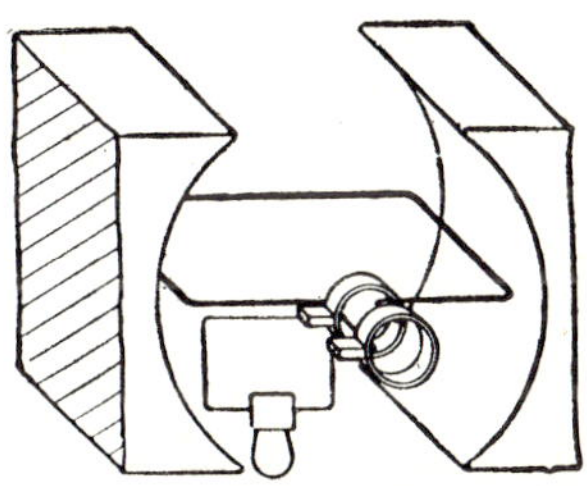

Fig. 5-3

The alternating e m f generated in a coil is impressed on an external circuit by means of two slip rings, one connected at each end of the coil (fig. 5-3). Stationary brushes resting on the rings connect the coil to the external circuit.

If the two slip rings are replaced by a single split ring (fig. 5-4a), called a commutator, the connection of the coil reverses at the same time that the induced e m f reverses. Therefore, the e m f impressed on the external circuit always remains in one direction. With a single coil, the voltage supplied will of course not remain steady, but will pulsate as indicated in fig. 5-4b. A pulsating voltage would be undesirable commercially, and furthermore, a single coil would utilize very little of the space available on the cylindrical iron core.

Lap Winding—The undesirable features of a single coil may be eliminated by placing a number of such coils on a cylindrical iron core (fig. 5-5a). If these are rotated in a clockwise direction, the voltages induced in the sides under the north poles are "in", and the voltages

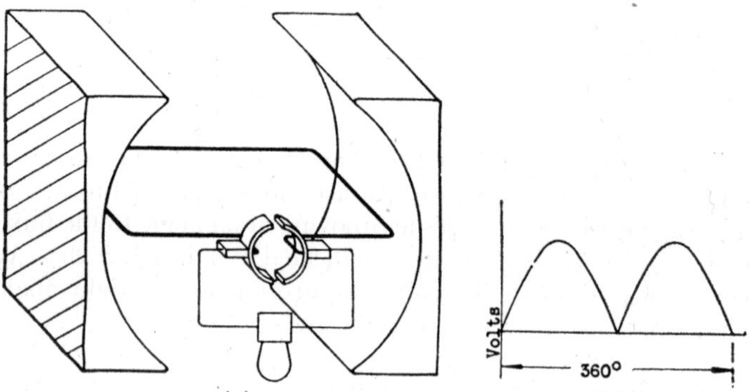

(a) (b)

Fig. 5-4

induced in the sides under the south poles are "out". These coils may be connected in such a manner that their induced voltages are additive from one brush to the other. A common method of doing this is illustrated in fig. 5-5b, which is known as a lap winding because the successive coils overlap. A complete view of this winding is shown in fig. 5-5c and a developed view in fig. 5-5d. One path and the direction of the voltage induced in it may be traced by starting with brush A and proceeding as follows: A-1-8-3-10-5-12-B. Another path may be traced from A as follows: A-4-13-2-11-16-9-B. The coil with sides 6 and 15 is short-circuited by brush A, and the coil with sides 7 and 14 is short-circuited by brush B. This is permissible because these coils are not cutting magnetism and have no voltage induced in them. Thus the winding has two paths, and the direction of the voltage in each tends to send the current from brush A to brush B. This is shown in fig. 5-5e. With reference to the winding itself, these voltages are equal and opposite, and no current flows through the two paths when no external load is connected to the brushes. With reference

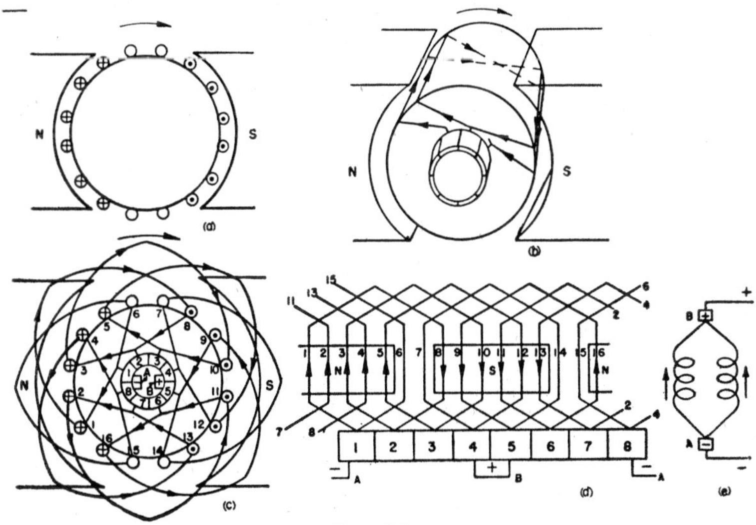

FIG. 5-5

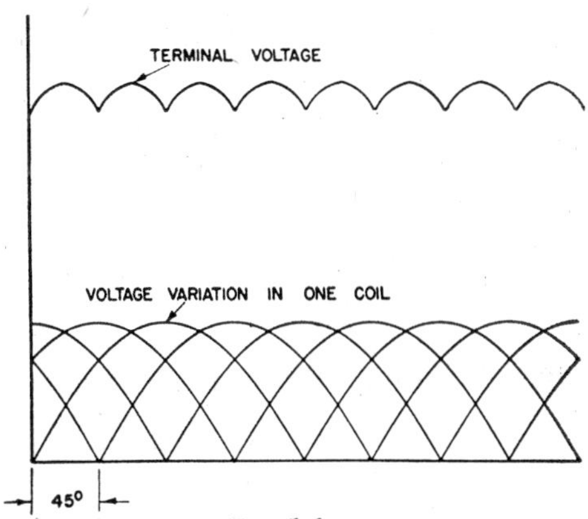

FIG. 5-6

to an external load, however, the voltages in the two paths act in the same direction, sending current through the external load. The current through each path will be half the load current.

The voltage induced in a path is the sum of the voltages induced in the coils of that path. When one coil moves past a brush or out of a path, another coil moves into the path. Fig. 5-6 shows how the voltage of one path is made up. Since the coils are displaced from each other by $360 \div 8 = 45°$, the voltage waves are displaced from each other by the same angle. These voltages are plotted separately. The generated voltage in a path at any instant is the sum of the individual voltages in that path at the same instant. This sum varies slightly, as indicated in fig. 5-6. The number of coils is much greater in most windings, and the voltage per path (or at the terminals of the machine) is nearly constant.

Armatures with only one winding, as shown in fig. 5-5, are said to be simplex wound. Those with more than one winding are said to be multiplex wound, and are to be considered later.

Pitch—If a coil spans from the center of one pole to the center of an adjacent pole, it has a full pitch. If the coil spans less than full pitch, it has a fractional pitch, which should not be less than .9 full pitch. If the pitch is too small, an appreciable reduction in generated e m f results. The fractional pitch type of winding is in general use, because it produces less sparking and because it requires less copper, which means lower armature resistance.

The most convenient way of expressing the span of a coil is in terms of the slots. To span from the center of one pole to nearly the center of an adjacent pole, a coil must span nearly ½ the distance around the armature in a 2 pole winding, ¼ the distance in a 4 pole winding, 1/6 the distance in a 6 pole winding, etc. Then for any number of slots S and poles P, the slot span is

$$y_s = \frac{S}{P} \text{ (preferably slightly less)} \qquad (5\text{-}1)$$

Double Layer Winding—When the coil sides are placed in one layer, as shown in fig. 5-5, it is difficult to arrange their ends so that they will pass each other. In order to avoid this difficulty, the coils are generally arranged in two layers, as shown in figs. 5-7 and 5-13a. Each coil has one of its sides at the top of a slot and the other at the bottom of some other slot. The coil ends will then lie side by side. Each coil is usually wound with the proper number of turns by a machine, and is then wrapped with varnished cambric or mica tape, either of which is covered with cotton tape. As far as connecting the coils is concerned, the number of turns per coil is immaterial. For simplicity, then, the coils in winding diagrams will be represented as having only one turn.

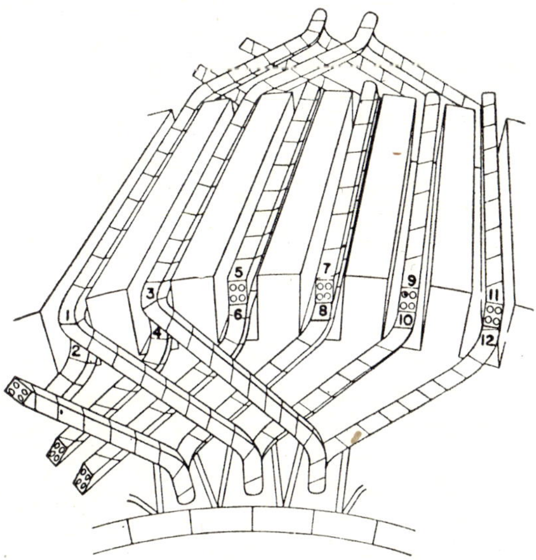

FIG. 5-7

Winding Pitch—In a two layer winding it is desirable to number the coil sides rather than the slots, in which case the rules established will apply to all windings of a given type. The coil sides are numbered as indicated in figs. 5-7 and 5-13a, those at the top of the slots being given odd numbers and those at the bottom being given even numbers. The number of coil sides spanned at the pulley end is called the back pitch, denoted by y_b (fig. 5-8). This back pitch equals the coil sides per slot (C_s) times the slot span (y_s) plus 1. Expressing this by means of an equation,

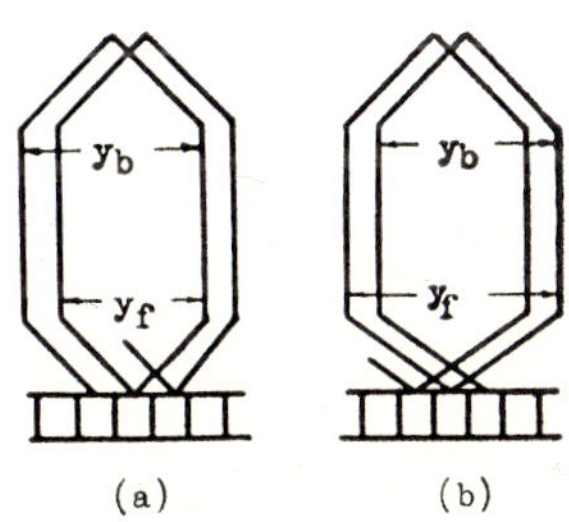

FIG. 5-8

$$y_b = C_s y_s + 1 \qquad (5\text{-}2)$$

The commutator pitch, denoted by y_c, is simply the number of segments spanned between the start end and the finish end of a coil. In a simplex lap winding, $y_c = 1$.

The number of coil sides spanned at the commutator end is called the front pitch, denoted by y_f. This span may be greater or less than y_b, but is never equal to it. If y_f is less, the winding is progressive, advancing in the clockwise direction (fig. 5-8a), and if y_f is greater, the winding is retrogressive, advancing in the counterclockwise direction (fig. 5-8b). In either case, y_b and y_f differ by 2. Therefore,

$$y_f = y_b \pm 2 \qquad (5\text{-}3)$$

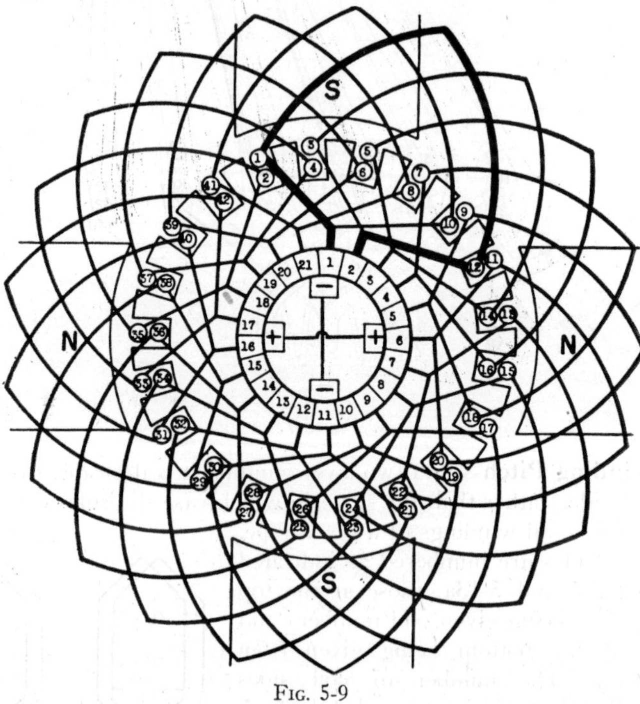

FIG. 5-9

When the negative sign is used, the winding is progressive, and when the positive sign is used, the winding is retrogressive. A retrogressive winding is seldom used because it requires more copper.

EXAMPLE—A 4 pole, simplex lap wound armature contains 21 slots and has two coil sides per slot. What is (a) the slot span? (b) The back pitch? (c) The commutator pitch? (d) The front pitch?

(a) $y_s = \frac{S}{P} = \frac{21}{4} = 5\frac{1}{4}$

Since a coil can span only a whole number of slots, in solving this problem the fraction ¼ is dropped, making $y_s = 5$.

(b) $y_b = C_s y_s + 1 = 2 \times 5 + 1 = 11.$

The back pitch of the first coil will therefore be from 1 to 12.

(c) As previously explained, for a simplex lap winding, $y_c = 1$.

(d) Considering a progressive winding, then

$$y_f = y_b - 2 = 11 - 2 \text{ or } 9$$

This winding is shown in figs. 5-9 and 5-10 (in fig. 5-10 the winding is rolled out into a plane).

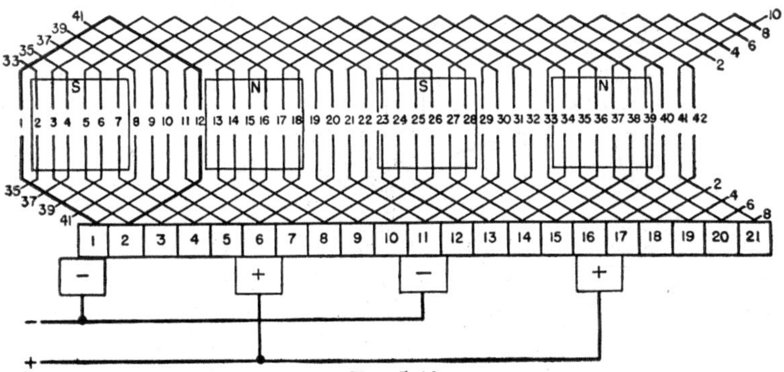

FIG. 5-10

Paths in a Simplex Lap Winding—In the four pole, simplex winding shown in fig. 5-12a (which is the same as fig. 5-9), there are

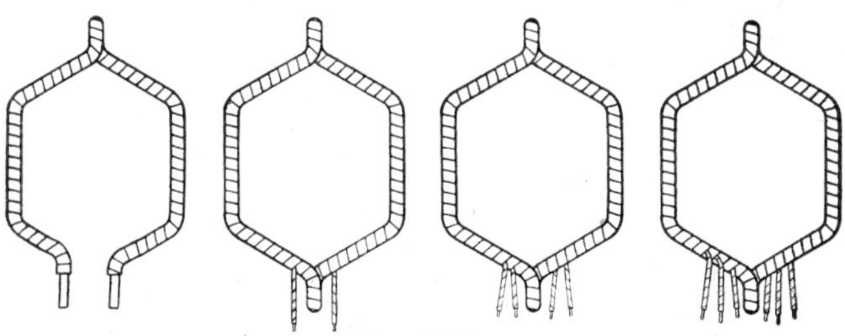

FIG. 5-11

four parallel paths between the negative and positive leads. One path is traced from the negative lead to brush A, through coils 1, to brush B, and then to the positive lead. A second path is traced from brush A, through coils 4, to brush D, to the positive lead. A third path is traced from the negative lead to brush C, through coils 2, to brush B, to the positive lead. A fourth path is traced from brush C, through coils 3, to brush D, and then to the positive lead. The connections of these paths are shown in a simplified manner in fig. 5-12b.

The winding shown in fig. 5-12a has 4 poles and 4 paths. A 6 pole winding has 6 paths, an 8 pole winding has 8 paths, etc. *In any simplex lap winding, then, there are always as many paths as there are poles.*

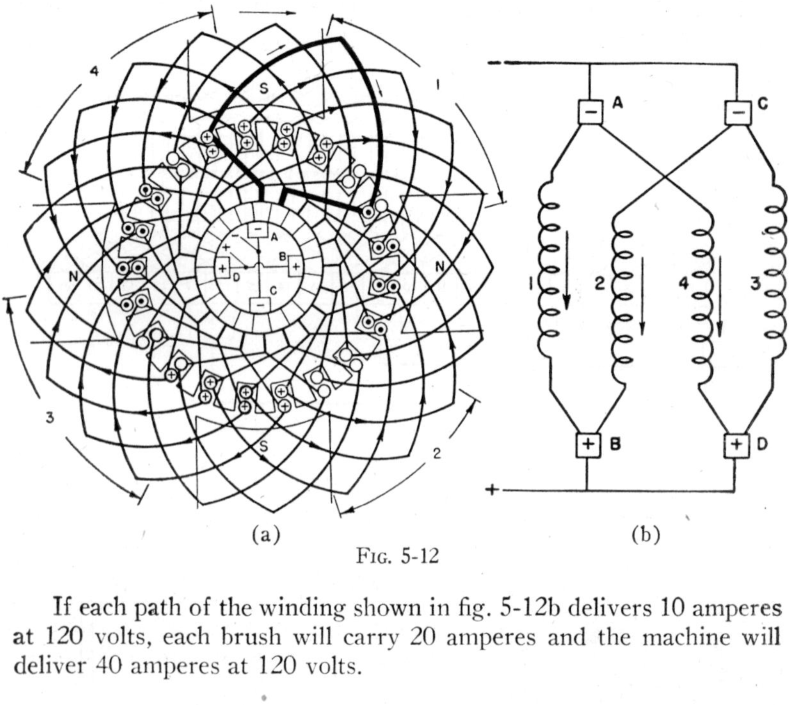

(a) (b)

FIG. 5-12

If each path of the winding shown in fig. 5-12b delivers 10 amperes at 120 volts, each brush will carry 20 amperes and the machine will deliver 40 amperes at 120 volts.

Multiple Coils—Although two coil sides per slot are used in many machines, in others a large number of slots would be required to keep the voltage between adjacent commutator segments at a low value. In some machines this would require small slots, in which case the ratio of the copper cross section to the slot cross section would be low. Also, the teeth roots would be so narrow that the teeth would be mechanically weak. This undesirable construction is eliminated by using less slots than commutator segments and placing more than two coil sides in each slot.

When more than two coil sides are to be placed in each slot, the armature is wound with multiple coils. A multiple coil consists of two or more coils taped together (fig. 5-11) and inserted in a slot as a unit, one side filling the upper half of one slot and the other side filling the lower half of another slot. These coil sides are numbered as shown in fig. 5-13b and are connected in the same manner as single coils.

Irrespective of the type of winding, there are always two coil sides connected to each segment, or an armature always has twice as many coil sides as segments. Then an armature with twice as many slots

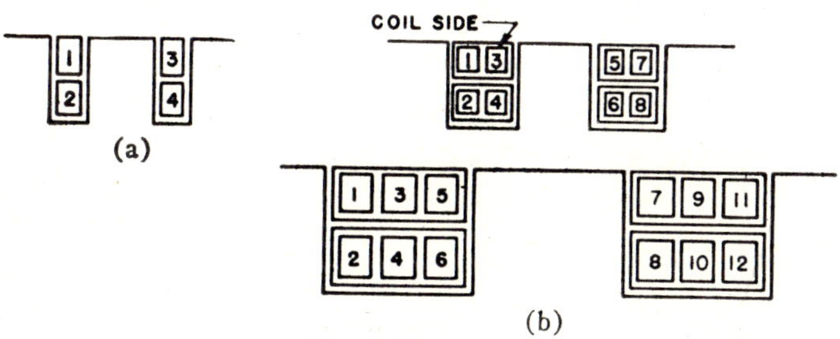

FIG. 5-13

as segments (fig. 5-5) must have one coil side per slot, one with the same number of slots as segments (fig. 5-9) must have two coil sides per slot, one with half as many slots as segments must contain four coil sides per slot, etc. The number of coil sides per slot may be determined by the following equation (s = the number of segments):

$$C_s = \frac{2s}{S}$$

EXAMPLE—A 4 pole simplex lap winding has 17 slots and 34 segments. What is (a) the back pitch? (b) the commutator pitch? (c) the front pitch?

$$\text{(a)}\quad y_s = \frac{S}{P} = \frac{17}{4} = 4\frac{1}{4}$$

$$C_s = \frac{2s}{S} = \frac{2 \times 34}{17} = 4$$

Making $y_s = 4$, then

$$y_b = C_s y_s + 1 = 4 \times 4 + 1 = 17$$

(b) $y_c = 1$

(c) Considering a progressive winding, $y_f = 17 - 2 = 15$.

This winding is shown in fig. 5-14.

FIG. 5-14

Position of Brushes—In the windings considered, the terminals of each coil connect to two segments midway between the coil sides. This method of connecting the coils is the most common because it requires the shortest coil leads. With such a winding, the brushes are placed in line with the pole centers, so that the coil voltages add in each path, and so that the brushes short-circuit coils only while they are in the neutral plane. If a coil is short-circuited while it is not in the neutral plane, a large current will circulate through it and the short-circuiting brush. Then a severe spark will occur at the instant that the brush breaks away from one of the two segments connected to the short-circuited coil.

In some machines the bearing-bracket arms are in line with the poles. In this case it is desirable to have each brush in line with a point midway between the pole tips, so that they are not hidden under the arms, where they are inaccessible. For this position of the brushes, the coils are connected to the segments as shown in fig. 5-15.

Before an armature is stripped for the purpose of rewinding it, the coil connections should be observed, because such connections depend upon the position of the brushes relative to the poles. If the

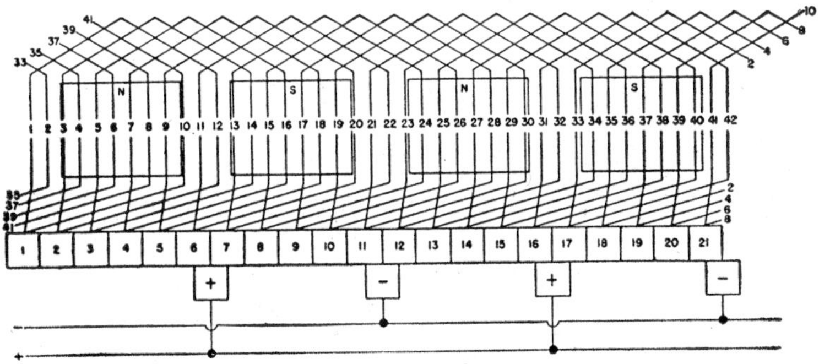

FIG. 5-15

coil leads are connected incorrectly, relative to the position of the brushes, sparking will occur unless the brushes are shifted. This cannot usually be accomplished.

Duplex Lap Winding—An armature that supplies a high current usually has several windings, each having an equal number of segments. Fig. 5-16 shows a 4 pole winding which occupies alternate slots and connects to alternate segments. In this case y_b is 13, since the first coil spans from 1 to 14. Coil side 14 connects back to coil side 5, making $y_f = 9$. Then the front pitch differs from the back pitch by 4 instead of by 2, which was the case in the simplex winding. Moreover, the commutator pitch is 2 instead of 1.

Since this winding occupies only alternate slots and connects to alternate commutator segments, a similar winding may be placed in the vacant slots and connected to the remaining segments. This second winding will obviously have the same back, front, and commutator pitches as the first. Although these two windings are insulated from each other, they are connected together by the brushes, each of which must cover at least two segments. Then each brush will connect to both windings at all times, permitting all the paths to supply current continuously. The two windings are therefore in parallel, and since each winding has four paths, the armature will have twice as many paths as there are poles. *The armature is said to be wound duplex*

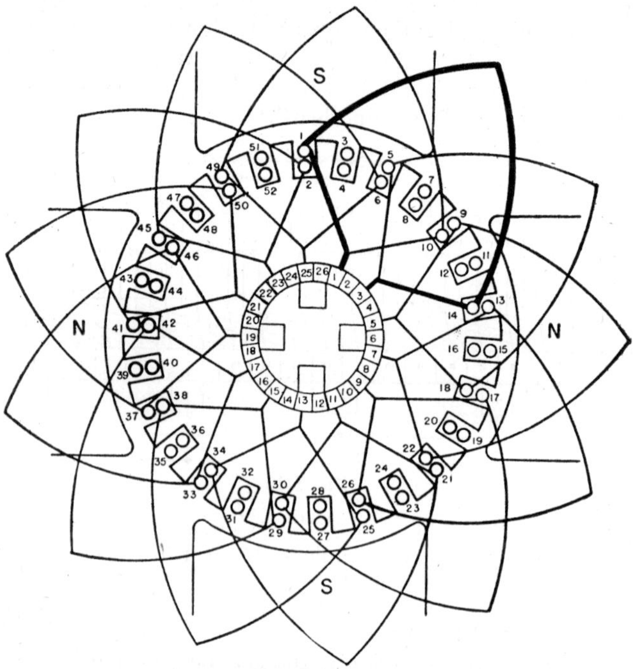

Fig. 5-16

because it has two windings, and is said to be double reentrant because each winding reenters upon itself.

Of course three windings could be placed on an armature core, making the armature triplex lap wound, etc. Regardless of the multiplicity of a winding, the number of paths is

$$b = mP \tag{5-5}$$

where m = the number of windings and P = the number of poles.

If the number of slots and segments in fig. 5-16 were one more or less than 26, the first winding would not close, and the complete winding would close only after the second winding was placed on the armature. This is illustrated by the 25 slot armature in fig. 5-17. It should be observed that the first winding does not return to segment 1, as does the winding in fig. 5-16, but terminates at 2. The second winding, shown with dotted lines, starts at segment 2 and terminates at segment 1. This duplex winding is single reentrant

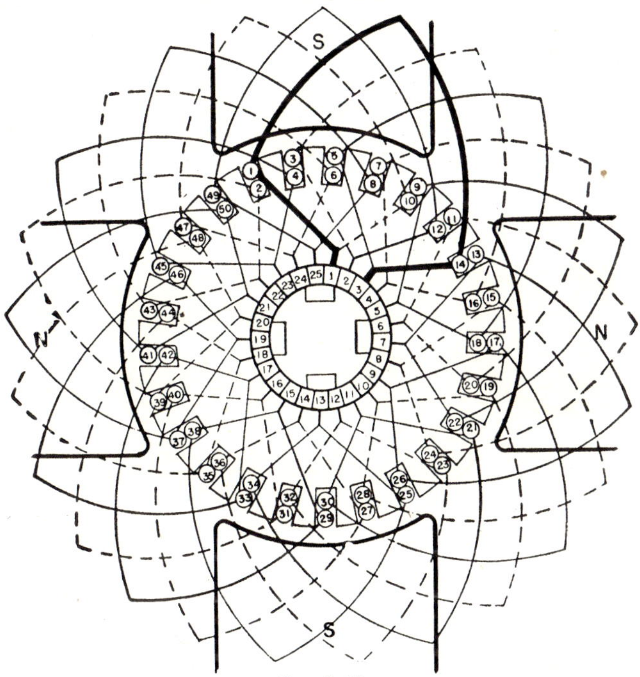

FIG. 5-17

because it returns to the starting point only after the armature is completely wound. As in fig. 5-16, the brushes must be wide enough to cover at least two segments. The difference between the windings of figs. 5-16 and 5-17 is illustrated in a simple manner in fig. 5-18. The two windings have the same number of paths and practically the same number of coils per path, and therefore are the same electrically.

In the foregoing it was shown that an armature with two windings is duplex, one with three windings is triplex, etc. *If all the winding on an armature must be traced to return to the starting point, it is single reentrant; if only half the winding is traced before return-*

ing to the starting point, it is double reentrant; if only a third is traced before returning to the starting point, it is triple reentrant; etc.

The degree of reentrancy in a lap winding is found by dividing the number of segments by the multiplicity. If this gives a whole number, the winding is multiple reentrant, and if not, the winding is single reentrant.

In a multiplex winding y_b is determined in the same manner as in a simplex winding (see equation 5-2), since in all cases this span should be slightly less than the distance from the center of one pole to the center of the next.

The front span y_f must differ from y_b by 2 in a simplex winding, by 4 in a duplex winding, by 6 in a triplex winding, etc. Then for any

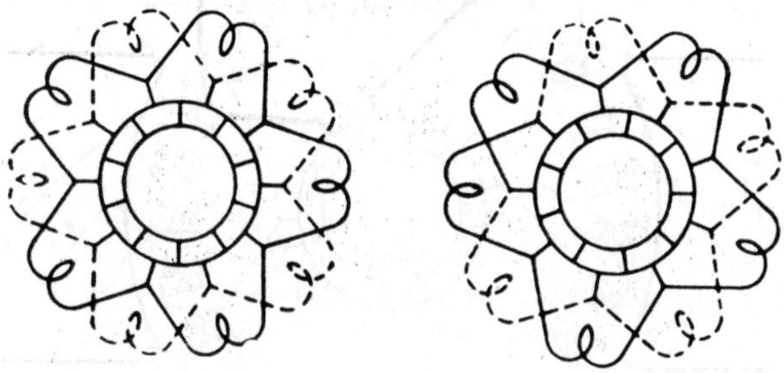

FIG. 5-18

multiplicity (m) the front pitch (y_f) must differ from the back pitch (y_b) by 2m. That is,

$$y_f = y_b \pm 2m \qquad (5\text{-}6)$$

For a simplex winding the commutator pitch (y_c) is 1, for a duplex winding it is 2, for a triplex winding it is 3, etc. Then for any multiplicity,

$$y_c = m \qquad (5\text{-}7)$$

EXAMPLE—A 4 pole, triplex lap winding has 24 slots and 24 segments. What is (a) the back pitch? (b) the commutator pitch? (c) the front pitch?

(a) $y_s = \frac{S}{P} = \frac{24}{4} = 6$

$y_b = C_s y_s + 1 = 2 \times 6 + 1 = 13$

(b) $y_c = m = 3$

(c) $y_f = y_b - 2m = 13 - 2 \times 3 = 7$

This 4 pole, triplex winding is shown in fig. 5-19.

If an armature is changed from simplex lap to duplex lap without making any other change, the numbers of paths is doubled and each path has half as many coils. The armature will then supply twice as much current at half the voltage. This principle is used when it is desired to reduce a voltage to half its value, as from 240 to 120 volts.

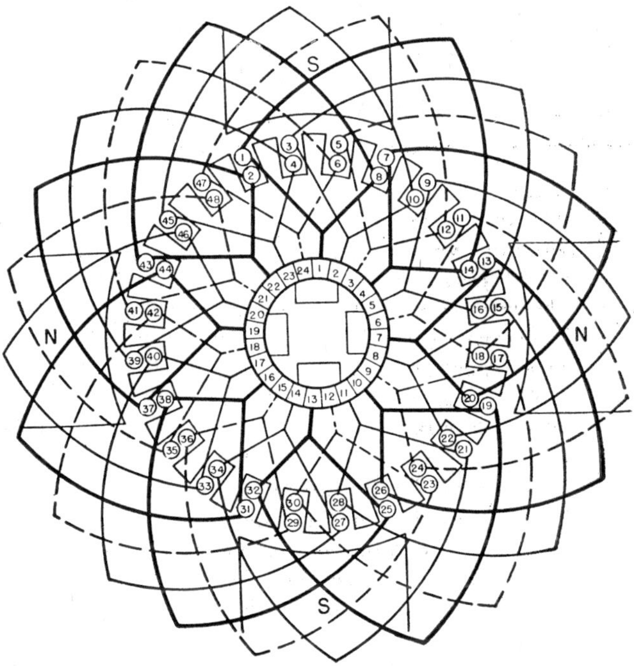

FIG. 5-19

A change from a simplex winding to a duplex winding is made merely by unsoldering the start end of each coil and connecting it one segment back of its previous location.

In the windings considered it will be noted that in every case the finish ends of adjacent coils, and also the start ends of adjacent coils, connect to adjacent commutator segments. This fact is utilized in connecting the coils to the commutator.

When placing a lap winding on an armature, a good procedure is as follows:

(1) Insert all the coils in the slots.
(2) Mark the two segments to which the ends of the first coil should be connected.
(3) Beginning with the "finish" end of coil 1, connect the "finish" ends of adjacent coils to adjacent segments.
(4) Beginning with the "start" end of coil 1, connect the "start" ends of adjacent coils to adjacent segments.

Requirements for Lap Winding—The most important requirements for lap winding may be summarized as follows:

(1) A lap winding can be placed on an armature with any number of slots (of course the number of commutator segments in a two layer winding must be equal to or a multiple of the number of slots).
(2) The brushes must be wide enough to cover at least m segments (where m is the multiplicity of the winding).
(3) There must be as many brushes as there are poles, and the brushes must be set so that they short-circuit only the coils cutting no flux.

Armature with Two Commutators—Some armatures have two independent windings in the same slots, and each of these windings has its own commutator, one commutator being placed at each end of the core. This system has the advantage that the two windings can be connected in series or in parallel, and that it can be adapted to the three wire system.

Equalizer Connections—Because of wear in the bearings, and for other reasons, the air-gaps in a generator become unequal, and therefore the flux in some poles becomes greater than in others. This causes the voltages of the different paths to be unequal. In a lap winding the paths connect to the negative line through the negative brushes, and connect to the positive line through the positive brushes. With unequal voltages in these paths, circulating currents will flow even if no current is supplied to an external load. If these currents are large, some of the brushes will be required to carry a greater current at full load than they were designed to carry, and this will cause sparking. To relieve the brushes of these circulating currents, points on the armature that are at the same potential are connected together by means of copper bars called equalizers. This is accomplished by connecting to the same equalizer the coils that occupy the same positions relative to the poles (fig. 5-20). The equalizers provide a low resistance path for the circulating current, keeping it from passing through the brushes, and thus reducing sparking.

Equalizers should be used only on windings in which the number of coils is a multiple of the number of poles. For best results each coil should be connected to an equalizer, but this is seldom done. Satisfactory results are obtained by connecting about every third coil to an equalizer. In order to distribute the connections to the equalizers equally, the number of coils per pole must be divisible by the connection pitch.

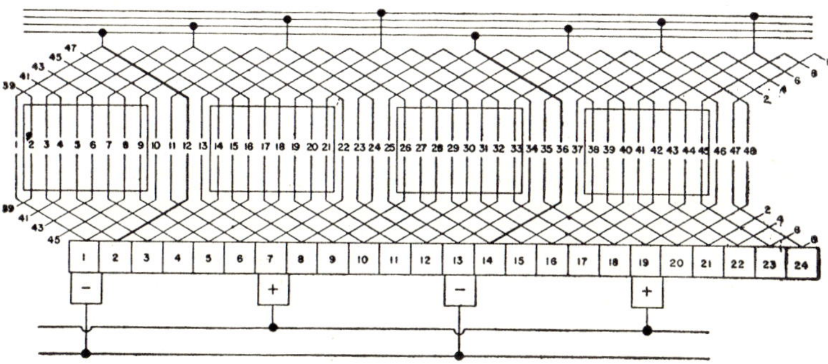

Fig. 5-20

Example—A 6 pole armature contains 54 coils. Suggest how equalizer connections should be made.

$$\text{Coils per pole} = \frac{54}{6} = 9$$

As the number 9 is divisible by 3, equalizer connections to every third coil will give good results.

Wave Winding—The essential difference between a lap winding and a wave winding is in the commutator connections. In a lap winding, the start end of the second coil connects to the finish end of the first, the start end of the third to the finish end of the second, etc.

(fig. 5-21). With such connections, the coil voltages add. These voltages will also add if the coils occupying approximately the corresponding position under all the poles are connected in series (see figs. 5-22 and 5-25). The winding must not close after it passes once around the armature, but, if it is a simplex winding, it must connect to a segment adjacent to the first, and the next coil must lie adjacent to the first, as indicated in fig. 5-22. This is repeated each time around until connections are made to all the segments and all the slots are occupied, after which the winding automatically returns to the starting point. If, after passing once around the armature, the winding connects to a

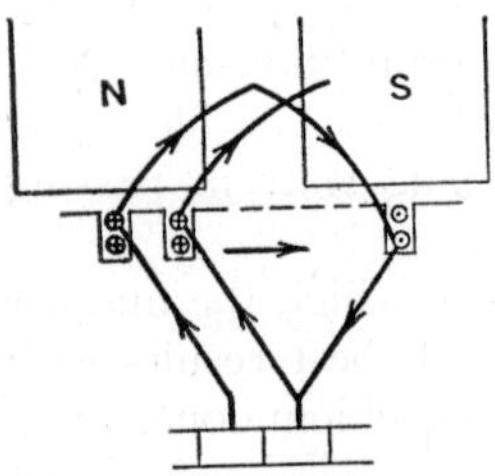

Fig. 5-21

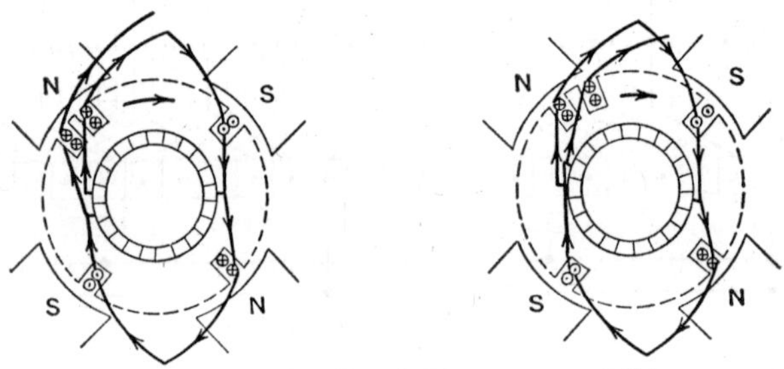

Fig. 5-22

segment to the left of the starting point (fig. 5-22a), the winding is retrogressive. If it connects to a segment to the right of the starting point (fig. 5-22b), it is progressive. This type of winding is known as a wave winding because it passes around the armature in a wavelike form.

A coil for a wave winding may have more than one turn (fig. 5-23), but the number of turns does not affect the coil connections, and for simplicity in the illustrations, only one turn per coil will be shown.

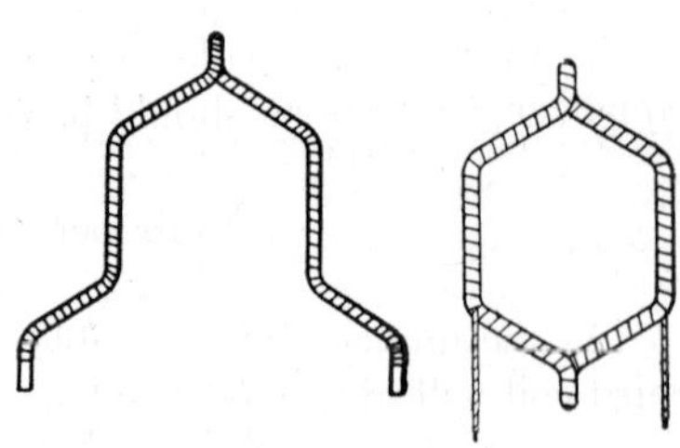

Fig. 5-23

In wave winding, as in lap winding, the number of coil sides spanned at the pulley end is known as the back pitch (y_b), the number of segments spanned at the terminals of the coils is known as the commutator pitch (y_c), and the number of coil sides spanned at the commutator end is known as the front pitch (y_f). This is indicated in fig. 5-24. In a wave winding, y_s and y_b are determined in the same way as in a lap winding.

In a simplex wave winding, the commutator pitch (y_c) times the number of coils that must be traced to pass once around the armature

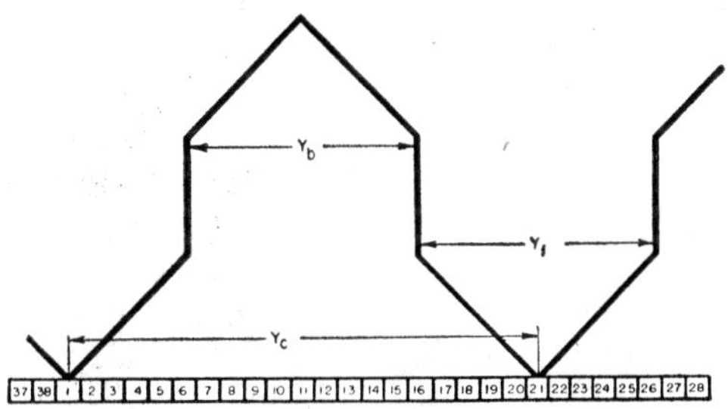

FIG. 5-24

(which is always P/2 coils) equals the total number of commutator segments (s) plus or minus 1 (see fig. 5-25). That is,

$$y_c \frac{P}{2} = s \pm 1 \tag{5-8}$$

or

$$y_c = \frac{2s \pm 2}{P} \tag{5-9}$$

where the minus sign indicates a retrogressive winding and the plus sign indicates a progressive winding.

In any winding the progress on the core must be exactly the same as on the commutator. This is illustrated in fig. 5-25, where the commutator pitch is 11 and the sum of the back and front span is 11 slots, which, when expressed in coil sides, equals the sum of the back and front pitch divided by 2. In a wave winding, then,

$$y_c = \frac{y_b + y_f}{2}$$

or

$$y_f = 2y_c - y_b \tag{5-10}$$

In a wave winding, y_b, y_c, and y_f may be equal.

EXAMPLE—An armature has 29 slots and 29 commutator segments. If it is to be wound simplex wave for 4 poles, determine (a) the back pitch, (b) the commutator pitch, and (c) the front pitch.

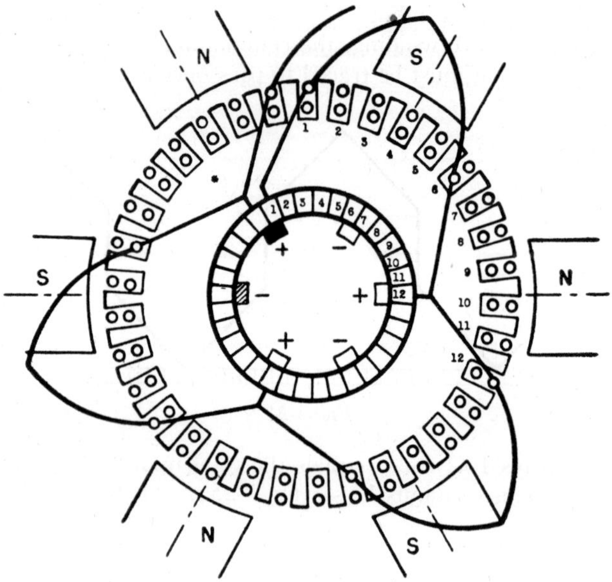

FIG. 5-25

(a) $y_s = \frac{S}{P} = \frac{29}{4} = 7\frac{1}{4}$

Dropping the fraction,

$$y_b = C_s y_s + 1 = 2 \times 7 + 1 = 15$$

(b) Considering a retrogressive winding,

$$y_c = \frac{2s - 2}{P} = \frac{2 \times 29 - 2}{4} = 14$$

(c) $y_f = 2y_c - y_b = 2 \times 14 - 15 = 13$

Fig. 5-26 shows this winding.

Paths in a Wave Winding—In fig. 5-26 one path (indicated by heavy lines) may be traced by starting with the negative brush A and proceeding through the coil sides as follows: A-1-16-29-44-57-14-27-42-55-12-25-40-53-10-23-38-51-8-21-36-49-6-19-34-47-4-B. Another path (indicated by light lines) may be traced by starting from the

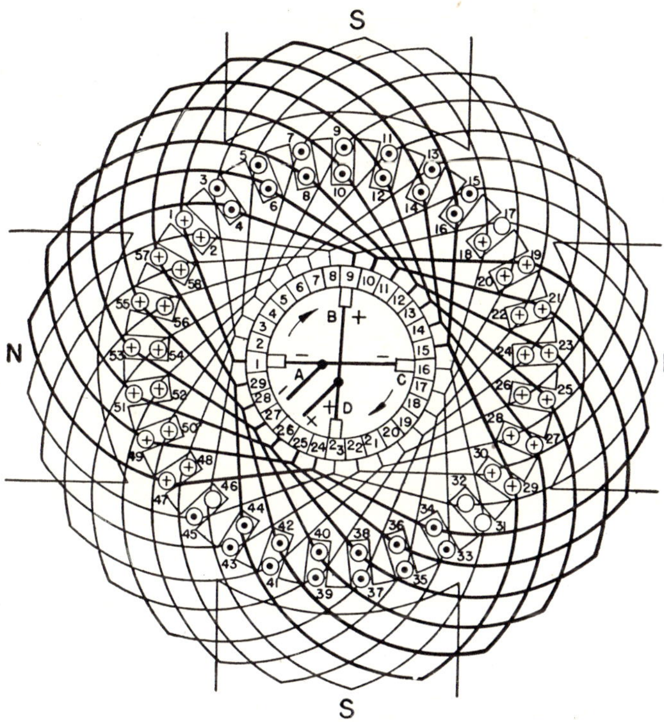

Fig. 5-26

negative brush C and proceeding as follows: C-18-3-48-33-20-5-50-35-22-7-52-37-24-9-54-39-26-11-56-41-28-13-58-43-30-15-2-45-D. The negative brushes short-circuit the coil having sides 31-46, and the positive brushes short-circuit the coil having sides 17-32. This is permissible, since these coils are not cutting flux. This winding, which is also shown diagrammatically in fig. 5-27a, has 4 poles and only two paths. A simplex winding with any other number of poles likewise has two paths. In any simplex winding, then,

$$b = 2 \tag{5-11}$$

The winding shown in figs. 5-26 and 5-27a has the same number of brushes as there are poles. In this winding, the current passes

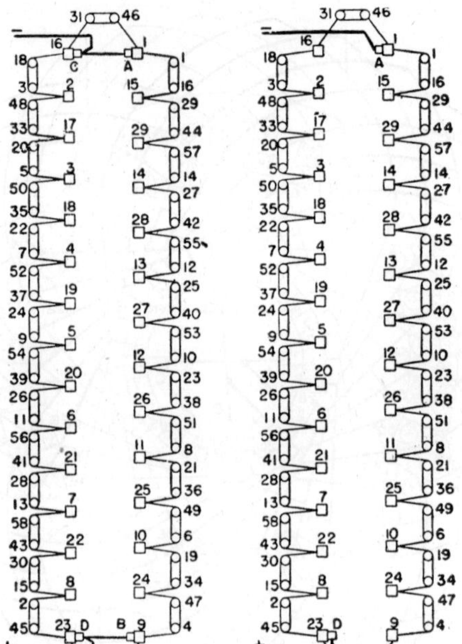

(a) (b)

FIG. 5-27

through the two paths, through the two positive brushes to the positive line, and back through the negative brushes. In a wave winding, however, the brushes of like polarity are short-circuited together by coils in the neutral plane. In figs. 5-26 and 5-27a, then, if brushes B and C are removed, the current will pass through the two paths, through the positive brush D to the positive line, and back through the negative brush A (fig. 5-27b). Therefore, in any wave winding, only one positive and one negative brush are required.

It should be realized, however, that if two brushes are used instead of four, each brush must carry twice as much current, and so must have twice the cross section. The thickness of a brush should not be increased, because it would short-circuit more coils. It must therefore be made wider, requiring a wider commutator and a longer motor. To avoid making the commutator wider, it is necessary to use the same number of brushes as there are poles.

The use of two brushes is a decided advantage in a few cases, such as crane and railway motors, in which access to the top and the bottom of the armature is difficult to provide.

Position of Brushes in a Wave Winding—The wave windings considered are required in machines in which the brushes are in line with the pole centers, which is the most common condition. Regardless of where the brushes are situated, the coils must be con-

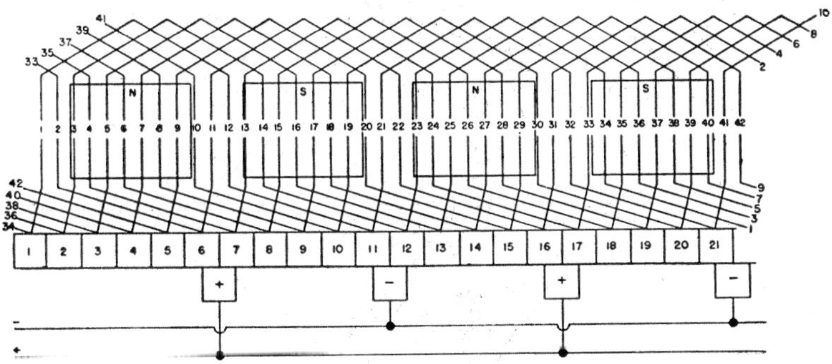

FIG. 5-28

nected to the commutator so that the coil voltages in each path will add, and so that the brushes will short-circuit coils only while they are not cutting flux, as explained in lap winding. If the brushes are in line with a point midway between the pole tips, the coils must be connected to the segments as shown in fig. 5-28.

Dummy Coils—Some armatures cannot be wave wound unless dummy coils are used. No dummy coils are needed if 2s ± 2 in equation 5-9 is a direct multiple of P. If it is not, the number of segments used must be reduced until it is a multiple. One dummy coil is required for each segment not used.

EXAMPLE—A 4 pole, simplex wave wound armature has 22 slots and 22 segments. Determine the number of dummy coils required, if any.

$$y_c = \frac{2s \pm 2}{4} = \frac{2 \times 22 \pm 2}{4}$$

$$y_c = 10\frac{1}{2} \text{ or } 11\frac{1}{2}$$

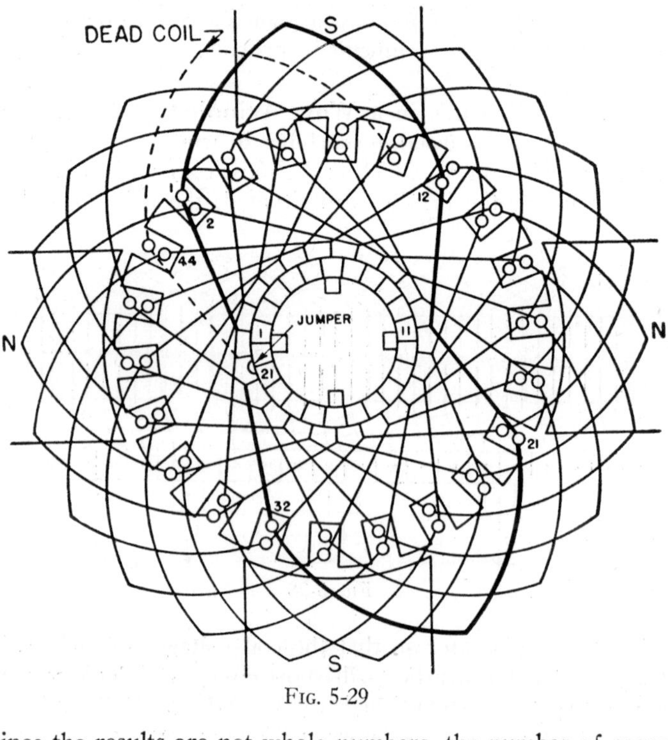

FIG. 5-29

Since the results are not whole numbers, the number of segments must be reduced. Considering 21 segments, then

$$y_c = \frac{2 \times 21 \pm 2}{4} = 10 \text{ or } 11$$

This means that the armature can be wound only if the number of segments and coils is reduced from 22 to 21. The coil to be omitted is inserted into the slots in the same way as the others, to make the armature dynamically balanced, but it is not a part of the armature circuit (fig. 5-29). One end of this coil is taped and the other end is connected to the unused commutator segment for the sake of appearance. Since only 21 segments are required, two are connected together and considered as one.

Multiplex Coils in a Wave Winding—A wave winding may have two or more times as many segments as slots, to keep the voltage between segments at a low value. The armature is then wound with multiple coils.

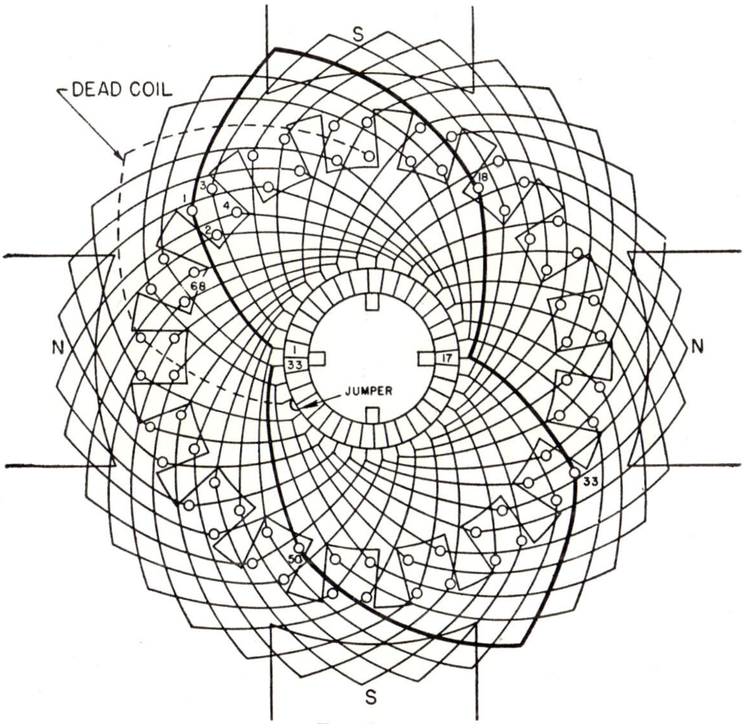

FIG. 5-30

EXAMPLE—A 4 pole, simplex wave wound armature has 17 slots and 34 segments. Determine (a) the back pitch, (b) the commutator pitch, and (c) the front pitch.

(a) $C_s = \frac{2s}{S} = \frac{2 \times 34}{17} = 4$

$$y_s = \frac{S}{P} = \frac{17}{4} = 4\frac{1}{4}$$

Dropping the fraction,

$$y_b = 4 \times 4 + 1 = 17$$

(b) $y_c = \frac{2s - 2}{4} = \frac{2 \times 34 - 2}{4} = 16\frac{1}{2}$

Since the result is not a whole number, 33 segments will be considered.

$$y_c = \frac{2 \times 33 - 2}{4} = 16$$

(c) $y_f = 2y_c - y_b = 2 \times 16 - 17 = 15$

This winding is shown in the diagram of fig. 5-30.

Multiplex Wave Winding—The definition given for multiplicity and reentrancy in lap winding also applies for wave winding. That is, in wave winding, multiplicity refers to the number of independent windings while reentrancy refers to the separately closed windings on the armature.

Fig. 5-31 shows one winding of a duplex wound armature. After tracing once around the armature, the connection is made two segments from the starting point. In a triplex winding, the connection would be made three segments from the starting point.

Then for any multiplicity, equation 5-8 must be expressed as

$$y_c \frac{P}{2} = s \pm m$$

or

$$y_c = \frac{2s \pm 2m}{P} \qquad (5\text{-}12)$$

This applies to all wave windings. A multiplex winding will have no dummy coils only if $2s \pm 2m$ is a direct multiple of P.

Paths in a Multiplex Wave Winding—The number of paths in a multiplex wave winding depends on the multiplicity and not on the number of poles. It was shown that a simplex wave winding always has two paths. A duplex wave winding is merely two simplex wave windings in parallel, and therefore has four paths. A triplex wave winding has six paths, etc. Then for any multiplicity (m), the number of paths is

$$b = 2m \qquad (5\text{-}13)$$

If the armature shown in fig. 5-31 has 24 slots and 24 segments, the winding would not fill the alternate slots and close, but would return to a segment to the left or the right of the one from which it started. In this case the winding will close only after the remaining alternate slots are filled. This is illustrated in fig. 5-32, in which one winding is shown dotted.

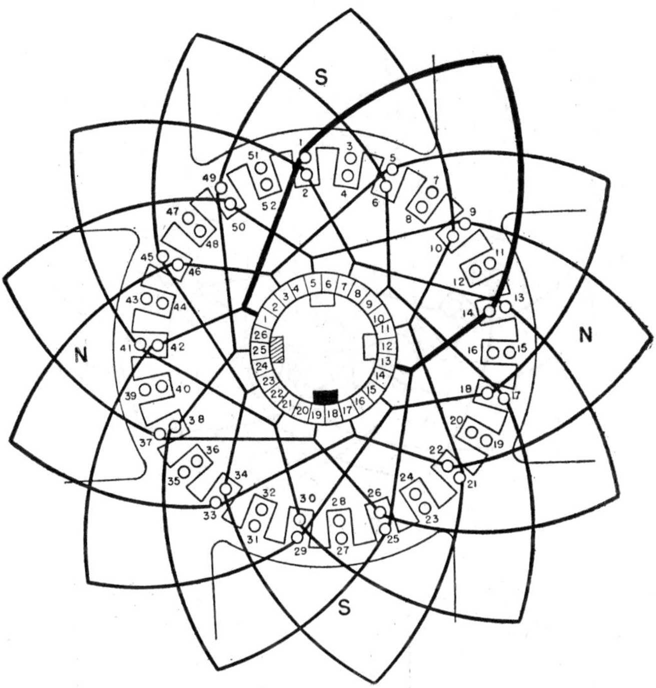

FIG. 5-31

Reentrancy in a Wave Winding—Fig. 5-33 shows a portion of a duplex wave winding, in which y_c is 20. In tracing through this winding it will be noted that it advances two segments each time it passes around the armature. If it arrives at 1, the winding will be double reentrant. If it skips segment 1, it will not close until a connection is made to all segments, in which case the winding will be single reentrant. To determine the degree of reentrancy in a wave winding, divide the commutator pitch by the multiplicity. If this gives a whole number, the winding is multiple reentrant; if this does not, it is single reentrant.

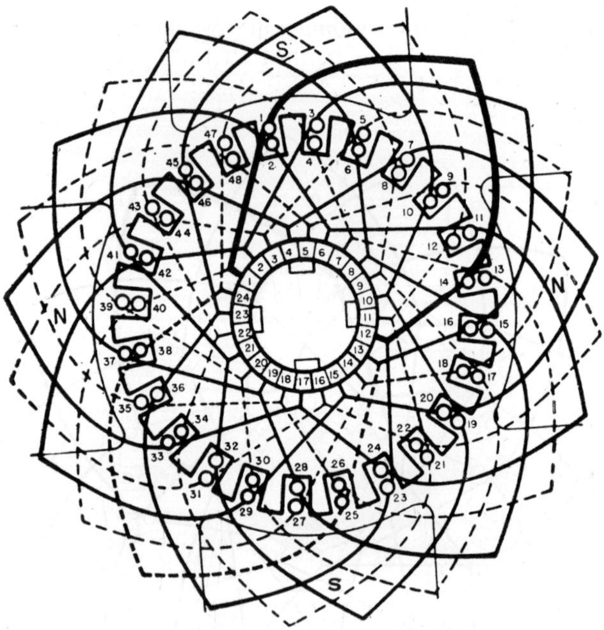

FIG. 5-32

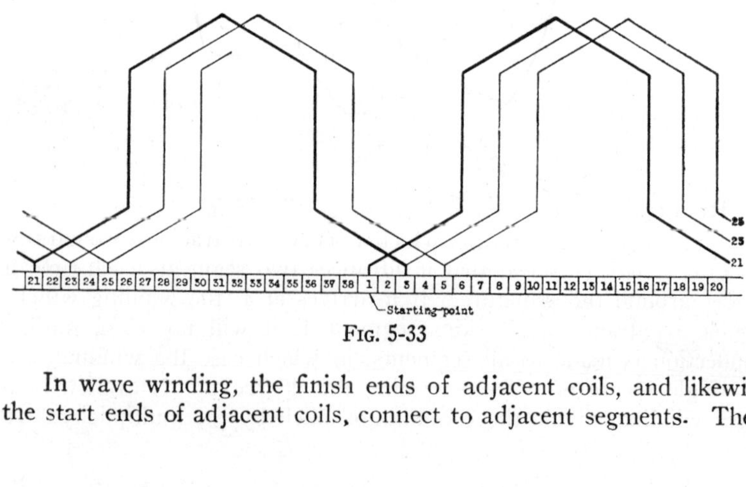

FIG. 5-33

In wave winding, the finish ends of adjacent coils, and likewise the start ends of adjacent coils, connect to adjacent segments. Then the procedure for inserting coils and the method of connecting them in wave winding is the same as in lap winding (see page 146).

Requirements for Wave Winding—Three important requirements for wave winding may now be listed as follows:

(1) An armature can be wave wound, without dummy coils, if (2s ± 2m) is a direct multiple of P.
(2) The brushes must be wide enough to cover at least m segments.
(3) Two brushes are necessary, but as many brushes as there are poles may be used, and they must be set so that they short-circuit only the coils cutting no flux.

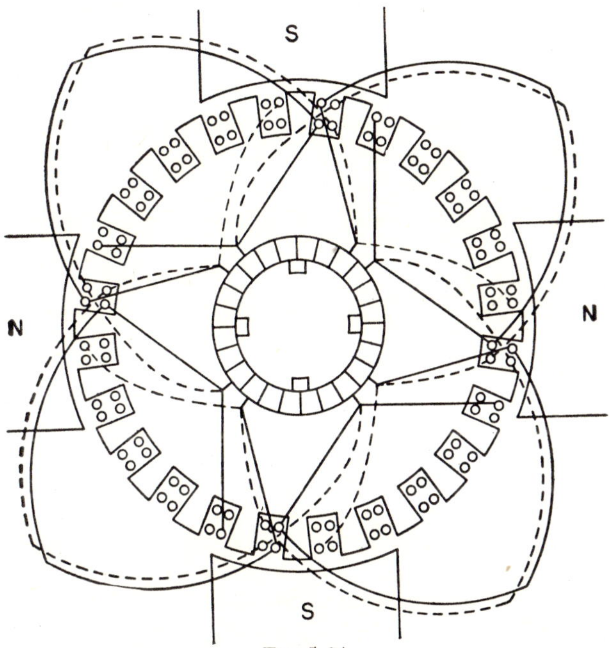

FIG. 5-34

Frogleg Winding—The frogleg winding consists of a lap winding and a wave winding placed in the same slots and connected to the same commutator segments. A part of a winding of this type is shown in fig. 5-34. The coils shown solid are a part of the lap winding and those shown dotted are a part of the wave winding.

This type of winding has the same industrial application as a lap winding with equalizers. The advantage of the frogleg type is that the wave winding serves as equalizer connections and also supplies voltage and current the same as a lap winding.

Selection of Type of Winding—If the current capacity of an armature is represented by I, the current per path or per conductor is I/b. The current per conductor fixes its size, because if less than a certain number of circular mils per ampere is used, the winding will overheat. In figuring an armature, if the size of wire determined is too large to be conveniently formed into a coil, a larger number of paths should be selected, so that the required size will be smaller. If the size determined is smaller than a size that can be conveniently formed into a coil, a smaller number of paths should be selected, so that fewer turns of larger wire will be required. Wire of larger size costs less per pound and requires less space for insulation. Also, in an armature with large size wire, the heat more readily leaves the copper, and therefore it operates at a lower temperature.

In general, a high current armature is lap wound to provide a large number of paths, and a low current armature is wave wound to provide a low number of paths. Usually 240 volt machines, up to about 75 horsepower, are simplex wave wound.

Generated Voltage—The amount of e m f induced in a conductor depends upon the rate at which it is cutting flux. When a conductor cuts flux at the rate of 10^8 magnetic lines per second, one volt is induced in it. By utilizing this fact it is possible to determine the e m f generated in an armature.

Let E_g = generated voltage (that is, the voltage across the terminals of the generator at no load), P = number of poles, Φ = flux per pole, N = speed of the armature in revolutions per minute, b = number of paths, and Z = total number of conductors on the armature. Since there are two conductors per turn, the total number of conductors on an armature (Z) equals two times the product of the number of turns per coil (t_c) and the number of coils (C). Thus,

$$Z = 2t_cC \qquad (5\text{-}14)$$

In one revolution each conductor cuts the flux per pole (Φ) times the number of poles (P), or $P\Phi$ lines. In one second the armature makes N/60 revolutions, and during this time each conductor cuts

$P \Phi N/60$ lines. To obtain the average voltage induced in each conductor, this expression is divided by 10^8, giving

$$\frac{P \Phi N}{10^8 \times 60}$$

In an armature the total number of conductors (Z) is divided into b paths. The total generated voltage (E_g) equals the e m f generated in one path. The number of conductors per path is Z/b. Then the voltage per path is

$$\frac{Z}{b} \times \frac{P \Phi N}{10^8 \times 60}$$

As the voltage per path is the same as the voltage at the terminals of the machine,

$$E_g = \frac{PZ \Phi N}{10^8 b 60} \qquad (5\text{-}15)$$

EXAMPLE—Determine the voltage generated in a 6 pole machine, running at 900 r p m, if its armature is simplex lap wound and has 300 active conductors. The flux per pole is 5×10^6 magnetic lines. In this winding the number of paths equals the number of poles. Substituting in equation 5-15,

$$E_g = \frac{6 \times 300 \times 5 \times 10^6 \times 900}{10^8 \times 6 \times 60} = 225 \text{ volts}$$

EXAMPLE—A 4 pole, 12 volt generator has a speed of 750 r p m. The flux per pole is 4×10^5 lines. If the armature is wound duplex lap and has 24 slots and 48 segments, find (a) number of coil sides per slot, and (b) number of turns per coil.

(a) $C_s = \frac{2s}{S} = \frac{2 \times 48}{24} = 4$

(b) Solving for Z in equation 5-15,

$$Z = \frac{10^8 b 60 E_g}{P \Phi N} = \frac{10^8 \times 8 \times 60 \times 12}{4 \times 4 \times 10^5 \times 750} = 480 \text{ conductors}$$

Solving for t_c in equation 5-14,

$$t_c = \frac{480}{2 \times 48} = 5 \text{ turns per coil}$$

Winding Small Armatures—Form wound coils are seldom used on small armatures, the coils being generally wound directly on the core by hand or machine. Machine winding is preferable, because the armature can be wound much faster. The coils may be wound on the core in many different ways, but they are always connected to the commutator as explained in lap and wave winding. Methods of winding the coils, and not methods of connecting them to the commutator, will therefore be discussed.

Loop Winding—A type of winding much used on small cores, known as loop winding, is shown in fig. 5-35a. For this winding, the armature must have the same number of bars as slots. After the first coil is wound, a loop is made, and then the next coil is started in the second slot (see fig. 5-35b). This is continued (see fig. 5-35c) until all the coils are wound.

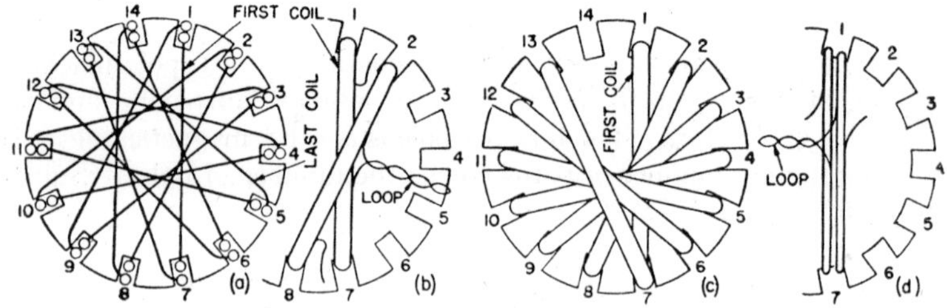

FIG. 5-35

If the armature has twice as many commutator segments as slots, the second coil is wound in the same slots as the first, with a loop between the two (see fig. 5-35d). The third is started in slot number two and the process is repeated. For a simplex winding the loops are connected to adjacent commutator segments.

This winding can be placed on a core with any number of slots. It is easily wound, either by hand or by machine. Its coils are not all alike, and consequently it is not electrically or mechanically balanced.

Full Pitch Split Winding—In a full pitch split winding, which is shown in fig. 5-36a, the number of slots must be even. Half of each coil is wound on one side of the shaft and the other half on the other side. The procedure in winding the coils is shown in fig. 5-36b. Sleeves should be used to indicate the start and finish ends. The second coil is started in the second slot, the third coil in the third slot, etc. After half the coils are wound, the bottom layer is complete. The top layer is started in the same slot as the finish end of the first coil, so that after the armature is wound, a start end and a finish end protrude from each slot.

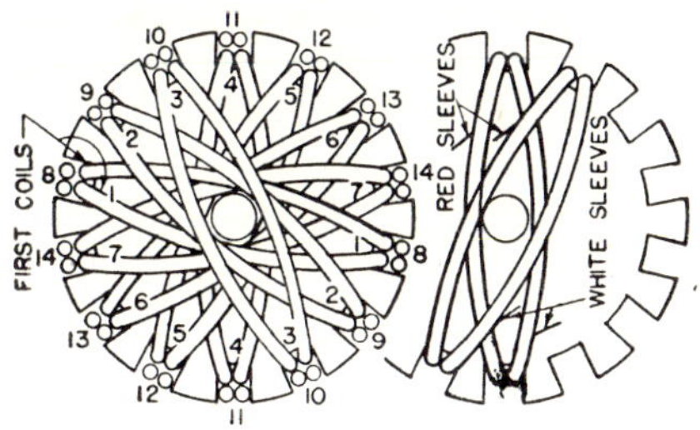

(a) (b)

FIG. 5-36

This type of winding is most suitable for an armature with coils that require a few turns of heavy wire. It is recommended for machines that have little winding space at the ends.

Fractional Pitch Split Winding—The coils in this type of winding, which is shown in fig. 5-37a, also require sleeves to distinguish

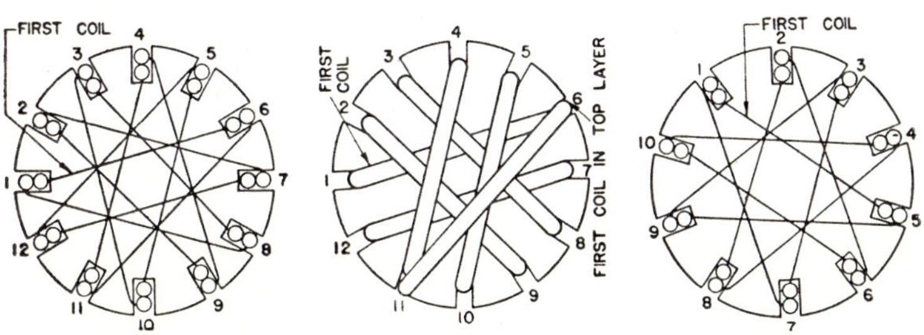

(a) (b) (c)

FIG. 5-37

the start from the finish ends. One coil is wound, and the armature is then turned 180 degrees and the next coil is started. The third coil is started in the third slot beyond the finish end of the second coil. After the third coil is wound, the armature is again turned 180 degrees and the next coil is started, etc.

The top layer is started in the second slot beyond the finish end of the last coil on the bottom layer (fig. 5-37b). To wind the remaining coils, the same scheme is used that was used in winding the bottom layer.

After all the coils are wound, it will be noted that each slot contains a start end and a finish end. In this winding, and also in the full pitch split winding, the last two coils are tied together to keep them from flying outward when the armature is rotating.

If the number of slots is not divisible by four but is divisible by two, the same general procedure is used; that is, the coils are wound on the core in pairs, proceeding in the clockwise direction until all are wound. Such a winding is shown in fig. 5-37c.

The advantage of split winding is that it gives a better mechanical and electrical balance than loop winding.

V Winding—A few steps in inserting the coils in the V winding shown in fig. 5-38a are illustrated in fig. 5-38b. The first coil will lie at the bottom of the slots. The following coils, except the last, will have one side at the bottom of a slot and the other side at the top of another slot. The last coil will have both sides at the top of slots. To eliminate complications which may arise when making commutator connections, sleeves should be used.

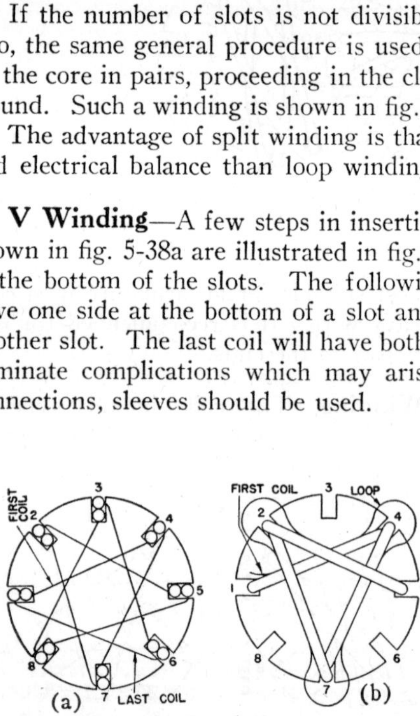

FIG. 5-38

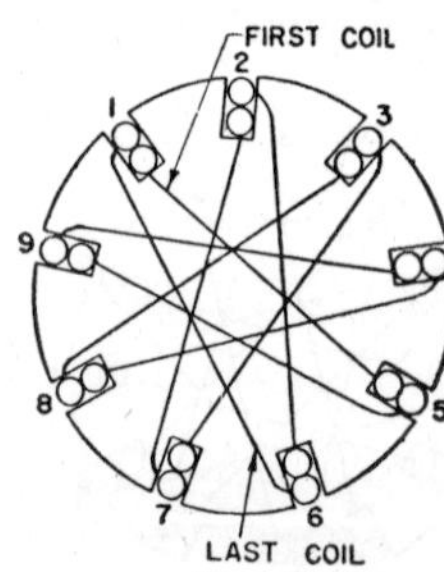

FIG. 5-39

A distinctive feature of this winding is that the slots are filled as the winding progresses.

The variation in coil lengths in this winding is about the same as in the split winding. Its advantage over the split winding is that it can be wound on a core with any number of slots. In fig. 5-38a is shown a winding with an even number of slots; in fig. 5-39 is shown one with an odd number.

Armature Testing—A method commonly used for testing armatures for short circuits and open circuits is known as the segment to segment test. This test may be made on an armature with any type of winding. The method of testing a simplex lap wound armature is illustrated in fig. 5-40a. A low D C voltage is impressed between two points diametrically opposite on the commutator.

Voltage measurements are made between adjacent commutator segments. To facilitate the rapid reading of these measurements, the terminals of the voltmeter are clamped close together by means of two non-conducting strips. Generally the voltage supplied must be adjusted so that the meter will indicate at some convenient value on the scale. If the armature has no faults, like readings will be obtained

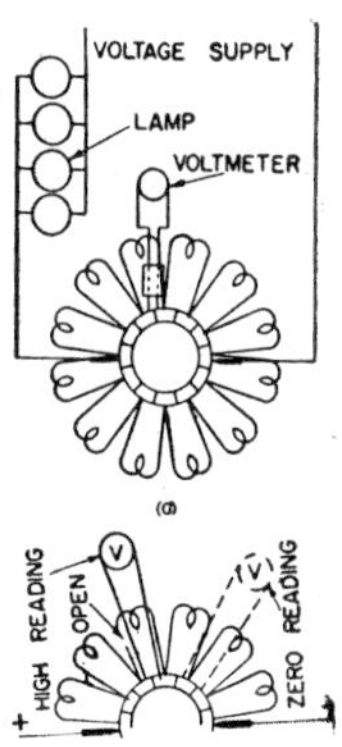

FIG. 5-40

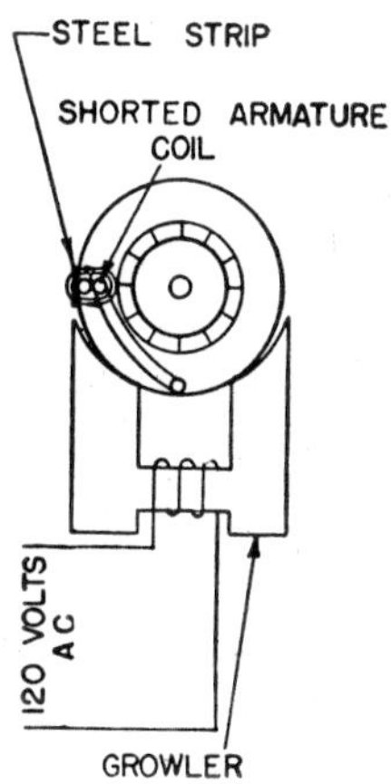

FIG. 5-41

between any two commutator segments. If little or no reading is obtained when the voltmeter terminals are placed on adjacent segments, a coil between these segments or the segments themselves are short-circuited. If zero readings are obtained between segments connected to coils in one path, except between two segments where the reading is high, the winding is open between these segments (see fig. 5-40b).

Growler Test—A coil with a few short-circuited turns cannot always be located by employing the segment to segment test. In this case the short circuit is located by means of a device commonly called a growler (so-named because of the growling noise it makes). Such a device is shown in fig. 5-41. When a growler is energized with alternating current, an expanding and contracting flux is established, which induces a voltage in the armature winding. If the winding has no faults, no current will flow in the armature coils. If a coil is short-circuited, a current will flow in the short-circuited turns. To locate the faulty coils, a steel strip such as a hacksaw blade is passed over the slots, so that when a coil side lies under the blade, the other coil side of the same coil lies between the pole faces of the growler. When one coil side of a short-circuited coil lies under a blade, the current in this coil sets up a flux which attracts the blade. If a part of the blade is held slightly away from the core, the blade will vibrate, making a noise, which indicates a short circuit.

Exercises

1. A 4 pole, lap wound armature has 28 slots and 28 segments. If the winding is progressive, what is (a) the back pitch? (b) the commutator pitch? (c) the front pitch?
2. Answer parts a, b, and c in problem 1 if the winding is retrogressive.
3. Make a diagram of the winding in problem 1.
4. How many paths has (a) an 8 pole, simplex lap winding? (b) a 10 pole simplex lap winding?
5. An 8 pole, 240 volt, simplex lap wound armature delivers 125 amperes. Determine (a) the voltage per path, (b) the current per path, and (c) the current per brush.
6. Using a cross section of 600 circular mils per ampere, what size wire is required for the armature of problem 5?
7. Determine the number of coil sides per slot in an armature that has (a) 21 slots and 42 segments; (b) 12 slots and 48 segments; (c) 20 slots and 10 segments.
8. Are one, two, or three wire coils required for an armature that has (a) 14 slots and 42 segments? (b) 20 slots and 20 segments? (c) 20 slots and 60 segments?
9. How many commutator segments may an 18 slot armature have?
10. Make a wiring diagram of a 4 pole, simplex lap winding with 18 slots and 36 segments. Place the brushes in line with the pole centers, number the coil sides, and show the slots.
11. Show a few turns of the winding of problem 10, considering each brush in line with a point midway between adjacent poles.
12. How many paths has an (a) 8 pole duplex lap winding? (b) a 10 pole triplex lap winding?
13. A 2 pole lap wound armature has 14 slots and 14 segments. Show a few coils of this winding if the armature is (a) simplex; (b) duplex; (c) triplex. Consider the brushes in line with the pole centers.

14. A 6 pole, duplex lap winding has 32 slots and 64 segments. Determine (a) the back pitch; (b) the commutator pitch; (c) the front pitch.
15. Show how a few coils of the winding in problem 14 are connected, considering the brushes in line with the pole centers.
16. Can the armature in problem 14 be wound (a) simplex? (b) triplex?
17. Is the armature in problem 14 single or double reentrant?
18. An 8 pole, triplex lap wound armature supplies a current of 250 amperes. Determine (a) the current per path, and (b) the current per brush.
19. How many brushes are required in a lap winding?
20. A 4 pole, 120 volt generator with a simplex lap winding is rated at 10 k w. What is (a) the voltage and (b) the current per path at full load?
21. What would be the terminal voltage and line current at full load if the armature of problem 20 is reconnected duplex.
22. What is the minimum number of segments that each brush must cover in a (a) duplex winding? (b) simplex winding? (c) triplex winding?
23. What changes must be made in the commutator connection of a generator with a simplex lap wound armature to change its terminal voltage from 240 volts to 60 volts?
24. A 4 pole motor has 24 slots and 24 segments. Determine the possible equalizer connection pitches.
25. The finish ends (or the start ends) of adjacent coils connect what number of segments apart in a (a) simplex lap winding? (b) duplex lap winding? (c) triplex lap winding?
26. How many paths has a simplex wave winding (a) with two poles? (b) with four poles? (c) with six poles?
27. (a) How many brushes are required in a wave winding? (b) Can any other number be used?
28. A 4 pole armature has 17 slots and 17 segments. If the armature is wound progressive simplex wave, what is (a) the back pitch? (b) the commutator pitch? (c) the front pitch?
29. Make a wiring diagram of the armature in problem 28, placing the brushes in line with the pole centers.
30. A 6 pole, simplex wave wound armature supplies a current of 200 amperes. Determine (a) the current per path, and (b) the current per brush (state the number of brushes used).
31. A 6 pole, progressive simplex wave winding has 40 slots and 80 commutator segments. Determine (a) the back pitch, (b) the commutator pitch, and (c) the front pitch.
32. Show the connections of a few coils of the winding in problem 31, considering the brushes in line with a point midway between the pole tips.

33. A 6 pole, 6 volt, 50 k w duplex lap wound generator is reconnected simplex wave. Determine (a) the terminal voltage, and (b) the current at full load.
34. Using a cross section of 600 circular mils per ampere, what size wire is required for the armature of problem 33?
35. Make a schematic diagram of a 6 pole, simplex wave winding (show six brushes).
36. How far apart do the finish ends (or the start ends) of adjacent coils connect on the commutator for (a) a simplex wave winding? (b) a duplex wave winding? (c) a triplex wave winding?
37. Are dummy coils required (a) in some lap windings? (b) in some wave windings?
38. An 8 pole, simplex wave winding has 50 slots and 100 commutator segments. What is the minimum number of dummy coils that this winding may have?
39. How many paths has (a) an 8 pole, duplex wave winding? (b) a 6 pole, triplex wave winding?
40. An 8 pole armature has 42 slots and 126 segments. If it is wound triplex wave determine (a) the back pitch; (b) the commutator pitch; (c) the number of dummy coils; (d) the front pitch; (e) the degree of reentrancy.
41. Can the armature in problem 40 be wound duplex wave?
42. What kind of machines are (a) lap wound? (b) wave wound?
43. How many paths has the winding shown in fig. 5-34 when it is completed?
44. How many coils has an armature with 34 slots and 68 segments?
45. How many conductors has a 20 coil armature if each coil contains 10 turns?
46. Determine the number of conductors on an armature which has 22 slots and 44 segments, if each coil has 4 turns?
47. A simplex winding on a two pole machine has 400 conductors. The flux per pole is 400,000 lines. What voltage is generated when the armature is running at 900 r p m?
48. An 8 pole generator, which runs at 850 r p m, has 10^6 magnetic lines per pole. How many conductors has the armature if it is wound duplex lap and supplies 240 volts?
49. If the armature in problem 48 has 21 slots and 21 segments, how many turns does each coil contain?
50. A 4 pole, duplex lap wound armature has 28 slots and 56 segments, and runs at 450 r p m. The flux per pole is 7×10^5 lines. If each coil contains 16 turns, what is the terminal voltage?
51. A 6 pole, 60 volt generator, which has a duplex lap wound armature, runs at 600 r p m. The flux per pole is 6×10^5 lines. If the armature has 100 coils, how many turns does each coil contain?

52. A duplex lap wound armature of a 4 pole generator delivers 100 amps at 12 volts. Each coil has 10 turns. (a) How many turns must each coil have if the armature is rewound simplex lap to generate 60 volts? (b) How much current will the armature deliver at full load?
53. A 120 volt, 6 pole generator has a triplex wave wound armature with 56 coils. Each coil has 4 turns and the flux per pole is 10^6 magnetic lines. What is the rated speed of the generator?
54. The 4 poles of a generator are each 8 square inches in cross section and have a flux density of 60,000 magnetic lines. The armature has 36 slots and 102 segments. If the armature is wound triplex lap, generates 240 volts, and runs at 1800 r p m, determine the turns per coil.
55. Give the advantages and disadvantages of winding the coils directly on the core instead of using form wound coils.
56. Make wiring diagrams of two pole, 8 slot armatures which are (a) loop wound; (b) full pitch split wound; (c) fractional pitch split wound; (d) V wound. Indicate the first and last coil in each.
57. Which windings stated in problem 56 can be placed on cores with any number of slots? Each of the other windings requires how many slots?
58. What is the chief undesirable feature of the full pitch split winding?
59. How are armature windings tested for short circuits and open circuits?

CHAPTER 6

GENERATORS

A modern D.C. generator with the main parts named is shown in fig. 6-1. These parts will be discussed in detail below.

Armature—The armature core is made of soft iron or mild steel stamping about .025 inch thick (fig. 6-2a). The core is laminated to cut down eddy currents, which would otherwise be excessive. In small

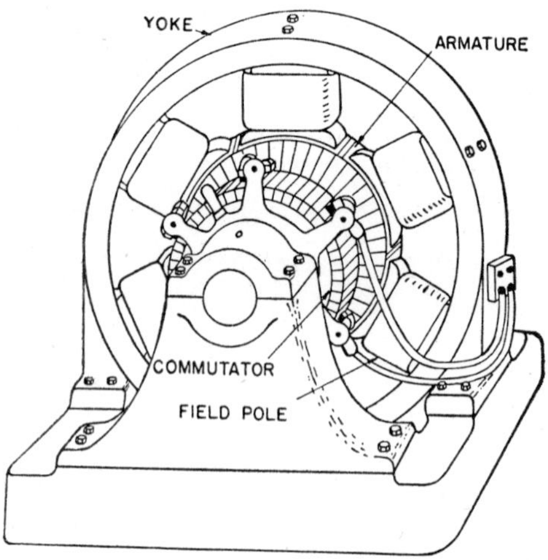

FIG. 6-1

machines these laminations are keyed directly to the armature shaft, while in large ones they are keyed to the spider, which in turn is keyed to the shaft. The slots of the armature core are first lined with fish paper then the coils are inserted into the slots. The coils are usually held in place by wooden or fiber wedges driven into the top of the slots. The sections of the coils not in the slots are held in place by band wires.

The commutator is made up of copper segments insulated from the supporting rings by mica cones (fig. 6-2b). These segments are held firmly in place by wedge shaped clamping rings drawn together by bolts in large machines and by a nut on the shaft in small machines. The leads from the armature coils are soldered into slits or into risers located at the ends of the segments.

Brushes—The purpose of brushes (fig. 6-2c) is to carry current from the commutator to the external circuit. They are usually made of blocks of graphite carbon, although in very low voltage machines brushes of copper gauze or metal patent compounds are sometimes used. The brushes should slide freely in their holder so as to follow any irregularity in the commutator. The brushes should bear upon the commutator with a pressure of about 1.5 lbs. per square inch of contact area. A suitable spring is provided on the brush holder to adjust the brush tension. To decrease the brush resistance the upper portion of the brush is copper-plated. This plating is connected to the brush holder by a pigtail made of twisted copper wire.

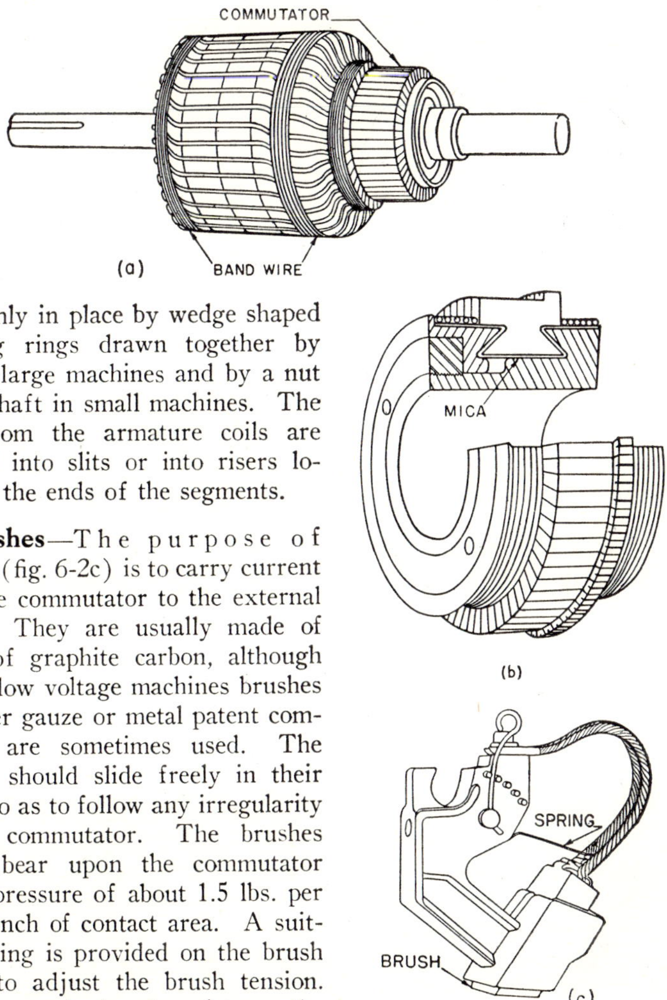

FIG. 6-2

Frame—The frame or yoke of a generator serves as a mechanical support for the machine and forms a part of the magnetic circuit connecting the field poles. In small machines, where weight is of little importance, the frame is made of cast iron and the feet form a part of the casting (fig. 6-3). In another type of construction the frame is made of a steel plate which has been rolled into a ring and then welded. The feet are then welded to this ring.

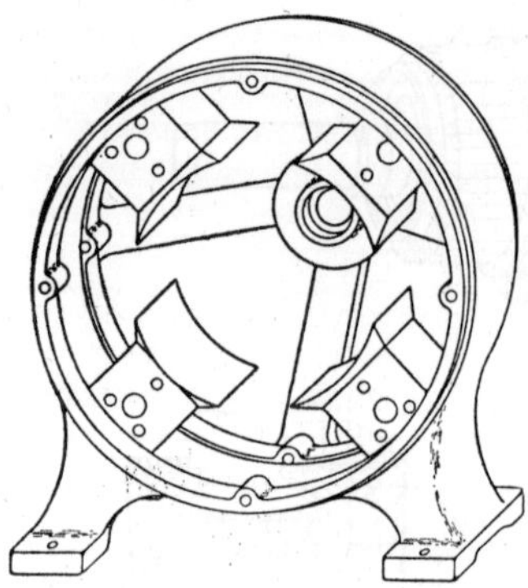

Fig. 6-3

Field Cores and Shoes—Field cores are made of cast steel, forged steel, or steel laminations. When cast or forged steel is used, the cores are usually made circular in cross section. Laminated cores, which are used almost universally except on very small machines, have a rectangular cross section and are held to the frame by bolts.

Field Excitation—In all generators, except in very small ones called magnetos, the field flux is produced by current flowing in coils placed on the pole pieces. The e m f required to force current through the field coils may be obtained either from a separate source, such as another generator, or from the armature of the generator itself. When the field is excited from a separate source the machine is said to be

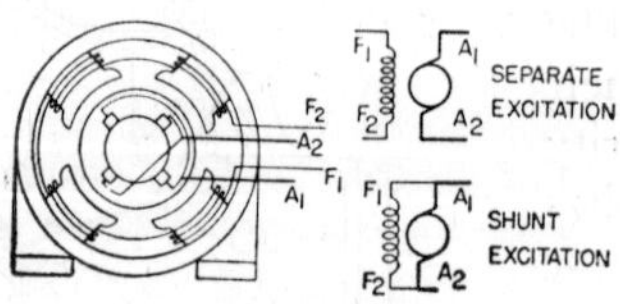

Fig. 6-4

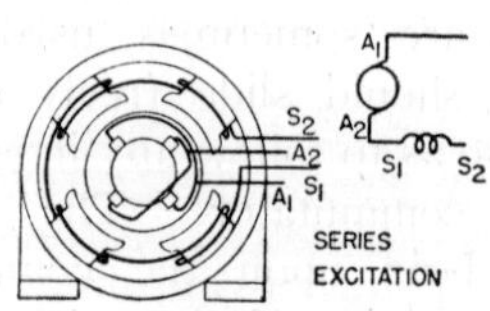

Fig. 6-5

separately-excited, and when the current is obtained from its own armature it is said to be self-excited. Self-excitation may be obtained in three ways—by shunt excitation, series excitation, or compound excitation.

In machines designed for either separate or shunt excitation (fig. 6-4), the field coils are made of a large number of turns of relatively small wire, thus providing a high resistance and taking a small current. A rheostat is usually included in the field circuit to vary the generated voltage.

In machines designed for series excitation, the field coils are made of a few turns of large wire, because these coils carry the total current of the machine (fig. 6-5).

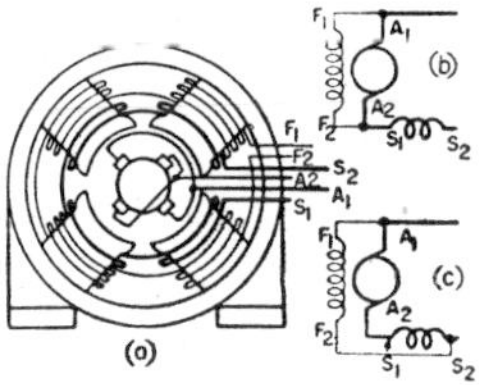

FIG. 6-6

A generator designed for compound excitation has both a shunt and a series winding placed on the field poles (fig. 6-6a). Ordinarily the series coils are connected to aid the shunt coils, and the machine is said to be cumulatively compounded. When the series winding is connected to oppose the shunt, it is differentially compounded. A compound machine is short shunted when the shunt field is connected directly across the armature fig. 6-6b and is long shunted when the shunt field is connected directly to the line (fig. 6-6c).

Effect of Armature Flux—When the field winding of a generator is energized and no current flows through the armature, the flux in the air-gap is uniformly distributed as shown in fig. 6-7a. At the instant that a coil is in plane ab, it is cutting no flux. Therefore no e m f is induced in it. The coil is said to be in the neutral plane.

The direction of the field flux and the direction of rotation of the armature in fig. 6-7b are the same as those in fig. 6-7a, except that the armature is delivering current to a load. The armature conductors to the left of the neutral plane carry current "in" and those to the right carry current "out". Not considering the magnetizing action of the field current, the conductors on the two sides of the neutral plane establish a flux downward, in a direction at right angles to the field poles. This direction is determined by the thumb rule.

In fig. 6-7c is shown the actual distribution of flux, which is established by the combined magnetizing action of the field and the armature current. At the upper tip of the north pole and at the lower tip of the south pole, the armature and field fluxes are in the same direction (see fig. 6-7a and 6-7b). Consequently, in these tips the field flux is strengthened as indicated in fig. 6-7c. At the lower tip of the north pole and at the upper tip of the south pole, the armature and field fluxed are in opposite directions. Hence, in these tips the field flux is weakened. Then the armature current distorts the field

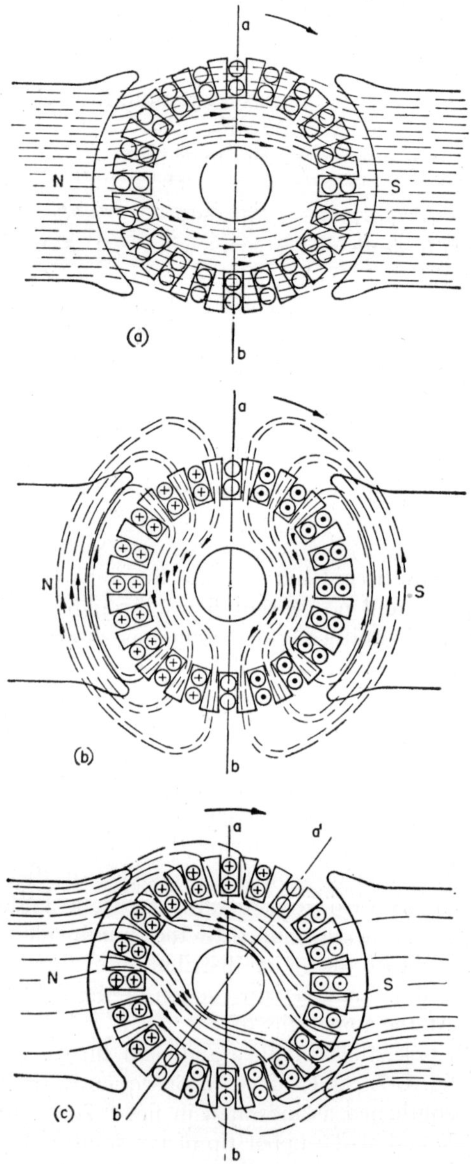

FIG. 6-7

flux in the direction of rotation. The flux on becoming distorted causes the neutral plane (the plane in which a coil cuts no flux) to shift from ab to position a′b′ in fig. 6-7c. *Note that this shift is in the direction of rotation of the armature.* Plane ab is known as the mechanical neutral plane and a′b′ is known as the electrical neutral plane.

If the brushes are set so that they short circuit the coils in plane ab (fig. 6-7a), as considered in armature winding, the coils in this plane will have e m fs induced in them since they are cutting flux. These e m fs will establish high currents in the short-circuited coils. As the segments at the forward ends of the short-circuited coils move away from the brushes the high current short circuits are opened and excessive sparking occurs at the commutator.

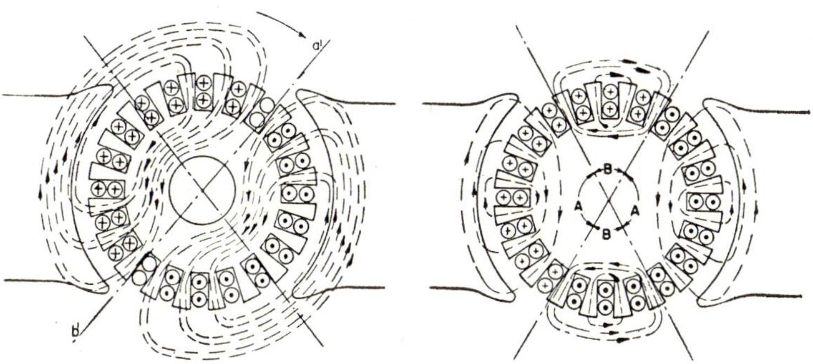

(a) (b)

FIG. 6-8

The sparking will be eliminated if the brushes are set so that the coils are short-circuited while they are in the electrical neutral plane (plane a′b′ fig. 6-7c). In this case the conductors to the left of a′b′ will carry current "in" and those to the right will carry current "out". The direction of flux will no longer be vertically downward, but will be downward and to the left (fig. 6-8a). The action of this flux is known as armature reaction. It may be considered as a resultant of two fields, one vertically downward as indicated in angles A (fig. 6-8b), called cross-magnetization, and the other to the left as indicated in angles B, called demagnetization. This component weakens the field flux, which is undesirable as will be shown later.

Even if the brushes were not shifted, the increased flux density in one tip of each pole causes an increase in the reluctance of these tips, and this causes a decrease in the field flux.

The twisting action of the field, and therefore the shifting of the neutral plane is proportional to the current through the armature. Then, with a changing armature current, which generally takes place in generators, the position of the neutral plane likewise changes. Then it is not possible to set the brushes in any one position and expect them to short circuit coils in the neutral plane. Consequently, shifting of brushes is not an entirely satisfactory method of eliminating sparking. This method is generally used only in small machines.

Armature Reaction in Multipolar Machines—In multipolar machines, armature reaction occurs in the same manner as in the bipolar machines described above. However, the diagram may appear

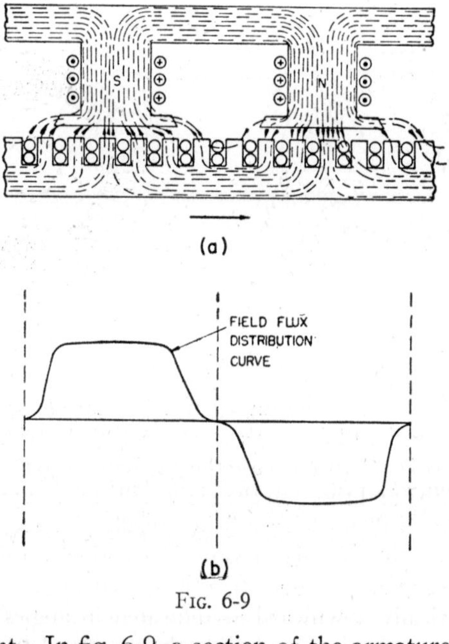

FIG. 6-9

slightly different. In fig. 6-9, a section of the armature and two field poles of a multipolar machine are shown, the armature being shown as a flat surface for convenience.

Fig. 6-9a, shows the distribution of the magnetic lines between two poles when current flows through the field winding only. The density of these lines at the surface of the armature is maximum under the

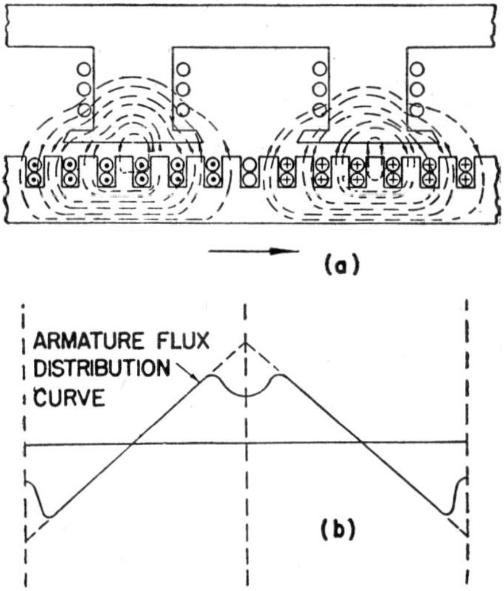

Fig. 6-10

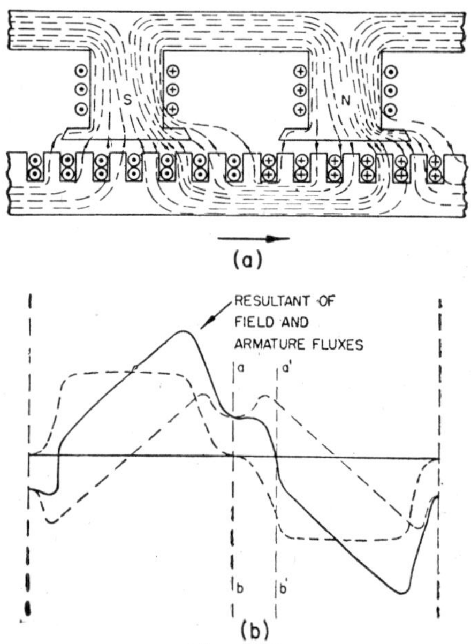

Fig. 6-11

central portion of each pole and decreases gradually toward the pole tips, then decreases rapidly beyond the pole tips, becoming zero midway between the poles. Although fig. 6-9a shows clearly the distribution of the lines of force in a section of the machine, the distribution of the useful flux at the surface of the armature may be more conveniently represented by a flux distribution curve as that shown in fig. 6-9b. In this curve note that the flux is substantially constant under the pole shoe, falls to zero at the neutral plane, and extends in the opposite direction under the next pole.

The magnetic field set up by the armature current when the brushes are in the mechanical neutral plane is shown in fig. 6-10a. The corresponding flux distribution curve is shown in fig. 6-10b. The conductors under the two poles may be thought of as forming a pancake coil with center midway between the two poles. If it were not for the relatively large air-space between poles, the flux distribution curve would be as indicated by the dotted lines (fig. 6-10b). However, the actual distribution is indicated by the solid line curve. Note that the maximum flux is under each pole tip.

The resultant flux distribution due to the combined action of the field and armature current is shown in fig. 6-11a (in this figure it is assumed that the iron has a constant permeability). By adding the flux curves of figs. 6-9b and 6-10b, the resultant distribution curve is obtained; this curve is shown in fig. 6-11b. It will be noted that the flux is distorted in the direction of rotation as in bipolar machines and that the neutral plane is shifted from ab to a′b′.

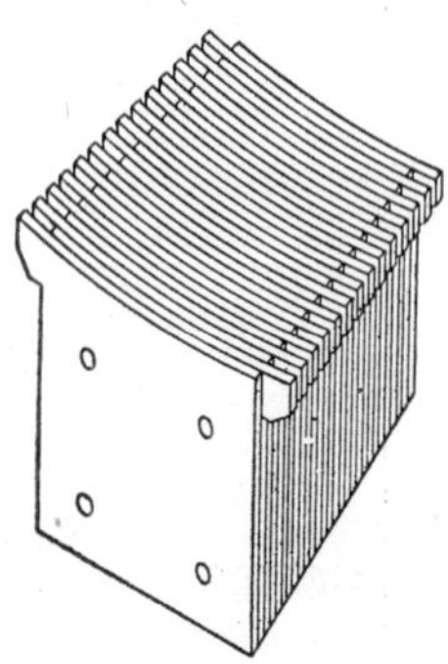

Fig. 6-12

Compensating Armature Reaction—Since the cross-magnetizing effect of the armature makes it necessary to shift the brushes with the load, it is often desirable to minimize the cross-magnetizing effect produced by the armature. One method of doing this is to build field poles of laminations having only one tip as shown in fig. 6-12. These, when stacked, are alternately reversed so as to leave space between laminations at the pole tips. The small cross-section of iron at the pole tips becomes saturated and thus the permeability is reduced. This has the effect of reducing flux distortion.

Another method of reducing the effect of the cross-flux is to place one or more slots in the pole face as shown in fig. 6-13. These slots produce a high reluctance for the armature flux, but have little effect on the field flux.

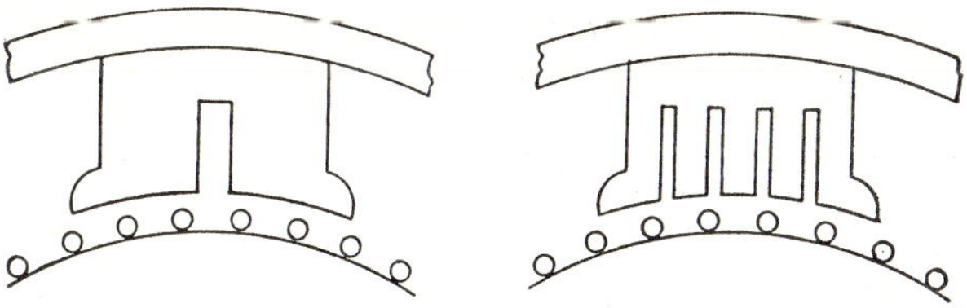

Fig. 6-13

The magnetizing effect of the armature may be neutralized by means of a compensating winding embedded in the pole faces as shown in fig. 6-14. These pole-face conductors are connected in series with the armature so as to carry the same current as the armature, but the connection is so made that every conductor embedded in the field

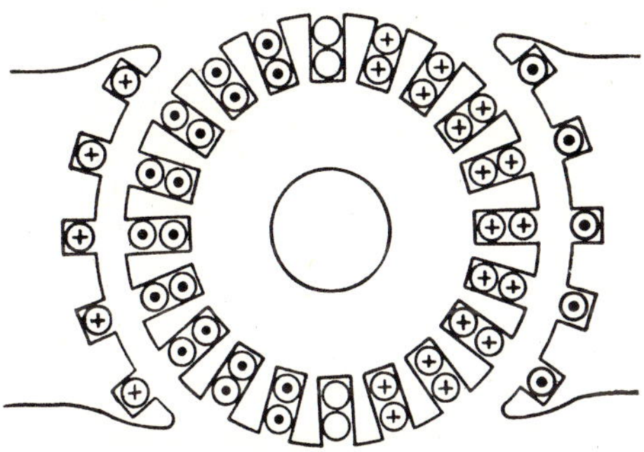

Fig. 6-14

poles carries current in the opposite direction to that of the adjacent conductor on the armature.

Compensating windings are not used on many machines, because of the cost of manufacturing them. Their greatest uses are on high speed and high voltage generators of large capacity.

Commutation—It was shown in chapter 5 that the current through a coil, rotating in a magnetic field, reverses as the commutator segments to which the coil ends are attached pass under the brush. The commutator serves also to maintain the current in one direction through the external circuit. The process of conducting the current from the coils to the brushes is shown in fig. 6-15 and is called commutation. In fig. 6-15a commutation of coil B has just started, because the brush is just beginning to make contact with segment B. Before this occurred, coil B was carrying the same current as all the other coils to the left of the brush. Assume this current to be 10 amperes. As soon as segment B makes contact with the brush, coil B is short-circuited;

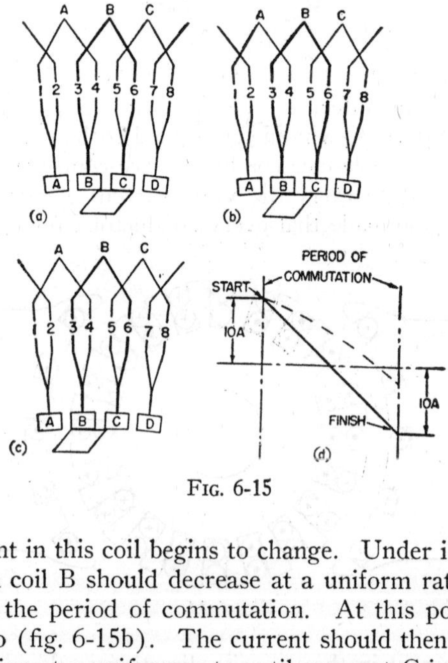

FIG. 6-15

and the current in this coil begins to change. Under ideal conditions, the current in coil B should decrease at a uniform rate until the coil is midway in the period of commutation. At this point the current should be zero (fig. 6-15b). The current should then increase in the reverse direction at a uniform rate until segment C is ready to break contact with the brush (fig. 6-15c). Now coil B is about to be cut into the armature circuit on the opposite side of the brush and must have 10 amperes flowing through it in the opposite direction. The current change in coil B under these ideal conditions is shown by the solid line in fig. 6-15d. The requirement for ideal commutation is that the coil, after being short-circuited by the brush, must have its current completely reversed and brought up to the initial value before it is again cut into the circuit. Then no current flows from segment C to the brush at the instant it breaks contact. If current is flowing between the brush and the segment when this break occurs, there will be sparking.

Securing the ideal conditions illustrated in fig. 6-15d was one of the outstanding difficulties in the early development of electrical machinery. In a coil undergoing commutation, it is difficult to reduce the current to zero and build it up to an equal value in the opposite direction during the brief interval that the coil is short-circuited by the

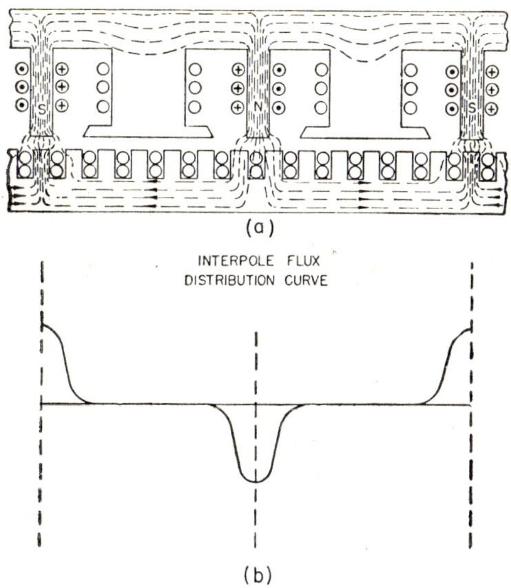

Fig. 6-16

brush. This difficulty exists because self-inductance prevents the current from changing rapidly. This results in coil B moving into the circuit to the right before the current has completely reversed and built up to 10 amperes. This condition is shown by the dotted line in fig. 6-15d. Note that the current must rise suddenly to 10 amperes when the break occurs, which causes sparking.

The early methods employed to eliminate sparking consisted of shifting the brushes (in the direction of rotation of the armature, in the case of a generator) until the short circuited coils were cutting a small amount of flux from the oncoming main poles. The induced voltage decreased the coil current to zero and built it up to the desired value in the opposite direction. This method is successful under steady load conditions, but, unfortunately, the strength of the required reversing field varies with the amount of current to be reversed, so that variable load conditions necessitate frequent shifting of brushes. Attempts have been made to use mechanical brush shifters actuated by solenoids in series with the armature, but these were not entirely satisfactory.

An important step was made when carbon brushes were substituted for the copper brushes first used. Having comparatively high

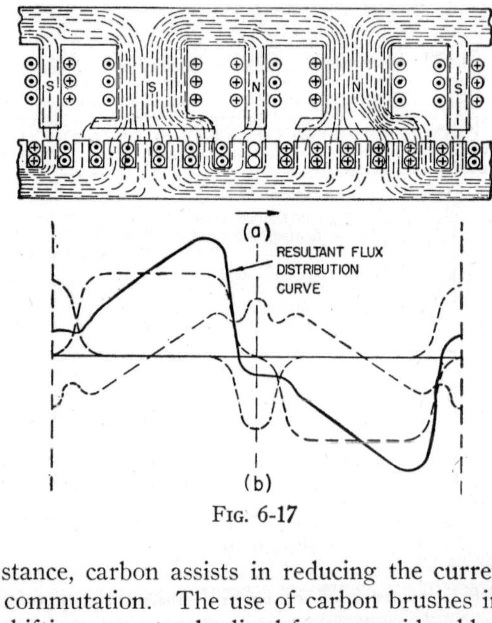

FIG. 6-17

contact resistance, carbon assists in reducing the current in the coil undergoing commutation. The use of carbon brushes in conjunction with brush shifting was standardized for a considerable period. The degree of shift was that required under average load conditions.

Commutating Poles—It was explained above that the voltage required to overcome the self-induced e m f in the short-circuited coil may be secured by shifting the brushes in the direction of rotation of the generator. Another way of securing the same effect is to leave the brushes in the mechanical neutral and to produce at this point the required commutating flux. This is accomplished by means of auxiliary poles called commutating poles or interpoles. These are placed midway between the main poles, as shown in fig. 6-16a. Interpoles are much narrower than the main poles and set up a flux in the proper direction and of sufficient magnitude to produce satisfactory commutation. The distribution of this commutating flux is shown by a curve in fig. 6-16b.

The resultant magnetic field in the air gap is then made up of three magnetic fields, namely, one due to the field current (fig. 6-9b), another due to the armature current (fig. 6-10b), and still another due to the current in the commutating winding (fig. 6-16b). The curves for these three fields and the resultant curve are shown in fig. 6-17b. In this figure it should be observed that the interpole flux neutralizes not only the field in the neutral plane due to armature reaction, but also produces the proper flux to generate an e m f in the short-circuited coil, equal and opposite to the e m f of self-inductance. This produces an ideal condition for commutation. Since armature reaction and the e m f due to self-inductance in the armature undergoing commutation are both proportional to the armature current, then the flux produced by the commutating poles must likewise be proportional to the armature current. This proportionality may be obtained by connecting the interpole winding in series with the armature, as shown in fig. 6-18. It should be noted that the coils of the interpoles are so connected to the brushes as to cause the commutating poles to have the same polarity as the main poles directly ahead of them.

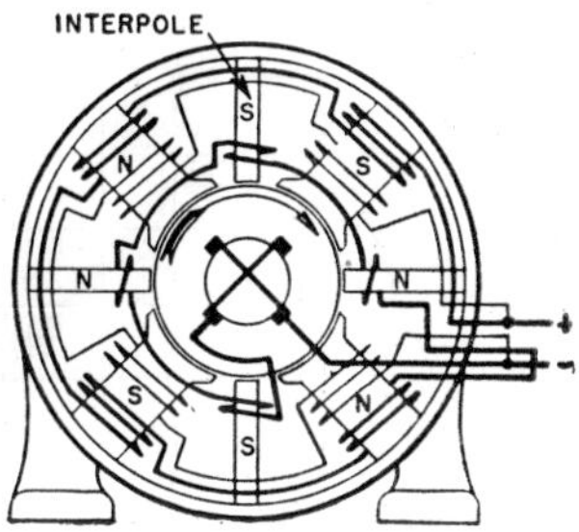

FIG. 6-18

The Saturation Curve—Suppose that equation 5-15 is written as

$$Eg = \left(\frac{PZ}{60b10^8}\right)\Phi N \qquad (6\text{-}1)$$

For any generator the quantity within the parenthesis is constant, and may be indicated by K. Then

$$Eg = K\Phi N \qquad (6\text{-}2)$$

This shows that the generated e m f in a machine is directly proportional to the flux per pole and to the speed. Consider the speed kept constant. Then the generated e m f is directly proportional to the flux. The flux is produced by the field ampere-turns, but as the turns remain constant, the flux depends upon the field current. It does not vary directly as the field current, however, because of the varying reluctance of the magnetic circuit. The curve in fig. 6-19a shows how the flux per pole varies with the field current. Due to residual magnetism, the curved part at a does not start at zero but at some value slightly greater. Beyond this part the curve to b is practically straight because the flux increases in proportion to the field current. At b the line begins to curve, for the magnetic circuit is becoming saturated.

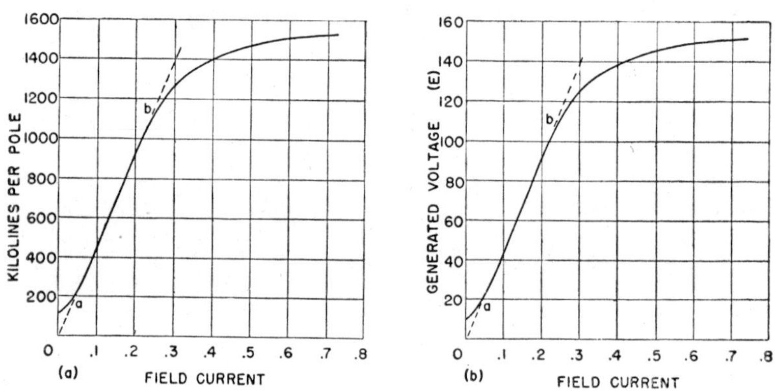

FIG. 6-19

Equation 6-2 shows that the generated e m f varies directly with the field flux. Hence, the generated e m f curve is the same as the flux curve, as shown in fig. 6-19b. Because the field poles become saturated, the voltage increases rapidly to b with a given change in field current. Shunt generators are designed to operate at a point slightly above b so that a slight change in speed produces an unnoticeable change in generated e m f. Compound generators are designed to operate slightly below b so that a few series field ampere-turns give the desired increase in voltage.

Separately Excited Generator—Suppose this type of generator is operated at a constant speed and with a constant field excitation. When it is unloaded, the terminal voltage will equal the generated e m f. When it is delivering current, the terminal voltage will be less than the generated voltage. This voltage decrease is due to two reasons. (1) As the flux per pole is reduced by the armature reaction,

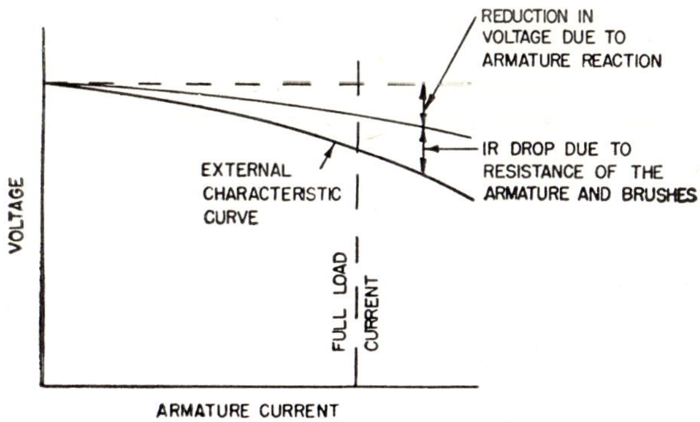

FIG. 6-20

the e m f generated by cutting this flux is also reduced. (2) The terminal voltage becomes less than the generated e m f by an amount equal to the IR drop of the armature winding and the brushes.

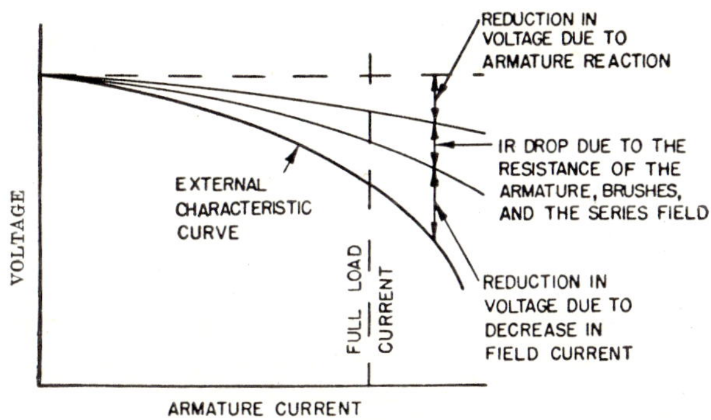

FIG. 6-21

These drops are indicated in fig. 6-20. Thus, when the load on a separately excited generator increases, its voltage decreases as indicated by the external characteristic curve.

Self-Excited Generator—The external characteristic curve of a self-excited generator is shown in fig. 6 21. When the machine is unloaded, the terminal voltage equals the generated e m f. As more and more current is delivered, the terminal voltage drops farther and farther from the no-load value. In this respect the self-excited generator behaves like a separately excited generator except that the rate at which the terminal voltage decreases is greater in the former than it is in the latter. This is due to the fact that there is another factor which is effective in lowering the voltage as the load increases. Figs. 6-20 and 6-21 show that in both generators the terminal voltage drops as the armature current increases, because of armature reaction and IR drop. In the self-excited generator there is a third factor—the drop in voltage due to the first two factors causes a decrease in the field current, and as a result less flux is produced. Then in this case the terminal voltage drops off more rapidly as indicated by the external characteristic curve.

Building Up of a Self-Excited Generator—If the resistance of the field circuit in a generator is kept constant, the current flowing through it is directly proportional to the terminal voltage. If a curve is plotted showing how the field current varies with the terminal voltage, a straight line will result. For example, if the resistance of the field circuit is 5 ohms, 2 amperes will flow at 10 volts, 4 amperes at 20 volts, 6 amperes at 30 volts, etc. This relation is shown by line a in fig. 6-22. Line b indicates the voltage-current relation when the resistance of the field circuit is 12.2 ohms, and line c indicates the relation when the resistance of the field circuit is 20 ohms. It should be noted that the higher the resistance of the field circuit, the greater the slope of the resistance curve.

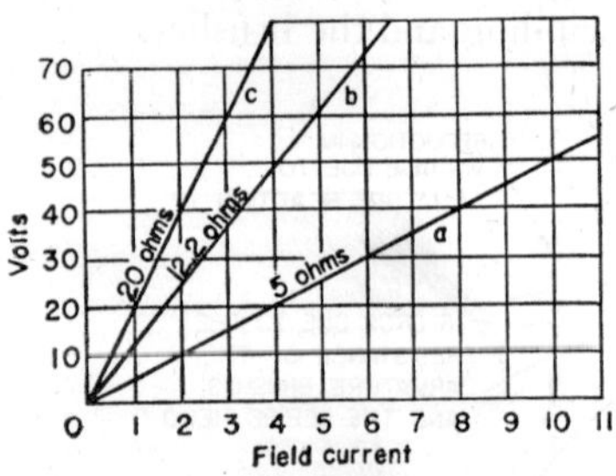

FIG. 6-22

Fig. 6-23 shows the saturation curve of the field drawn on the same graph and to the same scale as the resistance curve. In this particular case the resistance of the field circuit is 12.2 ohms. The voltage of the generator is built up as follows: The small amount of residual magnetism in the field is cut by the armature conductors, and a voltage of 6 volts is generated at the brushes. Since the field circuit is connected directly across the brushes, .6 of an ampere is sent through the field. But this field current (referring to the magnetization curve) causes 14 volts to be generated. This voltage (referring to the resistance curve) sends 1.1 amperes through the field circuit. As a result, 24 volts are generated, and so on. It may be seen from the curves that each value of field current produces a voltage in excess of its previous value, and this higher voltage in turn increases the field current (that is, the action is cumulative). The machine will continue to build up in this way until the 110 volt point is reached. At this point the field resistance curve crosses the saturation curve.

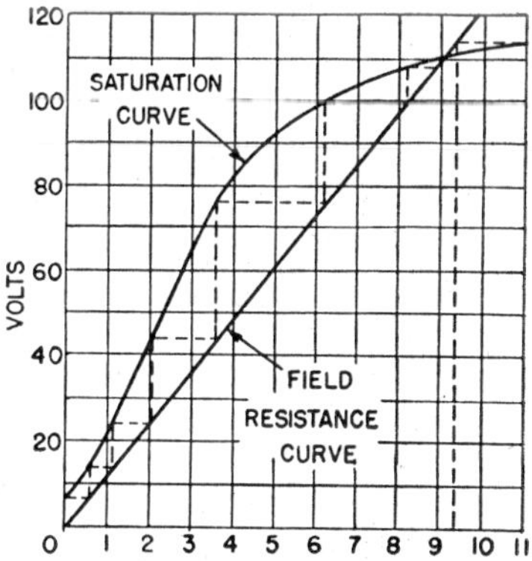

Fig. 6-23

The following illustration will show why the voltage will not rise higher than 110 volts. Assume that somehow 113 volts were generated. This voltage would produce a field current of only 9.3 amperes, and with this field current, only 111 volts would be generated. This voltage in turn would set up a smaller current, resulting in a smaller generated voltage. This procedure would continue until the terminal voltage would become 110 volts.

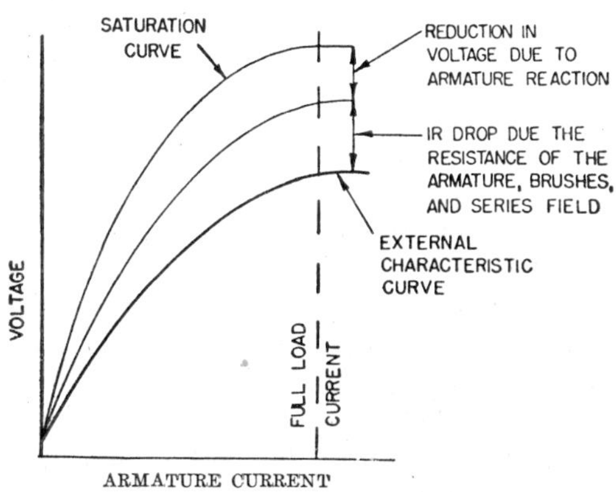

Fig. 6-24

Series Generators—The saturation curve of a series generator, which is shown in fig. 6-24, does not differ from that of a shunt machine. The external characteristic curve (fig. 6-24) is similar in shape to the saturation curve, but lies below it, for the following two reasons. The flux per pole is reduced, due to armature reaction, which causes a reduction in the generated voltage; and the resistance of the armature winding, field winding, and brushes causes an IR drop.

In the past, series generators were extensively used for series arc lighting. Even though arc lamps have been replaced to a great extent by series tungsten lamps, series generators are still used to supply power to these systems.

For proper operation, arc lamps and series tungsten lamps require a constant current. For this reason some kind of regulator must be connected to the line to prevent current fluctuation. A very simple type of regulator for this purpose is a carbon pile rheostat operated by a solenoid as shown in fig. 6-25. When the current in the external circuit increases, the pull on the solenoid also increases and the carbon pile is compressed. This has the effect of reducing the resistance of the rheostat so that more current is shunted from the series field coils. The flux in the field poles is therefore reduced, and the current drops until the line current reaches the value for which the pull of the solenoid is adjusted.

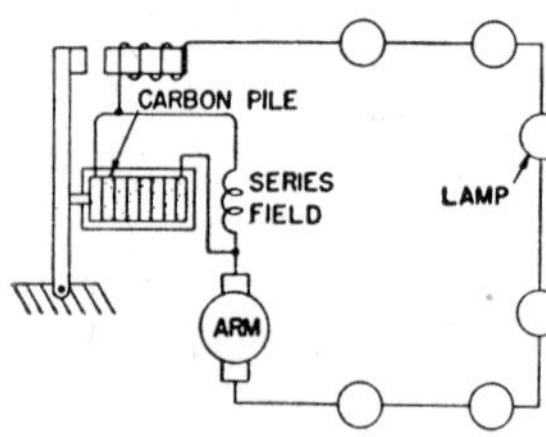

FIG. 6-25

As series generators are usually connected in a circuit that demands a constant current, they are sometimes called constant-current generators. It should be understood, however, that these machines are so-called not because they tend to maintain a constant current but because they are used in a circuit whose current must be kept constant, an external regulating device being used to keep it constant.

Compound Generator—In practice, power circuits require a constant voltage, especially lighting circuits. If a shunt generator, or even a separately excited generator, were used to supply power to these circuits, large variations in load would produce large voltage variations unless the field current was adjusted by a rheostat as the load changed. By using a compound generator, the desired voltage regulation may be automatically obtained without adjustment of the field rheostat. The curves in fig. 6-26 show some of the external characteristics that may be obtained from compound generators. The series winding in compound generators is usually connected to aid the magnetizing action of the shunt winding. A machine so connected is said to be cumulatively compounded. Its external characteristics depend upon the ampere-turns in the series field. By properly adjusting these ampere-turns the increase in terminal voltage due to the series field may be made to compensate for the drop in voltage due to the increased armature reaction and armature IR drop. In this case the series field also maintains a practically constant terminal voltage of the shunt field and

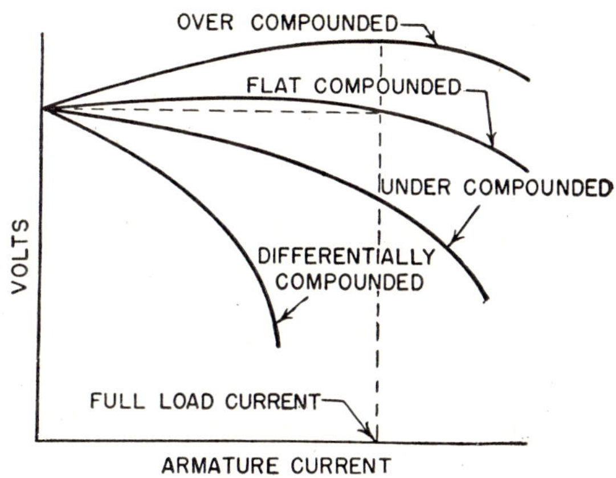

FIG. 6-26

the field current will not drop as the load increases. It is apparent that the three causes for voltage drop in a shunt generator are completely eliminated by the series field. The external characteristic of such a generator is shown in fig. 6-26. In this case the effect of the series field is to cause the generator to deliver the same terminal voltage at both no load and full load, and for this reason the machine is said to be flat-compounded. Actually, however, a flat characteristic curve is not exactly flat, because it is not possible to maintain a constant terminal voltage for all values of armature current below full load. The nearest approach is to adjust the series field ampere-turns so that the terminal voltage will rise slightly and then drop again, reaching the same voltage at full load as at no load. When the series ampere-turns are adjusted so that the terminal voltage of the generator is greater at full load than at no load, the machine is said to be over-compounded (see fig. 6-26). When the terminal voltage is less at full load than at no load, the machine is said to be under-compounded. Except those supplying power for welding, generators are seldom under-compounded. They are usually designed to be considerably over-compounded, and the degree of compounding is then regulated by shunting more or less current from the series field by means of a resistance called a diverter (fig. 6-27). For small machines the diverter is usually made of German silver ribbon or tungsten, and for large ones it is usually made of cast iron grids.

The resistance is ordinarily adjusted at the factory when the machine is tested and need not be changed unless it is desired to change the operating characteristics.

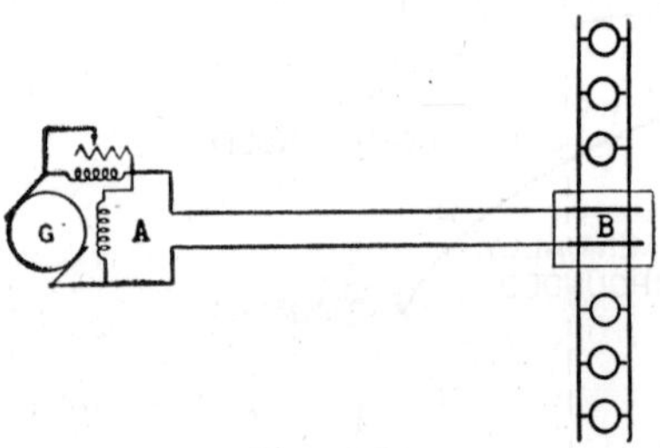

Fig. 6-27

If the terminal voltage at a generator remains constant, the voltage at a distant distribution point will decrease as the load is applied, because of the increased IR drop in the line wires. This will be made clear by considering fig. 6-27, which shows a generator at A supplying power to a group of lamps at B. If the voltage at A remains constant as the load increases, the voltage at B will fall, because this voltage is always equal to the voltage at A minus the IR drop in the line AB. If the series field ampere-turns are adjusted so that the voltage at A is increased just enough to equal the IR drop in the line AB, the voltage at B will remain constant.

The amount of over-compounding used in practice depends upon the service for which the machine is intended. For generators near the point of power consumption, the degree of over-compounding may be as low as 3 per cent, while for generators supplying power to a railway it may be as high as 10 per cent.

Factors Affecting the Building Up of a Self-Excited Generator—The e m f of a self-excited generator may fail to build up for one or more reasons. Chiefly among these reasons are: (a) the field circuit may be reversed, (b) the resistance of the field in a shunt generator or the resistance of the external load in a series generator may be too high, (c) the speed may be too low, (d) there may be insufficient residual magnetism.

The field must be connected to the armature so as to cause the field current to pass through the winding in a direction that will increase the flux, that is, produce building up action. If on the first trial the field connection is found to be wrong, the error may be remedied by reversing the field connection to the armature, or by reversing the direction of armature rotation if this is permissible. This latter case reverses the polarity of the generator, causing current to flow in the opposite direction through the external circuit.

Too high a resistance in the field circuit may be caused by neglecting to cut resistance out of the field rheostat, by poor connections in some part of the field circuit, or by poor contact between brushes and commutator due to grease or dirt. Due to these conditions the machine may completely fail to build up or may build up to a low value.

The starting of the building up of a generator depends upon speed as well as residual magnetism. A reduction in either of these may cause the machine to not generate a voltage. Occasionally it is necessary to overspeed a generator in order to start it to building up, after which it may be brought down to normal speed. This action compensates for a subnormal residual magnetism.

It is usually necessary to maintain a fixed polarity at the generator. If a generator builds up in reverse, which sometimes happens after a shut down, it becomes necessary to change the polarity back to the original condition. To do this, the residual magnetism must be reversed. This may be accomplished by sending a current through the field coils from an external source, so as to reestablish the residual magnetism in the proper direction. Occasionally in substations, a compound generator that is to be paralleled with others builds up with its polarity reversed. In this case the residual magnetism is reestablished in the proper direction merely by energizing the series field.

In power plants, power is usually supplied from several smaller units instead of from one large generator, for the following three reasons. (1) The efficiency of the power station may be maintained at a high point by varying the number of generators in service, so that the machines in use operate at approximately full load. (2) A generator that breaks down may be removed from the circuit for repairs without interrupting the power supply. (3) Generators may be added to the system as the power demand grows.

Shunt Generators in Parallel—The generators in power stations are connected in parallel through a common bus line, to which the feeders are also connected. This is shown in fig. 6-28a. The division of the load may be regulated easily in shunt generators, because

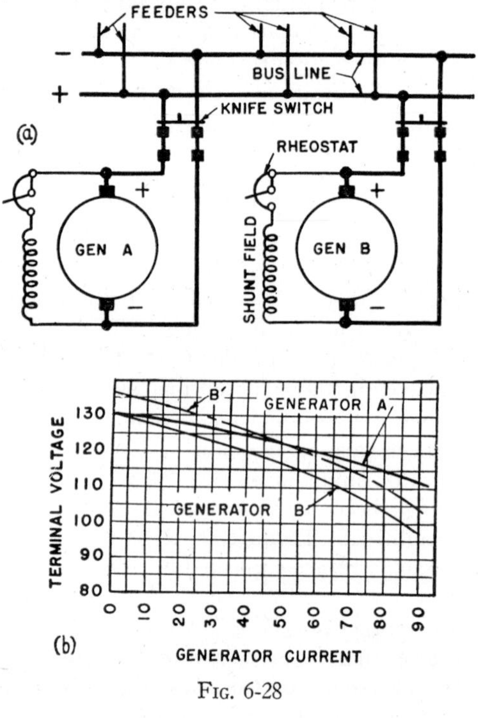

FIG. 6-28

of the drooping characteristics. The load may be shifted from one generator to another merely by manipulating a field rheostat.

Fig. 28b shows the characteristics of two shunt generators, A and B, which have their rheostats adjusted so that they supply the same e m f (130 volts) at no load. Generator B has a more drooping characteristic than generator A. If a load is connected to the bus line, causing A to deliver 60 amperes, B will deliver 40 amperes and the bus voltage will be 120 volts. If the load is increased so that A carries 80 amperes, B will carry 55 amperes and the e m f will be 115 volts.

Suppose it is desired to have each machine carry 50 amperes. This result is obtained by decreasing the field excitation of generator A or increasing that of generator B. If the latter is done, the voltage characteristic of generator B will be raised, as indicated by curve B′ (fig. 6-28). In this case the bus line e m f is 123 volts. This voltage

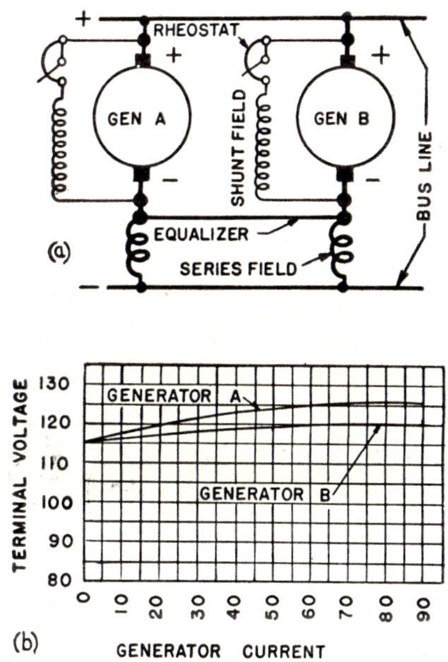

FIG. 6-29

can be raised, while keeping the currents equal, by increasing the field excitation of both generators. For generators of unequal capacities, the loads are generally divided in proportion to their ratings.

Compound Generators in Parallel—Under-compound generators also operate satisfactorily in parallel, but over-compounded generators will not operate satisfactorily in parallel unless their series fields are paralleled. This is accomplished by connecting the two negative brushes together, as shown in fig. 6-29a. The conductor used to connect these brushes is generally called an equalizer.

Suppose that an attempt is made to operate the two generators in fig. 6-29a in parallel without an equalizer (the external characteristics of the two machines are assumed to be as shown in fig. 6-29b). When generator A supplies 20 amperes, B supplies 60 amperes, and the total load is 80 amperes at 120 volts. If, for any reason, the current supplied by A increases slightly, say to 30 amperes, the current in the series field will increase and raise the generated e m f to 122 volts.

Since the total current is 80 amperes, the current of generator B must drop to 50 amperes, and as a result, its series field is weakened. This causes machine B to generate 119 volts. The result is that the load on generator A increases and that on generator B decreases. This will continue until in a very short time generator A will be driving B as a motor, in which case the current in the two machines will be in the direction shown in fig. 6-30. After machine B changes from a generator to a motor, the current in the shunt field will remain in the same direction, but the current in the armature and series field will reverse. Thus the magnetizing action of the series field opposes that of the shunt field. As the current taken by generator B increases, the demagnetizing action of the series field becomes greater and the resultant field becomes weaker. The resultant field will finally become zero, and at that time machine B will short circuit machine A, opening the breaker of either or both machines.

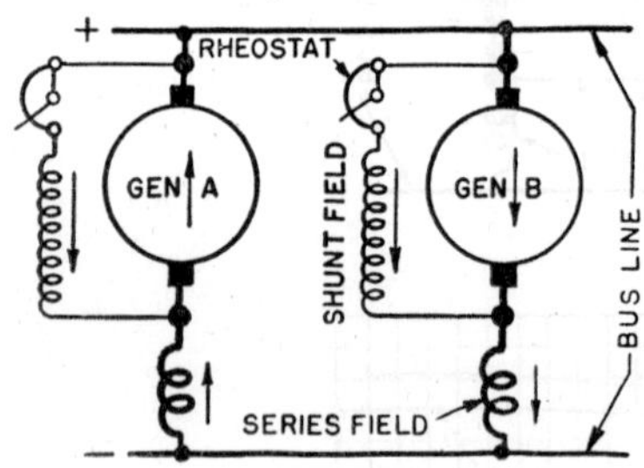

Fig. 6-30

When the equalizer is used, a stabilizing action exists and neither machine tends to take all the load. To consider this, suppose that the current delivered by generator A increases (fig. 6-29a). This increased current will not only pass through the series field of generator A, but also through the equalizer and series field of generator B. The voltage of both machines therefore increases, and generator B will take part of the load.

It was stated that the load on two shunt generators may be divided as desired merely by manipulating the field rheostats. It is usually desirable to have this division remain approximately constant at all loads. In compound machines of equal capacities, this takes place only if they have similar external characteristics and equal series field resistances. In machines of unequal capacities, the external characteristics must be similar, but the series field resistances must be inversely proportional to the capacities of the machines. If the latter condition does not exist, sufficient resistance should be connected in series with the proper series fields to bring about the desired condition.

Compound generator characteristics cannot be adjusted by means of series field diverters so that the loads on the machines divide as desired. In fig. 6-29a, suppose that the series field of generator A is shunted by a diverter. This diverter will also shunt the series field of generator B, since the bus line and equalizer have negligible resistance. In this case the diverter lowers the external voltage characteristics of the machine, but does not affect the load division.

Connecting a Generator to a Bus Line—Suppose it is desired to connect generator B of fig. 6-31 to a bus line. The generator is first brought up to its rated voltage and the negative and equalizer switches

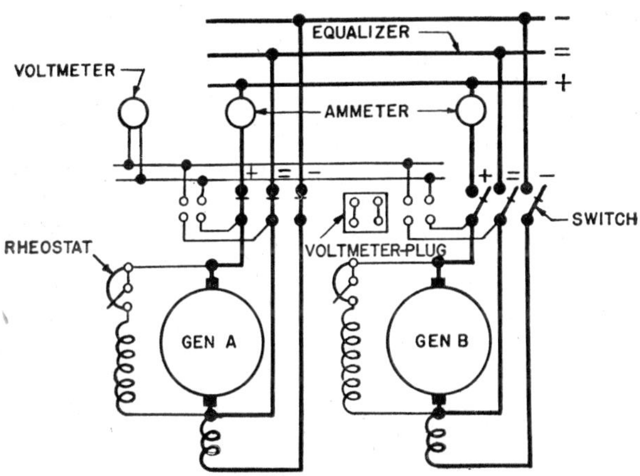

FIG. 6-31

are then closed, causing some of the load current to pass through the series field of machine B. The voltage of generator B is then adjusted to equal that of the bus line, and the positive switch is closed. This machine is then loaded by increasing its field strength.

To shut down a generator, its load is reduced to zero by weakening the shunt field, the positive switch is opened, and then the other two switches are opened.

Exercises

1. What is a suitable brush tension in a D C machine?
2. What is the purpose of a field rheostat?
3. Why are the field poles and armature of a direct current machine laminated?
4. Explain the ways in which the armature winding is held in the slots.
5. How does a series generator differ in construction from a shunt generator?
6. What are the detrimental results of armature reaction in a generator?
7. Name the different types of direct current generators.
8. With an increase in load, (a) why does the voltage of a shunt generator drop? (b) Why does the voltage of a series generator increase?
9. What conditions prevent a shunt generator from building up?
10. How are the external characteristics of compound generators changed?
11. Give at least one industrial application of generators with external characteristics as follows: (a) flat compound, (b) over compound, (c) differential compound.
12. A 250 volt, 20 k w, compound generator has a .5 ohm armature, a .3 ohm series field, and a 200 ohm shunt field connected long shunt. (a) Determine the current in the armature, shunt field, and series field at full load. (b) Determine the generated e m f at full load.
13. To obtain the same voltage at no load as at full load, the field current of a 120 volt, 4 k w, shunt generator must be changed from 1.5 to 2 amperes. The shunt field has 300 turns per pole. If this machine is to be equipped with a series field to give flat compound characteristics, how many turns will be required in this field?
14. (a) Give the procedure for connecting a generator with an equalizer to a bus line. (b) Give the procedure for disconnecting this generator from the bus line.
15. Two generators of 100 and 150 k w capacities are connected in parallel. If the series field resistance of the 100 k w generator is .0015 ohm, what must be the series field resistance of the other to properly divide the load?

CHAPTER 7

MOTORS

A GENERATOR is a machine for converting mechanical energy into electrical energy. The electric motor, on the other hand, converts electrical energy into mechanical energy. The structure of a D.C. generator is essentially the same as that of a D.C. motor. They usually differ slightly in detail of design because of different operating conditions. For example, the generator is usually of the open type; that is, the armature and field windings are exposed; whereas, the motor must often be partially or entirely enclosed to protect the winding from injury. However, the same machine may be run either as a generator or as a motor.

Principle of the Motor—Fig. 7-1a shows a uniform field between the opposite poles of two magnets. In fig. 7-1b is represented the cross-section of a conductor placed between the two poles and

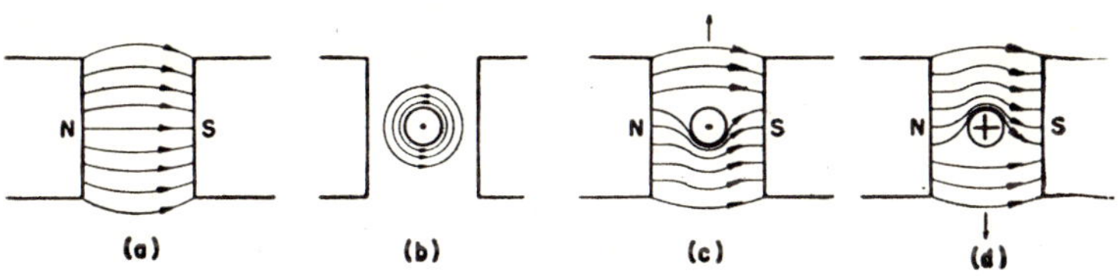

FIG. 7-1

carrying current toward the observer. Let us assume that the field due to the north and south poles has been temporarily removed. By applying the thumb rule (see Chapter 4) we find that the lines of force around the conductor are in a counter-clockwise direction.

Further, let us suppose that the conductor of fig. 7-1b is placed in the magnetic field of fig. 7-1a; the resultant magnetic field would then be as shown in fig. 7-1c. Above the conductor the lines of force are weakened because the two fields are opposite in direction and tend to destroy each other. Below the conductor the lines of force are strengthened because they are in the same direction. Magnetic lines, like rubber bands under tension, tend to straighten out, and, as a result, the conductor in fig. 7-1c is forced upward. If the current in the conductor is reversed as shown in fig. 7-1d, the direction of motion will be reversed, since, in this case, the lines above the conductor are strengthened, while those below are weakened.

The operation of an electric motor thus depends upon the principle that a conductor carrying current in a magnetic field tends to move at right angles to the field.

Fleming's Left Hand Rule—There is a definite relation between the direction of the lines in a magnetic field, the direction of

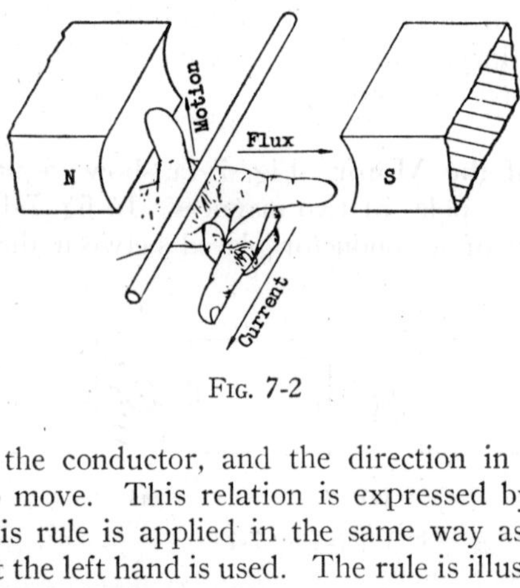

Fig. 7-2

the current in the conductor, and the direction in which the conductor tends to move. This relation is expressed by Fleming's left hand rule. This rule is applied in the same way as the right hand rule, except that the left hand is used. The rule is illustrated in fig. 7-2 and may be stated as follows:

Extend the thumb, the forefinger, and the middle finger at right angles to one another. Now turn the hand in such a position that the first finger points in the direction of the lines of force and the second finger in the direction of the current in the conductor; then the thumb will point in the direction of motion of the conductor.

Force Exerted on a Conductor—The force acting on a conductor carrying current while in a magnetic field is directly proportional to the field strength, the active length of the conductor, and the current flowing through it. The force produced may be computed by the equation

$$F = \frac{8.85\ BIl}{10^8} \qquad (7\text{-}1)$$

where F = force in pounds, B = flux density in lines per square inch, l = active length of the conductor in inches, and I = current in amperes.

EXAMPLE—Suppose that the conductor in fig. 7-2 is 12 inches long and is carrying a current of 50 amperes, and that the flux density is 70,000 lines per sq. in. What upward force acts on the conductor?

$$\text{The force on the conductor, } F = \frac{8.85 \times 70{,}000 \times 50 \times 12}{10^8}$$

$$= 3.6 \text{ lbs.}$$

Torque of a Motor—Since the armature of a motor is the same as that of a generator, the current from the supply line must divide

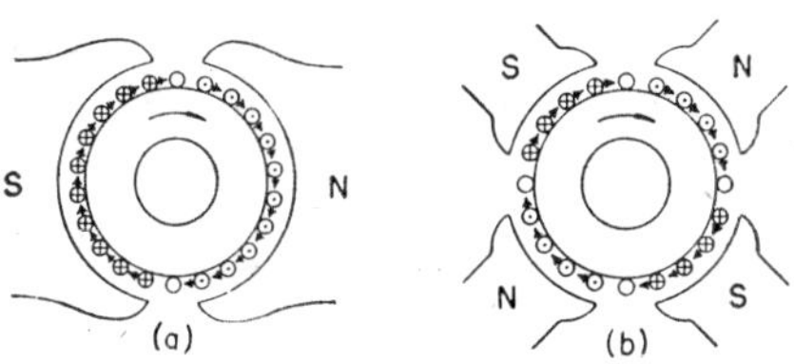

FIG. 7-3

and pass through the several paths of the armature. The currents through the conductors under a pole will all be in the same direction as shown in figs. 7-3a and 7-3b. In the bipolar machine of fig. 7-3a, the current flows toward the observer under the north pole and away from the observer under the south pole. The forces produced by the current in the conductors are therefore, all additive, as indicated by the arrows.

When a conductor moves from one side of a brush to the other, the current in that conductor is reversed, and, since it then comes under the influence of the next pole, which is of opposite polarity, the direction of the force remains the same. In a four pole machine (fig. 7-3b) the current flows in opposite directions under poles of unlike polarity so that the total force due to the current is the sum of the forces produced by the individual conductors.

The effect of these added forces in rotating the armature does not depend upon their magnitude (pounds pull) alone, but also upon the arm, that is, the radial distance at which they act. This effect is usually expressed as a product of the force and the length of the arm and is called torque. Expressed by an equation

$$T = Fr \qquad (7\text{-}2)$$

where T = torque expressed in pound feet, F = force in pounds, and r = radius in feet about which the added forces act.

EXAMPLE—In fig. 7-3a assume that the 20 active conductors are 10 inches long and 3 inches from the center of the armature. If the current through each conductor is 20 amperes and the air-gap flux density is 60,000 lines per square inch, find the torque developed.

$$\text{The force on each conductor, } F = \frac{8.85 \times 60{,}000 \times 20 \times 10}{10^8}$$

$$= 1.062 \text{ lbs.}$$

The torque developed by each conductor is 1.062 × 3/12 or .2655 lb. Then the torque produced by the 20 conductors is T = .2655 × 20 or 5.31 lb. ft.

Power Required to Drive a Generator—Whenever the armature of a generator is delivering current, a motor effect which opposes the motion is set up. This is readily explained by reference to fig. 7-4.

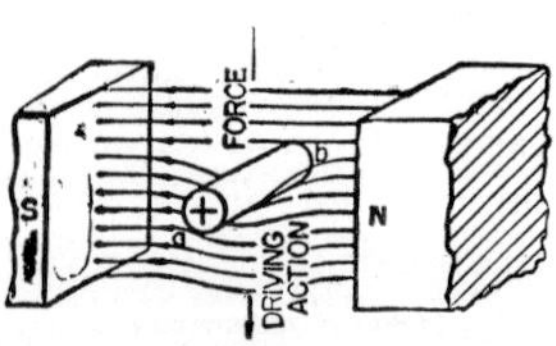

FIG. 7-4

Assume that ab represents a conductor on the surface of the armature. When ab is at rest, there is no force acting on the conductor. Just as soon as ab is moved downward, however, a current will flow in the direction a to b. The lines of force around the conductor, due to the current, will have a clockwise direction; hence, the lines above ab will be weakened and those below will be strengthened. This will produce an upward force that will oppose the driving action.

If the current in ab becomes greater, the motor effect will be increased, since the flux around ab will be increased. If ab is not connected to a load, there will be no opposing force, because there will be no magnetic field produced by the current through the conductor.

Thus little power is required to rotate a generator when the armature is delivering no current, but as the armature current increases, the strength of the field surrounding each conductor will increase in the same proportion. Hence the greater the amount of current delivered by a generator the greater the amount of driving power required. This is strictly in accordance with the law of conservation of energy, which may be stated as follows:

The energy output of a machine can never be greater than the energy input.

Counter Electromotive Force—We have just considered the motor effect in a generator. Likewise we shall see that every motor is subject to a generator effect.

Suppose that 230 volts are supplied to a 5 H P, 850 r p m motor. The armature resistance of this size motor is about .1 ohm. According to Ohm's law, the armature current will be 230 ÷ .1 or 2300 amperes. Of course such a high current would burn out the armature winding; then we suspect that something besides the armature resistance keeps the current at a reasonable value. We will now determine what this restraining action is.

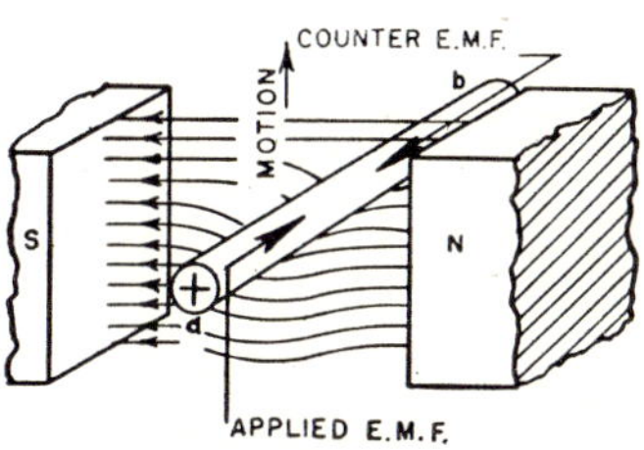

FIG. 7-5

Consider that a voltage is applied in the direction shown in fig. 7-5. A current will flow in the same direction, or from a to b, and according to Fleming's left-hand rule the conductor will be forced upward. On moving upward the conductor will cut lines of force, and according to Fleming's right-hand rule, an e m f will be induced in the opposite direction to the impressed e m f, or from b to a. This counter voltage always opposes the impressed voltage, and is called the counter (or back) voltage. It is directly proportional to the armature speed and field strength. The effective voltage across the armature equals the applied voltage E_a minus the counter voltage E_g. If an armature has R ohms resistance, the current flowing through it can be determined by the equation

$$I = \frac{E_a - E_g}{R} \qquad (7\text{-}3)$$

Suppose that the 5 H P motor considered previously has a counter-e m f of 228 volts at full load. Then the full load current is

$$I = \frac{230 - 228}{.1} = 20 \text{ amperes}$$

Torque and Power—Solving for E_a in equation 7-3,

$$E_a = E_g + IR$$

and multiplying both members of the equation by I, we have

$$E_aI = E_gI + I^2R \qquad (7\text{-}4)$$

The term E_aI represents the power supplied to the armature, and I^2R is the power lost as heat due to the resistance of the armature. The remainder of the power, E_gI, must therefore be the amount of mechanical power developed by the armature. Not all of this developed power is available at the shaft or pulley, for some of it is lost in overcoming friction of the bearings and brushes, windage resistance, hysteresis and eddy current in the armature core and pole faces. All these losses produce a mechanical drag on the armature. The remainder of the power, after subtracting these losses, appears as power available at the pulley or shaft for driving the external load. Nevertheless, in practice it is sufficiently accurate to assume that the power supplied by the pulley is E_gI watts.

Then the horsepower developed by the armature is

$$H\,P = \frac{E_gI}{746} \qquad (7\text{-}5)$$

But equation 2-25 shows that the horsepower may also be expressed as

$$H\,P = \frac{TN}{5252} \qquad (7\text{-}6)$$

Then it follows that the right-hand members of equations 7-5 and 7-6 are equal, that is

$$\frac{E_gI}{746} = \frac{TN}{5252}$$

Solving for T,

$$T = \frac{5252}{746} \times \frac{E_g}{N} \times I \qquad (7\text{-}7)$$

But when studying generators, we learned that

$$E_g = \frac{P\Phi ZN}{60b10^8}$$

Then

$$T = \frac{5252}{746} \times \frac{P\Phi ZN}{60b10^8N} \times I \qquad (7\text{-}8)$$

or

$$T = \left(\frac{PZ}{8.5b10^8}\right)\Phi I \qquad (7\text{-}9)$$

For a given motor the expression within the parentheses is constant. Then this equation may be written

$$T = K_1\Phi I \qquad (7\text{-}10)$$

where K_1 is a constant.

EXAMPLE—A simplex lap wound armature of a 4 pole motor has 140 active conductors, and there are 6×10^6 magnetic lines per pole. If at full load the armature runs at 1200 r p m and takes 6 amperes, determine (a) the torque produced, and (b) the horsepower developed.

(a) Substituting in equation 7-9, we have

$$T = \left(\frac{4 \times 140}{8.5 \times 10^8 \times 4}\right)6 \times 10^6 \times 6 = 5.93 \text{ pound-feet}$$

(b) Using equation 7-6 to find the horsepower,

$$H\,P = \frac{5.93 \times 1200}{5252} = 1.35 \text{ horsepower}$$

Armature Reaction in a Motor—In a generator the armature current flows in the direction of the induced e m f, whereas in a motor the armature current flows against the induced (or counter) e m f. It should therefore be expected that for the same direction of rotation and field polarity the armature flux of the motor will be in the opposite direction to that of the generator. Hence, instead of the main flux being distorted in the direction of rotation, as in the generator, it is distorted opposite the direction of rotation. The condition that exists in a generator is shown in fig. 7-6a, while fig. 7-6b shows the same machine operating as a motor. It should be observed that the direction of rotation and the polarity are unchanged, while the current is in the opposite direction. This causes the flux to likewise be distorted in the opposite direction.

As a result, in a generator the brushes must be shifted in the direction of armature rotation, whereas in a motor they must be shifted against the direction of rotation.

If it were not for the e m f of self-inductance the brush axis should be made to coincide with the neutral plane. Due, however, to the necessity of neutralizing this e m f, the brushes must be set slightly behind this neutral plane (fig. 7-6b). Thus in both the generator

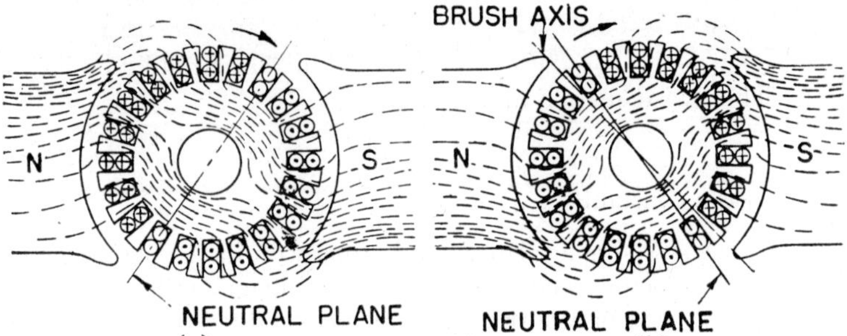

(a) FIG. 7-6 (b)

and the motor it is necessary to set the brushes beyond the neutral plane to neutralize the effect of self-inductance.

Commutating Poles—Commutating poles are just as important in motors as in generators. When a motor is not equipped with commutating poles the brushes must be shifted just as much as in a generator, if sparkless commutation is to be obtained.

Fig. 7-7a shows the condition that exists in a section of a multipolar machine. The corresponding flux distribution curves are shown in fig. 7-7b. The flux produced by the armature between the two poles is downward, and to oppose this flux, the commutating pole must be a south pole. *Therefore, in a motor the commutating poles must have the same polarity as the main poles directly back of them. This is the opposite of the corresponding relation in a generator.*

Direct current motors are divided into three principal types—shunt, series, and compound. They are named, as in generators, according to the way the field windings are connected.

Shunt Motor—In a shunt motor the field is connected directly across the line; and since the line voltage is steady, the field current and field flux are constant. If such a motor is unloaded, the retarding torque is small, being due only to windage and friction. The armature will develop a counter-e m f that will restrict the armature current to a value that will cause the motor to develop a torque equal to the resisting torque.

A motor must slow down when an external load is applied to it, because the small no-load current is just sufficient to produce a torque to overcome friction. This decrease in speed causes the generated voltage to decrease, since the field flux is constant, and more current

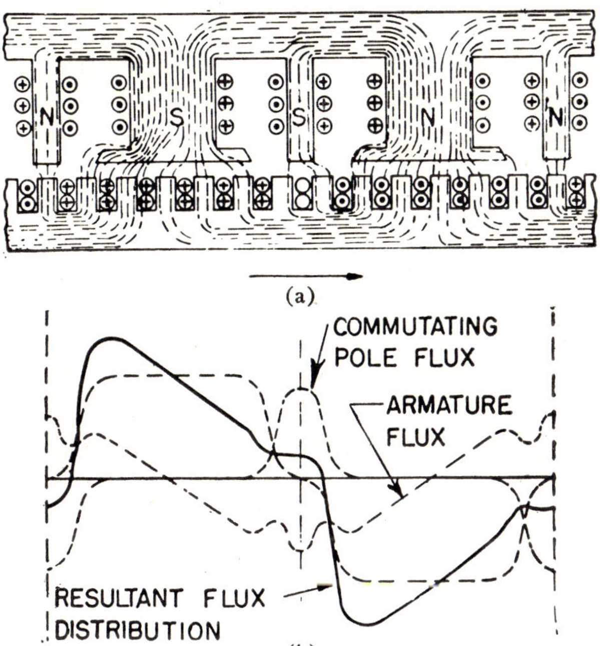

(a)

(b)

FIG. 7-7

flows through the armature (see equation 7-3). The armature current will continue to increase until a torque is developed to just equal that of the external load. When this point is reached, the armature current and speed will remain constant until the load is again changed. If the load is decreased, the torque developed will be greater than that of the external load. Then the speed will increase and the counter-e m f will decrease the armature current.

The foregoing discussion shows that the amount of current taken by a shunt motor is automatically regulated by the load placed upon it. This is true for all types of motors—shunt, series, and compound.

The industrial application for which a motor is best suited depends on the variation of the speed and the torque with the load.

An equation suited for investigating the speed variation in a motor may be obtained by substituting for E_g in equation 7-3, the expression $K\Phi N$ (which is given in equation 6-2). This gives

$$I = \frac{E_a - K\Phi N}{R}$$

Solving for N,

$$N = \frac{E_a - IR}{K\Phi} \qquad (7\text{-}11)$$

In a shunt motor, E_a, R, K, and Φ are practically constant, and I is the only variable. When the motor is carrying no load, the value of

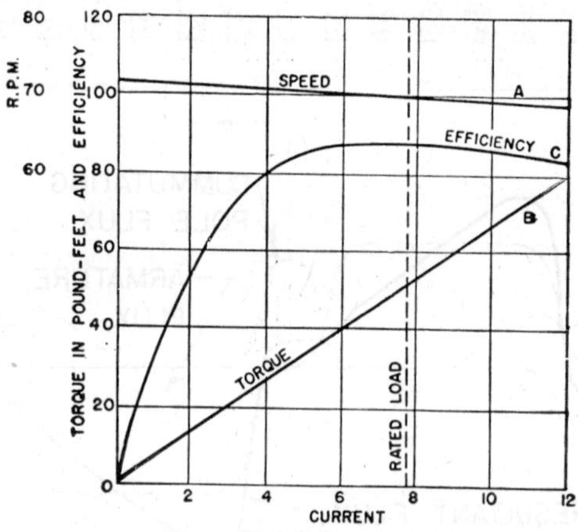

FIG. 7-8

I is small because the speed and therefore the counter-e m f are both at a maximum. In equation 7-11, then, IR (the voltage drop in the armature) is negligible compared to E_a. At full load, IR is ordinarily about 1/20 or 5 per cent of E_a, the actual value depending upon the size and design of the machine. Consequently, at full load, the speed determined by substituting in

$$\frac{E_a - IR}{K\Phi}$$

is about 95 per cent of the no-load value, as shown by curve A in fig. 7-8. This fall in speed is reduced slightly by armature reaction, which causes the flux to be less and the speed to increase correspondingly. In some cases, armature reaction is sufficient to cause the speed to remain nearly constant or even rise with an increase in load. For these reasons, a shunt motor is considered a constant speed motor, even though the speed usually decreases slightly with an increase in the load.

Speed Adjustment—The speed of a shunt motor can be changed by means of an adjustable resistance in the armature circuit or one in the field circuit (called a field rheostat). Inserting resistance in the armature circuit establishes a high power loss, and causes the speed to change considerably with a change in the load, as indicated

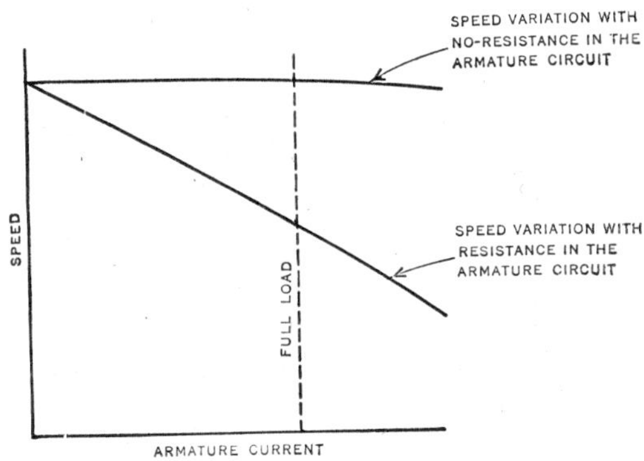

FIG. 7-9

in fig. 7-9. The most common method of changing the speed is by varying the resistance in the field circuit. Suppose that the resistance in the field circuit is increased. This causes the current through the field circuit to decrease, which in turn causes a reduction in the field flux. Thus fewer magnetic lines are cut by the armature and therefore less counter-e m f is generated. As a result, more current flows through the armature.

The increase in armature current is much greater than the decrease in field current, therefore the torque is greater and the armature increases in speed. In doing so, the counter-e m f increases and reduces the armature current. The speed increases until the developed torque becomes equal to the resisting torque, and then the motor again runs at a constant speed. Therefore, peculiar as it may seem, the speed of a shunt motor is increased by increasing the resistance of the field circuit. *The field of a shunt motor in operation should never be open because its speed will increase to an extremely high value.*

With this method of varying the speed, a motor will run at nearly constant speed for any setting of the rheostat (see fig. 7-10). This constant speed characteristic is desirable in many industrial applications in which the speed is adjusted at a definite value and it is important that it remain at approximately that value for load changes from no load to full load.

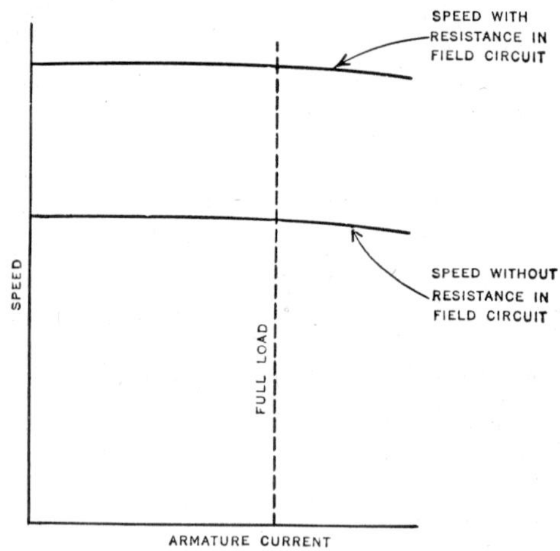

FIG. 7-10

When a lower speed in a shunt motor is desired, some of the resistance in the field circuit is cut out. This allows more current to flow through the field. The armature conductors then cut more magnetic lines, and a higher counter-e m f is produced. This cuts down the current (and torque) in the armature, and as a result the motor runs at a lower speed.

The relation between the torque developed by a shunt motor and the current flowing through its armature winding may be analyzed easily. As stated above, the field flux is practically constant. Then, from equation 7-10, the torque will vary directly with the armature current. This relation is shown by curve B in fig. 7-8. Note that when the armature current is 6 amperes, the motor develops a torque of 40 pound-feet, and when the armature current is 12 amperes the motor develops a torque of 80 pound-feet. That is, when the current doubles the torque doubles.

Curve C shows that as the load is increased the efficiency increases rapidly at light loads, reaches a maximum at a little over half load, and decreases gradually. This is due to the fact that at light loads the friction and copper losses are high when compared with the external load. Above full load the I^2R losses of the armature cause the efficiency to drop gradually.

The shunt motor is best suited for constant speed drives. It meets the requirements of a large range of industrial applications, such as the driving of machine tools, blowers, fans, and line shaftings.

Series Motors—Equation 7-11 may also be applied to a series motor if R is considered to include the resistance of the series winding. Hence R, and therefore the IR drop, is slightly greater than in a shunt motor. In a series motor the flux increases almost directly with the load. Then, since the numerator in the expression

$$\frac{E_a - IR}{K\Phi}$$

decreases appreciably while the denominator increases considerably with an increase in load, it should follow that the speed decreases very much as indicated by curve A in fig. 7-11. As full load is approached, the field becomes somewhat saturated. Thus the field flux does not increase as much for a given increase in armature current as it does at light loads, so that near full load the speed changes less rapidly.

If the load is removed from a series motor, the speed increases rapidly, and at very light loads a motor may run at far more than safe speeds. In very small motors this is not true, since friction offers enough load to restrain a motor from running at dangerous speeds. A large series motor should always be connected to the load by gears or other positive drives, and never used with a belt drive or any other drive where it is possible for the load to be entirely removed.

Equation 7-10 shows that the torque developed by any direct current motor varies with Φ and I. In a series motor the current through the armature also passes through the field. Then during the time the field is only moderately saturated Φ will be almost directly proportional to I, and the torque is therefore directly proportion to I^2. This may be expressed by means of the following equation:

$$T = K_2 I^2 \qquad (7\text{-}12)$$

where K_2 is a constant.

This relation is shown by curve B in fig. 7-11. When the armature current is 4 amperes, the motor develops a torque of 20 pound-feet, and at 8 amperes it develops a torque of 80 pound-feet. That is, when the armature current doubles the torque becomes four times

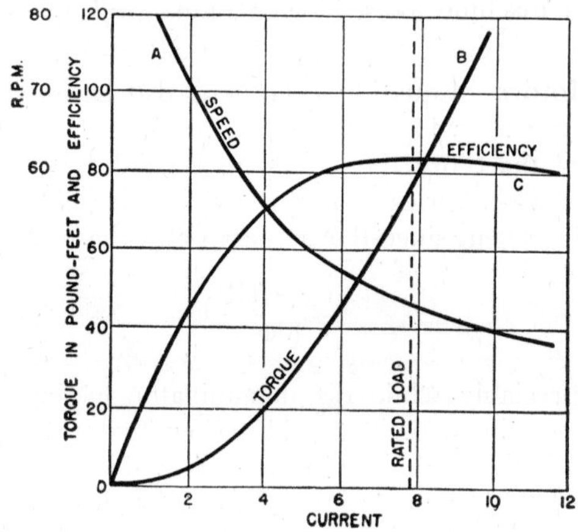

FIG. 7-11

as great. Thus the torque increases rapidly near and above full load. Such characteristics make the use of the series motor desirable where it is necessary to supply a large torque with a moderate increase in current. For this reason a series motor will start with a much heavier load than a shunt motor, and will accelerate with less current input.

Actually, however, saturation and armature reaction prevent the torque from increasing as rapidly as the square of the current, so that at very heavy loads the torque curve becomes almost straight.

Although the efficiency of a series motor varies about the same as that of a shunt motor (see curve C in fig. 7-11), the actual operating efficiency is usually considerably lower. This is due to the fact that a series motor seldom runs at its rated load, but is continually accelerating and decelerating.

Series motors are used chiefly for widely varying loads, where extreme speed changes are not objectionable, and where the operator is always present. They are used extensively for cranes and hoists, and for traction purposes. In these cases the variation of speed with the load is a favorable condition. For instance, in using a crane to lift a heavy load, it is generally desirable to proceed slowly, but in carrying a light load it is desirable to have increased speed for making rapid headway. The enormous torque of a series motor makes it very suitable for hoisting and work demanding frequent acceleration under heavy loads.

Compound Motors—As in compound generators, connections in compound motors are usually arranged so that the series winding aids

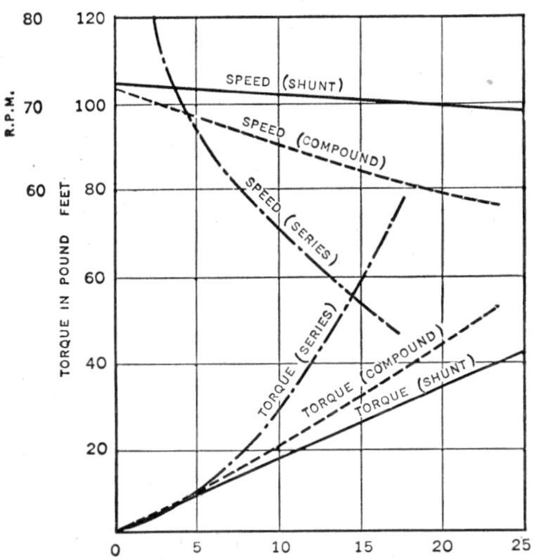

FIG. 7-12

the shunt winding. A motor so connected is said to be cumulatively compounded. If the series field is connected to oppose the shunt, however, the motor is said to be differentially compounded. The cumulatively compounded motor has the combined characteristics of the shunt and series motor, as may be seen in fig. 7-12. It has a definite no-load speed, and therefore may be run unloaded. The speed falls more rapidly than it does in shunt motors, because the field strength increases with the load. The method of speed control is the same as that used in shunt motors.

The current-torque curve of a compound motor is intermediate to the curves of the shunt and series motors (fig. 7-12). The field strength at zero armature current is due only to the shunt winding. As the load is applied, the series turns increase the flux, causing the torque for any given current to be greater than it would be for a straight shunt motor.

The speed and torque characteristics of the cumulatively compounded motor may approach those of the shunt or series motors as limits, depending upon the strength of the two fields. If the motor contains a light series winding, it will have better torque characteristics than a shunt motor, but will largely retain the good speed regulation or efficient speed control of the shunt motor. If the motor has a light shunt winding (one sufficient to keep the no-load speed within certain limits) and a heavy series winding, it will have the characteristics of a series motor, but will not run at a dangerous speed with light loads.

Compound motors are used for driving machines that demand a fairly constant speed and are subject to irregular loads or sudden applications of heavy loads, such as presses, shears, compressors, reciprocating tools, and elevators. They are also used where it is desired partially to protect the motor by causing it to decrease in speed under heavy loads.

To reverse the direction of rotation of any motor, the connections of the armature terminals alone, or the connections of the field winding (or field windings) alone, must be reversed. If both (or all) are reversed, the direction of rotation will remain unchanged.

Effect of a Flywheel—The power required by some kinds of machine tools is very irregular. In a punch press or stamping machine, for example, almost no power is required until the punch or die comes in contact with the metal to be punched or pressed. If the moving parts in such a machine are not very heavy, the current taken by the driving motor will vary widely, as shown by the solid curve in fig. 7-13. The motor selected to carry such a variation of current must be large enough to carry the greatest value of current without excessive sparking. If a considerably over-compounded motor is used (for instance about 30 per cent) with a heavy flywheel, the current through the armature will vary somewhat as indicated by the dotted curve in fig. 7-13. From this curve it can readily be seen that the driving motor may be much smaller when a flywheel is used (less than two-thirds as large).

The effect of a flywheel on a machine can be illustrated thus: At a (fig. 7-13) the load on the motor and flywheel suddenly increases and causes them to slow down. As the flywheel is slowed down, it is effective in supplying energy. That is, the electrical input to the motor will not be as great as the power demanded by the load, and as the flywheel slows down it assists the motor in carrying a part of the load. During this slowing down process, the current input to the motor must increase.

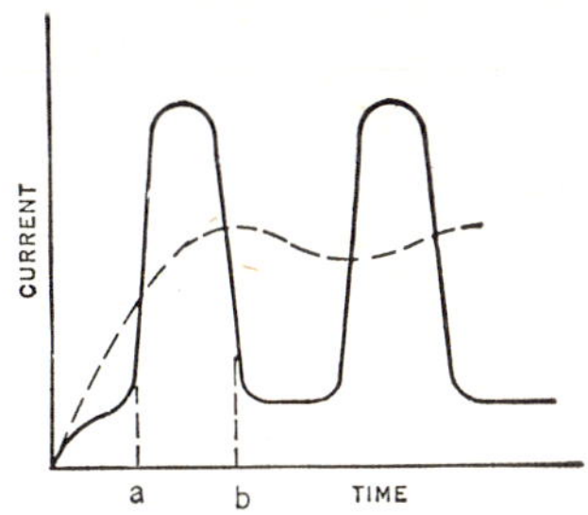

FIG. 7-13

At b the power required by the load is less than the input to the motor, so that the motor and flywheel begin to speed up and the current decreases. This procedure continues during the operation of the machine.

Exercises

1. What factors govern the strength of the counter-electromotive force?
2. What relation exists between the speed of a motor and the armature current?
3. Why does a motor take a very large current when starting?
4. What is the net voltage acting on an armature circuit?
5. Assume the wire in fig. 7-2 to be 8 inches long and to lie entirely within the magnetic field. The current carried by the conductor is 40 amperes, and the flux density is 60,000 lines per square inch. Find the upward force exerted on the conductor.
6. The current taken by an armature is 10 amperes, and its resistance is .2 ohm. If 120 volts are supplied, find the counter-e m f.
7. Find the horsepower developed by the motor in problem 6.
8. Determine the power in watts supplied to the motor in problem 6.
9. The armature of a 4 pole motor contains 48 coils of 6 turns each. The flux per pole is 2×10^6 magnetic lines. If the armature is simplex lap wound and the current taken by it is 40 amperes, what will be the torque?
10. If the motor in problem 9 is turning at 1200 r p m, what horsepower is being developed?

11. Name the different types of D C motors.
12. For what kind of service is a shunt motor adaptable?
13. (a) How does the speed of a shunt motor vary with the load? (b) How does the armature current vary with the torque?
14. How is the direction of rotation of a shunt motor reversed?
15. To vary the speed of a shunt motor, why is resistance inserted in the field circuit rather than in the armature circuit?
16. What will happen if the rheostat in the field circuit of a shunt motor becomes open?
17. A shunt motor connected across a 240 volt line takes 6 amperes and runs at 850 r p m. Its armature resistance is .2 ohm. At what speed will the motor run when the armature current is 7 amperes?
18. A 230 volt shunt motor runs at 900 r p m when taking a full load current of 10 amperes. The armature has a resistance of .5 ohm and the field has a resistance of 200 ohms. Find the resistance of the field rheostat required to raise the speed to 1200 r p m while the motor takes a full load current.
19. A 4 pole, 240 volt shunt motor takes 15 amperes at full load. The armature is simplex wave wound, has 180 active conductors, and has a resistance of .3 ohm. The flux per pole is 720,000 lines. What is the speed of the motor at full load?
20. What changes must be made in the armature winding of the motor in problem 19 so that it will run at half speed?
21. What effect will a decrease in the number of turns in the armature coils have on the speed of a motor?
22. A 4 pole shunt motor with a simplex wave wound armature runs at 1725 r p m. At what speed will this motor run if the armature is reconnected simplex lap?
23. What will be the speed of the motor in problem 22 if the armature is reconnected duplex lap?
24. It is desired to use a 240 volt, 4 pole, wave wound motor on 120 volts. If the horsepower and speed are to remain unchanged, what changes must be made?
25. When developing a torque of 20 lb.-ft., a 120 volt, 5 H P motor takes 10 amperes. What current will the motor take when developing a torque of 60 lb.-ft.? The field takes 1.5 amperes.
26. How is the speed of a series motor affected when a load is applied? When a load is removed?
27. What precaution should be taken in using series motors?
28. How does the flux vary with the load in a series motor? How does this affect the variation of the torque with the load?
29. For what general type of loads is the series motor most suited? Why?
30. How may a shunt, a series, and a compound motor be reversed?

31. Explain the speed and torque characteristics of a cumulatively compounded motor. What advantage has this motor over the series motor?
32. The armature of a 240 volt series motor has a resistance of .5 ohm and the series field has a resistance of .1 ohm. When the machine is taking 10 amperes, its speed is 1200 r p m. Neglecting armature reaction and assuming that the flux saturation curve is a straight line, determine the speed when the motor takes 30 amperes.
33. At full load a 220 volt series motor takes 40 amperes and runs at 980 r p m. If the armature has a resistance of .2 ohm and the series field .8 ohm, compute the speed of the motor when it is connected to a 150 volt line and takes 20 amperes.
34. A 220 volt, 4 H P series motor develops a torque of 12 lb.-ft. when the current is 10 amperes. What torque will it deliver when the current is 20 amperes?

CHAPTER 8

CONTROLLERS

In chapter 7 it was shown that the amount of current flowing through an armature may be determined by the equation

$$I = \frac{E_a - E_g}{R} \qquad (8\text{-}1)$$

where I = armature current, E_a = applied voltage, E_g = generated or counter-voltage, and R = armature resistance. R is very small in a D C motor, and at the instant of starting E_g is zero. Then if a motor is switched directly to a voltage E_a (its rated voltage), the armature current will be excessive. For instance, the armature resistance of a 5 H P, 230 volt motor is about .4 ohm. The current through the armature at the instant that the machine is connected directly to the line is 230 ÷ .4 or 575 amperes. The full load current of a motor of this size is about 20 amperes. Then the starting current is 575 ÷ 20

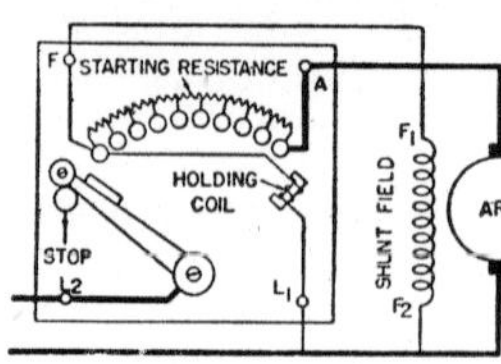

Fig. 8-1

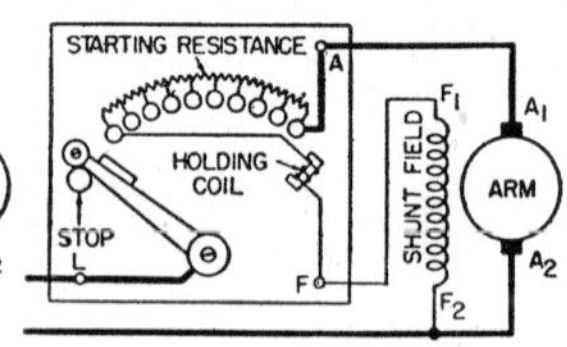

Fig. 8-2

or 28.75 times the full load current. This high current may damage the insulation by overheating it, and the motor and attached machinery may be damaged by the sudden jerk received when starting. A variable resistance should be connected in series with the armature, to reduce the starting current to a limit of about 1.5 times the full load current. As soon as the armature begins to rotate, a voltage, E_g, is generated, which increases with the speed. Then from equation 8-1 it can be seen that the armature current, I, decreases as the motor speeds up. Thus the starting resistance may be cut out gradually. This may be accomplished in a simple and practical way by a starting resistance arranged as shown in fig. 8-1.

This starting arrangement, usually called a starter, functions in the following manner: When the arm is raised to the first contact, the field is connected directly across the line and the starting resistance is placed in series with the armature. The starting resistance is gradually cut out of the armature circuit as the arm is rotated slowly. When the arm moves into contact with the electromagnet, all of the resistance is cut out and the armature is connected directly across the line. The electromagnet holds the arm in a running position as long as voltage is maintained. In case of a voltage failure, the magnet releases the arm, which is pulled back to the off position by a spring. Hence the motor will not be damaged when the voltage is restored. This starter may also be used for starting series or compound motors.

In some starters, the low voltage release magnet is designed to be connected in series with the shunt field (fig. 8-2). In such starters the electromagnet releases the arm if the field circuit is opened, a decided advantage, since, as a result, the armature comes to a standstill instead of racing. These starters are used only for starting shunt or compound motors.

Controllers—A controller is a device for regulating the operation of electrical equipment. A controller for electric motors is simply a starter which provides a convenient and safe means of performing several or all of the following functions: (a) starting and stopping, (b) accelerating and decelerating, (c) reversing, (d) regulating the speed, (e) dynamic braking or plugging.

Series motors and compound motors with heavy series windings have industrial applications that usually demand heavy service. In some of these applications the motor must be frequently started, stopped, and reversed, in addition to having its speed varied continually. The stopping is usually accomplished by dynamic braking (to be explained in detail later). The need of a controller for motors rendering such rugged service has been responsible for the development of two types of controllers: a face plate type and a drum type. Each of these types is operated by a single handle.

Face Plate Controller—An E C & M face plate controller is shown in fig. 8-3a. Segments are mounted in a circle on the face of this controller, inside of which two arc contacts (A and A′) are mounted. These contacts have renewable tips. Brushes resting on these contacts are attached to (but insulated from) the horizontal arm, which is actuated by a handle at the top of the controller. The two sets of brushes at each end of the arm are connected together (see fig. 8-3b), thus forming a circuit between the inner contacts and outer segments. The starting resistance, which is mounted at the back of the panel, is connected to the segments as indicated in fig. 8-3b.

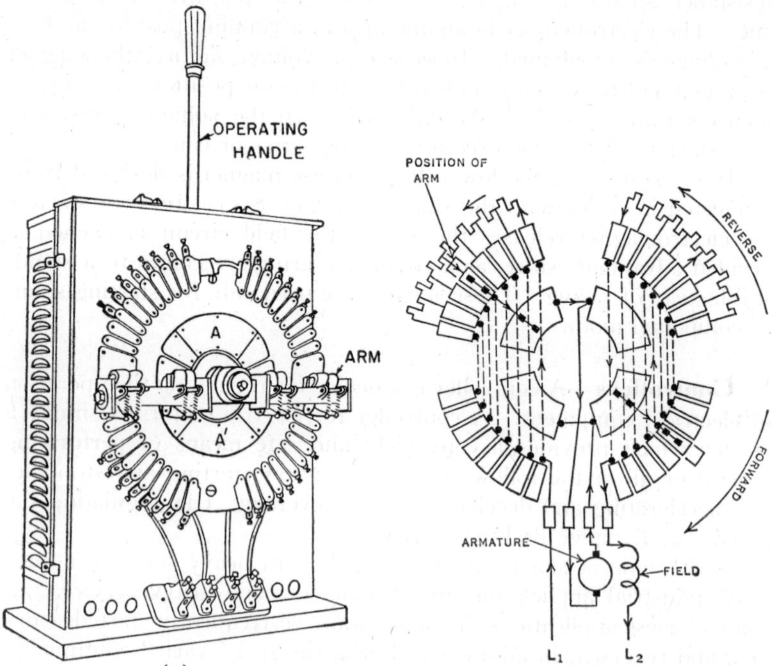

(a) (b)

FIG. 8-3

When the controller is in the "off" position, the brushes set on insulating pieces, and the operating handle is vertical.

When the handle is moved in the "forward" direction, sections of starting resistance are cut out in steps, thus causing the armature to accelerate. When the arm is in the position indicated in fig. 8-3b, the current flows in the direction indicated by the arrows on the wires.

When the arm is moved from the horizontal to the "reverse" direction, the current reverses in the armature, while the current in the series field remains in the same direction. This causes the motor to run in reverse.

Drum Controllers—One common type of drum controller is shown in fig. 8-4a. It consists essentially of a drum cylinder insulated from a central shaft to which an operating handle is keyed. Copper segments are attached to the drum, and are connected to or insulated

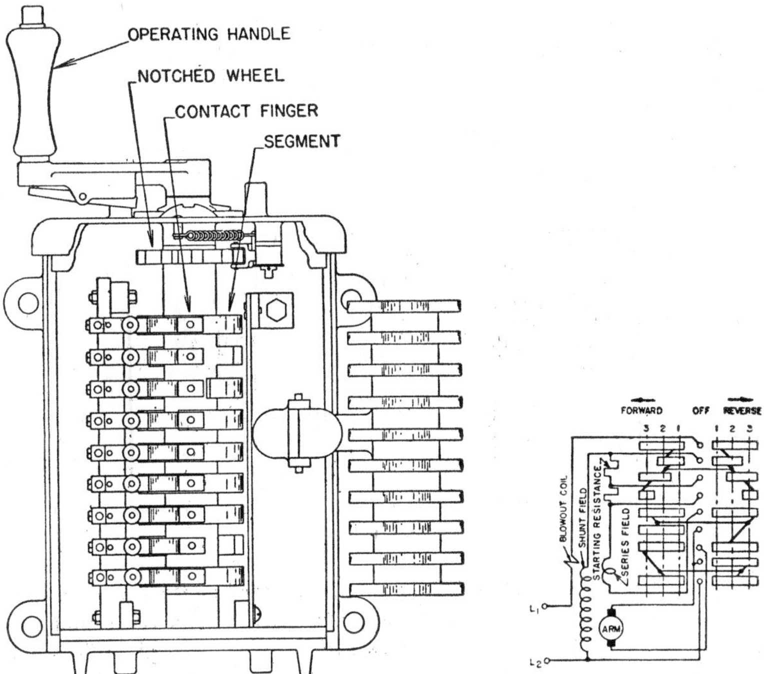

(a) (b)

FIG. 8-4

from one another as the case requires. In fig. 8-4a the segments are connected through the drum, as indicated in fig. 8-4b. A series of stationary fingers is arranged to contact with the segments. These fingers are insulated from one another but connect to the starting resistance and to the motor circuit. Just under the handle a notched wheel is keyed to the shaft. A spring forces a roller into one of the notches when contacts are properly made, thus indicating to the operator the correct position of the handle.

In fig. 8-4b, when the controller handle is moved forward one notch, the fingers are at position 1. The current flows from L_1, through all of the resistance, through the series field and armature, to L_2. This starts the motor. As the handle is turned further, the resistance is cut out of the armature circuit in steps and is inserted in the field circuit. When the handle reaches notch 3, all the resistance will be short-circuited out of the armature circuit. This will provide maximum speed.

The current through the armature, and therefore the motor, is reversed by moving the handle in the opposite direction from the "off" position.

Magnetic Blowout—Since controllers are frequently operated, the fingers and corresponding contacts would soon burn if means were

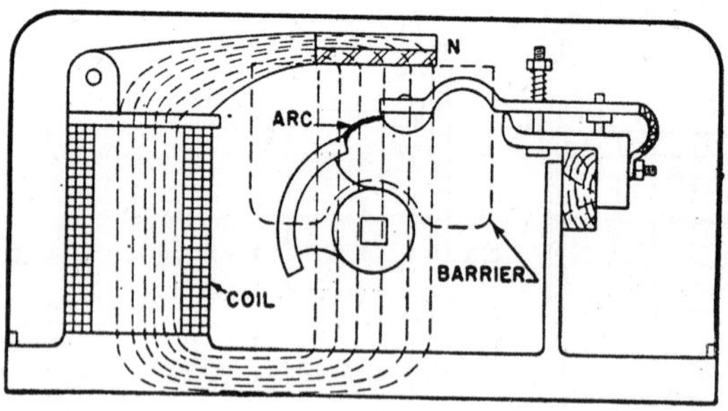

FIG. 8-5

not provided to extinguish the electric arc formed as the contacts open. This extinguishing effect is accomplished by means of a magnetic blowout, a diagram of which is shown in fig. 8-5. The plate N, which is over the fingers and hinged to the core of the blowout coil, extends the full length of the drum. This plate is lined with asbestos on the inner side, and from it, arc barriers project down between the fingers. The current through the blowout coil sets up a strong magnetic flux which passes across the contacts. When the contacts open the current forms an electric arc, which moves toward the reader (the direction is determined by Fleming's left-hand rule). In doing so the arc is drawn out to such an extent that it breaks. The coil of the magnetic blowout is usually in series with the line, so that the extinguishing action is in direct proportion to the size of the electric arc formed.

Cam Type Drum Controller—In a drum controller of the type discussed above, the fingers slide on segments, wearing the contacts. Furthermore, the blowout coil cannot extinguish the arc fast enough to prevent some burning action, because the contacts open slowly. These difficulties are overcome in a drum type controller having cams to actuate contacts, as suggested in fig. 8-6. The cams are insulated from one another by mounting them on a square micanized shaft, which is turned by an operating handle.

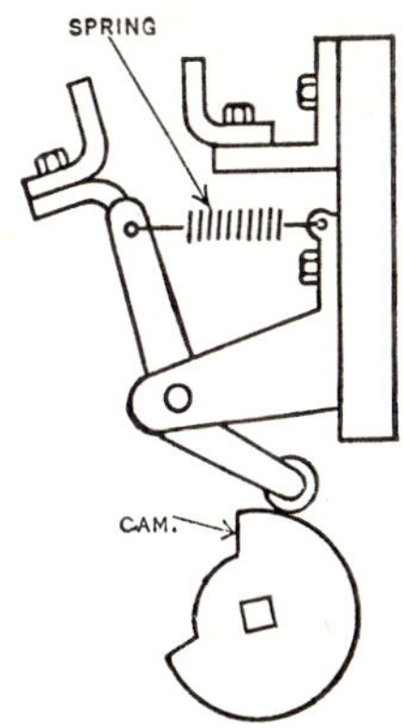

FIG. 8-6

Automatic Control—Automatic controllers are generally used because they have a number of decided advantages over the manual type. The operator may accelerate the motor too rapidly with the manual type controller, with the result that the motor may take excessive current. This causes fuses to blow or circuit breakers to open. Furthermore, the starting resistance may burn out.

With automatic controllers the starting resistance may be cut out at the maximum safe rate by means of magnetically operated switches. The operator need only press a button, and the magnetic switches start the motor and bring it up to speed by cutting out the starting resistance in steps. The motor is stopped merely by pressing the stop button.

Magnetic Switch—A Cutler Hammer magnetic switch is shown in fig. 8-7a. The essential parts of this switch are the magnet, consisting of an operating coil on an iron core, and the moving part or armature. A permanent air gap is usually made a part of the magnetic circuit to prevent the armature from sticking in the closed position after the holding coil has been de-energized. This gap may be of some non-magnetic material such as brass or bronze, as well as air.

Since line switches usually carry a heavy current, a large arc is formed when the main contact opens. This arc is extinguished by a magnetic blowout, which in this case is made as shown in fig. 8-7b. The arc formed after the contacts open is forced upward, drawn out, and broken.

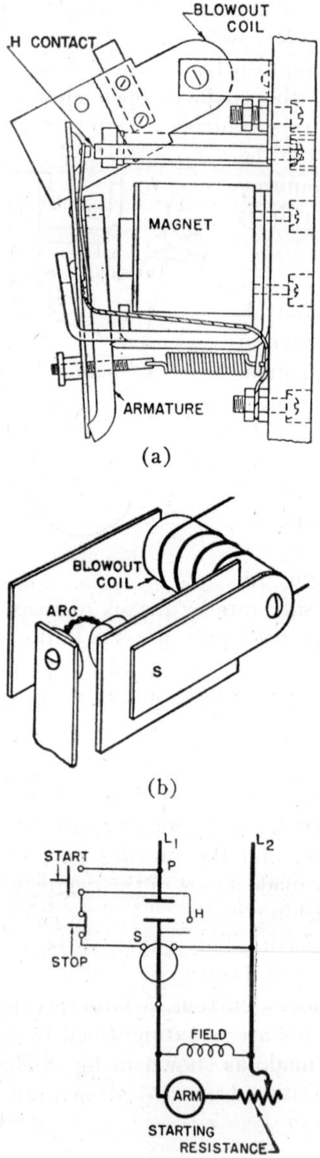

FIG. 8-7

Magnetic switches are provided with one or more auxiliary contacts which are connected in the controlling circuit. These contacts carry small currents and require no blowout magnets.

Magnetic switches are commonly placed in the line circuit as indicated in fig. 8-7c. The method by which the starting resistance is cut out is not shown, as this will be considered in detail later.

When the start button is pressed (fig. 8-7c), a circuit is formed from L_1, through the push button and the coil of the magnetic switch, to L_2. This closes the magnetic switch and power is supplied to the motor. The magnetic switch does not open after the finger is removed from the start button, because a new circuit is provided through which the closing coil is energized. This new circuit is from L_1, through the main contact, H contact, stop button, closing coil, to L_2.

The stop button is pressed to de-energize the closing coil. This allows the magnetic switch to drop out and the power to the motor is interrupted.

Such line magnetic switches are very much in use, for the following reasons:

(1) If a switch carrying a large current is to be operated frequently, less effort will be needed and there will be less distraction for the operator if he only pushes buttons instead of operating a heavy switch. Furthermore, the opening of large current-carrying switches subjects an operator to shocks, burns, or eye injuries.

(2) If the distance from the point of control is great, a considerable saving will be effected by having a magnetic switch near the

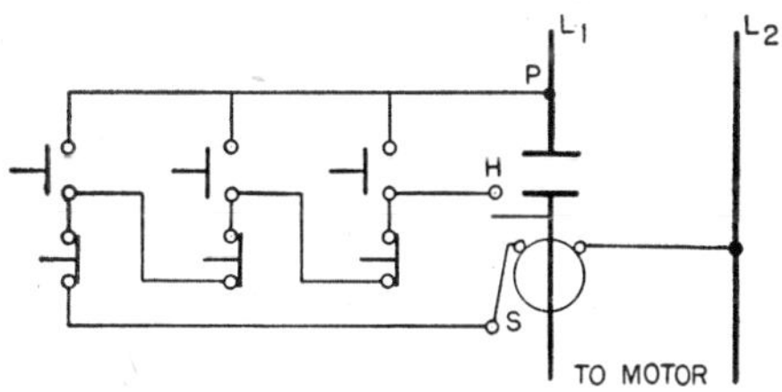

Fig. 8-8

motor to carry the large currents, with only small control wires extended to a push button near the operator.

(3) At times a motor must be controlled from several points, as in driving a conveyor. This may be accomplished easily by using a line magnetic switch (see illustration in fig. 8-8).

(4) A magnetic switch may be automatically actuated by a float switch, pressure or scale switch, limit switch, etc., instead of controlling it by a push button. This eliminates the necessity of an attendant. The force that such devices can exert is not always sufficient to operate switches mechanically.

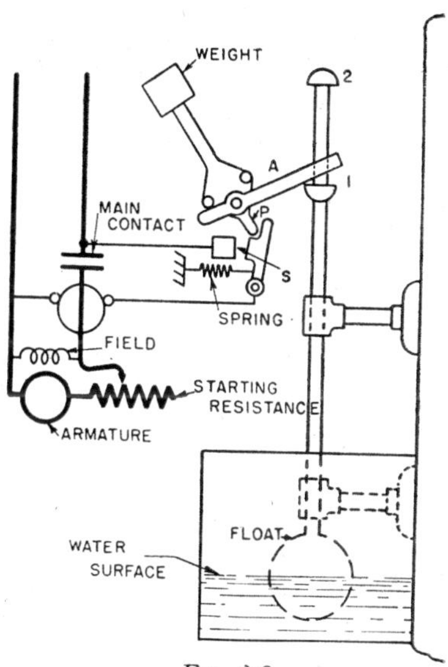

Fig. 8-9

A typical float switch control is shown in fig. 8-9. When the water raises the float above a certain level, knob 1 raises arm A and swings the weight to the left. Projector P opens auxiliary switch S. This de-energizes the holding coil and the magnetic switch opens, interrupting the power to the motor.

Knob 2 forces arm A downward when the water gets below a predetermined level. This swings the weight to the right and the spring closes switch S. This energizes the holding coil, which closes the magnetic switch, and the motor again starts.

The method shown in fig. 8-9 is used in keeping water within a certain level in a water storage tank. The motor starts when the water level is low and stops when it reaches a certain desired high level.

In order to maintain the pressure of motor driven pumps and compressors between certain specific limits, a pressure or scale switch is used in the pilot control. Fig. 8-10 shows a type of pressure switch

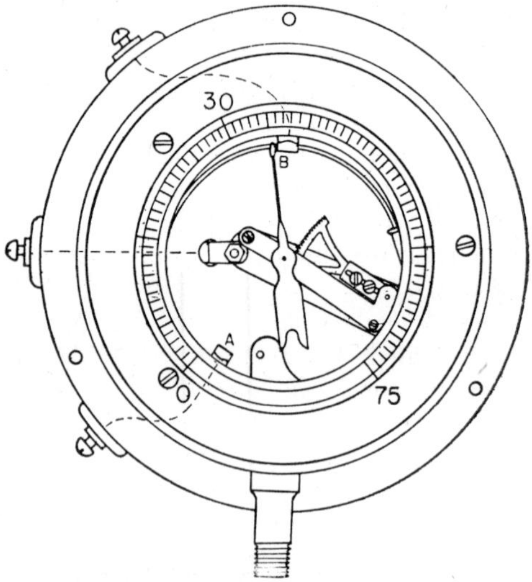

Fig. 8-10

that may be used to control the pressure of any gas or liquid. This switch may be connected so that the closing coil is de-energized when contact A opens, thus stopping the motor, or it may be connected so that the motor starts as soon as contact B is made.

To prevent the over-travel of doors, cranes, elevators, etc., limit switches are connected in the control circuit. Such switches not only stop the motor but provide means of starting it in the reverse direction. The method by which these switches are connected in the control circuit will be considered in discussing automatic starters.

A thermostat or thermocouple control is used for starting and stopping a motor when the temperature of a liquid or gas reaches a certain point. The thermostat control consists of two metals welded together, the two metals having different coefficients of expansion (fig. 8-11a). This bi-metal is fastened at one end so that the free end will bend slightly to one side under the influence of heat and make contact with a fixed post (fig. 8-11b), thus completing the circuit of the closing coil.

The magnetic switch in the motor circuit must be controlled indirectly when a thermocouple is used, as illustrated in fig. 8-12. When the temperature of the thermocouple reaches a certain point, sufficient current is established to close the midget relay. This completes the circuit of the coil of switch M, closing it and starting the motor.

The control of fig. 8-12 is used when it is desired to control a motor by a light, except that the thermocouple is replaced by a

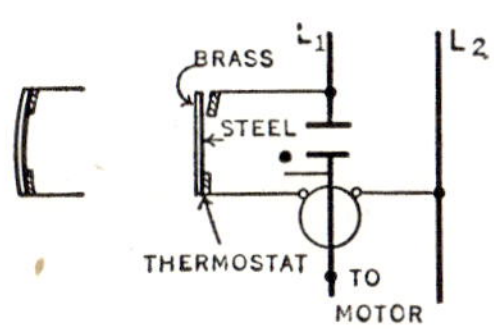

(a) (b)

Fig. 8-11

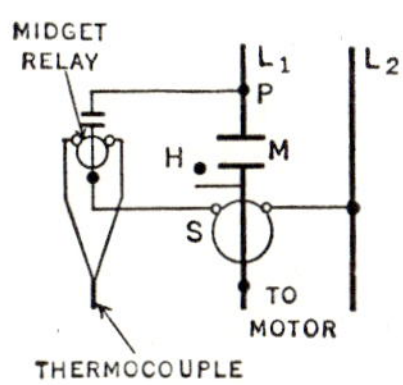

Fig. 8-12

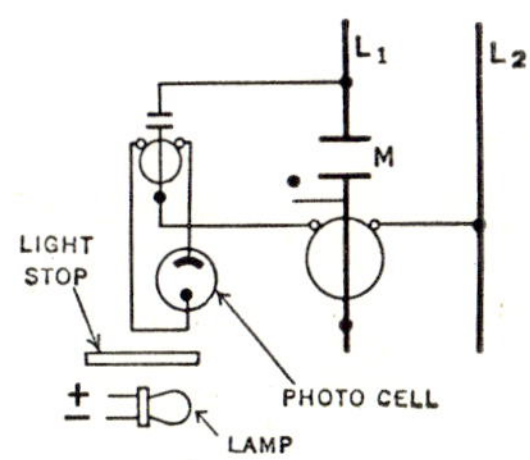

Fig. 8-13

photo cell (fig. 8-13). The controller shown in fig. 8-12 may be used when it is desired to control a motor with light, providing a photocell is used instead of a thermocouple (fig. 8-13). The cathode of the cell consists of a plate coated with a material that emits electrons when light strikes it. The anode, which consists of a metallic film through which light passes, is sprayed on the cathode. When light strikes the cell, electrons pass from the cathode to the anode and through the relay coil. This current causes the relay and switch M to close and start the motor.

(5) With the magnetic switch it is easy to provide low voltage and overload protection.

Low Voltage Protection—If the magnetic switch has a three wire control, as in fig. 8-7c, it automatically opens upon voltage failure, and remains open after the voltage is restored until the start button is again pressed. This is known as low voltage protection. Such protection is essential on all machines that may cause personal injury if they start unexpectedly.

In some cases as with motors driving blowers and pumps, it is desirable that the motor start automatically as soon as the voltage is back to normal. A motor provided with such a control is said to have low voltage release protection. This protection is obtained by using a two wire control, as shown in fig. 8-14.

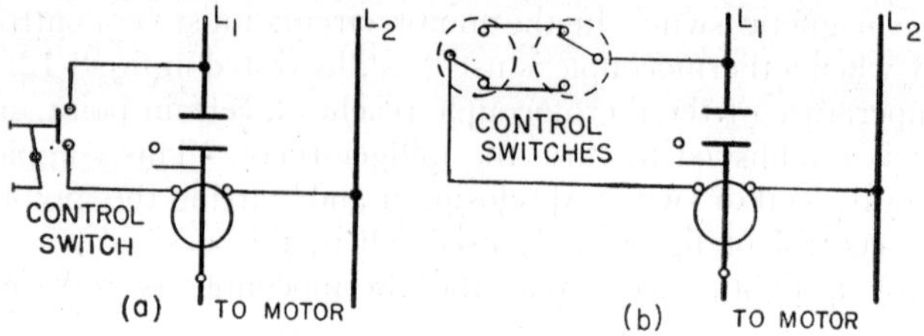

FIG. 8-14

Overload Protection—Overload protection may be obtained by means of fuses and magnetic or thermal overload relays. Fuses have commonly been used because of low cost, but the development of low price overload relays is making popular this type of overload protection, especially on large controllers.

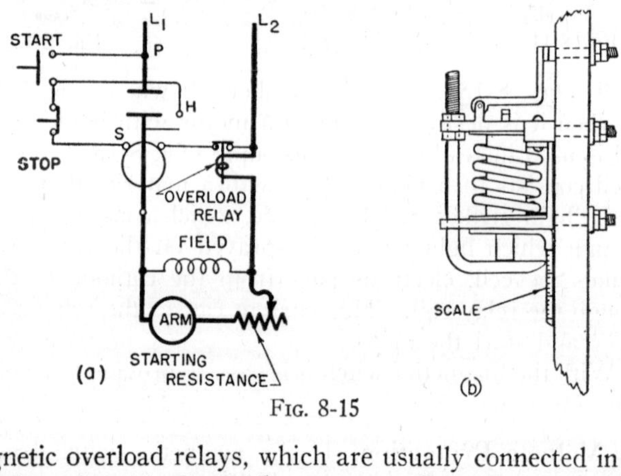

FIG. 8-15

Magnetic overload relays, which are usually connected in the control circuit as shown in fig. 8-15a, are generally made in two types. One type provides an instantaneous trip and the other a time delay trip. The instantaneous trip relay (fig. 8-15b) must be set so that it will not trip because of the starting current, but will trip only when the load current becomes abnormally high. The time delay trip is similar to the instantaneous trip except that a dash pot is added to provide the time delay feature. This dash pot prevents the relay from tripping because of momentary inrushes, but if a sustained overload occurs it will trip after a brief period.

The thermal overload relay is made somewhat like a thermostatic control as it consists of two metals of different coefficients of expansion welded together. This bi-metal is placed near a heating element through which the current to the motor passes (fig. 8-16). One end of the bi-metal is fixed and the other end is free to bend when the temperature changes. When a heavy overload persists long enough to raise the temperature of the motor to a dangerous value, the bi-metal is heated and bends. This opens the circuit at A, which deenergizes the coil of the magnetic switch.

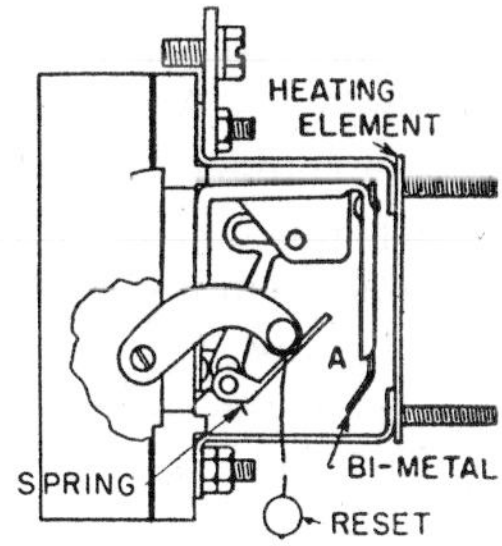

FIG. 8-16

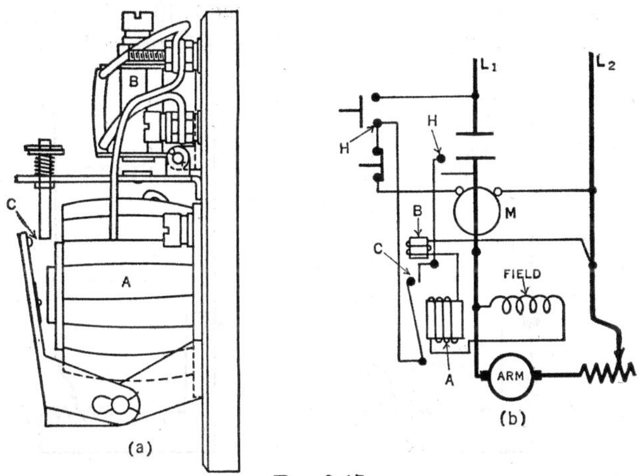

FIG. 8-17

(6) Protection against an open or short circuited shunt field may be provided with the magnetic switch. This protection, which is especially essential on grinding wheels, is effected with the use of a field relay (fig. 8-17a) connected in the circuit as shown in fig. 8-17b. Coils A and B, which are in series, are connected in series with the

shunt field. Auxiliary contact C is inserted between the H contact of the magnetic switch and the H terminal of the pushbutton. When the start button is pressed the magnetic switch is closed and current is supplied to the motor. A small current also passes through the field and coils A and B. The magnetizing force of coil A is sufficient to close contact C, which provides a circuit from L_1, through the H contact and coil M, to L_2. The flux produced by coil M holds the magnetic switch closed. The motor will not race if the field becomes open, because the flux of coil A becomes zero and contact C opens. This opens the circuit of coil M, and the magnetic switch opens.

If the field becomes shorted (or grounded), coil B produces a flux of sufficient strength to open contact C by attracting its armature. This again stops the motor from racing.

Motors should not be permitted to race, even for a short time, because at abnormally high speeds they are likely to damage themselves and the driven machinery by the action of centrifugal force.

Dynamic Braking—It is sometimes desirable to have a control that will cause a motor to come to a standstill quickly instead of coasting to a standstill after the stop button is pressed. This result

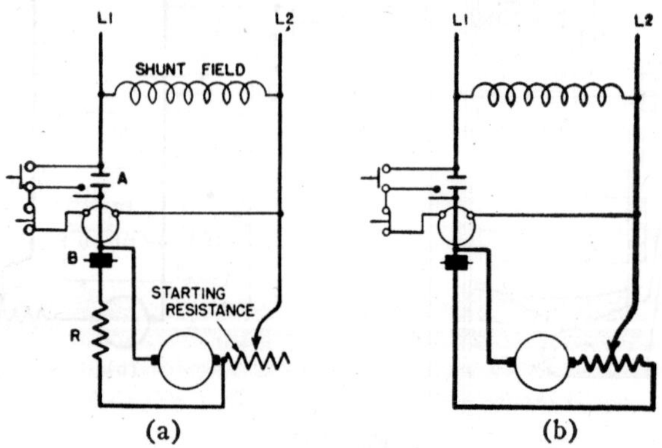

Fig. 8-18

is obtained by means of an action called dynamic braking. A connection for dynamic braking is shown in fig. 8-18a. Note that the shunt field is connected directly across the line so that it will remain energized at all times. While the motor is running normally, contact

A is closed and contact B is open, thus resistance R is not in the circuit. When the stop button is pressed, contact A opens and contact B closes. The armature is disconnected from the electrical source and is connected across resistance R. The armature, while slowing down, rotates in a strong magnetic field and therefore operates as a generator, establishing a high current through itself and through R. The armature current is in a magnetic field, and according to the motor rule, establishes a torque in a direction opposite to that of the direction of rotation of the armature, causing it to come quickly to a standstill.

The starting resistance is frequently used for dynamic braking as shown in fig. 8-18b. Some controllers equipped for dynamic braking are considered later.

It was previously mentioned that several methods may be employed to cut out gradually the starting resistance. These are: (1) counter electromotive force (usually written counter e m f) method, (2) current limit method, (3) time element method.

Counter e m f Method—A simple diagram illustrating the principle of the counter e m f method of cutting out one step of starting resistance is shown in fig. 8-19a. The magnetic switch, which is controlled by a pushbutton in the manner already explained, is represented at 1. A switch which serves to cut out the starting resistance is represented at 2. This switch is usually called an accelerating contactor, and one type is constructed as shown in fig. 8-19b.

The motor accelerates after switch 1 is closed, causing the counter e m f and therefore the terminal voltage of the armature to increase. Since the coil of contactor 2 is connected across the armature, its current will increase with the counter e m f. By properly adjusting the air gap of contactor 2, it will close at a fixed voltage, and thus short out resistance R which is in series with the armature. For a 240 volt machine, the contactor closes when the voltage at the brushes reaches about 200 volts.

In the simple arrangement of fig. 8-19a, contactor 2 will not open immediately when switch 1 is opened, but will be held closed by the counter e m f induced in the armature. The counter e m f must drop to about 20 per cent of the full load voltage in order to release the contactor. If the magnetic switch is opened and then closed again before the armature speed has reduced to 20 per cent of its full speed value, contactor 2 will remain closed. This will cause the armature to take a high current, and perhaps also severely jar the driven machinery. In order to eliminate this possibility, another contact is usually provided on the magnetic switch, which opens the circuit to the coil of contactor 2 as soon as switch 1 is opened. This is illus-

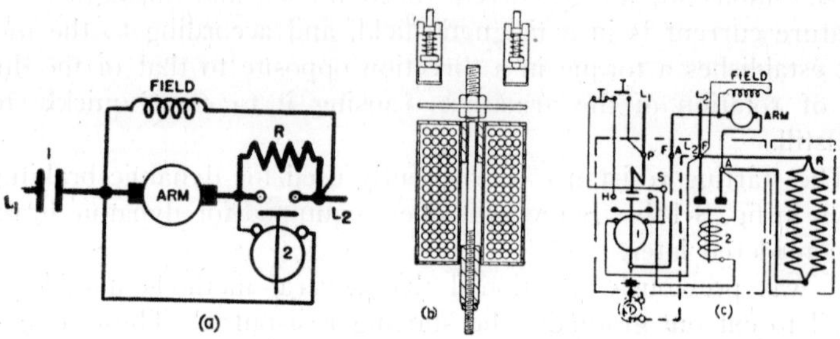

FIG. 8-19

trated in fig. 8-19c. The scheme shown in fig. 8-19a, however, is used fundamentally in all counter e m f starters.

If it is desired to equip the above starter for dynamic braking, the equipment indicated by dotted lines must be added.

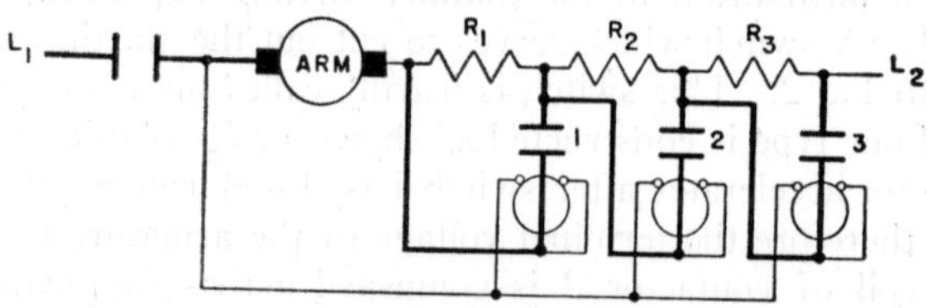

FIG. 8-20

It is possible to arrange several contactors to close at different armature voltages (fig. 8-20) so that the starting resistance may be cut out in several steps.

The scheme used in the counter e m f starter manufactured by the Cutler-Hammer Company is shown in fig. 8-21a. This starter differs from the one discussed previously in that it has only one auxiliary contact on the magnetic switch, and the accelerating contactor has a compound coil. When the start button is pressed (see fig. 8-21b), coil A is connected directly across the line, closing contactor 1. After the finger is removed from the start button, section B of the compound coil is connected in series with coil A. This reduces the power required to hold switch 1 closed, and the magnetizing action of section B aids section C, which closes contact 2 when the armature voltage reaches a selected value.

Coil A is deenergized when the stop button is pressed, allowing switch 1 to open, thus interrupting the current to the motor.

Counter e m f starters are objectionable where considerable variation in the line voltage exists. When the line voltage is above normal, the accelerating contactor closes too soon, and when far below normal the accelerating contactor at times does not close. This type of starter

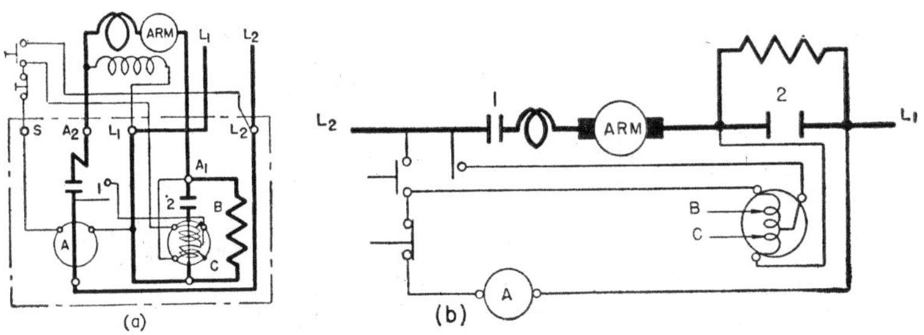

FIG. 8-21

has the advantage that it is simple in structure and requires few auxiliary contacts.

Current Limit Method—The two principal methods of obtaining current limit acceleration are by (1) series lockout contactors or (2) series relays.

(1) *Series Lockout Contactors*—The series lockout contactor is designed so that it will hold open when the current through its coil exceeds a set value, but will close promptly as soon as the current falls below this value. There are two general types of such contactors, one having a single coil and the other having two coils.

Single Coil Lockout Contactors—The single coil contactor depends for its operation upon the saturation of a portion of the magnetic circuit. A contactor of this type is shown in fig. 8-22a. Its operating characteristic is secured by providing two magnetic circuits, one through the yoke A and one through the tail piece B (fig. 8-22b). The flux through A tends to close the contactor and the flux through B tends to hold it open. The cross section of A is very small. Hence, when the coil current exceeds the set value, A becomes saturated and a large amount of flux passes through B, holding the contacts open. When the current falls below the set value, the flux passes almost entirely through A, because section A is no longer saturated and the air gap at the end of the tail piece provides considerable reluctance in the B circuit. Thus the contactor closes. In closing, the lockout air gap is increased and the air gap between the armature and coil is decreased. This causes the contacts to close rapidly, and then hold together with considerable force.

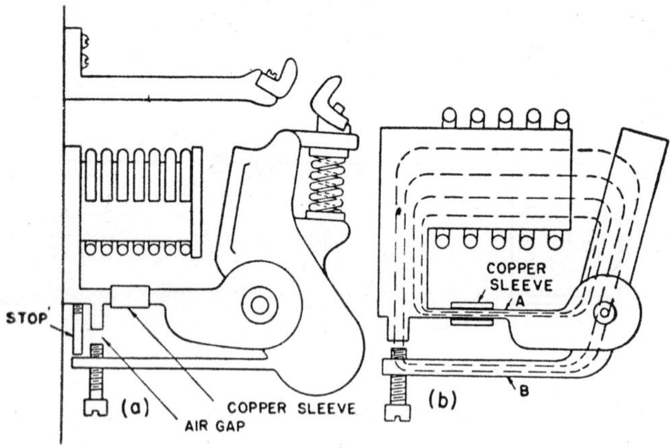

FIG. 8-22

The contactor may be adjusted to close at any suitable current value by changing the length of the air gap between the end of the tail piece and the yoke.

It may appear that the contactor will close while the coil current is increasing, because A must become saturated before the flux begins to build up in B. This difficulty is overcome by the copper sleeve around A. With a rising coil current a voltage is induced in the sleeve, causing a flow of current in the sleeve. This current sets up a flux which prevents the main flux from passing through A, thus causing it to first build up in B.

A series lockout contactor, which operates fundamentally on the same principle as that of fig. 8-22a, is shown in fig. 8-23. With a high coil current, considerable flux passes through B, holding the contactor open. With a current below the set value, little flux passes through B, thus permitting the contactor to close. The yoke of fig. 8-23, as that of fig. 8-22, must be surrounded by a copper sleeve.

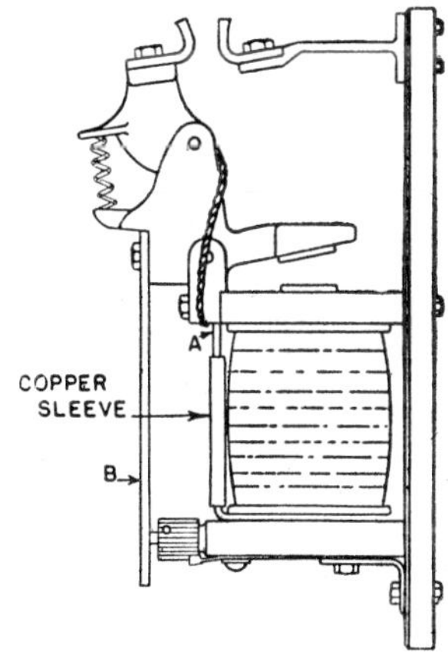

FIG. 8-23

A simple starter with series lockout contactors is shown in fig. 8-24. At the instant that the magnetic switch is closed by pressing the start button, the shunt field is connected directly across the line, and all of the resistance is connected in series with the armature. Contactor 1 closes as soon as the counter e m f reduces the armature current to the value at which contactor 1 is set to function. This cuts out R_1 and inserts the coil of contactor 2 in the armature circuit. As the speed of the motor increases further, the current decreases sufficiently to allow contactor 2 to close, thus connecting the armature directly across the line.

Fig. 8-25 shows a series lockout contactor different in form from those previously discussed. It is of the plunger type. The coil is of heavy wire, because it must carry the armature current, and surrounds a brass tube, within which the plunger P moves freely. The stem of plunger P extends into the soft iron tube S, which may be screwed in or out to vary the length of air gap G. A circular brass plate is attached to the top of the plunger by means of a non-magnetic pin. This contacts with two brushes when the plunger is forced up. A short iron tube, which extends nearly to the head of the plunger, surrounds the non-magnetic pin and thus establishes another air gap A. The coil is enclosed in an iron case C, providing an external path of low reluctance for the returning flux.

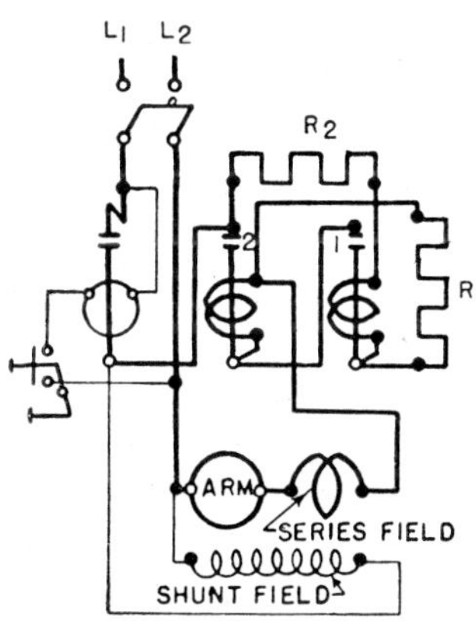

FIG. 8-24

The operation of this contactor will be considered in connection with the simplified diagram shown in fig. 8-26. When a current flows through the coil, the magnetic lines pass through air gap A and tend to raise the plunger. If the current is high, two magnetic paths are formed at the lower end of the plunger. One path of the flux enters the stem at right angles to the direction in which the plunger moves, thus making this flux ineffective in producing a downward pull. Because the stem becomes saturated, another path of flux is set up through air gap G (fig. 8-26a). The flux through this path and the weight of the moving parts prevent the plunger from moving upwards.

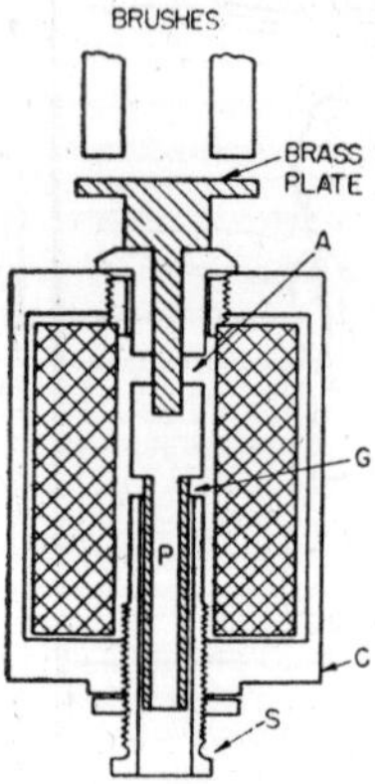

FIG. 8-25

When the coil current drops below a set value, the stem will no longer be saturated because less flux will be established. Thus nearly all the flux will enter the stem and little will pass through air gap G (fig. 8-26b). As a result, the upward pull due to the flux through air gap A will exceed the downward pull, which in this case is only the weight of the moving parts, and the plunger will move upward and close the contacts (fig. 8-26c).

In fig. 8-25, air gap G is lengthened if plug S is screwed out, allowing the contactor to close at a larger value of current. Air gap G

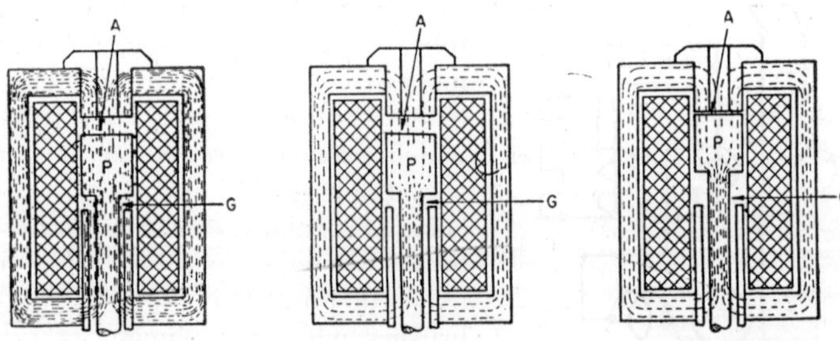

FIG. 8-26

is shortened if S is screwed in, allowing the switch to close at a smaller value of current.

The wiring diagram of a starter with E C & M type series contactors is shown in fig. 8-27. After the magnetic switch closes and the start button is released, R is inserted in series with the closing coil.

This resistance serves to reduce the power supplied to the closing coil, as little power is required to hold the magnetic switch closed.

After the magnetic switch is closed, the shunt field is connected directly across the line, and current flows from L_1, through the series field and armature, through the entire starting resistance and the coil of contactor 1, to L_2. Although current flows through the coil of contactor 1, it will not close until the counter e m f has reduced this current to the value for which it is set. When contactor 1 does close, R_1 is short-circuited and the current flows through the series field and armature, through R_3, R_2, and the coils of contactors 2 and 1, to L_2. The increase in current due to the closing of contactor 1 causes con-

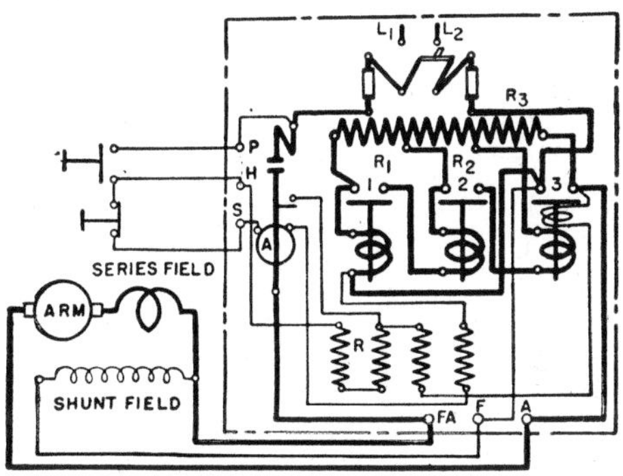

FIG. 8-27

tactor 2 to be locked out. Contactor 2 closes as soon as the counter e m f has reduced the current to a value for which it is set, shorting out R_2 and inserting the coil of contactor 3 in the circuit. Contactor 3 closes as soon as the current drops to a value for which it is set, shorting out R_3 and connecting the series field and armature directly across the line. Contactor 3, on closing, short circuits its closing coil and those of the other contactors. Contactors 1 and 2 drop open, but 3 is held closed by the small shunt coil near the top of the contactor. This shunt coil is just strong enough to hold 3 closed after the series coil has closed it.

The motor is stopped merely by pressing the stop button.

Double Coil Lockout Contactor—Another type of current limit contactor is shown in fig. 8-28a. It has two coils which are connected in series with the armature, the upper coil closing the contactor and the lower one locking it open. The iron surrounding the closing coil is worked near saturation, so that the closing force is strong at a low current and increases slightly as the current increases. The iron surrounding the lower coil is not saturated, so that the pull exerted by this coil increases rapidly with an increasing current. The pull of each coil varies with the current in the manner indicated by the curves in fig. 8-28b. When the current exceeds the value at A, the lockout pull is stronger and the contactor remains open. As the counter e m f

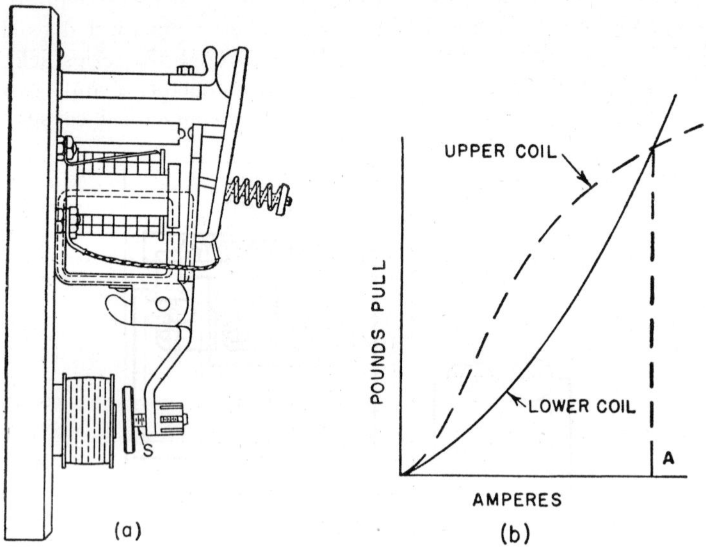

FIG. 8-28

reduces the current, the lockout force decreases rapidly until the closing force predominates when the current decreases below A, closing the contactor. The contactor operates at a current which is set by adjusting screw S.

The diagram of a starter that employs this type of contactor is shown in fig. 8-29. At the instant that the magnetic switch is closed, a high current flows from L_1, through coils A and B, through all the starting resistance, through the armature and series field, to L_2. At the same time the shunt field is connected across the line. Contactor 1 closes as soon as the current reduces to a value for which this contactor is set, shorting out B and R_1, and providing a current circuit through the closing coil A and through coils C and D of contactor 2. Contactor 2 closes as soon as the current again reduces sufficiently, shorting out coil D and providing a circuit through coils C, E, and F.

All the starting resistance is cut out as soon as contactor 3 closes, connecting the motor across the line.

Another type of double coil, current limit type contactor is illustrated in fig. 8-30. It has a shunt coil for closing the contactor and a series coil under the tail piece for locking it open. The shunt coil exerts a constant pull because it is connected across the line. The series coil, being in series with the armature, produces a pull that varies with the armature current. When the motor starts, the high current through the series coil causes a lockout force to exceed by far the closing force. The lockout force decreases as the current reduces, until finally the closing force predominates and closes the contactor. This contactor is set to close at a given current by adjusting the set screw located at the center of the series coil.

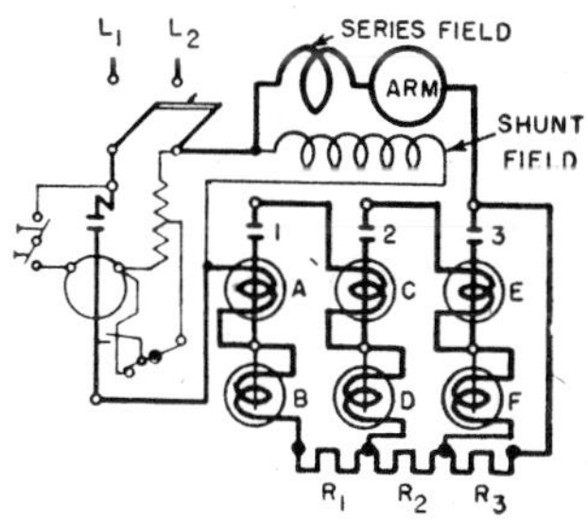

FIG. 8-29

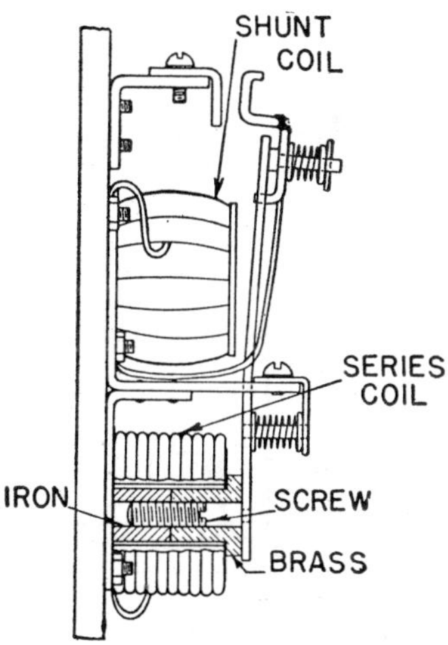

FIG. 8-30

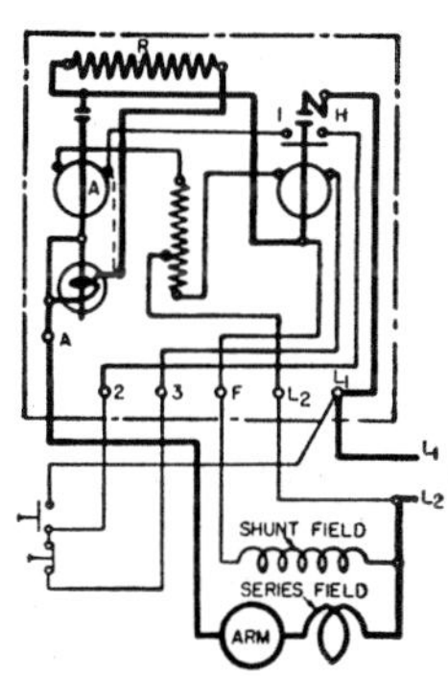

FIG. 8-31

The wiring diagram of a starter, manufactured by the Monitor Controller Company, which employs this type of contactor is shown in fig. 8-31. In this starter note that the magnetic switch has two auxiliary contacts, serving the same purpose as those on the magnetic switch of the counter e m f starter shown in fig. 8-19c. Contact H maintains the circuit of the holding coil after the push button is released. Contact 1 disconnects coil A from across the armature so that the accelerating contactor will open immediately after the magnetic switch opens. This eliminates the possibility of disconnecting the motor, then throwing it across the line before the armature stops rotating. This is explained more in detail in the discussion on counter e m f starters. When the magnetic switch is closed, the high current flows from L_1, through the magnetic switch, resistance R, and series coil, through the armature and series field, to L_2. The contactor closes as soon as the current decreases to the value for which the contactor is set. This shorts out resistance R, connecting the armature with the series field directly across the line.

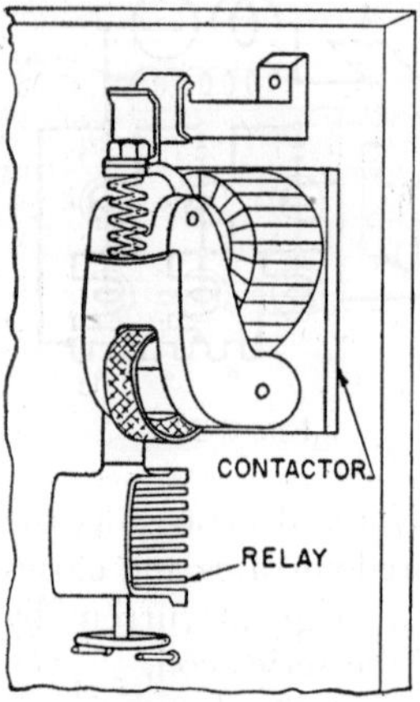

FIG. 8-32

The main disadvantage of this starter is that the accelerating contactor connects the motor directly across the line if the starting resistance is open. To overcome this difficulty, the shunt coil may be connected across the brushes as shown by the dotted lines with contact 1 removed.

(2) *Series Relays*—A starter employing series relays is generally more complicated than the other current limit types, because, in addition to the contactors, a relay must be provided for each step of resistance, and electrical or mechanical interlocks are required.

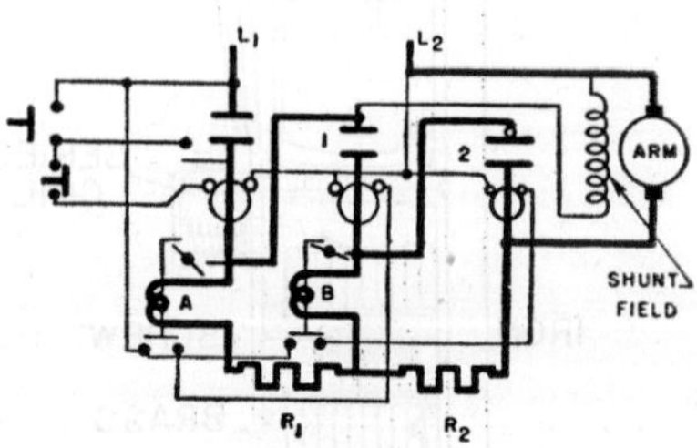

FIG. 8-33

Fig. 8-32 shows a contactor and relay. The relay is normally closed by force due to gravity. However, the magnetic force produced by the coil holds it open until the current has decreased to the set value. The relay is sometimes mechanically interlocked with the contactor so that it is held open until the contactor closes.

A starter employing series relays is shown in fig. 8-33. At the instant that the magnetic switch is closed, a high current flows from L_1, through the magnetic switch and relay A, through all the resistance and armature, to L_2. At the same time, the magnetic switch releases the plunger of relay A so that its contacts may close. However, the high current through this relay holds its contacts open until the current drops to a value for which it is set to close. The contacts of relay A close as soon as this current value is reached, causing contactor 1 to close because the circuit of its closing coil has been completed. This shorts out relay A and R_1, providing a current path through contactor 1 and relay B. Relay B closes as soon as the current reduces to a value for which it is set to operate, completing the circuit of the coil of contactor 2. Contactor 2 closes, shorting out relay B and R_2, connecting the motor across the line.

General Characteristics of Current Limit Starters—Current limit starters are popular because of their simplicity. These starters require practically no interlocks (with the exception of those with series relays), and they are not likely to cause trouble. They are easily adjusted, even for extreme conditions. However, current limit starters may be undesirable for loads that are extremely heavy at starting, because their contactors may fail to short out all of the starting resistance. Furthermore, some of these starters are not suitable for handling very light loads, because the contactors will not remain closed.

Dynamic Braking and Reversing—Some industrial operations require the use of controllers that provide dynamic braking and reversing. The kind of operating switch required by one type of such a controller is shown in fig. 8-34a. The sequence in which the contacts are made is shown in fig. 8-34b. The line wires are connected to the + and − terminals. When the handle is in the "reverse" position, H and A_1 are positive while F and F_1 are negative (see fig. 8-34b). On moving the handle to the "forward" position, F, F_1 and H, A_1 exchange polarities. If an armature is connected between A_1 and F_1 (through a starting resistance), as indicated in fig. 8-34c, the current through it, and therefore the direction of rotation, will reverse as the handle is shifted from "reverse" to "forward." Hence, this switch is fundamentally a reversing switch, and may be connected in conjunction with any starter.

If the handle is brought from a "running" to a "drift" position, the armature is disconnected from the line, and the motor is brought to a standstill by the load.

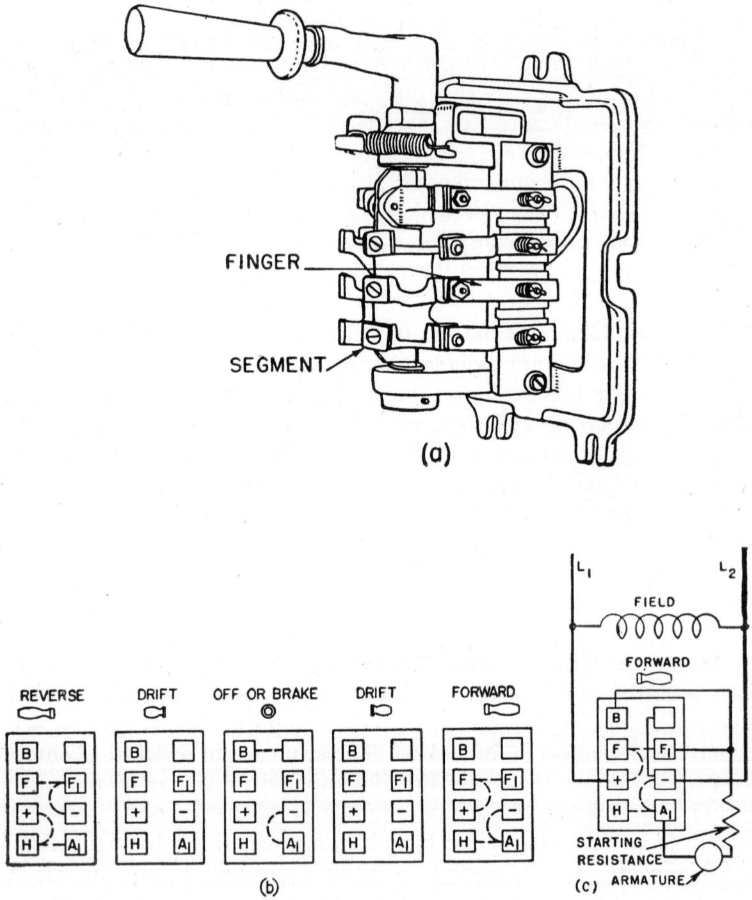

FIG. 8-34

If the handle is brought from the "running" to the "off" or "brake" position, contacts are made which connect terminals A_1 and B (fig. 8-34b), placing the starting resistance directly across the armature terminals. The rotation of the armature under the influence of the excited shunt field causes a voltage to be generated, sending current through the starting resistance. As a result, the machine functions as a loaded generator and develops a retarding torque which rapidly slows down the motor, bringing it to a standstill.

Fig. 8-35 shows a reversing controller manufactured by the E C & M Company. In this controller the starting resistance is automatically cut out of the armature circuit as explained in the discussion relating to fig. 8-27.

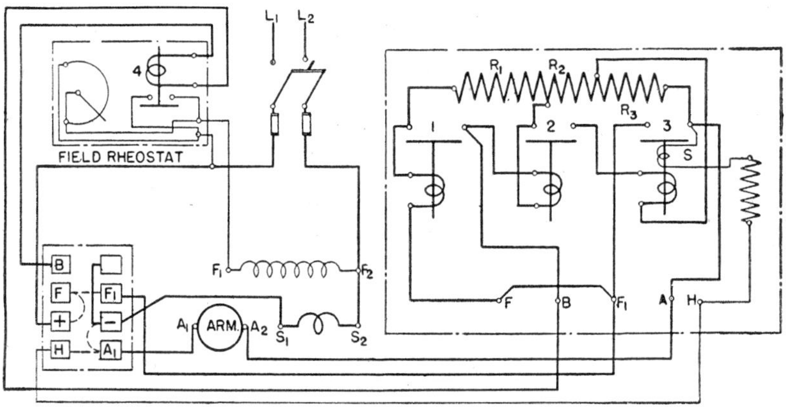

FIG. 8-35

The shunt field is connected across the line, through the field rheostat, when the knife switch is closed. With the control switch in the "forward" position, the main circuit is from L_1 to the + terminal and F_1 on the operator switch, through all the starting resistance and armature, through A_1 and the − terminal on the control switch, through the series field to L_2. Contactors 1, 2, and 3 close in sequence as the armature current decreases, connecting the motor across the line. The other contactors open as soon as contactor 3 closes, because their coils are short-circuited. Contactor 3 is held closed by the shunt coil S.

Dynamic braking is effected by bringing the handle of the control switch from the "forward" to the "off" position. Segments in the control switch connect terminals A_1 and B together, placing sections R_2 and R_3 of the starting resistance across the brushes. This circuit provides dynamic braking action, causing the motor to rapidly slow down. As it slows down, the generated voltage decreases and the current through the resistance therefore decreases, tending to reduce the braking torque. The contactors, on closing, increase the braking action as the motor slows down. When the current drops to the value for which contactor 2 is set, this contactor closes and shorts out R_2, leaving R_3 in the circuit. This increases the current and the braking torque. Contactor 3 closes when the current drops to the value for which it is adjusted, but this does not short out additional resistance. Relay 4 short-circuits the field rheostat, assuring a full field during the dynamic braking period.

Dynamic braking likewise occurs when the handle is brought from the "reverse" to the "off" position.

Other Hookups of the Operator Switch—In the discussion relating to fig. 8-34, it was suggested that an operator switch may,

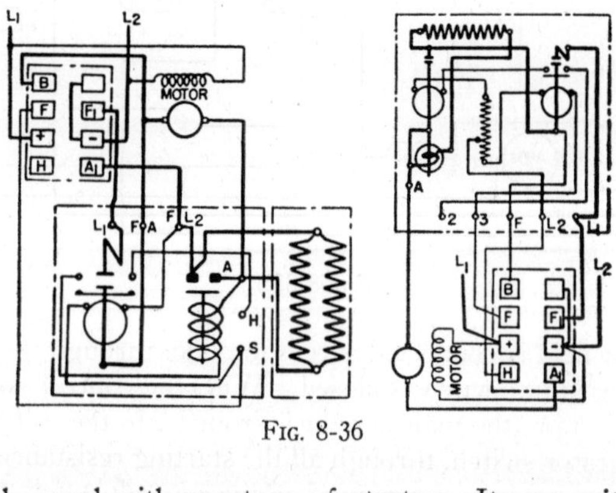

FIG. 8-36

in general, be used with any type of starter. Its use with two types of starters previously discussed is illustrated in fig. 8-36.

Time Element Method—If a motor, equipped with a type of starter previously discussed, is so heavily loaded that it does not start when all the starting resistance is in the armature circuit, the accelerating contractors will not short-circuit the starting resistance. In such an instance the starting resistance is likely to burn out unless the overload protection opens the load circuit. If a motor is equipped with a time element controller, the starting resistance is cut out in a definite time, regardless of the load, as long as this load does not exceed the value for which the overload is set.

There are three important methods of obtaining definite time limit acceleration. These are: (1) By dash pot, or by a pendulum arrangement similar to the escapement of a clock; (2) By a pilot motor; (3) By inductive time limit.

Dash Pot Timing—Fig. 8-37a shows a starter with an accelerating mechanism of the air dash pot type (an oil dash pot could have been used). In this starter the time of acceleration may be varied over a wide range by adjusting the air inlet valve. This valve is usually adjusted so that the motor will accelerate properly under the most severe load conditions. The time of acceleration for light loads is therefore longer than is necessary.

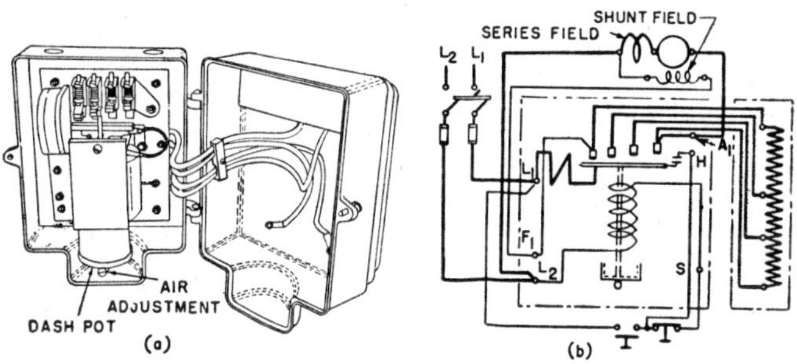

FIG. 8-37

The wiring diagram of this starter is shown in fig. 8-37b. The solenoid causes the bar mounted on the plunger to engage successively with four fingers. A step of resistance is cut out as each finger is engaged. The dash pot prevents the solenoid from driving the plunger up too rapidly, as this would cause the motor to be accelerated too fast.

The starter functions in the following way: After the main switch is closed and the start button is pressed, a circuit is completed from L_1, through the pushbutton and solenoid, to L_2. The current through the solenoid causes the plunger to move slowly upward, the motion being retarded by the dash pot. First, the horizontal bar closes contact H, causing the solenoid to remain energized after the start button is released. Contact H and all the fingers are under spring tension: thus the horizontal bar continues to move upward, cutting out the remaining two steps of starting resistance and connecting the motor to the line.

The coil circuit is opened when the stop button is pressed, and the horizontal bar drops instantly from the fingers.

The main disadvantage of this starter is that the dash pot sometimes gives trouble. If an air dash pot is used, the air valve is likely to become clogged with dust and thus change the time of acceleration. Furthermore, the leather dries if the air dash pot is not oiled occasionally, with the result that little or no vacuum is produced when the plunger moves upward. The timing of oil dash pots varies considerably with changes in temperature, and no oils are available which entirely eliminate this undesirable feature.

Time Element Starter with Pendulum Timing—A starter provided with a pendulum type timing mechanism is shown in fig. 8-38a. Its principle of operation is much the same as that of the dash pot starter (fig. 8-37). It has a multicontact magnetic switch consisting of one line contact and three accelerating contacts, all operated by a single solenoid. When the solenoid is energized, the line contact closes at once but the accelerating contacts close in sequence. each after a definite time interval. The timing mechanism, which governs the rate at which the accelerating contacts close, consists of a pendulum arrangement with an escapement similar to that used in a clock (see fig. 8-38b). This mechanism is connected to the accelerating fingers through a ratchet and pawl. When the start button is pressed (fig. 8-38c), the solenoid closes the line contact and also tries to close the accelerating fingers, which are prevented from closing by the timing mechanism. However, the escapement permits the gears to rotate at a rate which may be regulated by the adjusting nuts (fig. 8-38b). When this nut is screwed out, the escapement compels the gears to rotate more slowly, thus increasing the accelerating period. Hence the pawl is permitted to move slowly, allowing the accelerating contacts to close in sequence, each after a definite time interval. The last contact, on closing, connects the motor directly across the line.

Overload protection is provided by means of a hand-reset temperature overload relay (this type of relay was shown in fig. 8-16). The bimetal bends when the current through the heater, which is in series with the armature, exceeds a given value, opening the circuit to the solenoid. This allows the contacts to drop open instantly, interrupting the power to the motor. After the bimetal has cooled, the overload relay may be reset by means of a cord.

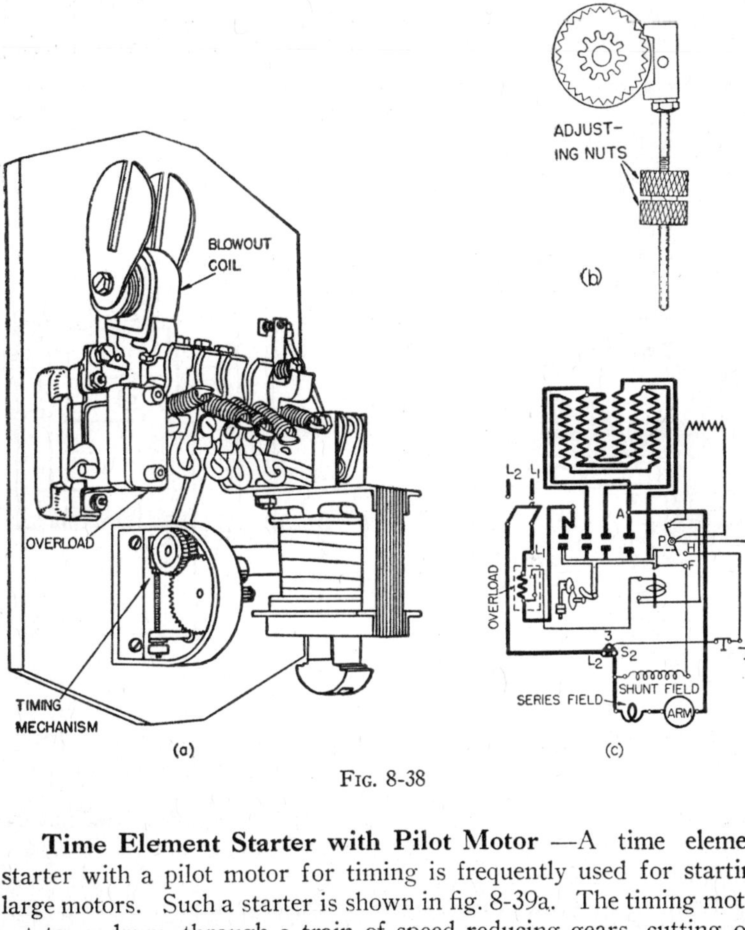

FIG. 8-38

Time Element Starter with Pilot Motor —A time element starter with a pilot motor for timing is frequently used for starting large motors. Such a starter is shown in fig. 8-39a. The timing motor rotates a drum, through a train of speed reducing gears, cutting out the starting resistance in steps. With this method of cutting out the starting resistance, it is possible to get a very long timing which is accurate and independent of disturbing conditions. Furthermore, the acceleration is smooth under all load conditions, and can be varied by changing the speed of the timing motor.

The wiring diagram of the starter, with a developed view of the drum, is shown in fig. 8-39b. The ring A keeps all the segments connected to line L_1. When the magnetic switch M is closed, by pressing the "start" button, power is supplied to the motor, with all the starting resistance in the armature circuit, the shunt field being

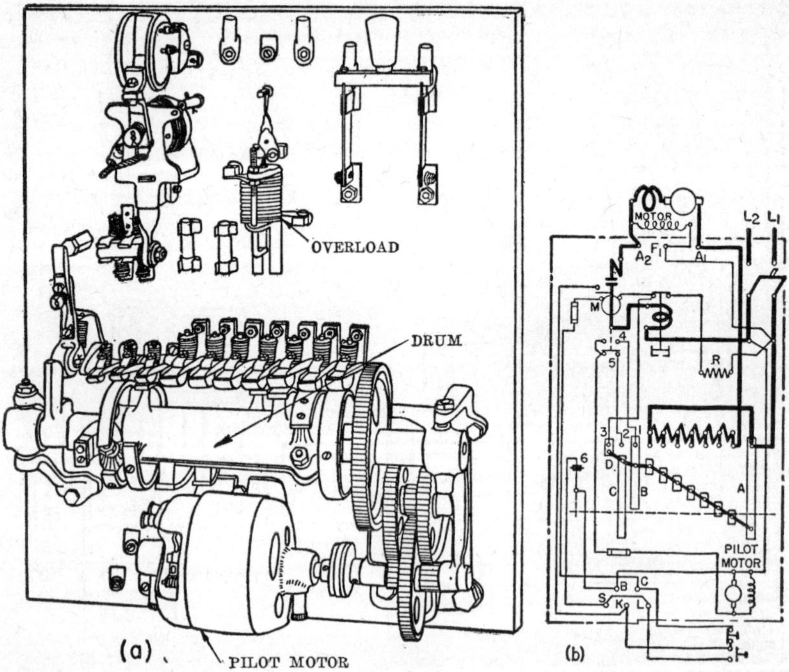

FIG. 8-39

connected directly across the line. At the same time, a circuit is formed from L_1, through ring A, segment B, finger 1, auxiliary contact 4, and pilot motor, to L_2. Thus power is supplied to the pilot motor, which turns the drum. The drum, on rotating, cuts out the starting resistance in steps, and forms a contact between finger 2 and segment C. As soon as all the starting resistance is cut out, segment B runs from beneath finger 1, stopping the pilot motor. The main motor thus remains connected directly across the line.

When the "stop" button is pressed, or when the overload trips the magnetic switch M, auxiliary finger 5 on the magnetic switch M completes the pilot motor circuit through finger 2 and segment C, causing the timing motor to rotate the drum back to the starting position. At this point segment C runs from beneath finger 2 and the pilot motor again stops.

After switch M is closed and the drum starts to rotate, the contact between finger 3 and segment D opens, inserting R in series with the closing coil. This reduces the holding power.

Contacts 6, which are operated by a cam on the drum, and are held closed when the drum is in the "off" position, are in series with the closing coil. This feature makes it impossible to start the motor unless all the starting resistance is in the armature circuit.

Inductive Time Element Starter—Fig. 8-40 shows a controller of the inductive time limit type. The accelerating contactor (see fig.

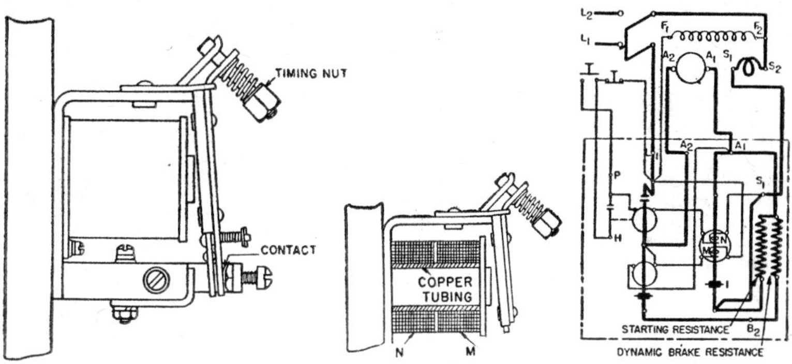

(a) (b) (c)

FIG. 8-40

8-40a) provides a definite time delay, which is adjustable from approximately one-half to four seconds. A heavy copper tubing is on the magnetic core of the contactor, as well as two coils which are known as a neutralizing coil and a main coil. The neutralizing coil N (figs. 8-40b and 8-40c) is connected in series with the coil of the magnetic switch, and the main coil M is connected across the contacts of the magnetic switch.

When the knife switch is closed, a current flows from L_1 (fig. 8-40c), through coil M, armature, starting resistance, and series field, to L_2. The magnetizing action of the current through coil M holds the accelerating contactor open against an adjustable spring pressure.

When the magnetic switch is closed, by pressing the "start" button, coil M is short-circuited and its current reduces to zero. Because the magnetic path of the contactor is a closed circuit of hard steel, the flux set up by coil M would not decrease very much if it were not for coil N. This coil produces a magnetizing action in a direction that will reduce the residual magnetism to zero. The current in coil N (and therefore the neutralizing action) increases gradually, because of the copper sleeve surrounding the core (Fig. 8-40b). This sleeve also prevents the flux from decreasing rapidly. Thus the flux reduces from maximum to zero at a comparatively slow rate. It follows, then, that contactor 1 will close as soon as the residual magnetism decreases sufficiently, connecting the motor across the line. The time delay of the contactor can be varied by changing the spring pressure with the adjusting nut.

Limit Switches—An operator switch is frequently used in connection with equipment in which a definite travel should occur. In

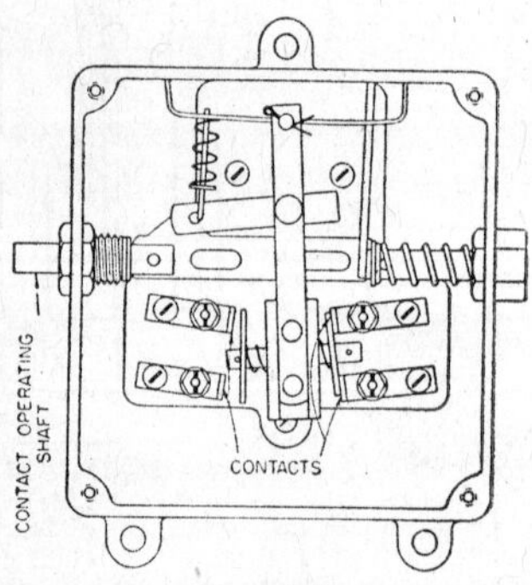

Fig. 8-41

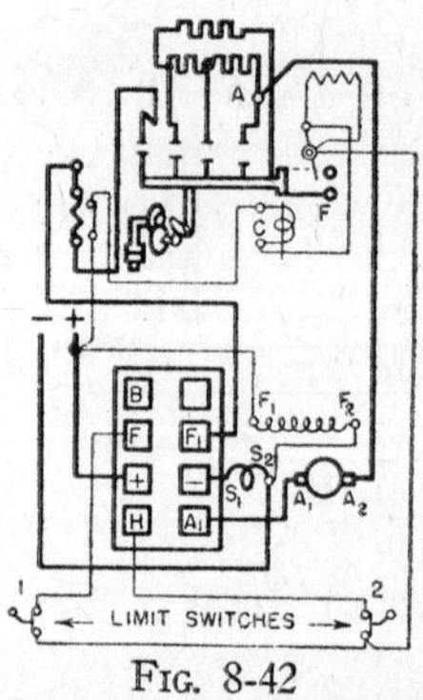

Fig. 8-42

such a case an operator cannot be expected to stop the driven equipment just within the safe limits. In fact, in most cases it is desirable to stop the motor automatically just before the driven equipment exceeds the limit of travel. This is accomplished by using a limit switch; one type, manufactured by Mackworth G Rees, Inc., being shown in fig. 8-41.

Fig. 8-42 shows a control system using a General Electric controller (fig. 8-38), an operator switch (fig. 8-34), and two limit switches placed at the limits of travel. The terminal of coil C, which normally connects to the heater of the overload, is transferred to the + line, and the other terminal is connected to the limit switches. When the handle is in the "reverse" position (see fig. 8-34b), the circuit of coil C is completed to the negative through limit switch 1. With this position of the operator switch, a short-circuit exists between F and H, having opposite polarity. This is remedied by removing the segment that makes contact with finger H in this position of the switch. Coil C, on becoming energized, closes the accelerating contacts, causing the motor to run in the "reverse" direction. The motor runs until the driven equipment reaches its limit and trips switch 1. This opens the circuit of coil C, which allows the main contacts to open, and stops the motor.

When the handle is shifted to the "forward" position, the circuit of coil C is completed to − through limit switch 2, although F and H are again shorted. This is remedied by removing the segment in contact with finger F. On again becoming energized, coil C closes the main contacts, causing the motor to run in the forward direction. The motor runs until the driven equipment trips switch 2.

Time-Current Relay—Sometimes a combination of the above methods of acceleration is used in one controller. For instance, a popular type of hoisting controller employs relays whose time of closing is a function of time and current, and for this reason they are generally called time-current accelerating relays. These relays are very simple in structure (see fig. 8-43). They have only one moving part, an aluminum sleeve S, which carries a contact bar B that rests on stationary contacts C when it is in its lowest position. The relay coil is connected in the armature circuit, and the bar B (when making contact) completes the circuit of a contactor coil. The relay functions as follows: The current rises as soon as the circuit containing the relay coil is completed, causing a flux to build up rapidly in the relay. As this flux builds up it passes through sleeve S and induces a voltage in it, setting up a high current. This current establishes a magnetic field in the opposite direction to that established by the coil. These two opposing

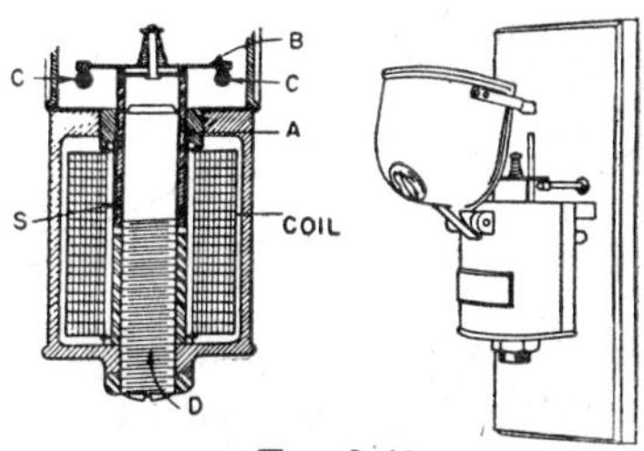

FIG. 8-43

fields cause the sleeve to move upward. The repelling force would cause the sleeve to move with considerable speed if it were not for the fact that the sleeve passes through the magnetic field as it starts to move upward, setting up a second current in the sleeve. This second current (called eddy current) causes a retarding action, and combines with the force of gravity to allow the sleeve to move slowly upward for a short distance. The sleeve starts to fall after reaching its highest point, and in so doing it again cuts flux, setting up eddy currents which retard the downward motion. This retarded motion of the sleeve provides the time of acceleration, which for maximum current is as high as 2.75 seconds.

Since the relay coil current governs the flux strength, it is apparent that a lower current will cause the sleeve to move upward a shorter distance, so that contacts C will be open for a shorter period. The contacts of such a relay are capable of carrying only a small current. Therefore they are not suitable for carrying the current of the main circuits of a controller, but may be used to complete the circuit of contactor coils.

The time of acceleration of the relay may be varied by regulating the threaded core D. When the core is screwed out, the air gap A increases, reducing the flux. Less current is then set up in the aluminum sleeve. The sleeve then rises a shorter distance, thus reducing the accelerating time.

Hoisting—The connections in a hoisting controller during the hoisting period are usually as shown in fig. 8-44a. The series brake

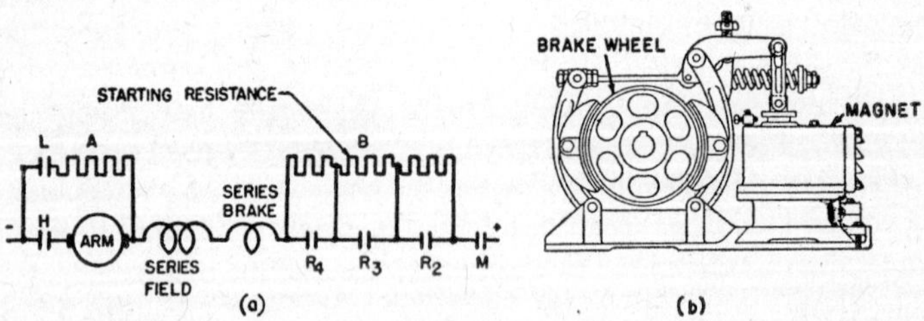

FIG. 8-44

(one type of which is shown in fig. 8-44b) is mounted on one end of the armature shaft and locks the motor when no current flows through its coil. This brake holds the load when the motor is disconnected from the power line. The contactors L, H, R_2, R_3, R_4, and M are all operated in proper sequence by means of a master switch. Contactors L, H, and M close when the handle of the master switch is moved to the first hoisting position. The main circuit is from −, through contactor H, armature, field, brake, starting resistance B and contactor M, to +. Some current is also shunted around the armature, through resistance A. The armature receives little current with these connections, giving a very low speed.

Moving the handle to the second hoisting position opens contactor L and removes the shunting resistance A. This increases the armature current, and therefore increases the motor speed.

Moving the handle to the successive hoisting positions closes contactors R_2, R_3, and R_4 in consecutive order, connecting the motor across the line.

The motor is thus operated as a series machine during hoisting. Five hoisting speeds are secured, varying from a very low speed for the first position of the handle to the highest speed when the motor is connected directly across the line.

Lowering—When a crane load is being lowered, it will overhaul the motor if it is heavy enough to overcome friction. Usually a load of about one-third the rated capacity of the crane is enough to overcome the friction of the hoisting machinery. Any load exceeding this would cause a motor to be overhauled if means were not provided to prevent this. Overhauling is prevented by connecting the motor as a shunt machine (a shunt machine runs at an approximately constant speed at all loads, as discussed in Chapter 7). If a load tends to overhaul a shunt-connected motor, the machine prevents this by automatically becoming a generator, supplying a counter-torque, and returning power to the supply lines. When the load is not heavy enough to overcome friction, the motor furnishes the necessary extra power to lower the load.

A wide lowering speed range is secured by changing the field strength, just as the speed is changed in any shunt motor.

The connections during the lowering period are shown in fig. 8-45. Contactors L, S, R_2, R_3, R_4, and M close when the handle is moved to the first lowering position. This connects the motor as a shunt machine, with the series brake in the field circuit. The circuit is from −, through contactor L, through resistance A, after which it divides into two circuits. One of these is through the field, brake, contactors R_4, R_3, and R_2 (which completely short-circuits resistance B), through contactor M, to +. The other circuit is through the armature, resistance C, contactors S and M, to +. The current through the first circuit sets up a flux in the field and causes the friction brake to release. The current through the second circuit flows through the armature in the proper direction to lower the load. These connections provide the greatest field strength and the lowest armature current. Thus the motor runs at the lowest speed.

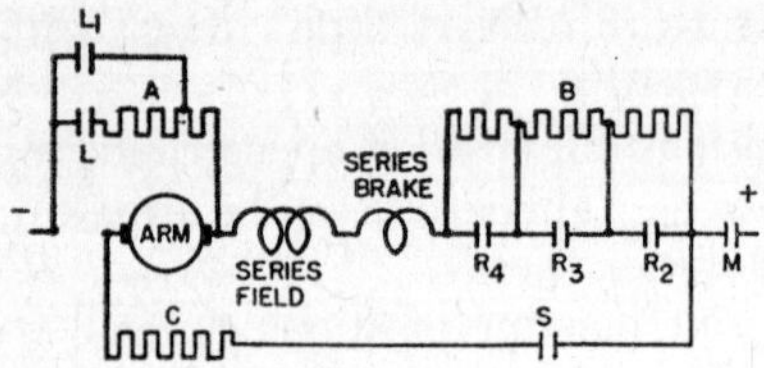

Fig. 8-45

Moving the handle to the two next lowering positions opens contacts R_4 and R_3, inserting two steps of resistance in the field circuit. This weakens the field and causes the motor speed to increase.

Moving the handle to the fourth lowering position closes contactor L_1. This shorts out a section of resistance A, causing an increase in armature current and motor speed.

Moving the handle to the last position opens contactor R_2, placing the last step of resistance in the field circuit. This provides the highest lowering speed.

Stopping—Fig. 8-46 shows the connections when the handle is in the "off" position. In this case only contactor E is closed. This connects the armature, series field, and resistance C in a closed circuit. If for some reason the load should overhaul, the motor would function as a series generator, forcing current through resistance C. This would cause the machine to develop a counter-torque, allowing the load to descend at a low speed. Thus a safety feature is provided, which is effective if the friction brake fails to hold the load.

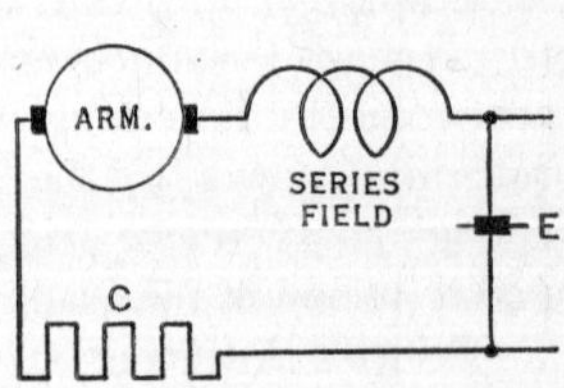

Fig. 8-46

The structure of a controller, which functions as explained above, is shown in fig. 8-47. Its wiring diagram is shown in fig. 8-48. (In these figures the resistors and contactors have the same designations as those of the preceding figures.)

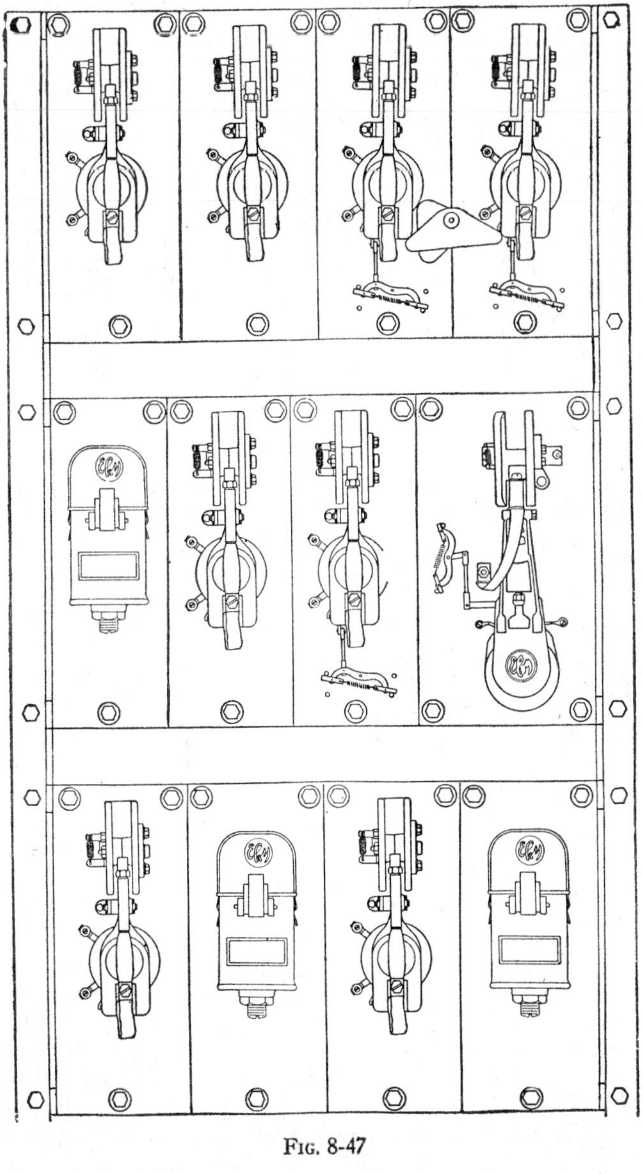

FIG. 8-47

In fig. 8-48 it will be noted that accelerating relays are provided, which are connected in the armature circuit. These are of the time-current type, and each closes a circuit to a coil of an accelerating contactor as soon as the current falls to a value for which the relays are set to operate. *In other words, these relays provide time-current limit acceleration during hoisting.*

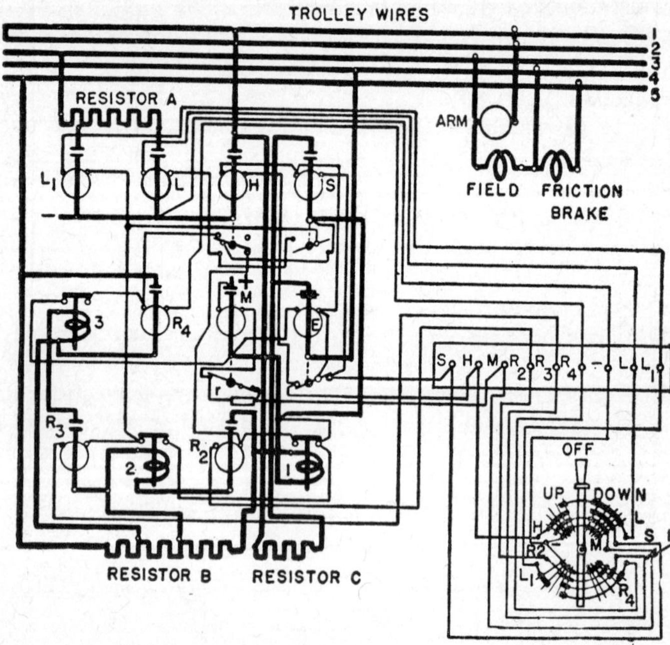

Fig. 8-48

Note further that all contactors except E are designed to close when their coils are energized. Contactor E is designed to be held closed by a spring, and opens only when its coil is energized.

Hoisting—In fig. 8-48, the following circuits are made when the handle of the master switch is moved to the first hoisting position:

A circuit is made from −, through the master switch to terminal M, through the coil of contactor M, to +. Contactor M thus closes, inserting resistance r in series with the closing coil and preventing the coil from overheating and reducing the holding power.

A second circuit is formed from −, through the master switch to terminal H, through the coil of contactor E, to +. The coil of contactor H is also connected to the same control circuit wire, but is short-circuited by the auxiliary contacts of contactor E. Then contactor E must open before the coil of contactor H can become energized, thus closing this contactor.

A third circuit is formed from −, through the master switch to terminal L, through the coil of contactor L, through the auxiliary contacts of the closed contactor H to the bottom of contactor S, and then through contactor M to +. Contactor L thus closes.

Contactors M, H, and L are now closed, and E is open (see fig. 8-44a). It was explained that H could not close until E had opened, and L could not close until H had closed. Without such an interlocking system the controller would not be reliable. For instance, suppose that L and M in fig. 8-44a are closed, and H is open. Current would then flow through the magnetic brake and release it, but no current would flow through the armature. This would allow the load to overhaul the motor.

The lowest hoisting speed is provided with contactors M, H, and L closed and E open (see fig. 8-44a).

When the handle is moved to the second hoisting position, all the circuits remain the same, except the one including the coil of contactor L. This circuit is opened, removing the armature shunting resistance A and thereby increasing the speed of the motor.

Moving the handle to the third position adds a circuit, which is from − on the master switch to terminal R_2, through the coil of contactor R_2 to a contact of the time-current relay 1. As soon as the current through this relay drops to a value for which it is set, the coil circuit of contactor R_2 is completed. This causes R_2 to close, short-circuiting a section of the starting resistance and again increasing the motor speed. After R_2 closes, relay 2 is inserted in series with the armature.

Moving the handle to the fourth position closes contactor R_3 as soon as the current drops to the value for which relay 2 is set. This shorts out a second step of resistance and inserts relay 3 in the armature circuit.

Moving the handle to the last position closes contactor R_4 as soon as relay 3 operates. This connects the motor directly across the line, giving the highest speed.

Lowering—The following circuits are made when the handle is moved to the first lowering position:

A circuit is made from − on the master switch to terminal M, through the coil of contactor M, to +. Contactor M closes, inserting r in the control circuit.

A second circuit is from −, through the master switch to terminal R_4, through the coil of contactor R_4 and one auxiliary contact of H to the bottom of contactor S, through contactor M, to +. Thus R_4 closes. Note that this circuit can be formed only when contactor H is open.

A third circuit is from −, through the master switch to terminal R_2, through the coil of contactor R_2 and one auxiliary contact of H to the bottom of contactor S, through contactor M, to +. Thus R_2 closes. As in the case of R_4, contactor R_2 can close only while contactor H is open.

A fourth circuit is from −, through the master switch to terminal S, through an auxiliary contact and coil of contactor E, through contactor M, to +. Contactor E opens, inserting the coil of contactor S in series with that of E, causing contactor S to close. Thus contactor S cannot close until E opens.

A fifth circuit is from −, through the master switch to terminal L, through the coil of contactor L, through closed auxiliary contacts of S, through the bottom of contactor S, through contactor M, to +. This circuit, which cannot be established until contactor S is closed, closes contactor L.

A sixth circuit is from −, through the master switch to terminal R_3, through the coil of contactor R_3, through one closed auxiliary contact and to the bottom of contactor S, through contactor M, to +. This circuit, which also cannot be established until contactor S is closed, closes contactor R_3.

Contactor E is now open and contactors L, S, R_2, R_3, R_4 and M are closed (see fig. 8-45). This provides the lowest speed for lowering.

When the handle is moved to the second lowering position, the coil circuit of contactor R_4 is opened. Contactor R_4 on opening inserts a section of resistance B in series with the field, increasing the speed of the motor.

Moving the handle to the third position opens the coil circuit of contactor R_3. R_3 on opening inserts another section of resistance B in the field circuit, providing a weaker field and a higher speed.

When the handle is moved to the fourth position, a circuit is made from −, through the master switch to terminal L_1, through the coil of contactor L_1, through one contact of S to the bottom of contactor S, through contactor M, to +. Contactor L_1 closes, shorting out part of resistance A, causing more current to flow in the armature and thereby giving a higher speed.

Moving the handle to the last lowering position opens the coil circuit of contactor R_2. R_2 on opening inserts the remaining portion of resistance B in the field circuit, providing the weakest field strength and giving the fastest lowering speed.

Off—When the handle is in the "off" position, the − terminal is disconnected (in the master switch) from all the contactor coils. Then all the contactors except E will be open. Contactor E establishes the connections shown in fig. 8-46, providing the dynamic brake action explained in connection with this figure.

Need for a Duplicate Trolley Wire—In fig. 8-48 note that one terminal of the armature is connected to trolley wires 1 and 2 by two separate collectors. If only one collector and trolley wire were used, the collector might slip off the trolley wire when a load was being lowered. Current would then flow through the series field and friction brake, but none would flow through the armature. The friction brake would then release and a heavy load would descend rapidly, for the motor would supply no counter-torque. With the use of two trolley wires, however, one collector will maintain the circuit even though the other should slip off.

Ward-Leonard System—A simplified diagram of the Ward-Leonard system is shown in fig. 8-49, in which a separate generator G supplies power to motor M to be controlled. The generator is driven by a constant speed alternating current motor, indicated by A. The fields of motor M and generator G are connected to a constant voltage source and their armatures are connected together. No starting resistance is required in the armature circuit of the motor. The motor is brought up to speed by gradually increasing the generator voltage from zero, which is done by increasing the excitation of the generator field. For a given armature voltage, the motor speed remains practically constant, being a shunt machine. With this type of control, the motor may be operated at any speed up to its maximum speed, which is reached when the maximum voltage is impressed on the armature.

The motor is reversed by reducing the generated voltage to zero then increasing it with opposite polarity. This is done merely by turning rheostat R.

With this control system, the motor may be brought to a standstill quickly, merely by rapidly reducing the generator voltage. When the

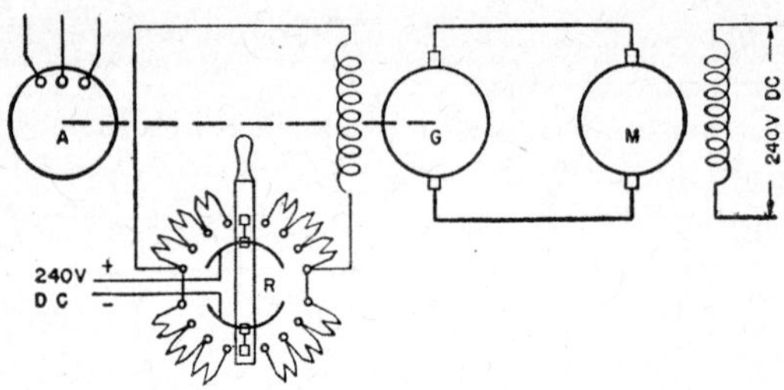

FIG. 8-49

generator voltage is reduced below the counter e m f of the motor, this counter e m f sends current through the generator armature, establishing dynamic braking. While this takes place, generator G operates as a motor, driving motor A, which returns power to the line.

The Ward-Leonard system of control is often used in connection with large motors, such as are used in steel mills, and with elevators when a direct current supply is not available.

Exercises

1. Why is a starter necessary on D C motors?
2. Draw a sketch showing how a series motor is connected to the starter of fig. 8-1.
3. What is the purpose of the holding magnet?
4. What two methods are used to connect the holding coil? What are the relative advantages and disadvantages of these methods?
5. What would be the effect of placing the starting resistance in series with the entire shunt motor instead of in the armature circuit?
6. What is the difference between a starter and a controller?
7. A 240 volt motor, which has a .3 ohm armature and a 120 ohm field, takes 50 amperes at full load. How much resistance must the starter contain to limit the starting current to 50 per cent above full load?

8. Why are cam type drum controllers more durable than the other types? Why are these types not in more general use?
9. Explain why a counter-electromotive force starter is so named.
10. Describe briefly the different types of current limit contactors.
11. In a current limit type controller, when the last contactor closes it short-circuits all the contactors and is held closed by a small auxiliary coil. Why is this desirable?
12. What is meant by dynamic braking, and how is it accomplished?
13. Why are the shunt fields in fig. 8-36 connected directly across the lines?
14. Make a diagram showing how the starter of fig. 8-37 may be controlled by an operator switch.
15. For what industrial applications should time element controllers be used?
16. Make a wiring diagram of a control system that contains a Ford dash pot type starter, an operator switch (see fig. 8-34), two limit switches, and a compound motor.

CHAPTER 9

GEOMETRY AND TRIGONOMETRY

In the study of alternating current, the equations developed are made clearer by the use of straight lines and angles. In dealing with lines and angles, it is necessary to be familiar with certain fundamentals in geometry and trigonometry. In the consideration of this work, symbols are used, which are as follows: angle ($\angle$); triangle ($\triangle$); parallel to ($\|$); and perpendicular to ($\perp$).

In this chapter all lines are considered in the same plane and are assumed to be straight unless otherwise specified.

Angles—An angle is formed when two lines are drawn from the same point as shown in fig. 9-1. In this figure suppose that line AC is rotated from coincidence with AB about A in the counter clockwise direction. The size of the angle formed depends upon the amount of rotation of line AC. When the line makes one complete revolution, it is considered as having rotated 360 degrees. Then, when the line AC has rotated one quarter of a revolution, it will have rotated 90 degrees, and is said to be perpendicular to AB, or the two lines are said to form a right angle. In the following a right angle is indicated by a small arc as ∟. When AC rotates half a revolution, it will have rotated 180 degrees. The two sides now lie in a straight line and for this reason the angle formed is called a straight angle.

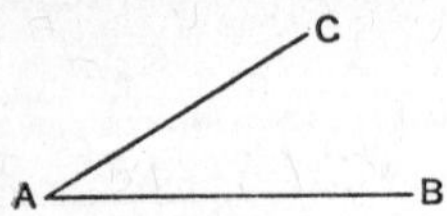

FIG. 9-1

Then an angle is measured in a unit called a degree. A degree is divided into 60 minutes and a minute is divided into 60 seconds. In problems pertaining to electricity, the answers obtained will be sufficiently accurate if the seconds are neglected.

An angle should not be expressed in the form 32.275°, but the decimal part should be converted to minutes, which is done by multiplying this part by 60. That is $.275° = .275 \times 60$ or 16.5′. The decimal part may be dropped. Then $32.275° = 32°\ 16'$.

Definitions—In the study of geometry and trigonometry, the meaning of certain words should be clearly understood. These are defined in the following;

(*a*) *Acute angle*—An angle smaller than a right angle.

(*b*) *Obtuse angle*—An angle larger than a right angle.

(*c*) *Supplementary angles*—Two angles whose sum is 180 degrees, one is said to be the supplement of the other.

(*d*) *Triangle*—An area bounded by three lines.

(*e*) *Right angle triangle*—A triangle with one 90 degree angle.

(*f*) *Oblique triangle*—A triangle with no 90 degree angle.

(*g*) *Isosceles triangle*—A triangle with two equal angles.

(*h*) *Equilateral triangle*—A triangle with three equal sides.

(*i*) *Equal triangles*—Triangles that have the same shape and size.

(*j*) *Similar triangles*—Triangles that have the same shape but are not the same size.

(*k*) *Hypotenuse*—The side of a right triangle opposite the right angle.

(*l*) *Legs*—The two sides of a right triangle that form the right angle.

(*m*) *Parallel lines*—Lines that do not cross even when extended indefinitely.

(*n*) *Parallelogram*—A four sided figure with opposite sides parallel.

(*o*) *Rectangle*—A parallelogram with right angles.

(*p*) *Square*—A parallelogram with equal sides and right angles.

(*q*) *Rhombus*—A parallelogram with equal sides and no right angles.

Axiom—A statement so evident that it requires no proof is known as an axiom. Axioms frequently used are given in the following:

(1) Things equal to the same thing are equal to each other.

(2) In an equation, a quantity may be substituted for its equal.

(3) If equals are added to, subtracted from, multiplied by, or divided by equals, the results are equal.

Theorem—A theorem is a statement which can be proven. In the following a proof is given for only those theorems that are least evident.

THEOREM 1—If two lines intersect, the opposite angles are equal.

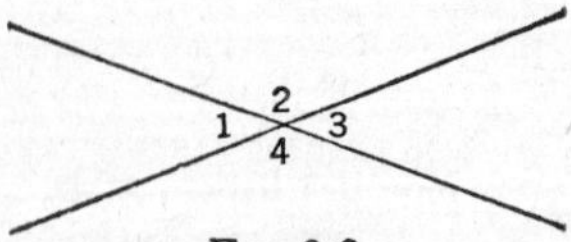

FIG. 9-2

PROOF—In fig. 9-2, $\angle 1 + \angle 2 = 180°$

$\angle 3 + \angle 2 = 180°$

Then $\angle 1 + \angle 2 = \angle 2 + \angle 3$ (Axiom 1)

Subtracting $\angle 2$ from both sides leaves $\angle 1 = \angle 3$

Coincide—When one figure is placed on another and every point on one falls on a corresponding point on the other, the figures are said to coincide. Two triangles are equal if they can be made to coincide throughout. Theorems 2 and 3 can be proven in this way.

THEOREM 2—Two right triangles are equal if

(a) Their corresponding legs are equal.

(b) The hypotenuse and leg of one equal respectively to the hypotenuse and leg of the other.

(c) The hypotenuse and acute angle of one equals respectively to the hypotenuse and acute angle of the other.

(d) The leg and acute angle of one equal respectively to the leg and acute angle of the other.

THEOREM 3—Two triangles are equal if

(a) Their corresponding sides are equal.

(b) Two angles and included side of one equal respectively to two angles and included side of the other.

(c) Two sides and included angle of one equal respectively to two sides and included angle of the other.

Corollary (*Cor.*)—A statement whose truth is made evident by a theorem is called a corollary.

COR. 1—If two triangles are equal, their corresponding angles are equal.

If two parallel lines are cut by a third line (called a transversal), angles 1 to 8 (fig. 9-3) are formed, which bear the following names: 1, 2, 7, and 8 are exterior angles; 3, 4, 5, and 6 are interior angles; 3 and 6, as well as 4 and 5, are alternate interior angles; 1 and 8 as well as 2 and 7 are alternate exterior angles; 1 and 5, 2 and 6, 3 and 7, as well as 4 and 8 are interior exterior angles.

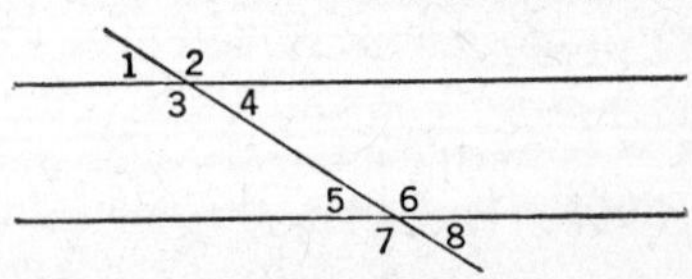

FIG. 9-3

THEOREM 4—If two parallel lines are cut by a transversal,
(a) The alternate interior angles are equal.
(b) The alternate exterior angles are equal.
(c) The exterior interior angles are equal.

Proof of 4a—In fig. 9-4, AB and CD are two parallel lines cut by line MN. Line GI is drawn ⊥ AB and through the midpoint of EF.

Then in △ HGF and HIE,

FH = HE
(H was selected at midpoint).

∠GHF = ∠IHE
(Theorem 1)

Then △ HGF = △HIE
(Theorem 2 c)

Therefore ∠GFH = ∠IEH
(Cor 1)

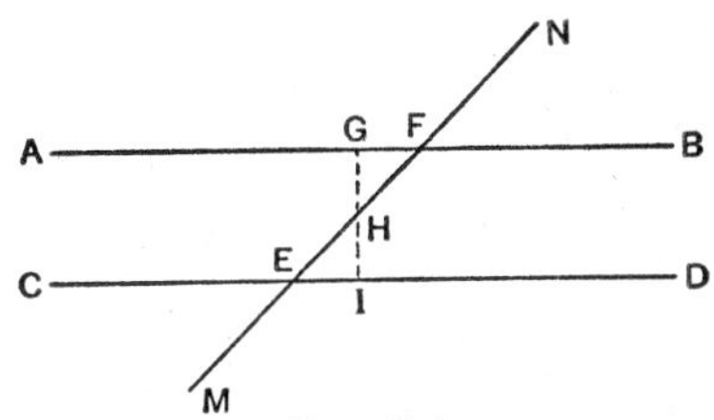

FIG. 9-4

THEOREM 5—Two angles are equal if their sides are parallel, right side to right side and left side to left side (as observed from vertices).

PROOF—In fig. 9-5, AB (left side of ∠B) is ∥ DE (left side of ∠E), and BC (right side of ∠B) is ∥ EF (right side of ∠E). Side ED is extended to H.

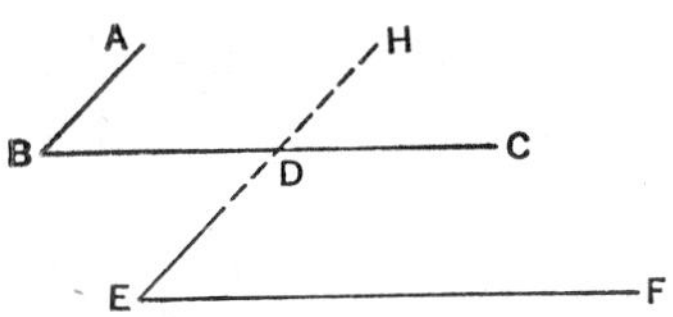

FIG. 9-5

∠B = ∠HDC (Theorem 4 c)
∠E = ∠HDC (Theorem 4 c)
Therefore ∠E = ∠B (Axiom 1)

THEOREM 6—The sum of the interior angles of any triangle equals 180 degrees.

PROOF—A line DE is drawn through B (fig. 9-6) ∥ AC. Side AB is prolonged to G and side CB to F.

∠2 = ∠5 (Theorem 1)
∠1 = ∠6 (Theorem 4 c)
∠3 = ∠4 (Theorem 4 c)
∠4 + ∠5 + ∠6 = 180° (straight angle)

Therefore

∠1 + ∠2 + ∠3 = 180° (Axiom 1)

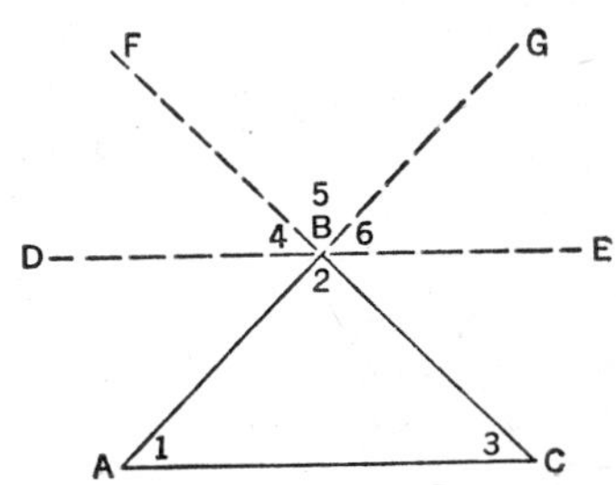

FIG. 9-6

COR. 2—The sum of two acute angles of a right triangle equals 90°.

COR. 3—When two angles of a triangle are equal to two angles of another triangle, the third angles are equal.

THEOREM 7—Two angles are equal if their sides are perpendicular, right side to right side and left side to left side.

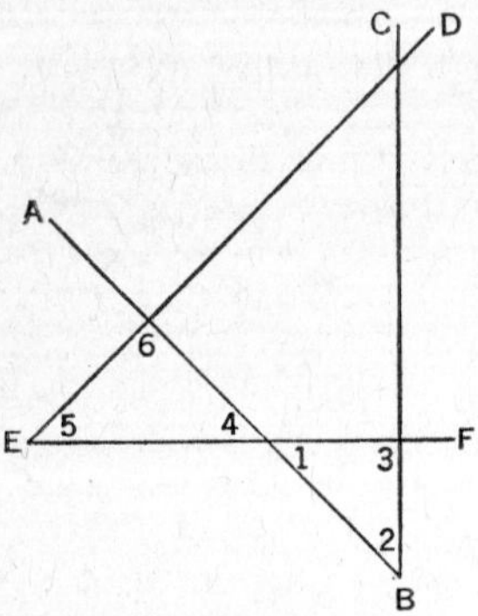

FIG. 9-7

PROOF—In fig. 9-7, AB (left side of $\angle B$) is $\perp ED$ (left side of $\angle E$), and BC (right side of $\angle B$) is $\perp EF$ (right side of $\angle E$)

$$\angle 1 + \angle 2 + \angle 3 = 180°$$

$$\angle 4 + \angle 5 + \angle 6 = 180°$$

$$\text{Then } \angle 1 + \angle 2 + \angle 3 = \angle 4 + \angle 5 + \angle 6$$

$$\angle 1 = \angle 4 \text{ (Theorem 1)}$$

$$\angle 3 = \angle 6 \text{ (both 90°)}$$

$$\text{Therefore } \angle 2 = \angle 5 \text{ (axiom 3)}$$

Exercises

1. What is the supplement of 20° 10′?
2. Determine the number of degrees and minutes in an angle whose supplement is 130° 6′.
3. How many degrees has each angle of an equilateral triangle?
4. If one acute angle of a right triangle is 26° 42′, what is the value of the other acute angle?
5. If two angles of a triangle are 30° 21′ and 42° 56′, what is the value of the third angle?

Determine the lettered angles in the following figures.

(6)

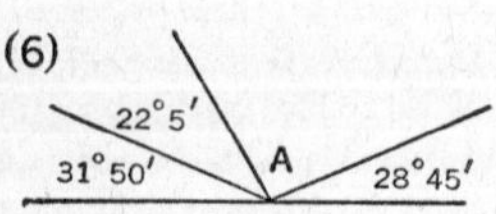

(7)

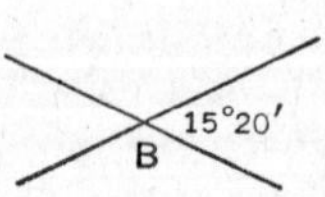

(8)

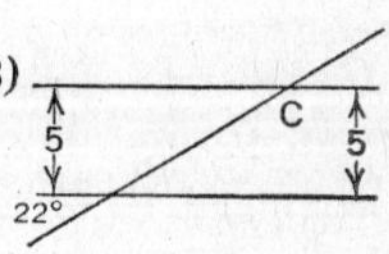

(9)

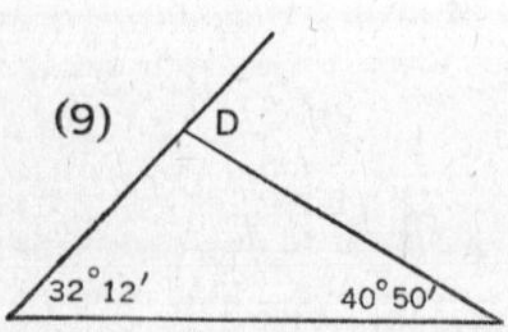

(10)

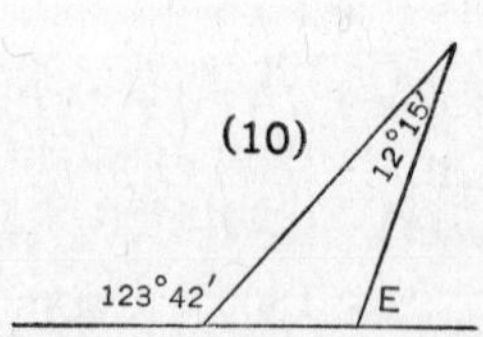

THEOREM 8—In a parallelogram,
(a) Either diagonal drawn forms two equal triangles
(b) The opposite sides are equal.
(c) The opposite angles are equal.
(d) The two diagonals bisect each other.

Proof of 8a and 8b—In fig. 9-8 AB is || DC and AD is || BC, (by definition).

Diagonal AC is drawn, forming two triangles.

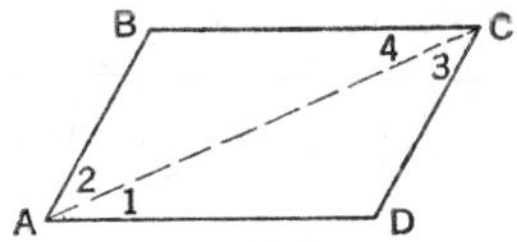

FIG. 9-8

$\angle 1 = \angle 4$ (Theorem 4 a)

$\angle 2 = \angle 3$ (Theorem 4 a), and since AC is common to both triangles, $\triangle ABC = \triangle CDA$ (Theorem 3 b)

Then $AB = DC$ and $AD = BC$.

THEOREM 9—The diagonals of a rhombus
(a) are $\perp$ each other
(b) bisect each other
(c) bisect the angles.

THEOREM 10—If three or more parallel lines intercept equal lengths on one transversal, they intercept equal lengths on any other transversal.

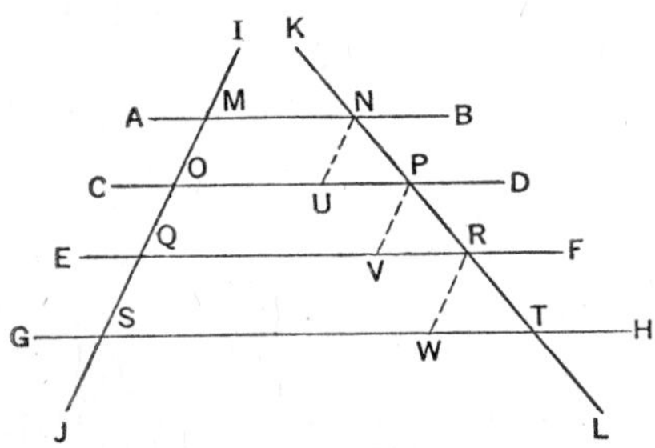

FIG. 9-9

PROOF—In fig. 9-9 parallel lines AB, CD, EF, and GH intercept equal lengths MO, OQ, and QS on the transversal IJ.

Lines NU, PV, and RW are drawn || IJ.

Then NU is || PV is || RW, and are equal (Theorem 8b).

$\angle NUP = \angle PVR = \angle RWT$, and $\angle UNP = \angle VPR = \angle WRT$. (Theorem 5).

Then $\triangle NUP = \triangle PVR = \triangle RWT$ (Theorem 3 b)

Therefore, $NP = PR = RT$ (Converse of Theorem 3 a)

Theorem 11—A line drawn through two sides of a triangle and parallel to the third side, divides the first and second side proportionally.

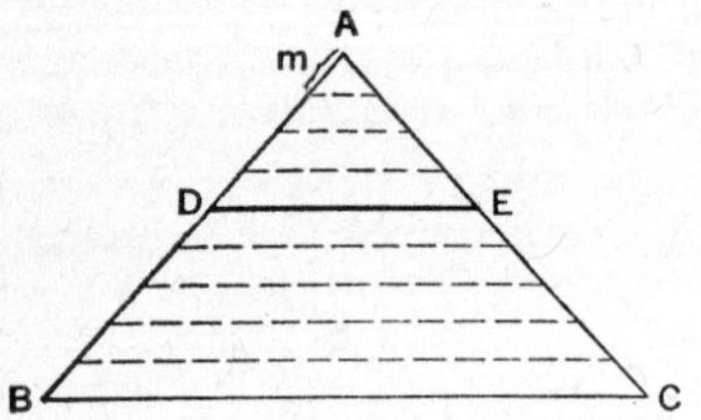

Fig. 9-10

Proof—In fig. 9-10, DE is $\parallel$ BC. Suppose that distance m is a unit of measure. In this case it is contained 4 times in AD and 5 times in DB. Then $\frac{AD}{DB} = \frac{4}{5}$

Lines are drawn parallel to BC, through the points of division along AB. Then AC will be divided into equal parts (Theorem 10); there will be 4 equal parts in AE and 5 in EC. Then $\frac{AE}{EC} = \frac{4}{5}$

$$\text{Since } \frac{AD}{DB} \text{ and } \frac{AE}{EC} \text{ each equals } \frac{4}{5},$$

$$\text{then } \frac{AD}{DB} = \frac{AE}{EC} \text{ (Axiom 1).}$$

Theorem 12—Two triangles are similar if two angles of one are equal to two angles of the other.

Theorem 13—If two triangles are mutually equiangular, their corresponding sides are in proportion.

Proof—In fig. 9-11, $\angle A = \angle A_1$, $\angle B = \angle B_1$ and $\angle C = \angle C_1$.

$\angle A$ is placed on $\angle A_1$ so that the two angles coincide. Then $\triangle ABC$ will take the position A_1DE. Then DE is $\parallel$ B_1C_1 (converse to theorem 4 c).

$$\text{Therefore } \frac{A_1D}{A_1B_1} = \frac{A_1E}{A_1C_1} \text{ (Theorem 11) and } \frac{AB}{A_1B_1} = \frac{AC}{A_1C_1}$$

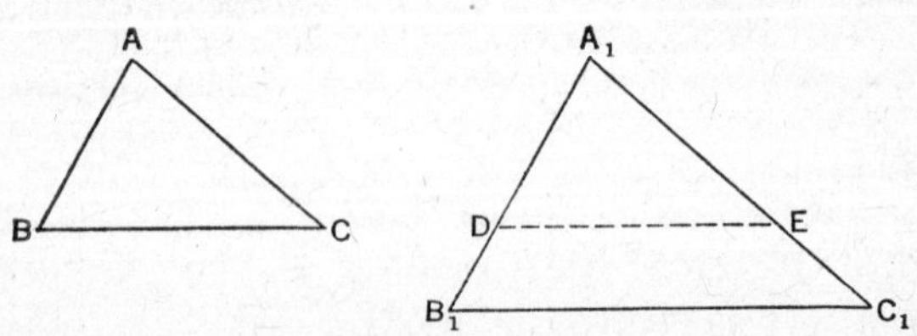

Fig. 9-11

Cor. 4—Two right triangles are similar if an acute angle of one is equal to an acute angle of the other.

COR. 5—Two triangles are similar if the corresponding sides are parallel or perpendicular.

THEOREM 14—If in any right triangle a perpendicular is drawn from the vertex of the right angle to the hypotenuse, the two triangles formed are similar to each other and to the original triangle.

PROOF—Fig. 9-12 is a right triangle in which BD is drawn from $B \perp AC$.

$\angle 1 = \angle 1_1$ and $\angle 2 = \angle 2_1$ (Theorem 7)

$\triangle ABD$ is similar to BCD is similar to ACB (Theorem 12)

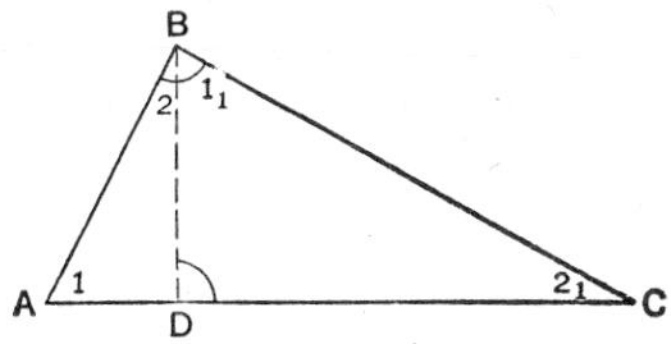

FIG. 9-12

$$\text{Then } \frac{AB}{AD} = \frac{BC}{BD} = \frac{AC}{AB} \text{ (Theorem 13)} \qquad (1)$$

$$\text{Also } \frac{AB}{BD} = \frac{BC}{CD} = \frac{AC}{BC} \qquad (2)$$

$$\text{and } \frac{AD}{DB} = \frac{BD}{DC} = \frac{AB}{BC} \qquad (3)$$

THEOREM 15—The square of the hypotenuse of any right triangle is equal to the sum of the squares of the other two sides.

Equation 1 (Theorem 14) shows that $\frac{AB}{AD} = \frac{AC}{AB}$

Multiplying both members by $AD \times AB$ gives $\overline{AB}^2 = AD \times AC$ (1)

Similarly, in equation 2 (Theorem 14), $\frac{BC}{CD} = \frac{AC}{BC}$. (2)

Then $\overline{BC}^2 = CD \times AC$

Adding (1) and (2) gives $\overline{AB}^2 + \overline{BC}^2 = AD \times AC + CD \times AC$

$= AC(AD + CD)$ (3)

But $AD + CD = AC$ (see fig. 9-12).

Then (3) becomes $\overline{AB}^2 + \overline{BC}^2 = \overline{AC}^2$

If $AB = 3$ and $BC = 4$, then $AC = 5$.

This is shown in fig. 9-13, which makes this theorem more evident.

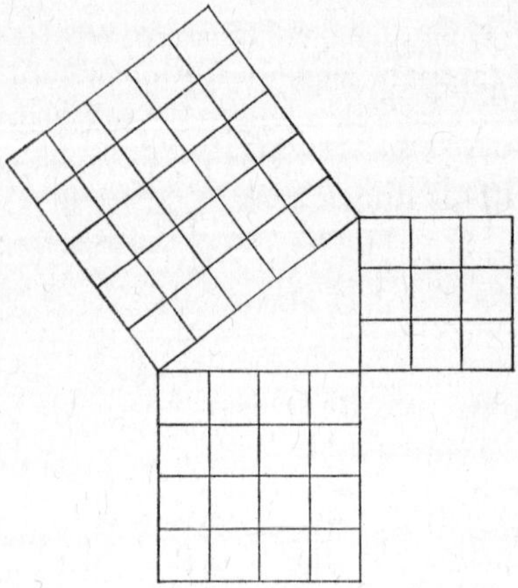

FIG. 9-13

Exercises

1. In fig. 9-14 are $\angle 1$ and $\angle 2$ equal?
2. In fig. 9-15, determine angles A, B, and C.

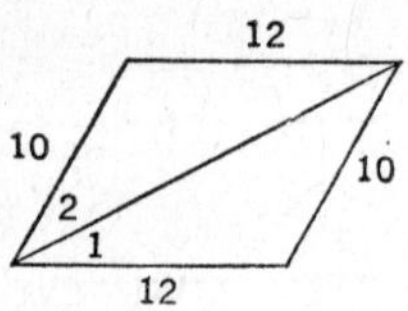

FIG. 9-14

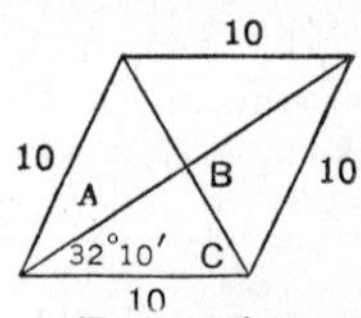

FIG. 9-15

Determine the value of the lettered sides and angles in the following figures:

3.

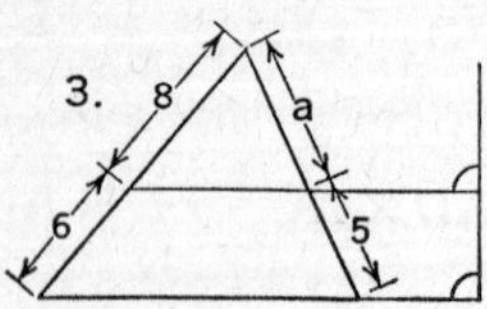

4.

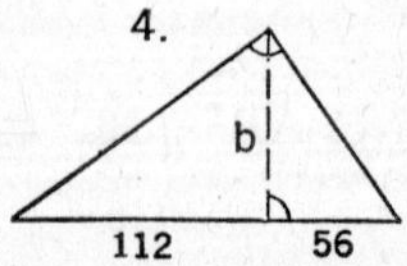

5.

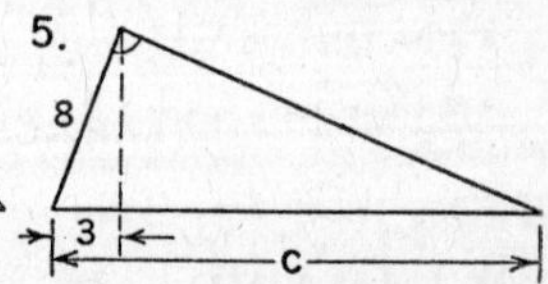

6.

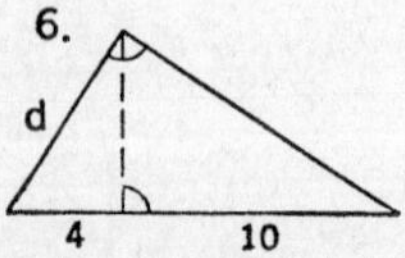

7.

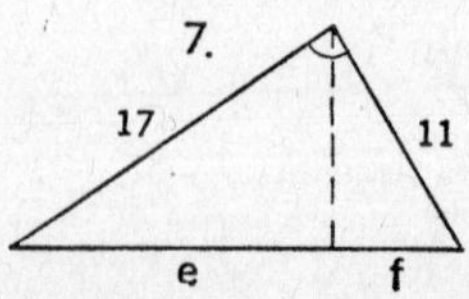

8.

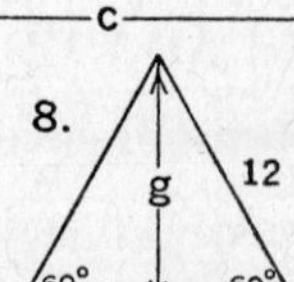

9. If two legs of a right triangle are 10 and 20, what is the length of the hypotenuse?
10. If the length of the hypotenuse of a right triangle is 16 and the length of one leg is 12, what is the length of the other leg?

TRIGONOMETRY

In fig. 9-16 is shown an angle of 30 degrees. If from three points on line AB, lines are drawn $\perp$ AC, three right triangles are formed, which are similar (Theorem 12). Hence, a ratio of two sides of one triangle is equal to the ratio of the corresponding sides of the other triangles, and for a given angle (which in this case is 30°) this ratio equals a constant.

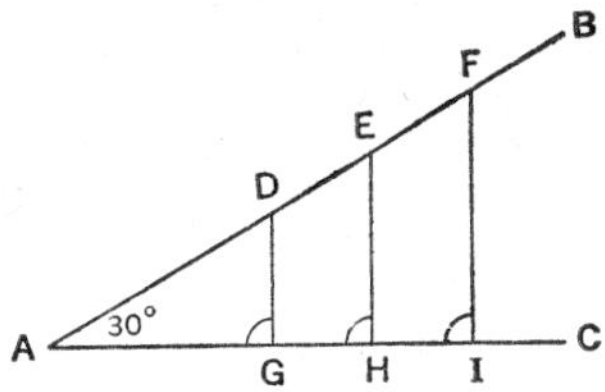

FIG. 9-16

$$\text{That is, } \frac{DG}{DA} = \frac{EH}{EA} = \frac{FI}{FA} = .5$$

$$\frac{GA}{DA} = \frac{HA}{EA} = \frac{IA}{FA} = .866$$

$$\text{and } \frac{DG}{GA} = \frac{EH}{HA} = \frac{FI}{IA} = .5774$$

This illustrates that for a given angle A, the ratio of two sides has the same value regardless of the length of the sides. If angle A is made smaller or larger, the equal ratios will change, but for any given angle they will remain the same. Relative to any angle there are six possible ratios, which are called trigonometric functions of the angle. In the right triangle ABC, (fig. 9-17) AB is the hypotenuse, BC is the side opposite $\angle$A, and AC is the side adjacent $\angle$A. The ratios are named as follows:

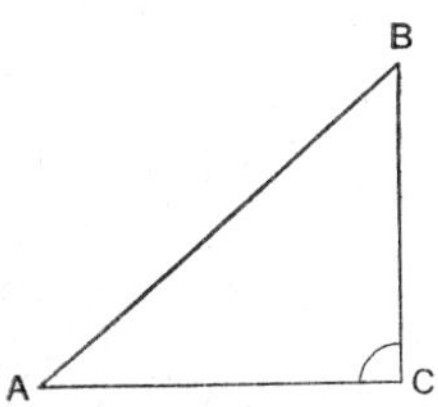

FIG. 9-17

$$\frac{BC}{AB} = \frac{\text{side opposite } \angle A}{\text{hypotenuse}} = \text{sine of } \angle A\text{, abbreviated sin A}$$

$$\frac{AC}{AB} = \frac{\text{side adjacent } \angle A}{\text{hypotenuse}} = \text{cosine of } \angle A\text{, abbreviated cos A}$$

$$\frac{BC}{AC} = \frac{\text{side opposite } \angle A}{\text{side adjacent } \angle A} = \text{tangent of } \angle A\text{, abbreviated tan A}$$

$$\frac{AC}{BC} = \frac{\text{side adjacent } \angle A}{\text{side opposite } \angle A} = \text{cotangent of } \angle A\text{, abbreviated cot A}$$

$$\frac{AB}{AC} = \frac{\text{hypotenuse}}{\text{side adjacent } \angle A} = \text{secant of } \angle A\text{, abbreviated sec A}$$

$$\frac{AB}{BC} = \frac{\text{hypotenuse}}{\text{side opposite } \angle A} = \text{cosecant of } \angle A\text{, abbreviated csc A}$$

Relative to any angle, these six ratios are expressed as follows:

$$\frac{\text{side opposite the angle}}{\text{hypotenuse}} = \text{sine of the angle}$$

$$\frac{\text{side adjacent the angle}}{\text{hypotenuse}} = \text{cosine of the angle}$$

$$\frac{\text{side opposite the angle}}{\text{side adjacent the angle}} = \text{tangent of the angle}$$

$$\frac{\text{side adjacent the angle}}{\text{side opposite the angle}} = \text{cotangent of the angle}$$

$$\frac{\text{hypotenuse}}{\text{side adjacent the angle}} = \text{secant of the angle}$$

$$\frac{\text{hypotenuse}}{\text{side opposite the angle}} = \text{cosecant of the angle}$$

These equations should be committed to memory because they are used throughout the study of trigonometry and alternating current electricity.

Trigonometry Tables—Since the ratio of the sides of a right triangle is a constant for a given angle, the value of the different ratios have been determined for angles expressed in degrees and minutes, ranging from 0 to 90 degrees. These values are given in tables of trigonometric functions, and are usually arranged as shown in table 9-I.

Function of angles from 0° to 45° have their heading at the top of the pages, and minute columns at the left. To find the function of an angle between 0 and 45°, say cos 30° 7′, reference is made to the page in the trigonometry tables that has 30° written at the top. The value of the cos 30° 7′ will be found in the column below the heading "cos" and in the line corresponding to 7′ in the minute column at the left. The value is .865 (see table 9-I).

Table 9-I

30°

M	Sin	Cosine	Tan.	Cotan.	Secant	Cosec.	M
0	.50000	.86603	.57735	1.7320	1.1547	2.0000	60
1	.50025	.86588	.57774	.7309	.1549	1.9990	59
2	.50050	.86573	.57813	.7297	.1551	.9980	58
3	.50075	.86559	.57851	.7286	.1553	.9970	57
4	.50101	.86544	.57890	7274	.1555	.9960	56
5	.50126	.86530	.57929	1.7262	1.1557	1.9950	55
6	.50151	.86515	.57968	.7251	.1559	.9940	54
7	.50176	.86500	.58007	.7239	.1561	.9930	53
8	.50201	.86486	.58046	.7228	.1562	.9920	52
9	.50226	.86471	.58085	.7216	.1564	.9910	51
10	.50252	.86457	.58123	1.7205	1.1566	1.9900	50
11	.50277	.86442	.58162	.7193	.1568	.9890	49
12	.50302	.86427	.58201	.7182	.1570	.9880	48
13	.50327	.86413	.58240	.7170	.1572	.9870	47
14	.50352	.86398	.58279	.7159	.1574	.9860	46
15	.50377	.86383	.58318	1.7147	1.1576	1.9850	45
16	.50402	.86369	.58357	.7136	.1578	.9840	44
17	.50428	.86354	.58396	.7124	.1580	.9830	43
18	.50453	.86339	.58435	.7113	.1582	.9820	42
19	.50478	.86325	.58474	.7101	.1584	.9811	41
20	.50503	.86310	.58513	1.7090	1.1586	1.9801	40
21	.50528	.86295	.58552	.7079	.1588	.9791	39
22	.50553	.86281	.58591	.7067	.1590	.9781	38
23	.50578	.86266	.58630	.7056	.1592	.9771	37
24	.50603	.86251	.58670	.7044	.1594	.9761	36
25	.50628	.86237	.58709	1.7033	1.1596	1.9752	35
26	.50653	.86222	.58748	.7022	.1598	.9742	34
27	.50679	.86207	.58787	.7010	.1600	.9732	33
28	.50704	.86192	.58826	.6999	.1602	.9722	32
29	.50729	.86178	.58865	.6988	.1604	.9713	31
30	.50754	.86163	.58904	1.6977	1.1606	1.9703	30
31	.50779	.86148	.58944	.6965	.1608	.9693	29
32	.50804	.86133	.58983	.6954	.1610	.9683	28
33	.50829	.86118	.59022	.6943	.1612	.9674	27
34	.50854	.86104	.59061	.6931	.1614	.9664	26
35	.50879	.86089	.59100	1.6920	1.1616	1.9654	25
36	.50904	.86074	.59140	.6909	.1618	.9645	24
37	.50929	.86059	.59179	.6898	.1620	.9635	23
38	.50954	.86044	.59218	.6887	.1622	.9625	22
39	.50979	.86030	.59258	.6875	.1624	.9616	21
40	.51004	.86015	.59297	1.6864	1.1626	1.9606	20
41	.51029	.86000	.59336	.6853	.1628	.9596	19
42	.51054	.85985	.59376	.6842	.1630	.9587	18
43	.51079	.85970	.59415	.6831	.1632	.9577	17
44	.51104	.85955	.59454	.6820	.1634	.9568	16
45	.51129	.85941	.59494	1.6808	1.1636	1.9558	15
46	.51154	.85926	.59533	.6797	.1638	.9549	14
47	.51179	.85911	.59572	.6786	.1640	.9539	13
48	.51204	.85896	.59612	.6775	.1642	.9530	12
49	.51229	.85881	.59651	.6764	.1644	.9520	11
50	.51254	.85866	.59691	1.6753	1.1646	1.9510	10
51	.51279	.85851	.59730	.6742	.1648	.9501	9
52	.51304	.85836	.59770	.6731	.1650	.9491	8
53	.51329	.85821	.59809	.6720	.1652	.9482	7
54	.51354	.85806	.59849	.6709	.1654	.9473	6
55	.51379	.85791	.59888	1.6698	1.1656	1.9463	5
56	.51404	.85777	.59928	.6687	.1658	.9454	4
57	.51429	.85762	.59967	.6676	.1660	.9444	3
58	.51454	.85747	.60007	.6665	.1662	.9435	2
59	.51479	.85732	.60046	.6654	.1664	.9425	1
60	.51504	.85717	.60086	1.6643	1.1666	1.9416	0
M	Cosine	Sin	Cotan.	Tan.	Cosec.	Secant	M

59°

Functions of angles between 45 and 90 degrees have their headings at the bottom of the pages and the minute columns at the right. To find the function of an angle between 45° and 90°, say tan 59° 12′, reference is made to the page that has 59° at the bottom. The value of tan 59° 12′ will be found in the column above the heading "tan" and in the line corresponding to 12′ in minute column at the right. The value is 1.6775 (see table 9-I).

To find an angle corresponding to a given value, find the value in the column corresponding to the function. If the function is at the top of the column, read the angle at the top and the minutes from the left column. If the function is at the bottom of the column, read the angle at the bottom and the minutes from the right column. If a value cannot be found corresponding to a function, select the nearest value. This is near enough if considering problems pertaining to electricity.

EXAMPLE—Find ∠A if tan A = 1.6758. The value 1.6758 cannot be found in the column headed "tan" (see table 9-I), then the nearest value will be selected, which is 1.6753. This value corresponds to 59° 10′.

Exercises

1. Find the value of the following:

   (a) sin 45° (d) cos 78° 12′
   (b) cos 10° 5′ (e) sec 5° 32′
   (c) tan 36° 20′ (f) csc 46° 23′

2. If cos A = .8784, determine

   (a) sin A (b) cot A (c) sec A

3. If sin B = .94561, determine

   (a) cos B (b) tan B (c) sec B

4. The value .85660 is (a) the cotangent of what angle? (b) The tangent of what angle? (c) The cosine of what angle?

Solution of Right Triangles—An unknown in a right triangle may be determined, with the use of trigonometry tables, if any two sides or one side and an acute angle are given. When two sides are given, use one of the above trigonometric equations that involves the two given sides and the unknown angle to be determined. When an acute angle and a side are given, use an equation that involves the given angle, given side, and the unknown side to be determined. In either case, the equation will contain only one unknown, which is readily determined.

EXAMPLE—Find $\angle A$ and $\angle B$ in fig. 9-18.

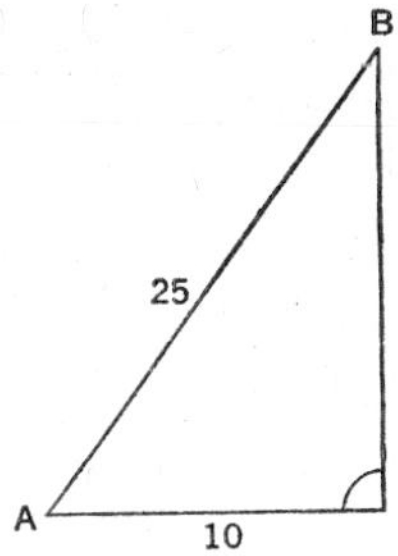

FIG. 9-18

$$\frac{10}{25} = \cos A$$

$$\cos A = .4$$

$$\angle A = 66° \ 25'$$

$$\angle B = 90° - 66° \ 25' = 89° \ 60' - 66° \ 25'$$

$$= 23° \ 35'$$

EXAMPLE—Find sides a and b in fig. 9-19.

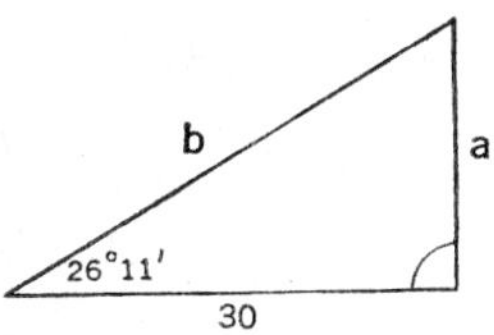

FIG. 9-19

$$\frac{a}{30} = \tan 26° \ 11'$$

$$a = 30 \tan 26° \ 11' = 30 \times .4917 = 14.751$$

$$\frac{b}{30} = \sec 26° \ 11'$$

$$b = 30 \sec 26° \ 11' = 30 \times 1.1143 = 33.429$$

Exercises

Determine the values of the lettered sides and angles in the following figures:

1.

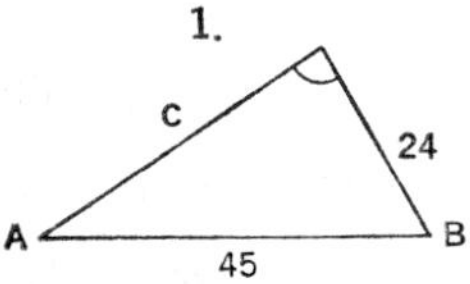

2.

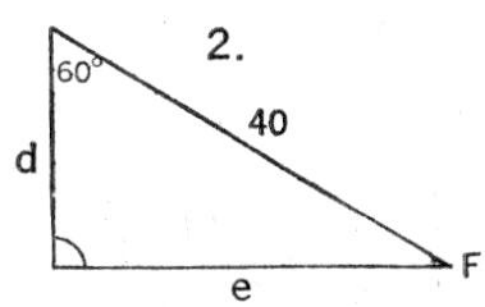

3.

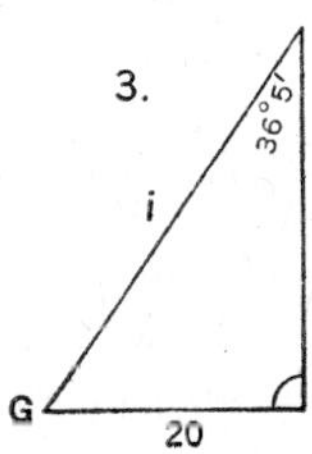

4.

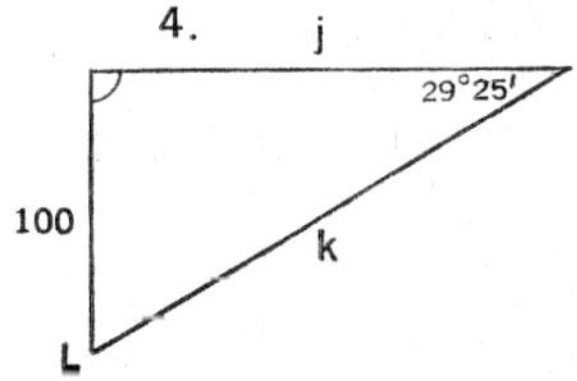

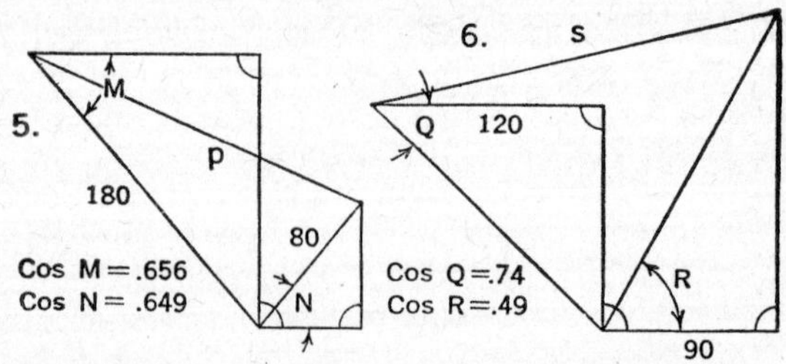

7. What length line is necessary to establish service between a residence connection 15 feet above the ground and a distributing transformer atop a 30 foot pole, if the distance on the ground from the house to the pole is 100 feet (allow an extra 3% for sag)?
8. A vertical service run requires one 18 inch offset. If the maximum bend allowed in the conduit is 60 degrees, how much additional conduit must be cut to allow for the offset?

Oblique Triangles—In an oblique triangle, the unknown values may be determined if the values given in any one of the following four cases are known.

(1) Two angles and any side.
(2) Two sides and an angle opposite either side.
(3) Two sides and an included angle.
(4) Three sides.

Law of Sines—Problems classified as cases 1 and 2 may be solved by the "law of sines," which may be stated as follows:

A side of a triangle is to the sine of the angle opposite that side, as any other side is to the sine of the angle opposite that side.

PROOF—In fig. 9-20a is shown a triangle with three acute angles, and in fig. 9-20b is shown a triangle with two acute angles and one

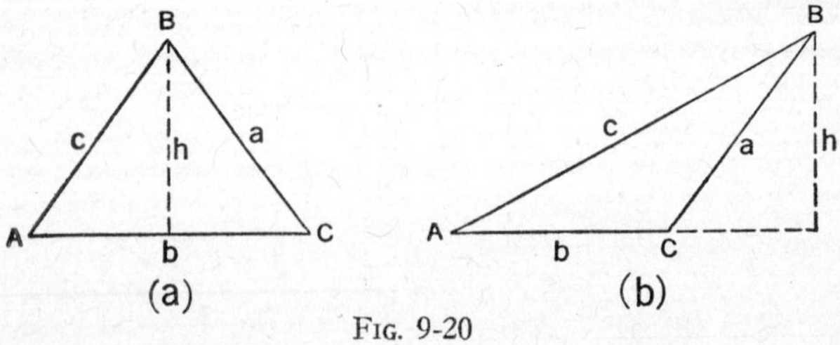

FIG. 9-20

obtuse angle. In these figures and those that follow, each angle and the side opposite it are denoted by the same letter. Altitudes h are drawn from B ⊥ b forming right triangles. Since fig. 9-20 includes an obtuse angle, its supplementary angle will be considered and will be denoted by C.

Then, in either figure,

$$\frac{h}{a} = \sin C \text{ and } \frac{h}{c} = \sin A \quad \text{or } h = a \sin C \text{ and } h = c \sin A$$

Then, $a \sin C = c \sin A$ and consequently

$$\frac{a}{\sin A} = \frac{c}{\sin C}$$

Example—Determine the value of side a in fig. 9-21.

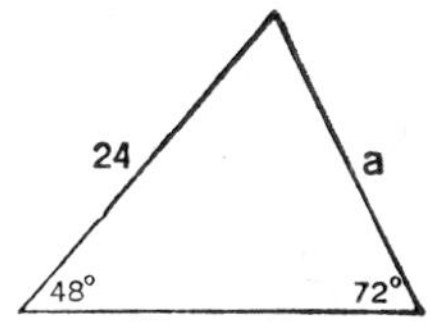

Fig. 9-21

$$\frac{a}{\sin 48°} = \frac{24}{\sin 72°}$$

$$a = \frac{24 \sin 48°}{\sin 72°} = \frac{24 \times .743}{.951} = 18.75$$

Example—Determine the value of b in fig. 9-22.

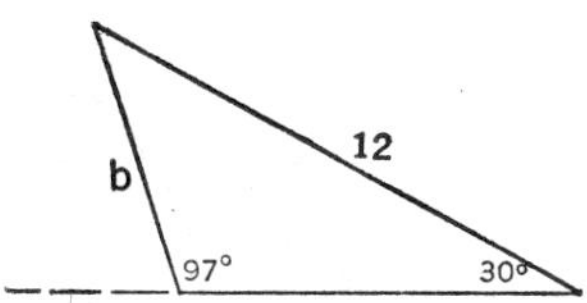

Fig. 9-22

Since the 97° angle exceeds 90° its supplementary angle must be used, which is 180 − 97 or 83°. Then

$$\frac{b}{\sin 30°} = \frac{12}{\sin 83°}$$

$$b = \frac{12 \sin 30°}{\sin 83°} = 6.045$$

When two sides and an angle opposite one of them is given, the construction of two triangles is sometimes possible. Fig. 9-23 shows ⚠ ABC and ABD each with sides 4 and 5, and the angle opposite 4 is 30° in either case.

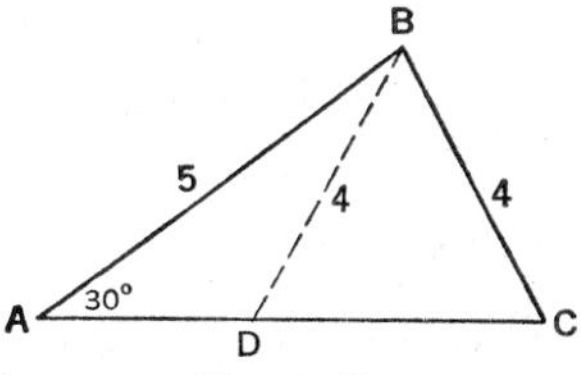

Fig. 9-23

$$\frac{\sin C}{5} = \frac{\sin 30°}{4}$$

$$\sin C = \frac{5 \sin 30°}{4} = .625$$

$$C = 38° 41'$$

The acute ∠D also = 38° 41′. Then the obtuse ∠D = 180° − 38° 41′ = 179° 60′ − 38° 41′ = 141° 19′.

Exercises

Determine the value of the lettered sides and angles in the following figures:

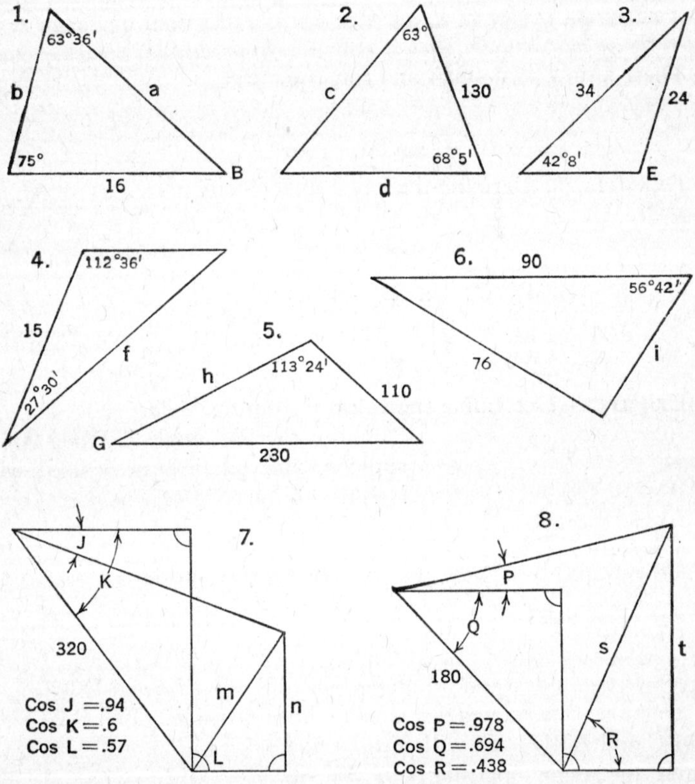

Law of Cosines—Problems classified as cases 3 and 4 may be solved by the law of cosines, which may be stated as follows:

(*a*) *The square of the side opposite an angle less than 90° equals the sum of the squares of the other two sides minus twice the product of these two sides and the cosine of the included angle.* (*b*) *The square of a side opposite an angle greater than 90° equals the sum of the squares of the other two sides plus twice the product of these two sides and the cosine of the supplement of the included angle.*

PROOF—In fig. 9-24a is shown a triangle with three acute angles, and in fig. 9-24b is shown a triangle with two acute angles and one obtuse angle. Altitudes h are drawn from $B \perp b$ forming right triangles. Since fig. 9-24b includes an obtuse angle, its supplementary angle will be considered and will be denoted by A.

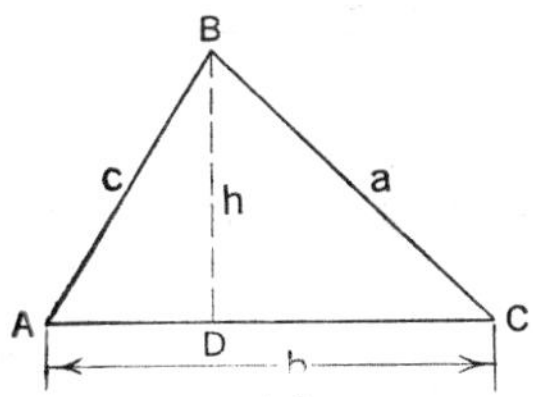

(a)

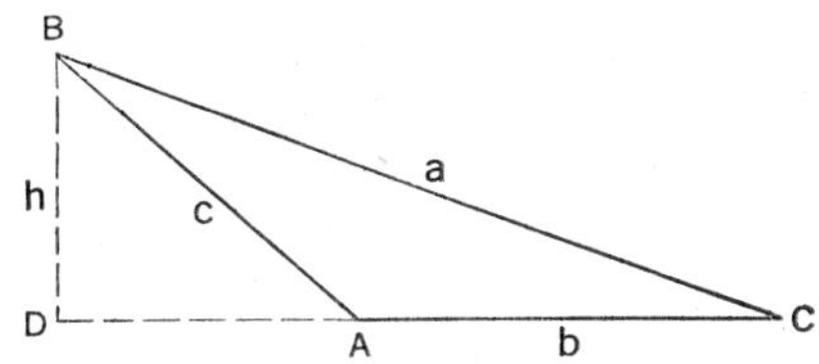

FIG. 9-24 (b)

Then in either case

$$h^2 = a^2 - \overline{CD}^2 \text{ and } h^2 = c^2 - \overline{AD}^2$$

Since the right members both equal h^2, then

$$a^2 - \overline{CD}^2 = c^2 - \overline{AD}^2 \quad \text{or} \quad a^2 = c^2 - \overline{AD}^2 + \overline{CD}^2 \quad (1)$$

In fig. 9-24a, $CD = b - AD$ or $\overline{CD}^2 = b^2 - 2b \times AD + \overline{AD}^2$

Substituting in equation (1),

$a^2 = c^2 - \overline{AD}^2 + b^2 - 2b \times AD + \overline{AD}^2$ Cancelling gives

$$a^2 = b^2 + c^2 - 2b \times AD \qquad (2a)$$

$$\frac{AD}{c} = \cos A \text{ or } AD = c \cos A$$

Substituting for AD in equation (2a) gives

$$a^2 = b^2 + c^2 - 2bc \cos A$$

In fig. 9-24b, $CD = b + AD$ or $\overline{CD}^2 = b^2 + 2b \times AD + \overline{AD}^2$

Substituting in equation (1),

$$a^2 = c^2 - \overline{AD}^2 + b^2 + 2b \times AD + \overline{AD}^2 \text{ or } a^2 = b^2 + c^2 + 2b \times AD \qquad (2b)$$

$$\frac{AD}{c} = \cos A \text{ or } AD = c \cos A$$

Substituting for AD in equation (2b) gives

$$a^2 = b^2 + c^2 + 2bc \cos A$$

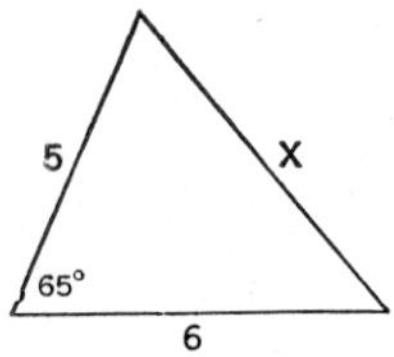

FIG. 9-25

EXAMPLE—Determine the length of side X in fig. 9-25.

Since X lies opposite an angle less than 90°, part "a" of the law of cosines applies. Then $X^2 = 5^2 + 6^2 - 2 \times 5 \times 6 \cos 65° = 35.64$

$$X = \sqrt{35.64} = 5.97.$$

EXAMPLE—Determine the length of side Z in fig. 9-26.

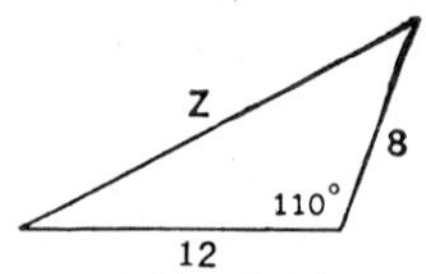

FIG. 9-26

Since Z lies opposite an angle greater than 90°, part "b" of the law of cosines applies. Then $Z^2 = 8^2 + 12^2 + 2 \times 8 \times 12 \cos 70° = 273.67$

$$Z = \sqrt{273.67} = 16.54$$

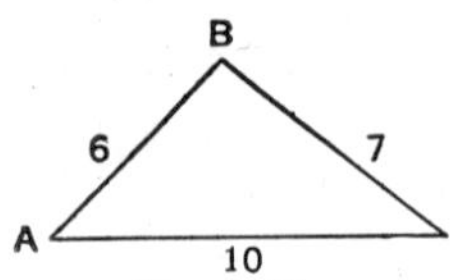

FIG. 9-27

EXAMPLE—Determine $\angle A$ and $\angle B$ in fig. 9-27. Obviously, $\angle A$ is less than 90°, then part "a" of the law of cosines applies.

$$7^2 = 6^2 + 10^2 - 2 \times 6 \times 10 \cos A$$

$$\cos A = .725, \text{ or } A = 43° \ 32'$$

From appearance, we cannot tell if angle B is greater or less than 90°. Suppose we assume that B is less than 90°, thus applying part "a" of the law of cosines.

$$10^2 = 6^2 + 7^2 - 2 \times 6 \times 7 \cos B$$

$$\cos B = -.17851$$

The negative sign indicates that the assumption is incorrect, and therefore angle B is greater than 90°. Letting S represent the supplementary angle and applying part "b" of the law of cosines,

$$10^2 = 6^2 + 7^2 + 2 \times 6 \times 7 \cos S$$

$$\cos S = .17857 \text{ or } S = 79° \ 43'$$

$$B = 180 - S = 180 - 79° \ 43' = 100° \ 17'$$

Exercises

Determine the values of the lettered sides and angles in the following figures:

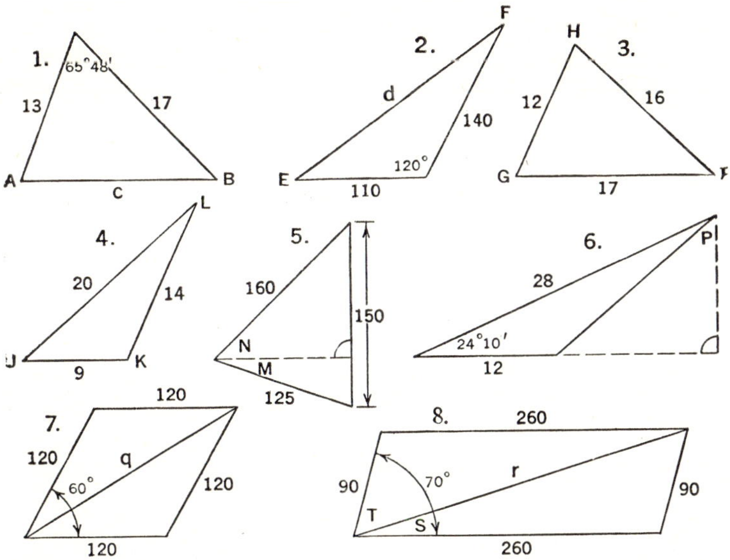

Vectors—In order to express some quantities completely, it is necessary to indicate their direction as well as their magnitude. This is done by means of a straight line drawn in the proper direction

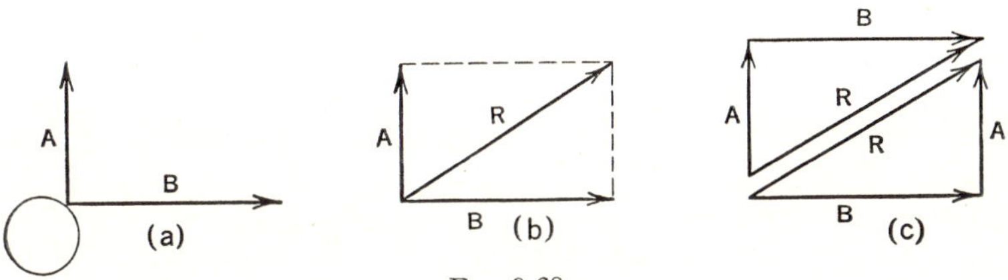

FIG. 9-28

having a length proportional to the amount of the quantity. An arrow head is placed at one end of the line to indicate the direction of the quantity. In the study of alternating current it is frequently necessary to find the resultant of two or more vectors, which represent electrical quantities. But, at the present time, the methods of finding resultants are more easily made clear by means of examples in which forces are considered.

Suppose that two forces A and B act at right angles on a body (fig. 9-28a). These two forces are equivalent to a single force, which is called the resultant. The resultant is the diagonal of the rectangle formed by using the two vectors as sides (fig. 9-28b). The diagonal divides the rectangle into two equal triangles and either of these shows the relation between the forces and the resultant (fig. 9-28c).

EXAMPLE—Forces of 30 and 40 pounds act on a body at right angles. Determine (a) the resultant and (b) the angle it makes with the 40 pound force.

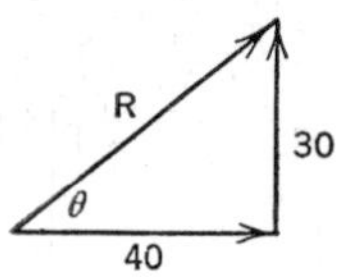

FIG. 9-29

(a) $R = \sqrt{30^2 + 40^2} = 50$ (see fig. 9-29)

(b) $\text{Tan}\ \theta = \dfrac{30}{40} = .75$ or $\theta = 36°\ 52'$

Components of Vectors—Since a resultant is the equivalent of two vectors at right angles, so two vectors at right angles are equivalent to a single vector, and therefore any single vector may be replaced by two at right angles. This is called replacing a vector by its components. If two vectors are displaced by some angle other than 90°, their resultant may be obtained by replacing one vector by its components, one in the same direction as the second vector and one at right angles to the second.

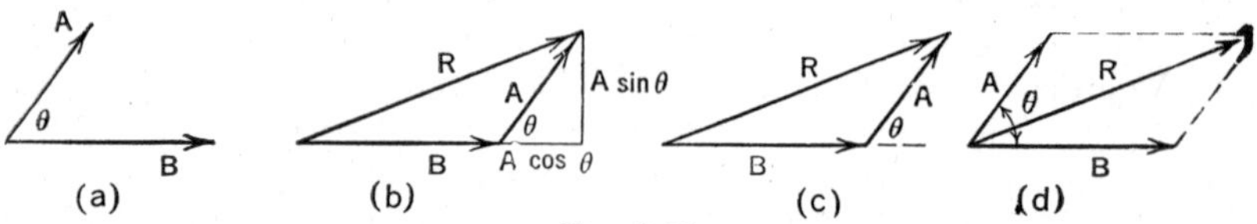

FIG. 9-30

Suppose it is desired to find the resultant of two vectors A and B that are $\theta°$ displaced (see fig. 9-30a). Vector A may be replaced by its components, and these are combined with B to give the resultant (fig. 9-30b).

EXAMPLE—Forces of 6 and 8 pounds 30 degrees apart (fig. 9-31a) act on a body. Find (a) the resultant and (b) the angle it makes with 8 pound force.

(a) The horizontal component is $8 + 6 \cos 30° = 8 + 5.196 = 13.196$ (fig. 9-31b).

The vertical component is $6 \sin 30° = 6 \times .5 = 3$.

The resultant is $R = \sqrt{13.196^2 + 3^2} = 13.53$ pounds.

$$\text{(b)}\quad \sin A = 3 \div 13.53 = .22173$$

$$A = 12°\ 49'$$

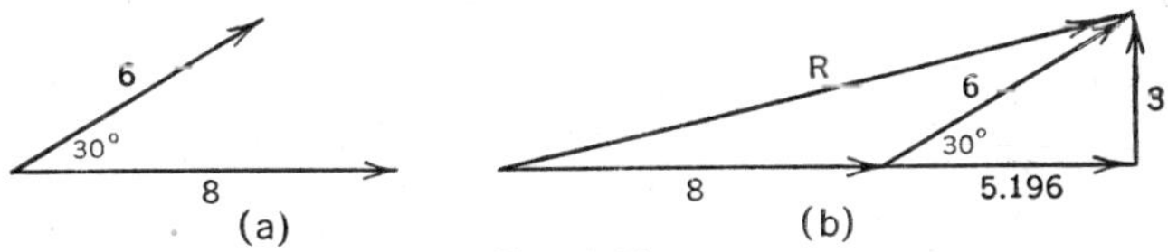

FIG. 9-31

Triangle Method—From fig. 9-30b a second method known as the triangle method may be deduced for finding the resultant. The two vectors, whose resultant is to be found, are placed, without changing their direction, so as to form two sides of a triangle. The third side is the resultant (fig. 9-30c).

Parallelogram Method—The resultant is also the diagonal of a parallelogram formed by using the two vectors as sides (fig. 9-30d). This method of determining the resultant is known as the parallelogram method.

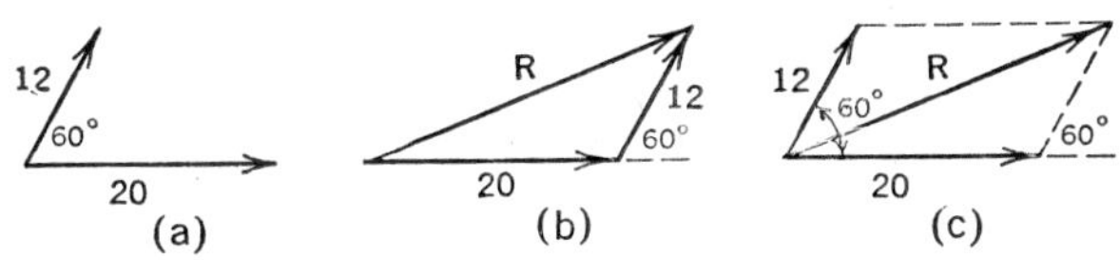

FIG. 9-32

When the resultant of two vectors is obtained by the triangle or parallelogram method, the law of cosines is used.

EXAMPLE—Find the resultant of the two vectors shown in fig. 9-32a.

In finding the resultant by the triangle method (fig. 9-32b) or the parallelogram method (fig. 9-32c),

$$R^2 = 20^2 + 12^2 + 2 \times 20 \times 12 \cos 60° = 400 + 144 + 240 = 784$$

$$R = \sqrt{784} = 28$$

Resultant of Several Vectors—The most convenient method of finding the resultant of more than two vectors is to find the horizontal and vertical component of each, then find the resultant horizontal and the resultant vertical component, which are then combined.

EXAMPLE—Find the resultant of the vectors shown in fig. 9-33a.

The resultant horizontal component is $8 + 10 \cos 30° + 12 \cos 60° = 22.66$ (fig. 9-33b).

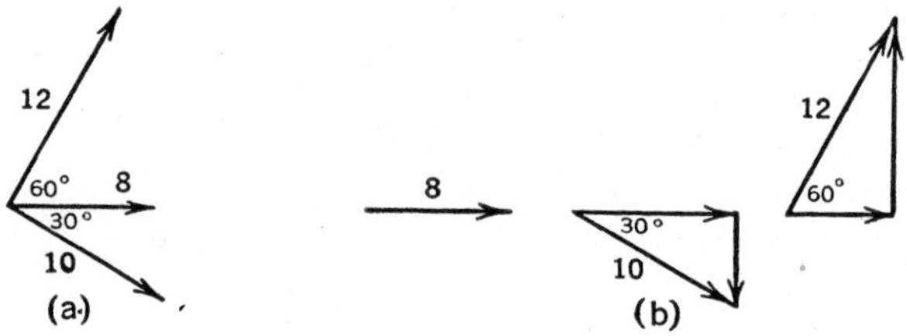

FIG. 9-33

The resultant vertical component is $12 \sin 60° - 10 \sin 30° = 5.39$. The resultant is $\sqrt{22.66^2 + 5.39^2} = 23.29$.

The resultant of several vectors may also be obtained by finding the resultant of two, then combining this resultant with a third, etc. until the final resultant is obtained.

Exercises

1. Name the three methods of finding the resultant of two vectors.
2. Find the resultant of a 10 pound force and a 30 pound force acting at right angles to each other.
3. What horizontal and vertical force added together give a resultant of 200 pounds acting at an angle of 70° 22′ with the horizontal?

Find (a) the resultant of the vectors shown below and (b) determine the angle that this resultant forms with the horizontal.

4.

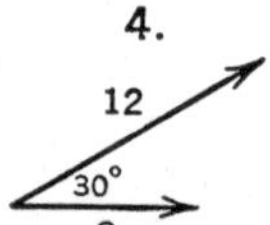

5.

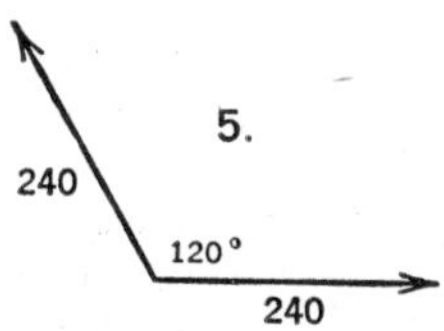

6.

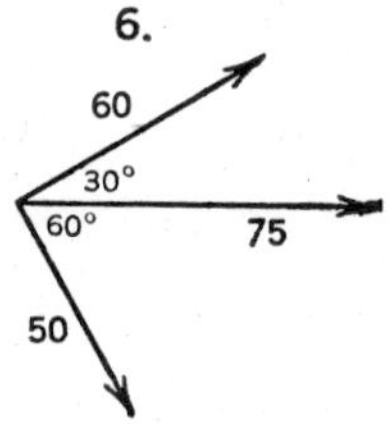

7.

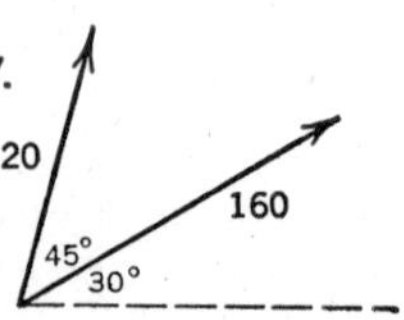

8.

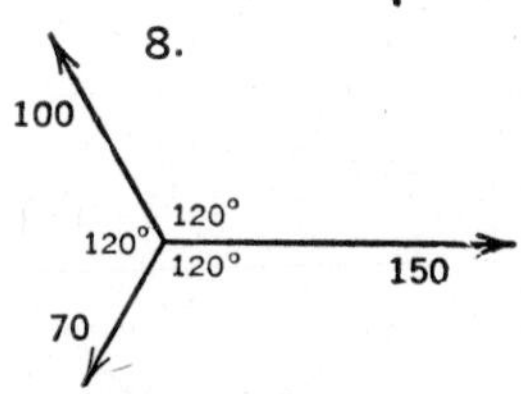

CHAPTER 10

ALTERNATING CURRENT

It was shown in chapter 5 that the voltage generated in a coil reverses as the coil sides pass from pole to pole, and that the commutator maintains a fixed polarity with reference to the external circuit. No commutator is used in an alternating current generator (see fig. 10-1a), and consequently, the alternating voltage produced is impressed directly upon the external circuit. This voltage (or current) not only reverses in direction, but also varies in magnitude at different instances, as shown by the curve of fig. 10-1b. This curve shows that the voltage increases from zero to a maximum point,

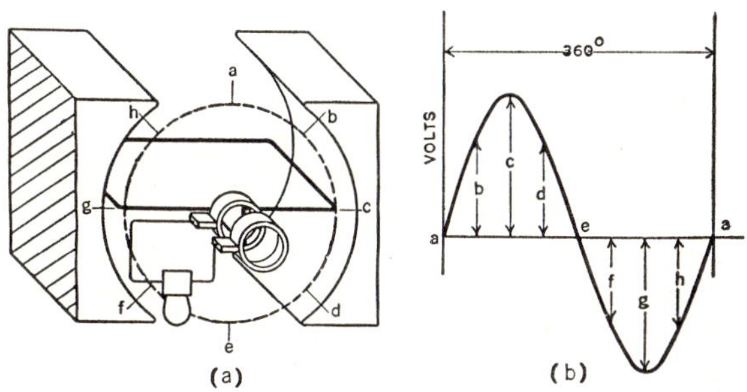

Fig. 10-1

decreases through zero to a maximum point in the opposite direction, and then falls to zero again. When such a change in voltage has taken place, a cycle is said to have been completed. From then on the cycle is merely repeated.

The number of cycles occurring per second is called the frequency. The usual frequencies in this country are 25 and 60 cycles. The 25 cycle frequency is less common, because it produces a perceptible flicker when it is used for lighting. Furthermore, 25 cycle machines of a given rating are larger and more expensive than 60 cycle machines.

Hydraulic Analogy of Alternating Current—The alternating characteristic of alternating current can be illustrated very well by the hydraulic analogy of Fig. 10-2. When the piston P is moved toward A, water is forced out at A and sucked in at C, and thus flows through the pipe line in the direction ABC. After the piston has reached the end of the stroke, it moves back toward C. The water is then forced out at C and sucked in at A, and flows around the pipe

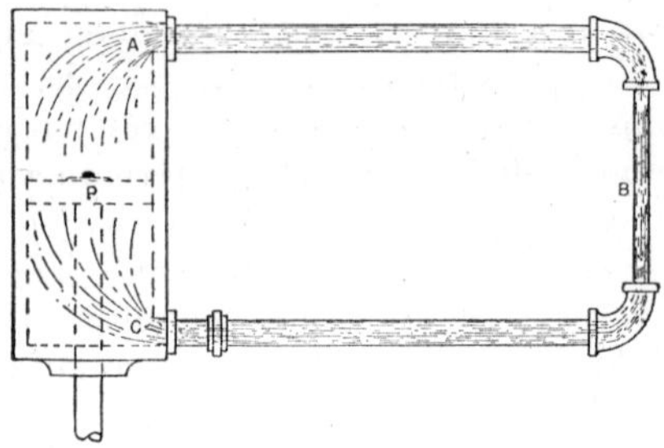

Fig. 10-2

line in the direction CBA. As the piston is moved back and forth, it sets up an alternating pressure which causes the current of the water to change its direction at the end of each stroke. Thus the current of water will alternate in direction, and the rate of flow will vary widely.

Advantages of Alternating Current—A large percentage of the electrical plants generate power in the form of alternating current. As alternating current generators have no commutators, they may be built for greater capacities and higher voltages than direct current generators. With alternating current, moreover, the voltage may be raised and lowered economically by means of transformers. A transformer operates at a high efficiency and requires little attention, because it contains no moving parts.

In most cases, electrical energy must be used at low voltages, because that is safer. When energy is to be used at a distance of more than a mile or two from the generator, it must be transmitted at a high voltage. The present practice is to use at least 1000 volts per mile. In this way the energy is transmitted at a low current value, which means that smaller conductors are required and the line losses are small. To increase the efficiency, then, the voltage is stepped up (by means of one transformer) at the generator end of the line, and afterwards stepped down to a safe value by means of another transformer at the point where the power is to be used.

Direct current voltages cannot be raised or lowered on a large scale without using machines with commutators, and the permissible voltage of a commutator is low. Hence, direct current cannot be transmitted as efficiently as alternating current.

The high transmission efficiencies obtainable with alternating current make it desirable to generate electrical energy in large quantities at a single station and to distribute the power over a large territory. The use of large boilers, automatic stokers, steam turbines, etc., which are found in large stations, makes possible a high over-all efficiency. Another reason for desiring alternating current is that it is required for induction motors, which are cheaper than direct current motors in first cost and in maintenance. Induction motors have no commutators and the squirrel cage type has a very simple rotating unit which requires practically no maintenance.

Until recently it was thought that direct current was essential for some classes of work. While this is true of electroplating and battery charging, in most other industrial applications alternating current is rapidly replacing direct current. Chief among these cases are applications requiring variable speed motors. For instance, the speed of squirrel cage induction motors (a very common type of A C motor) cannot be varied when a constant frequency is supplied. Where it is desired to change the speed on certain production machines, conveyors, etc., standard squirrel cage motors cannot be used. Recent developments have made possible slight speed changes in this type of motor, by the use of electronic controls.

Generation of a Sine Wave of Voltage—To investigate the true shape of the alternating current voltage curve discussed above, suppose in fig. 10-3 that 10^8 magnetic lines per inch emanate from the poles shown and that the coil side rotates at the rate of one inch per second. This flux distribution is selected because a conductor cutting 10^8 magnetic lines per second has one volt induced in it.

When a coil side moves one inch at a (fig. 10-3a), it is traveling parallel to the magnetic field, and no voltage is induced in it. When it moves one inch at c (fig. 10-3b), it is traveling at right angles to the flux. It then cuts 10^8 magnetic lines, and a maximum of one volt is induced in it. When the coil side moves one inch at b (fig. 10-3c), where the plane of the coil forms an angle of 45° with the horizontal, the coil side is traveling at an angle of 45° with the vertical. This relation can easily be proved by geometry. In moving one inch at this angle, the coil side generates an e m f equivalent to that which would be generated if the coil side were moving the distance ad horizontally, for the same amount of flux would be cut. The distance ad $= 1 \times \sin 45° = .707$ inch. At this instant the coil side cuts .707 of 10^8 magnetic lines per second, and .707 of a volt is therefore induced.

In fig. 10-3d assume that the number of magnetic lines per inch is such that a maximum voltage E_m is induced when the coil side moves one inch at c. Now, when the coil side moves one inch elsewhere such as at e, it has the effect of moving $(1 \times \sin \theta)$ inch horizontally. The voltage induced will be $(1 \times \sin \theta)$ of E_m. This means that the instantaneous voltage $e = (1 \times \sin \theta)\ E_m$ or

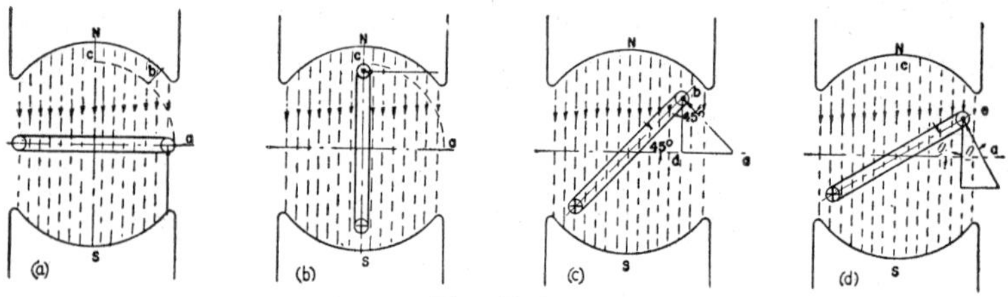

FIG. 10-3

$$e = E_m \sin \theta \qquad (10\text{-}1)$$

where $\theta =$ angle formed by the coil with the horizontal (or the angle formed by the coil side with the vertical); $E_m =$ maximum induced voltage; and e = voltage induced at instant under consideration.

EXAMPLE—If a maximum of 2 volts is induced in a coil rotating in a magnetic field, determine the voltage induced when the coil forms an angle of 20 degrees with the neutral plane.

Using equation 10-1, $e = 2 \sin 20° = 2 \times .342 = .684$ volt.

If a sine wave of e m f is impressed upon a fixed resistance, the alternating current produced will also have a sine form and may be represented by the equation

$$i = I_m \sin \theta \qquad (10\text{-}2)$$

Since the voltage and the current in an alternating current circuit are always changing in value, the question naturally arises as to how these may be measured. The permanent-magnet type of direct current meter cannot be used since the torque would reverse each half cycle, giving an average indication of zero. It is sometimes useful, however, to know the average value over half a cycle, which evidently is not zero. This average value may easily be obtained by plotting a number of equally spaced instantaneous values for one-half a cycle and finding their average. A more accurate way of making this computation is by means of calculus. The result of either method will show that the average value of alternating voltage is .637 times the maximum value (see fig. 10-4). Expressing this by an equation:

$$E_{av} = .637\ E_m \text{ or } \frac{2}{\pi} E_m \qquad (10\text{-}3)$$

Since the current also changes according to the sine curve, then

$$I_{av} = .637\ I_m \text{ or } \frac{2}{\pi} I_m \qquad (10\text{-}4)$$

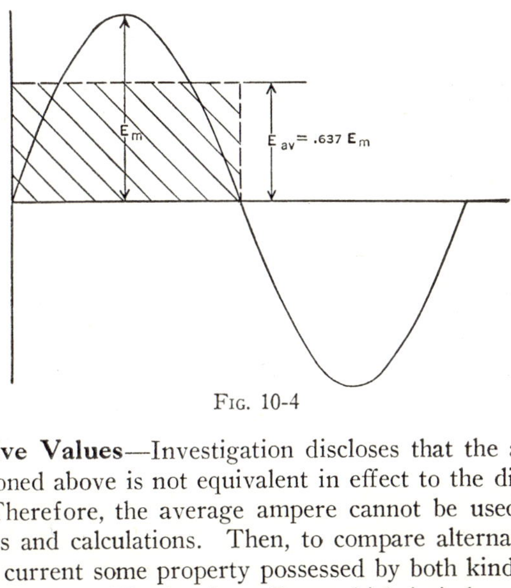

FIG. 10-4

Effective Values—Investigation discloses that the average ampere mentioned above is not equivalent in effect to the direct current ampere. Therefore, the average ampere cannot be used in making comparisons and calculations. Then, to compare alternating current with direct current some property possessed by both kinds of current must be employed. The heating effect, which is independent of the direction of flow, is a convenient base for comparison. Hence, the unit of alternating current is that current which will produce the same heating effect as a unit of direct current. On this basis an alternating current is said to have an effective value of one ampere if it produces heat in a certain resistance at the same average rate as heat is produced in the same resistance by one ampere of direct current.

For instance, suppose two calorimeters (containers for measuring the heat produced) are arranged so that one measures the heat generated per unit of time by an alternating current flowing through a resistance R. The other measures the heat generated per unit of time by a direct current flowing through an equal resistance R.

Consider that the alternating current is so regulated that it produces the same amount of heat per unit time as the direct current. Then from the above definition

$$I_{dc} = I_{ac}$$

where I_{dc} = amperes of direct current; I_{ac} = effective value of alternating current.

By direct current, the heat in B T U generated, $H = .000948\ I_{dc}^2Rt$.

By alternating current, the heat generated, $H = .000948\ (\text{av}\ i^2)Rt$. where i = instantaneous value of currents and t = time in seconds.

But the heat generated by the direct current equals the heat generated by the alternating current.

Therefore $.000948\ I_{dc}^2Rt = .000948\ (\text{av}\ i^2)Rt$

Then $I_{dc}^2 = \text{av}\ i^2$,

or $I_{dc} = \sqrt{\text{av}\ i^2}$.

But $I_{dc} = I_{ac}$.

Thus $I_{ac} = \sqrt{\text{av}\ i^2}$

This means that the effective value of alternating current =

$$\sqrt{\text{average of squares of instantaneous values.}}$$

The method of obtaining the effective value of current when the maximum is known is illustrated in Fig. 10-5. The curve a b c is a sine curve with a maximum value of $2\sqrt{2}$ amperes. Instantaneous values at regular intervals over the first half of the cycle are squared. For example, the instantaneous value x y = 2 amperes. Hence, its squared value x z = 4. The current squared curve, a d c, is plotted as shown. Its maximum obviously must be $(2\sqrt{2})^2$ or 8. The i^2 values form a curve which lies entirely above the zero axis, because the square of the negative values is positive. This new curve forms a sine wave which has a frequency twice that of the original wave. The average of this i^2 wave is 4 as shown by the dotted line, because the area above the dotted line will just match the shaded area below it. Then the effective value is $\sqrt{4}$ or 2. The ratio of the effective to the maximum current is $\frac{2}{2\sqrt{2}}$ or .707. From this illustration it should be realized that

$$I_{eff} = .707\ I_m = \frac{I_m}{\sqrt{2}} \tag{10-5}$$

By using equation 10-5, we may calculate the effective value directly without the necessity of determining the instantaneous values as described above.

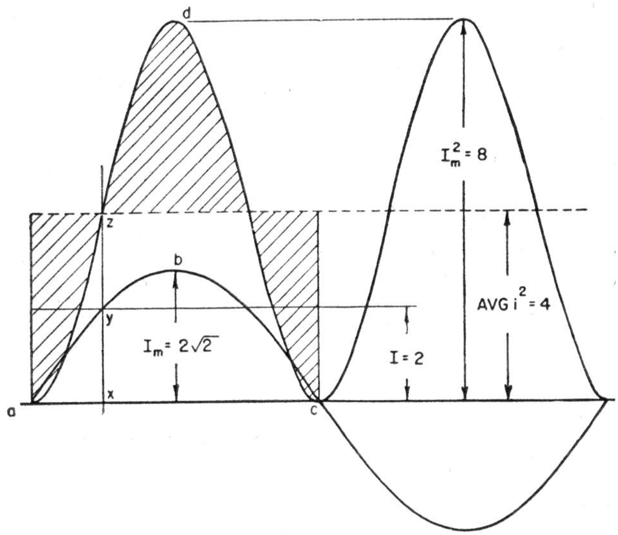

FIG. 10-5

The effective value of an alternating voltage is the square root of the average of the squares of the instantaneous values. Hence,

$$E_{eff} = .707E_m = \frac{E_m}{\sqrt{2}} \tag{10-6}$$

When an alternating current or voltage is specified, it is always the effective value that is meant unless there is a definite statement to the contrary.

Vector Representation of Alternating Voltage or Currents—The change which occurs in the value of an alternating voltage (or current) during a cycle need not be represented by a curve plotted

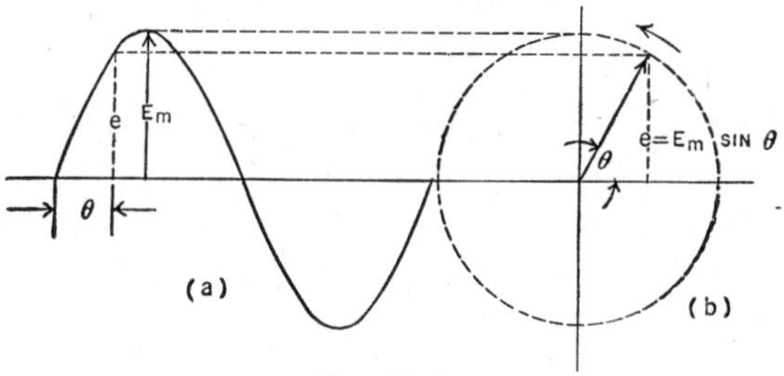

FIG. 10-6

as shown in fig. 10-1b, but may more easily be represented by vectors, as in fig. 10-6b. The vector E_m is made with its length equal to the maximum height of the sine curve (fig. 10-6a), and is rotated in the counter-clockwise direction at a speed in revolutions per second equal

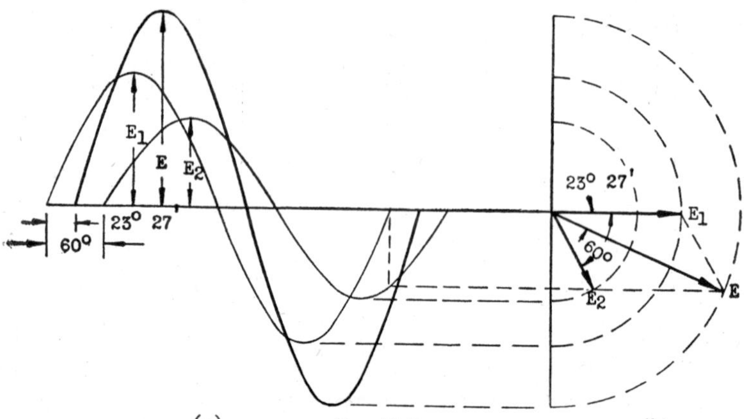

(a) FIG. 10-7 (b)

to the frequency of the voltage. That is, if the voltage in fig. 10-6a has a frequency of 60 cycles, the vector E_m in fig. 10-6b must make 60 revolutions per second. The vertical projection of the rotating vector E_m determines points upon the sine curve, because its projection equals $E_m \sin \theta$, which from equation 10-1 gives the instantaneous value of the voltage.

Assume that it is desired to add voltages E_1 and E_2 with effective values of 150 and 100 volts respectively, E_2 lagging E_1 by 60 degrees. From equation 10-6 these voltages will have maximum values of 212 and 141.4 volts. The addition may be performed by plotting the waves of the two voltages (E_1 and E_2, fig. 10-7a) and adding at equally spaced distances their instantaneous values, and with the sums obtained, plotting a new curve (E). This new curve, if determined accurately, will be found to have a maximum value of 308 or an effective value of 218 volts, and will lag E_1 by 23° 27′. Thus the sum of two voltages with effective values of 150 and 100 volts and differing in phase by 60 degrees is 218 volts.

This problem may be solved without using such a lengthy process of plotting waves and adding instantaneous values. These waves may be entirely represented by rotating vectors spaced 60 degrees apart, as indicated by E_1 and E_2 in fig. 10-7b. By adding these two vectors and completing the parallelogram, the third vector E is obtained. This vector will be found to have a value of 308 volts, the same value as the maximum of the resultant wave in fig. 10-7a. Furthermore, if a wave is plotted using E as a rotating vector, it will be found to coincide with wave E (fig. 10-7a), which was obtained by the addition of the instantaneous values of E_1 and E_2.

Since effective values are merely .707 times the maximum values, the same kind of diagram shown in fig. 10-7b, with vectors .707 times as long, would represent the effective values of E_1, E_2, and E. This is illustrated in fig. 10-8. Note that the 150 and 100 volts are represented 60 degrees apart, the 100 volt vector lagging. The resultant E is obtained by completing the parallelogram.

Its value is obtained as follows:

$$E = \sqrt{150^2 + 100^2 + 2 \times 150 \times 100 \times \text{Cos } 60°} = 218 \text{ volts}$$

EXAMPLE—At what speed in revolutions per minute must the field poles in fig. 10-9 rotate to give a voltage of 60 cycle frequency?

The e m f passes through one cycle when two poles pass a conductor, or for one revolution two cycles are produced. Then the field poles must make 30 revolutions per second or 1800 revolutions per minute in order to generate a voltage of 60 cycles per second.

EXAMPLE—To what extent would an electrician be shocked if he came in contact with a 440 volt line?

He would be subject to the maximum voltage, or

$$E_m - \frac{440}{.707} = 622.35 \text{ volts}$$

EXAMPLE—If the maximum value of an alternating current is 48 amperes, what is its effective value?

$$I = .707 \times 48 = 33.94 \text{ amperes}$$

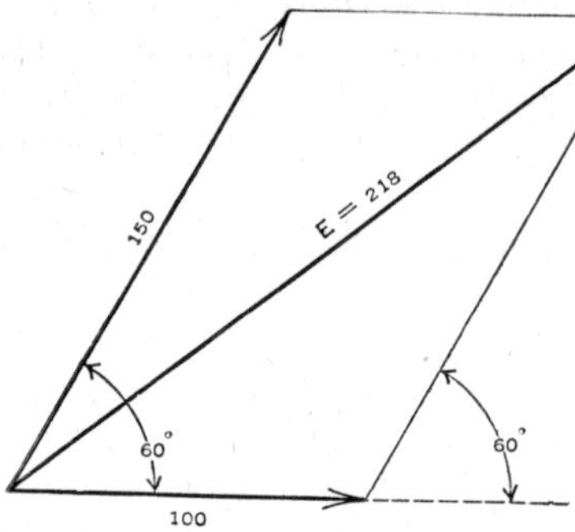

FIG. 10-8

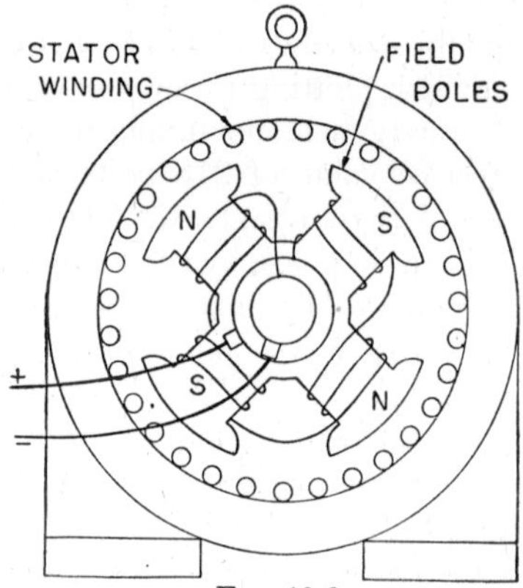

FIG. 10-9

Self-Inductance—Referring to fig. 10-10a, when the switch is closed at a, a magnetic field is produced, which increases as the cur-

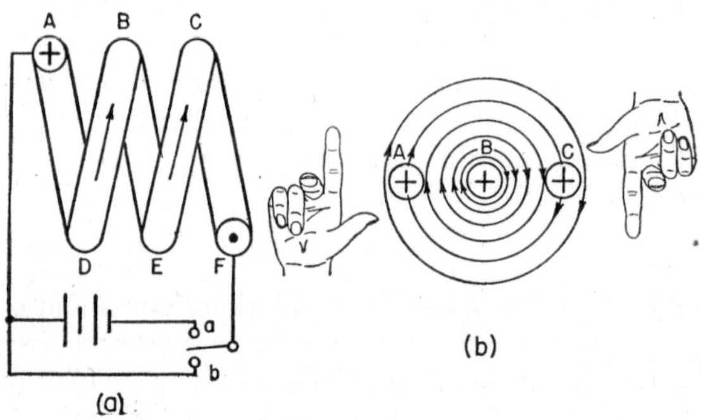

FIG. 10-10

rent increases. The magnetic loops start at the centers of the conductors and expand to their ultimate positions. As they expand, they cut all the conductors, which they ultimately encircle, and therefore induce a voltage in them. The flux expanding from cross section B (fig. 10-10a) is shown in fig. 10-10b. As this flux expands, it cuts across sections A and C. Cross section A is cut by flux in the upward direction, moving toward the left. The direction of the induced voltage in A may be determined by considering that the flux surrounding B is stationary and assuming that A is moved toward the right. On applying the right-hand rule, it is found that an e m f is induced outward in A. In like manner, a voltage is induced outward in conductors C and B. An e m f is induced inward in cross sections D, E, and F. *Then an e m f is induced by an increasing flux, in a direction opposite to the applied e m f.* After the current becomes constant, an e m f is no longer induced. The effect of an induced voltage is to cause the

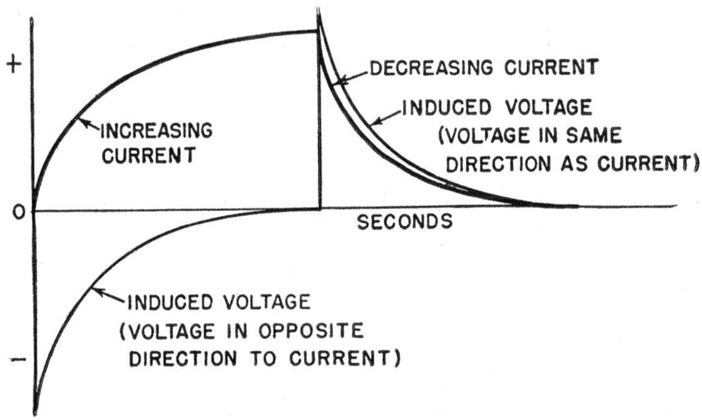

FIG. 10-11

current to increase at a slow rate, instead of increasing instantly to its constant value (see fig. 10-11). Its constant value may be found by Ohm's law ($I = E/R$).

If the switch is now opened at a and instantly closed at b, the current will decrease and the magnetic field will contract. When magnetic lines contract, they remain closed loops which shrink until they become points at the centers of the conductors, and then vanish. While the magnetic lines are contracting, they cut the turns in the opposite direction and therefore induce an e m f in the same direction as the current. The current will then decrease at a slow rate, instead of decreasing instantly to zero (see fig. 10-11).

These relationships are expressed by Lenz's law, as follows: *When a current in an inductive circuit is increasing, the induced e m f opposes the applied e m f, or tends to keep the current from increasing; and when the current is decreasing, the induced e m f reverses and acts in the same direction as the applied e m f, or tends to maintain the current in the circuit.*

Computation of Induced Voltage—We learned that a conductor cut by 10^8 magnetic lines per second has one volt induced in it. Then if the flux surrounding a conductor changes from Φ_1 to Φ_2 lines in t seconds, the conductor will be cut at the rate of

$$\frac{\Phi_2 - \Phi_1}{t}$$

lines per second. The average voltage induced in it equals

$$\frac{\Phi_2 - \Phi_1}{10^8 t}$$

For a coil consisting of T turns, the average voltage generated in it is

$$E_{av} = T\,\frac{\Phi_2 - \Phi_1}{10^8 t} \qquad (10\text{-}7)$$

EXAMPLE—A flux of 2×10^6 lines links a coil of 300 turns. What e m f will be induced in it if the flux decreases at a uniform rate to zero in .2 of a second?

$$E_{av} = \frac{300(2 \times 10^6)}{10^8 \times .2} = 30 \text{ volts}$$

Inductance in an Alternating Current Circuit—Any change in the current flowing through a coil causes the flux to change and thus induce a voltage in it. If an alternating current is supplied to a coil, an e m f will be continuously induced in it, because such a current is continuously changing. Suppose that an alternating current is supplied to the coil shown in fig. 10-12a, which is assumed to have a negligible resistance. The current is limited by the induced voltage. In fig. 10-12b, at a, c, and e, the current I is changing at the greatest rate and therefore the induced e m f at these instances is at a maximum. At b and d, however, the current is at its maximum value and is not changing. Then the induced e m f at these instances is zero. During the interval from a to b, the current is increasing, and is in the positive direction. According to Lenz's law, the induced e m f tends to oppose this change, and hence is in the opposite or negative direction, as indicated by line a′b. From b to c the current is decreasing, and the induced e m f tends to keep it flowing. This induced e m f is in the same, or positive, direction, as shown by line bc′. From c to d the current is increasing in the negative direction. Then the induced e m f remains positive, as indicated by line c′d. From d to e the current is again decreasing. Then the induced e m f tends to keep the current flowing and must be in the negative direction, as shown by line de′.

To force this current (I) through the coil, the applied voltage (E_a) must at each instant be equal and opposite to the induced voltage (E_i).

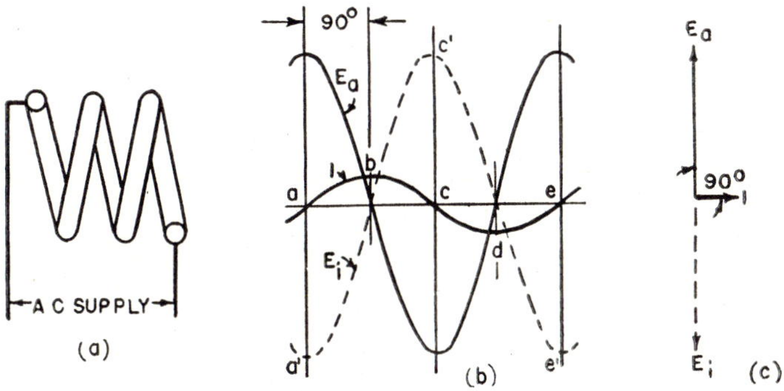

Fig. 10-12

This is shown vectorially in fig. 10-12c. If the applied voltage varies as a sine wave, the current and the induced voltage will likewise vary as a sine wave.

For reasons that will become evident later, we are primarily concerned, in any circuit, with the applied voltage and the current. *This discussion shows that in an inductive circuit with a negligible resistance, the current lags the applied voltage by 90 degrees.*

Computation of Inductance—Apparently, if a coil has many turns of large wire and much iron of a high permeability, only a small current must flow through it to establish flux sufficient to induce a voltage equal to the applied voltage. Such a coil is said to have a high inductance. The unit of inductance is called the henry, and is denoted by L. *A coil is said to have an inductance of one henry if* *the current on changing at the rate of one ampere per second establishes a change in flux that induces an average of one volt.* This relation may be expressed as

$$E_{av} = L\frac{I_2 - I_1}{t} \qquad (10\text{-}8)$$

where E_{av} = average induced e m f in volts, I_1 = initial current in amperes, I_2 = final current in amperes, t = time in seconds during which the current changes from I_1 to I_2, and L = inductance in henries.

EXAMPLE—The field of a D C generator has an inductance of 25 henries. If the field current is 20 amperes and is interrupted in .05 of a second, determine the average e m f induced in the winding.

$$E_{av} = \frac{25(20)}{.05} = 10{,}000 \text{ volts}$$

Inductive Reactance—The effect of inductance in a coil carrying alternating current may be expressed more conveniently in another manner. Consider a current of frequency f and of maximum value I_m, flowing through a coil of negligible resistance. During one cycle the current makes four changes, as indicated in fig. 10-13. (1) The current rises from zero to maximum in the positive direction; (2) it drops from maximum to zero; (3) it rises from zero to maximum in the negative direction; (4) it drops from maximum to zero. In each of the four changes, the current goes from zero to maximum, or vice versa, and during each of these changes, which take place in a quarter of a cycle, the same average voltage is generated. The time for one cycle is 1/f and for 1/4 cycle is 1/f × 1/4 or 1/4f. During 1/4 cycle, the current changes from zero to I_m or from I_m to zero, and for one of these changes, equation 10-8 becomes

$$E_{av} = L\frac{I_m}{t} \qquad (10\text{-}9)$$

Since $t = \frac{1}{4f}$,

$$E_{av} = L\frac{I_m}{\frac{1}{4f}} = 4fLI_m \qquad (10\text{-}10)$$

But

$$E_{av} = \frac{E_{eff}}{1.11} \text{ and } I_m = \frac{I_{eff}}{.707} \qquad (10\text{-}11)$$

Substituting 10-11 into 10-10,

$$\frac{E_{eff}}{1.11} = \frac{4fLI_{eff}}{.707}$$

or

$$E_{eff} = 6.28fLI_{eff} \qquad (10\text{-}12)$$

Since $6.28 = 2\pi$, this equation is more generally written as

$$E_{eff} = 2\pi fLI_{eff} \qquad (10\text{-}13)$$

The expression $2\pi fL$ is generally represented by X_L, which is called the inductive reactance. Then

$$X_L = 2\pi fL \qquad (10\text{-}14)$$

and equation 10-13 becomes

$$E = X_L I \qquad (10\text{-}15)$$

The subscripts are omitted, because when quantities are written without subscripts, effective values are understood.

Since the applied and induced voltages are equal in an inductive circuit with negligible resistance, E (in equations 10-13 or 10-15) may represent the applied voltage. Then equation 10-15 is similar to Ohm's law, when it is written as $E = RI$. The inductive reactance (X_L) is expressed in ohms, since it is the quantity by which the current is multiplied to give the voltage.

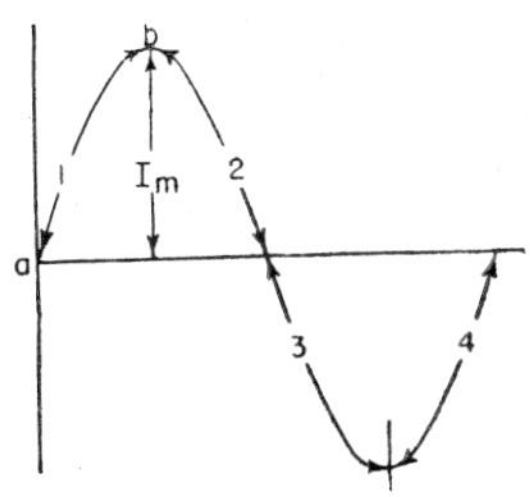

FIG. 10-13

EXAMPLE—A coil with negligible resistance takes 2 amperes when it is connected to a 120 volt, 60 cycle supply. Compute the inductance of the coil.

Substituting in equation 10-15,

$$X_L = \frac{120}{2} = 60 \text{ ohms}$$

Substituting in equation 10-14,

$$L = \frac{X_L}{2\pi 60} = \frac{60}{377} = .159 \text{ henry}$$

This shows how the inductance of a coil is determined from current and voltage readings. An equation will now be developed to show the factors upon which the inductance of a coil depends.

Equation 10-7 shows that

$$E_{av} = T\frac{\Phi_2 - \Phi_1}{10^8 t}$$

In an A C circuit, however, the change in flux per second is

$$\frac{\Phi_2 - \Phi_1}{t} = \frac{\Phi_m}{t}$$

Then, equation 10-7 becomes

$$E_{av} = \frac{T\Phi_m}{10^8 t} \qquad (10\text{-}16)$$

From equation 4-2,

$$\Phi_m = \frac{I_m T}{\mathcal{R}} \qquad (10\text{-}17)$$

Substituting equation 10-17 into equation 10-16,

$$E_{av} = \frac{I_m T^2}{10^8 t \mathcal{R}} \qquad (10\text{-}18)$$

Equation 10-9 may be written

$$L = \frac{tE_{av}}{I_m} \qquad (10\text{-}19)$$

Substituting equation 10-18 into equation 10-19,

$$L = \frac{T^2}{10^8 \mathcal{R}} \qquad (10\text{-}20)$$

This equation shows that the inductance of a coil is not a factor which depends upon the current and voltage. It shows that the inductance varies directly as the square of the number of turns and inversely as the reluctance.

For a core that has no air gap, equation 10-20 may be expressed as

$$L = \frac{T^2 \mu A}{10^8 l} \qquad (10\text{-}21)$$

EXAMPLE—A 400 turn coil is wound on a sheet steel ring with a cross section of 6 square inches and a length of 16 inches. If the flux density is 74,000 lines per square inch, determine the inductance of the coil.

Fig. 4-22 shows that 10 ampere-turns per inch are required at this flux density. Then the permeability (μ) is 74,000 ÷ 10 or 7400. Substituting in equation 10-21,

$$L = \frac{400^2 \times 7400 \times 6}{10^8 \times 16} = 4.44 \text{ henries}$$

Another equation useful in determining the inductance of a coil with an iron or steel core and no air gap may be developed by substituting for E_{av} in equation 10-19, the expression given in equation 10-16. This gives

$$L = \frac{T\Phi_m}{10^8 I_m} = \frac{TB_m A}{10^8 I_m} \qquad (10\text{-}22)$$

Non-Inductive Coil—It is sometimes desirable to wind a coil so that it will have a high resistance and no inductance. This is accomplished by winding the coil with two wires connected together at the start end, as shown in fig. 10-14. The current through such a coil is in opposite directions through adjacent turns, and consequently, the magnetizing action of half the turns just neutralizes the magnetizing action of the other half. No flux is therefore produced, and the coil is non-inductive.

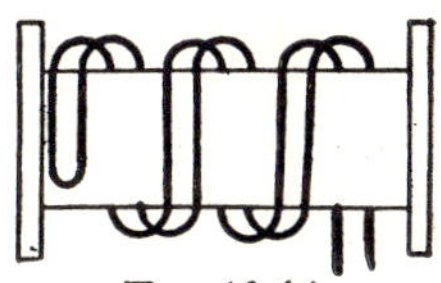

FIG. 10-14

Condensers—A condenser consists of two conductors separated by an insulating material called a dielectric (see fig. 10-15). Each atom in the dielectric consists of protons and an equal number of electrons. About half these electrons are with the protons in the central part of the atom, called the nucleus. The other half are called external electrons. These are so strongly attracted to the nucleus that they cannot be torn loose by a moderate electric pressure, but they can move to a certain extent. When they are moved a force is set up between electrons and nucleus which tends to bring them back to their original position. While the dielectric in a condenser is made up of an enormous number of atoms, only one atom is shown in fig. 10-15, in which only a few external electrons are indicated.

When the switch (fig. 10-15) is closed at A, the outer electrons are repelled by the negative plate and attracted by the positive plate. They will then move to new positions as indicated by the dotted circles. Of course, external electrons in all the atoms in the dielectric will move to new positions, and while they are moving, there will be an electric current through the circuit. Then electrons will flow out of the positive plate and into the negative plate. This will occur for only an instant, dying out as soon as the electrons in the dielectric reach a certain position, which depends on the voltage. After this no flow occurs as long as the switch is closed.

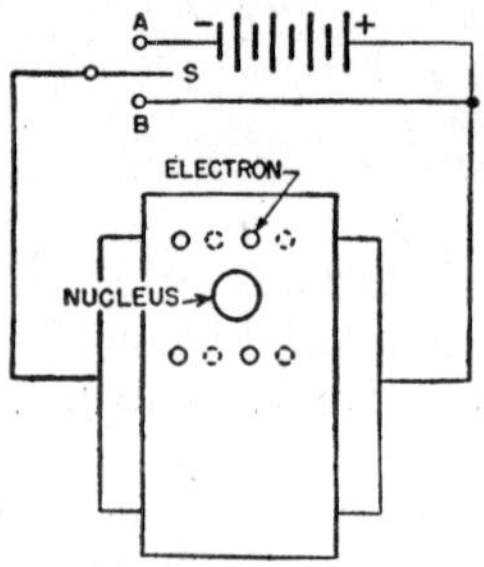

FIG. 10-15

If the switch is opened, the electrons that entered the negative plate cannot flow back, and therefore they will hold the electrons in the dielectric in their new position. In such a case the condenser is said to be charged.

If the switch is closed at B, the condenser is short-circuited and the force between the electrons and nucleus will cause the electrons to return to their normal position. While they are returning, there will be an electronic flow through the circuit in the opposite direction.

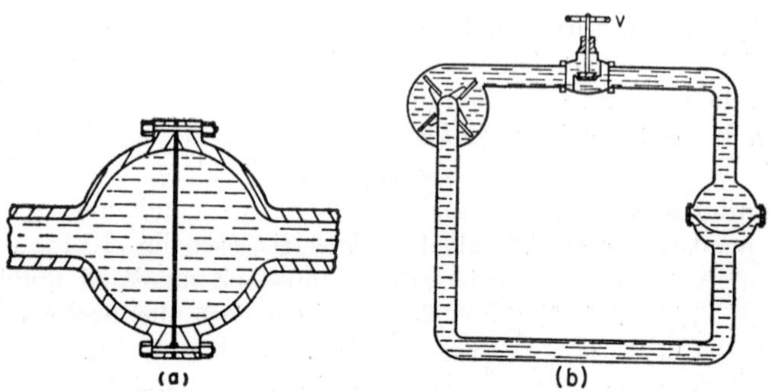

FIG. 10-16

Thus, a current flows in one direction when a potential is applied, and in the opposite direction when the potential is removed and the condenser shorted.

An electric condenser has an excellent hydraulic analogy in a rubber diaphragm stretched across a pipe as shown in fig. 10-16a. If this pipe is connected to a centrifugal pump as shown in fig. 10-16b, and the pump is started, there will be a flow of water in the pipe while the diaphragm is being stretched. After this, the diaphragm will remain stretched and no water will flow as long as the pump is running. If the valve is closed, which is equivalent to opening the switch in fig. 10-15, the water cannot flow back, even if the pump is stopped. Suppose that the pump is stopped, instead of the valve being closed. Due to the elasticity of the diaphragm, there will be a flow of water back through the pump.

Capacity—The size of an electric condenser is usually indicated by stating its capacity, which is measured in farads. *A condenser is said to have a capacity of one farad when a change of one volt per second across it produces an average current of one ampere.* But the farad is much too large a unit for practical purposes, so the microfarad (one millionth of a farad) has come into use.

The definition of a farad can be expressed by an equation, as follows:

$$I_{av} = C\frac{E_2 - E_1}{t} \qquad (10\text{-}23)$$

I_{av} = average current in amperes, C = capacity in farads, E_1 = initial voltage, E_2 = final voltage, and t = time required for the voltage to change from E_1 to E_2.

EXAMPLE—A condenser has a capacity of 20 microfarads. Find the average current that will flow if the voltage is raised from 0 to 240 volts in .1 of a second.

$$20 \text{ microfarads} = \frac{20}{10^6} \text{ farads}$$

$$I_{av} = \frac{20}{10^6} \times \frac{240}{.1} = .048 \text{ ampere}$$

If the voltage rises from 0 to E at a uniform rate in t seconds, Equation 10-23 may be expressed as follows:

$$I = C\frac{E}{t} \qquad (10\text{-}24)$$

But the quantity of electricity put into a condenser is It. Therefore,

$$It = C\frac{E}{t}t \qquad (10\text{-}25)$$

But in chapter 1 it was shown that $It = Q$. Substituting Q for It in Equation 10-25, then,

$$Q = CE \qquad (10\text{-}26)$$

This may also be expressed in the following way:

$$C = \frac{Q}{E}$$

EXAMPLE—Determine the number of coulombs of electricity stored in a 250 microfarad condenser when the voltage is raised from 0 to 250.

$$Q = \frac{250}{10^6} \times 250 = .0625 \text{ coulomb}$$

Thus, the capacity of a condenser can be defined in two ways. (1) The capacity is that factor by which the rate of voltage change must be multiplied to give the current. When a change of one volt per second causes one ampere to flow, the condenser is said to have one farad capacity. (2) The capacity is the charge per volt. When one volt will put a charge of one coulomb on a condenser, it has the capacity of one farad.

Condensers in Series—Condensers connected in series (fig. 10-17) have a voltage across the combination equal to the sum of the voltages across the individual condensers. Thus,

$$E = E_1 + E_2 + E_3 \qquad (10\text{-}27)$$

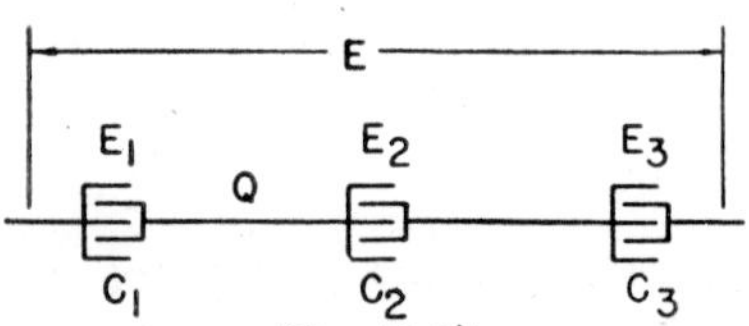

FIG. 10-17

Under the action of the voltage E, a charge Q is sent into the condensers. As they are in series, the same charge must be sent into each one, just as the same current is sent through resistances in series. Then, from Equation 10-26,

$$E = \frac{Q}{C}, \quad E_1 = \frac{Q}{C_1}, \quad E_2 = \frac{Q}{C_2}, \text{ and } E_3 = \frac{Q}{C_3} \qquad (10\text{-}28)$$

Substituting 10-28 into 10-27,

$$\frac{Q}{C} = \frac{Q}{C_1} + \frac{Q}{C_2} + \frac{Q}{C_3}$$

Dividing by Q,

$$\frac{1}{C} = \frac{1}{C_1} + \frac{1}{C_2} + \frac{1}{C_3} \qquad (10\text{-}29)$$

Thus, in a series circuit the reciprocal of the combined capacity equals the sum of the reciprocals of the capacities of the separate condensers.

EXAMPLE—Compute the capacity of three condensers of 2, 5, and 10 microfarads, when connected in series.

$$\frac{1}{C} = \frac{1}{2} + \frac{1}{5} + \frac{1}{10} = \frac{8}{10} \text{ or } C = 1.25 \text{ mf.}$$

Condensers in Parallel—When condensers are connected in parallel, the effect is the same as increasing the number of plates, the capacity of the circuit being increased. Let us consider three condensers (fig. 10-18) joined in parallel across a supplied voltage (E). In such a circuit the total charge is

$$Q = Q_1 + Q_2 + Q_3 \qquad (10\text{-}30)$$

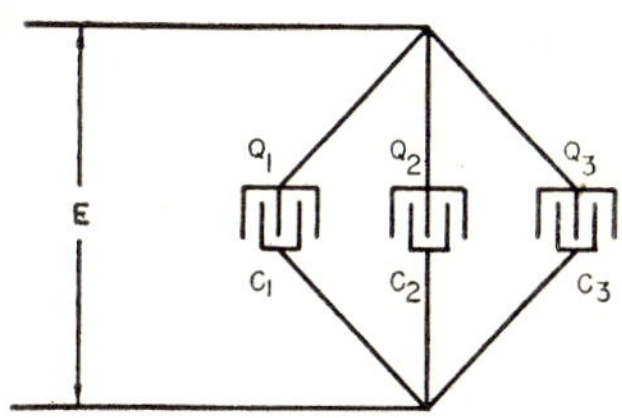

FIG. 10-18

But $Q = CE$, $Q_1 = C_1E$, $Q_2 = C_2E$, and $Q_3 = C_3E$ (10-31)

Substituting 10-31 into 10-30,

$$CE = C_1E + C_2E + C_3E$$

Dividing by E,

$$C = C_1 + C_2 + C_3 \qquad (10\text{-}32)$$

In a parallel circuit the total capacity equals the sum of the capacities of the separate condensers.

EXAMPLE—Three condensers, having capacities of 10, 25, and 50 microfarads, are connected in parallel. Determine their combined capacity.

$$C = 10 + 25 + 50 = 85 \text{ microfarads}$$

Condensers in A C Circuits—It was shown that a condenser is charged when a voltage is applied to its terminals and discharged when the condenser is shorted. If an alternating e m f is applied to a condenser, it will be alternately charged and discharged and an alternating current of the same frequency will flow in the circuit.

Suppose that an alternating e m f is applied to the condenser in fig. 10-19a. According to Equation 10-26, the charge on the plate at any instant is proportional to the voltage E and will therefore be in phase with it. That is, for zero voltage there will be zero charge and for maximum voltage there will be maximum charge. This is shown in fig. 10-19b (curve Q representing the charge).

Since the voltage is alternating, the condenser is charged first in one direction and then in the other, and an alternating current will flow. During the time from a to b, the voltage is increasing and the

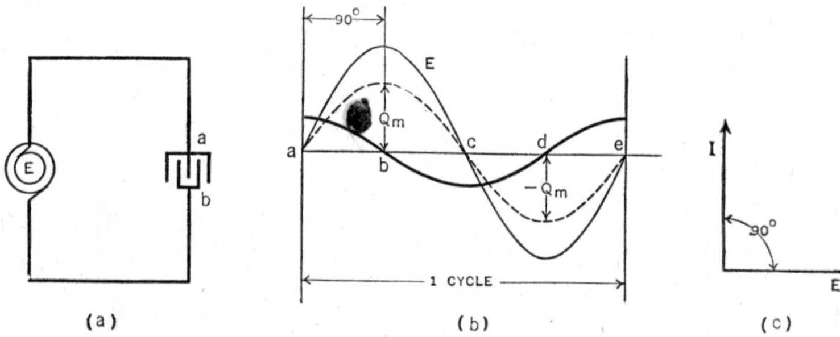

Fig. 10-19

condenser is being charged. This means that both the voltage and the current are from b, through the generator to a (fig. 10-19a). Curve E indicates that the greatest change in voltage occurs at a, and according to Equation 10-23, the maximum current will flow at that time. At b the rate of change in voltage is zero, and zero current flows, even though the charge in the condenser is at a maximum. Furthermore, during the time the impressed e m f is increasing, the voltage and the current will flow in the same direction. This gives the portion of the current curve between a and b. Between b and c the voltage is decreasing, and therefore the charge on the condenser is decreasing. This means that the condenser must discharge. To do this, the current must flow in the opposite direction to the impressed voltage; hence, the current flow between b and c is in the opposite direction to the voltage and must be represented below the horizontal line. Between c and d the condenser is being charged in the opposite direction, but nevertheless, the current will again be in the same direction as the voltage. Between d and e the condenser again discharges, and the current will then be in the opposite direction to the voltage. An examination of the curves shows that the current leads the condenser voltage by ninety degrees, which is represented vectorially in fig. 10-19c.

Computation of Capacity Reactance—During the interval of time from a to b (fig. 10-20), or one-fourth of a cycle, the voltage changes from zero to E_m. For this voltage change, Equation 10-23 becomes

$$I_{av} = C\frac{E_m}{t} \qquad (10\text{-}33)$$

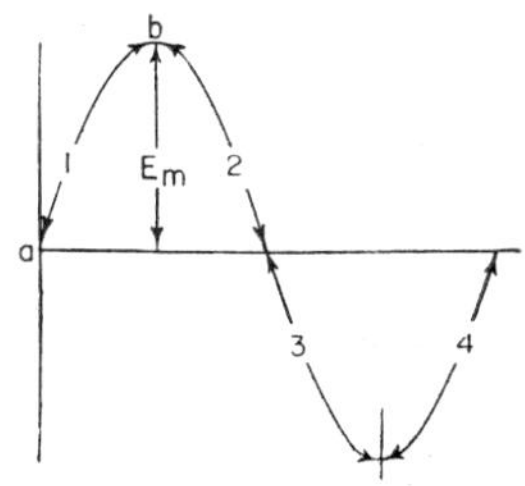

FIG. 10-20

Fig. 10-20 shows that the voltage across a condenser changes four times per cycle. The time required for each change can be found in the following way. Let f = the number of cycles per second (frequency). Then 1/f = the time required for one cycle. Each change occurs in one-fourth of a cycle, therefore $1/f \times 1/4 = 1/4f$, which is the time (t) required for each change. That is,

$$t = \frac{1}{4f} \qquad (10\text{-}34)$$

Substituting for t in Equation 10-33,

$$I_{av} = C\frac{E_m}{\frac{1}{4f}} = 4fCE_m \qquad (10\text{-}35)$$

But

$$I_{av} = \frac{I_{eff}}{1.11} \text{ and } E_m = \frac{E_{eff}}{.707}$$

Substituting in 10-35,

$$\frac{I_{eff}}{1.11} = \frac{4fCE_{eff}}{.707}$$

Dropping the subscripts and transposing,

$$E = \frac{1}{2\pi fC} I \qquad (10\text{-}36)$$

The expression $\frac{1}{2\pi fC}$ is usually represented by X_C. That is,

$$X_C = \frac{1}{2\pi fC} \qquad (10\text{-}37)$$

Equation 10-36 then becomes

$$E = X_C I \qquad (10\text{-}38)$$

EXAMPLE—A condenser of 25 microfarad capacity is connected to a 240 volt, 60 cycle supply. Compute the current that will flow.

$$25 \text{ microfarads} = \frac{25}{10^6} \text{ farad}$$

$$X_C = \frac{10^6}{2\pi \times 60 \times 25} = 106 \text{ ohms}$$

$$I = \frac{240}{106} = 2.26 \text{ amperes}$$

Calculation of Condenser Capacity—Condensers are usually made up of two long strips of tin foils separated by an insulator and rolled together as shown in fig. 10-21. The strips extend out at the ends to which the terminal connections are made. The capacity of a condenser may be computed by the equation:

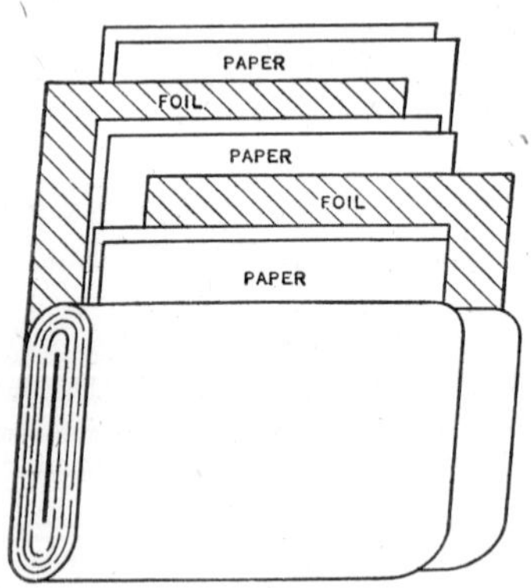

FIG. 10-21

$$C = \frac{2246 \text{ KA}}{10^{10}t} \qquad (10\text{-}39)$$

where C = capacity in microfarads, K = dielectric constant (see table 10-I), A = active area of one tin foil strip in square inches, and t = distance between the strips in inches.

EXAMPLE—Determine the capacity of a condenser that has two strips of tin foil, each 30 feet long and 4 inches wide. The strips are separated .005 inch by paraffin paper.

TABLE 10-I—DIELECTRIC CONSTANT AND DIELECTRIC STRENGTH OF VARIOUS MATERIALS

Material	Dielectric constant (K)	Dielectric strength	Material	Dielectric constant (K)	Dielectric strength
Air	1	86	Mica	4.5-6	1000-3000
Aluminum oxide layer	8-12		Oil (pyranol)	4.2	340
			(transformer)	2.2-2.6	250
Bakelite	4-8	450	Paper (beeswaxed)	2.5-3.5	1800
Beeswax	1.85	250	(manilla)	.15- .2	500-750
Cermics (titanium ox-			(paraffin)	2-2.4	850-1500
ide base)	90		Paraffin wax	1.8-2.2	310
Ebonite	2-3.5	1100-2200	Porcelain	4.5-6.5	300-1200
Fiber	5-8	51	Tantalum oxide		
Glass (plate)	7-9	750-1000	layer	11.5	
Glycerine	40		Varnished cambric	3.5-5.5	1000-1300
Gutta percha	3.9	325			

Since both sides of one strip are active, $A = 2 \times 30 \times 12 \times 4 = 2{,}880$ square inches. Considering $K = 2.2$ (see table 10-I) and substituting in equation 10-39,

$$C = \frac{2246 \times 2.2 \times 2880}{10^{10} \times .005} = .2846 \text{ microfarad.}$$

Dielectric Strength—If the voltage impressed on a condenser is raised to an extremely high value, it will exert a tremendous pressure on the electrons forcing them through the dielectric. The electrons pass through the dielectric in the form of an arc and puncture it. In paper, varnish cambric, etc., the arc actually burns a hole. The ability of a material to resist being punctured is called its dielectric strength, and is expressed in volts per mil (1 mil = .001 inch) in table 10-I.

Power in A C Circuits—Thus far, the voltage and current relations in the three fundamental alternating current circuits (resistance alone, inductance alone, and capacity alone) have been discussed. Before combining such circuits by connecting them in series or in parallel the power taken by each will be considered.

The power taken by a direct current circuit equals the product of the current flowing and the voltage impressed. At any instant, the power taken by an alternating current circuit equals the product of the current at that instant and the voltage at that instant. For example, fig. 10-22 shows a current wave, I, in phase with a voltage wave, E. This occurs when a circuit contains resistance only. To obtain the power at each instant the amperes and volts are multiplied together. With these products a new curve P may be plotted. Assume that at instant a, 2 amperes flow (indicated by ab) at a 3 volt pressure (indicated by ac). The power at this instant is found by obtaining their product which gives 6 watts. Other points are found similarly and the curve P is plotted. The power curve is positive during the first half cycle because both the current and voltage are positive. During the second half cycle both the current and voltage are negative, hence, their product will again be positive.

It should be noted that this power curve is a sine wave having double the frequency of the current or voltage.

The maximum value of voltage and current waves being E_m and I_m respectively, it follows that the maximum value of the power wave is E_mI_m. The average of the power wave is $\frac{E_mI_m}{2}$, since the portion above the dotted line will just fill in the unshaded areas below the dotted line. Then we may write average power

$$W = \frac{E_m I_m}{2} = \frac{E_m}{\sqrt{2}} \times \frac{I_m}{\sqrt{2}}$$

now $\frac{E_m}{\sqrt{2}}$ and $\frac{I_m}{\sqrt{2}}$ equal the effective values of voltage and current

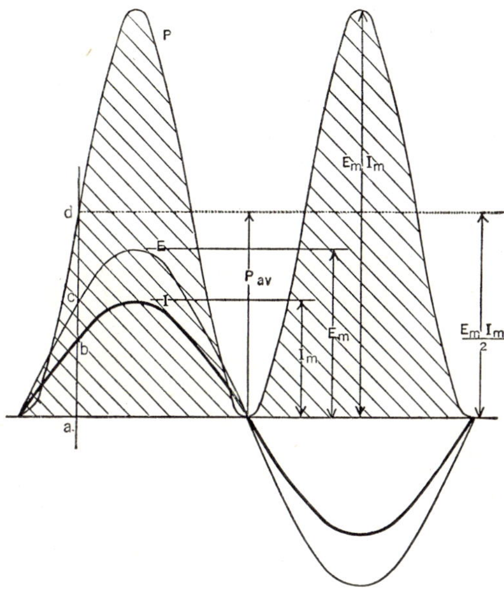

FIG. 10-22

respectively. Therefore the average power in a circuit containing only resistance equals the effective value of current times the effective value of voltage. Expressed by a formula

$$W = EI \tag{10-40}$$

Positive and Negative Power—In Fig. 10-23a is shown a battery supplying power to a resistance, R. The e m f of the battery, which is 6 volts, sends current through the external circuit from the positive terminal, through the resistance, R, and back to the negative terminal. That is, as far as the external circuit is concerned, both the voltage and current act in the same direction. Under this condition, the current is positive and the power delivered to the resistance is positive.

Fig. 10-23b shows the 6 volt battery with a generator connected across its terminals; the positive terminal of the generator is connected to the positive terminal of the battery. If the e m f of the generator equals that of the battery, obviously no current will flow.

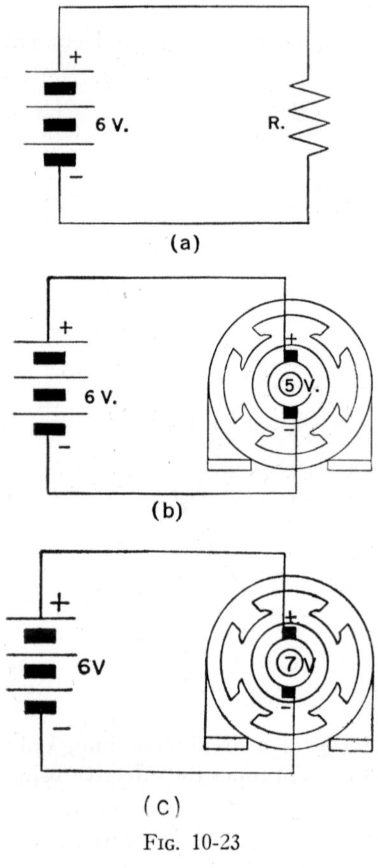

Fig. 10-23

If the generator e m f is lower than the battery e m f, current will flow from the positive terminal of the battery to the positive terminal of the generator (fig. 10-23b). In this case, as before, the current acts in the same direction as the battery voltage and the power is positive. The battery is delivering energy to the generator and tends to drive it as a motor. As far as the voltage, current, and power of the battery are concerned, the conditions are the same as those of fig. 10-23a, which contains resistance only across the battery terminals.

Suppose that the generated e m f is adjusted to exceed the battery e m f (fig. 10-23-c). Current will then flow from the positive terminal of the generator to the positive terminal of the battery. The battery e m f remains in the same direction, but the current is reversed, now being negative. As the battery is being charged, it is receiving energy. But since the battery is considered to be the source of energy, we may say that it is delivering negative power. Looking at it in another way, the current and voltage of the battery are in opposition, that is, they are opposite in sign, and thus the power is negative.

The purpose of this discussion is to show that in any device capable of delivering and receiving energy the power is positive when the current and voltage are in the same direction and negative when they are in opposite directions.

Power in an Inductive Circuit—Fig. 10-12 shows that in an inductive circuit with negligible resistance the current lags the impressed voltage by 90 degrees. This condition is illustrated again in fig. 10-24. To determine the power wave, the voltage and current at each instant are multiplied together, as in fig. 10-22. At a, b, c, d,

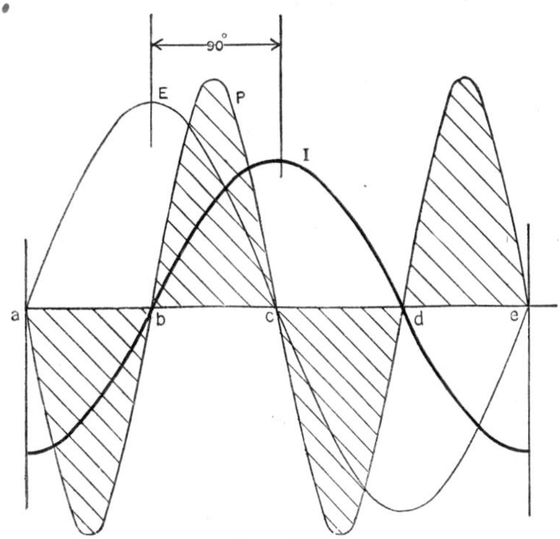

FIG. 10-24

and e in fig. 10-24, either the voltage or the current is zero. The power at each of these instants must then be zero. Between a and b the voltage is positive and the current is negative; thus they are in opposition. The current is flowing against the voltage, as suggested in fig. 10-23c. This means that the coil is delivering energy to the source of supply, just as the generator in fig. 10-23c delivered energy to the battery, which was considered to be the source. Between b and c the voltage and the current are both positive, because they are acting in the same direction. Thus the power during this time must be positive. The source is supplying power to the inductance. Between c and d the voltage is negative and the current is positive. As they are in opposition, the power is again negative. Between d and e the voltage and current are both negative. As they are acting in the same direction, the power is again positive. The power curve P is a sine curve with double the frequency of the current or voltage. The positive area of the power curve equals the negative area, so that all the energy received by the coil for the source of power is returned to the source. The average power input to the coil is then zero, and a wattmeter connected in the circuit would read zero.

Power in a Capacitative Circuit—Fig. 10-19 shows that in a condenser circuit the current leads the impressed voltage by 90 de-

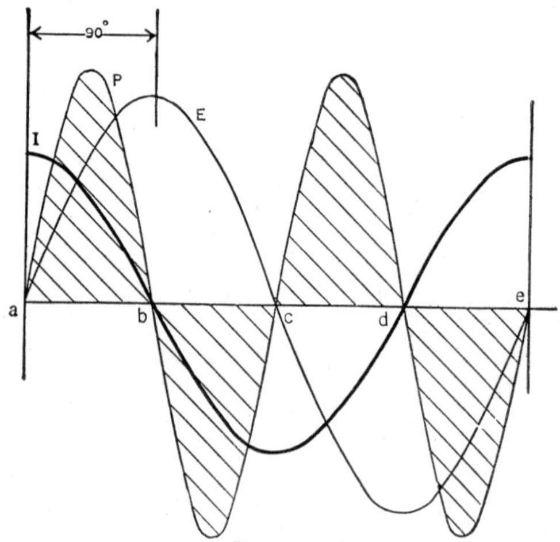

FIG. 10-25

grees. This condition is illustrated again in Fig. 10-25. As in Fig. 10-24, at a, b, c, d, and e either the voltage or current is zero. Hence the power at each of these instants is zero. Between a and b the voltage and the current are both positive because they are acting in the same direction; hence, the power is positive. Between b and c the voltage is positive and the current is negative; hence, the power is negative. Between c and d, the voltage and the current are both negative, and between these points the power is positive. Between d and e the voltage is negative, the current is positive, and the power is negative. The power curve P obtained is a sine curve with double the frequency of the current or voltage. The positive area of the power curve equals the negative area, so that for each cycle the average power is zero.

Power in a Circuit Containing Resistance and Inductance—It was shown above that when the voltage and current are in phase, their product gives the power; and that when the voltage and current are 90 degrees out of phase, the power is zero. When an inductive circuit contains resistance, the current is neither in phase nor 90 degrees out of phase with the voltage, but differs in phase by an angle which lies between these two limits. Such a case

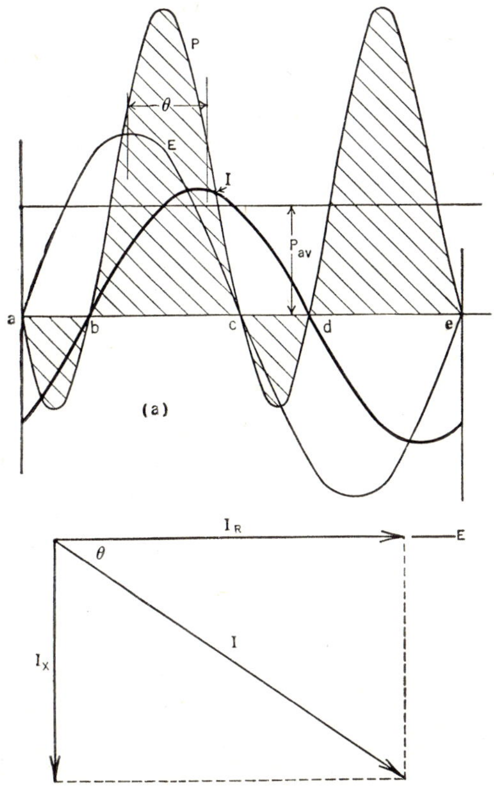

(b)

Fig. 10-26

is illustrated in Fig. 10-26a. The power curve P is the product at each instant of the volts and the amperes. Between a and b and between c and d the voltage and the current are in opposition, thus the power is negative. During these intervals the external circuit is delivering energy to the source of power. Between b and c and between d and e the voltage and the current are in the same direction, thus the power is positive. During these intervals the source of power is delivering energy to the external circuit. The power curve, P, as before, is a sine curve with double the frequency of the current or voltage, but the area above the horizontal line exceeds the area below it. Hence, there is more positive energy than negative energy, which means that more energy is supplied from the source than is returned to it.

By means of a vector diagram (fig. 10-26b) it is possible to determine the average power supplied to the external circuit.

Let I represent the effective current, lagging θ degrees behind the voltage. Consider this current made up of two currents, I_R and I_X. The current I_R is in phase with the voltage and the current I_X is 90 degrees out of phase with the voltage. Since I_R is the only current that takes energy from the source, the average power supplied to the external circuit equals the effective voltage times the effective current in phase with the voltage. This is expressed by an equation as

$$P = EI_R$$

But

$$I_R = I \cos \theta$$

Therefore

$$P = EI \cos \theta \qquad (10\text{-}41)$$

This is the general equation for the average power consumed by an electric circuit. *Cos θ is called the power factor because it is the factor by which the apparent power (EI) must be multiplied in order to obtain the true power P.*

EXAMPLE—A single phase motor takes 3 amperes at 120 volts. What is the power factor if the wattmeter reads 180 watts?

$$\text{Power factor} = \cos \theta = \frac{P}{EI} = \frac{180}{120 \times 3} = .5 \text{ or } 50\%$$

Exercises

1. How many revolutions must a 6 pole alternator turn to generate 5 cycles of voltage?
2. In 3 seconds, how many cycles of voltage are generated in an 8 pole alternator rotating at 360 r p m?
3. Find maximum voltage at the terminals of a condenser connected to a 110 volt supply.
4. An e m f has a maximum value of 115 volts. Find average value and effective value.

5. After passing through 30 degrees of its cycle, an e m f has a value of 45 volts. This voltage has what effective value?
6. An alternating e m f has a value of 110 volts. What is the instantaneous value of the voltage when it has reached 57 degrees of its cycle?
7. How many degrees and minutes of its cycle has an e m f of 440 volts reached when its instantaneous value is 40 volts?
8. An alternating current, which has passed through 30 degrees of its cycle, has an instantaneous value of 10 amperes. Determine its effective value.
9. A 20 k w furnace is supplied with 55 volts. Find maximum value of supplied current.
10. Two machines connected in parallel take a current of 3 and 20 amperes, which are 70 degrees out of phase. Determine the total current taken.
11. A loop of one turn is linked with 10^6 lines of flux. The flux is reduced to zero in .1 second. What is the average induced voltage?
12. The inductance of a certain generator field is 40 henries. The current in the field is changed from 1.25 amperes to 0.75 ampere in 2 seconds. What is the induced voltage?
13. What is the reactance of a coil of 5 henries connected to a 60 cycle supply?
14. What will be the reactance if the coil in Problem 13 is used in a telephone circuit with a frequency of 800 cycles?
15. The choke coil of fig. 10-27 contains 1000 turns and is wound on a sheet steel core. Determine the inductance, in henries, when .3 ampere flow through it.
16. What will be the inductance of the coil in fig. 10-27 if the current is increased to .9 ampere?

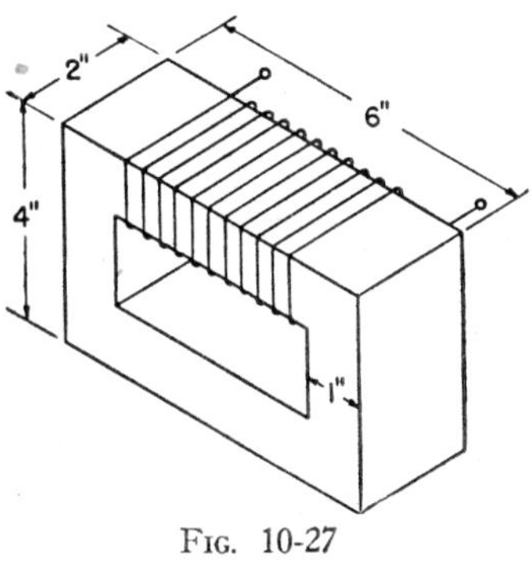

Fig. 10-27

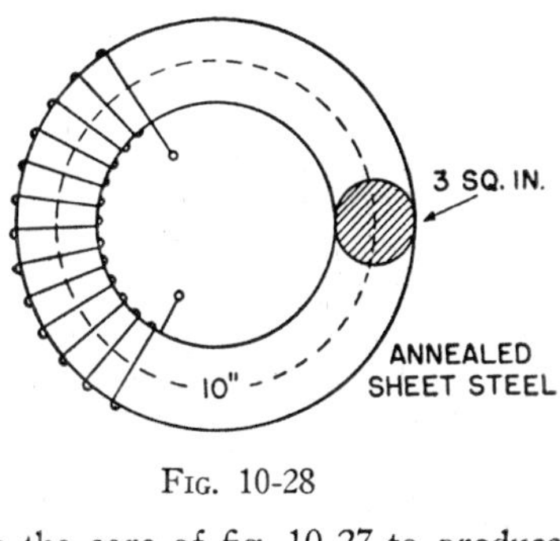

Fig. 10-28

17. How many turns must be wound on the core of fig. 10-27 to produce a choke of .75 henry? $\mu = 4500$.

18. Consider that the core in fig. 10-28 is capable of carrying 300,000 magnetic lines. Determine the number of turns that the coil must contain to have an inductance of 3 henries.
19. Determine the inductance of the coil in Problem 18 when .275 ampere flow through it.
20. How much electricity is stored in a 10 microfarad condenser connected to a 110 volt D C supply?
21. What voltage will store .002 coulomb of electricity in a 2 microfarad condenser?
22. Find the combined capacity of a 2 microfarad, a 3 microfarad, and a 5 microfarad condenser connected in series.
23. The combined capacity of two condensers connected in series is 16 microfarads. If the capacity of one is 26 microfarads, what is the capacity of the other?
24. Of three condensers connected in parallel, one has a capacity of 2 microfarads and another has a capacity of 4 microfarads. If the combined capacity of all three is 12 microfarads, find the capacity of the third.
25. How much electricity is stored in a 1 microfarad, a 2 microfarad, and a 4 microfarad condenser connected in parallel across a 100 volt line?

26 & 27. Find the resultant capacity in each of the following diagrams:

26.

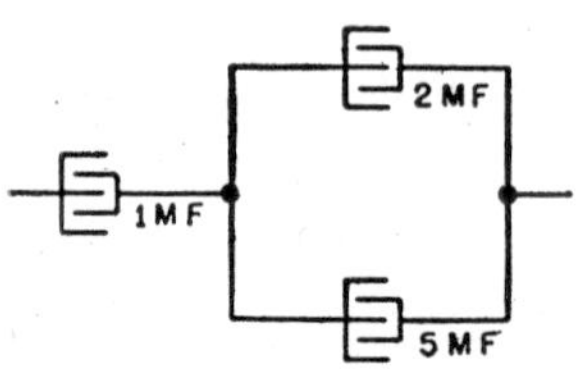

27.

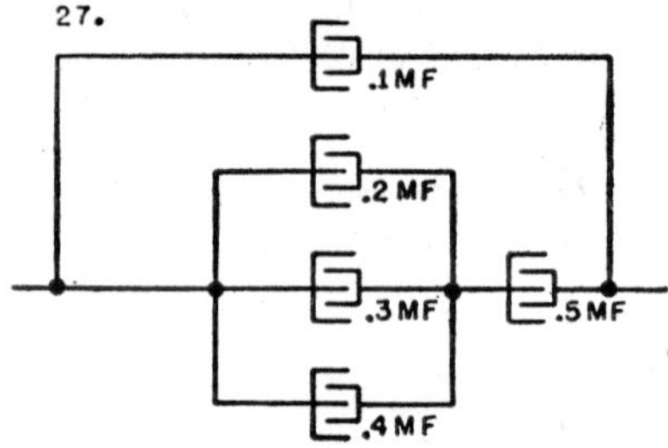

28. How much current will a 1 microfarad condenser take from a 220 volt, 60 cycle line?
29. At what frequency will 110 volts force 50 milliamperes through a 10 microfarad condenser?
30. Determine the capacity of a condenser that consists of two strips of tin foil each 50 feet long and 3 inches wide separated by beeswaxed paper .003 inch thick ($K = 3$).
31. Is power absorbed by a resistance? By a pure inductance? By a pure capacitance?
32. The current taken by a coil is 10 degrees out of phase with the supplied voltage. What is the power factor of the coil?

33. An appliance connected to a 240 volt line takes 10 amperes. If the current and voltage are 80 degrees out of phase, determine the power taken.
34. A coil takes 300 watts from a 120 volt supply. If 3 amperes flow through the coil, determine its power factor.
35. A motor, operating at a power factor of 70 per cent, takes 5 amperes at 120 volts. Determine the power taken.
36. What is the power factor of a 440 volt, single phase motor when the current is 3 amperes and the power taken is 1000 watts?
37. What maximum power can a 5 k v a, single phase alternator supply when the power factor is 90%?
38. Determine the line current in Problem 37 if the supplied voltage is 110.
39. At full load, a 3 H P, 110 volt, single phase motor operates at 60% power factor and has an efficiency of 70%. Determine (a) the power input, and (b) the line current.

CHAPTER 11

A C CIRCUITS

Resistance and Inductance in Series—Thus far we have considered how a circuit containing resistance only, inductance only, or capacity only, behaves when an alternating voltage is impressed upon it. When a circuit contains resistance and inductance together (as in a coil), it is convenient to consider it as a resistance connected in series with a pure inductance. To understand the relations of the quantities involved in such a circuit, Fig. 11-1a will be considered. As this is a series circuit, the current at any instant must have the same value at all portions of the circuit. Hence, only one current vector is needed. This current vector is used for reference only, and is laid

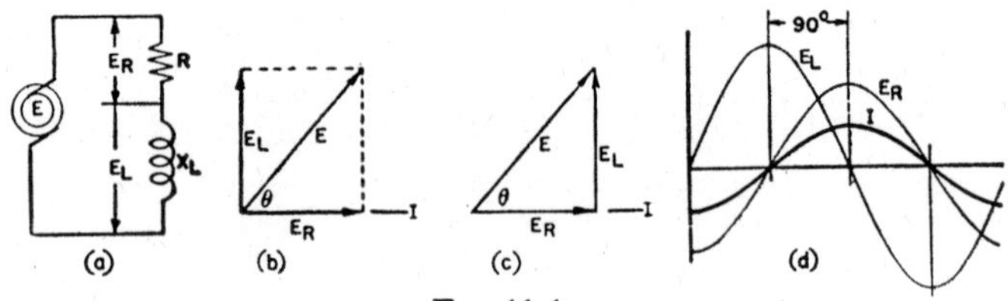

FIG. 11-1

off horizontally (fig. 11-1b). The phase relation of the voltages in the various parts of the circuit is expressed with reference to the current vector. The voltage E_R across resistance R (which equals IR) is in phase with the current. Therefore, the vector representing this voltage must be laid off along the current vector. The voltage E_L across the inductance L (which equals IX_L where $X_L = 2\pi fL$) must be laid off in such a direction that the current will lag it by 90°. Then it must be drawn vertically upward, because the vector rotation is always considered counterclockwise. Since E_R and E_L are 90° out of phase, E must be equal to their vector sum. The parallelogram method of adding the voltages is shown in fig. 11-1b, and the triangle method, which is the simplest, is shown in fig. 11-1c. In either case,

$$E = \sqrt{E_R^2 + E_L^2} = \sqrt{(IR)^2 + (IX_L)^2}$$

$$E = I\sqrt{R^2 + X_L^2} \qquad (11\text{-}1)$$

The expression $\sqrt{R^2 + X_L^2}$ is denoted by Z and is called impedance. Hence,

$$E = IZ \qquad (11\text{-}2)$$

Z, as well as R and X_L, must be expressed in ohms, since it is the factor by which the current must be multiplied to give the voltage.

The voltage and current relations of fig. 11-1a may also be represented by sine curves as shown in fig. 11-1d.

The cosine of the angle between the line voltage and the line current is the power factor. This may be determined as follows:

$$\text{Cos}\,\theta = \frac{E_R}{E} = \frac{IR}{IZ} = \frac{R}{Z} \qquad (11\text{-}3)$$

To illustrate how the above formula may be applied to a circuit containing resistance and inductance, the following example will be considered.

Example—A coil containing 10 ohms resistance and .04 henry inductance is connected to a 120 volt, 60 cycle line. Determine: (a) the impedance of the coil, (b) the current that flows, (c) the voltage required to force the current through the resistance, (d) the voltage required to force the current through the inductance, (e) the power factor, (f) the power taken.

$$X = 2\pi fL = 2 \times 3.14 \times 60 \times .04 = 15 \text{ ohms}$$

(a) $Z = \sqrt{R^2 + X_L^2} = \sqrt{10^2 + 15^2} = 18$ ohms

(b) $I = \frac{E}{Z} = \frac{120}{18} = 6.67$ amperes

(c) $E_R = IR = 6.67 \times 10 = 66.7$ volts

(d) $E_L = IX_L = 6.67 \times 15 = 100$ volts

(e) $\text{Cos}\,\theta = \frac{R}{Z} = \frac{10}{18} = .56$ or 56 per cent

(f) $W = EI \cos\theta = 120 \times 6.67 \times .56 = 448.22$ watts

Resistance and Capacity in Series—Fig. 11-2a shows a resistance (R) connected in series with a capacity (C), and the combination connected across an E volt supply. Suppose we consider the relation of the quantities involved in this circuit.

Since this is a series circuit, only one current vector is needed (fig. 11-2b), and it is laid off horizontally. The voltage E_R across the resistance R (which equals IR) is in phase with the current. Therefore, the vector representing this voltage must be laid off along the current vector. The voltage E_C across the capacity C (which equals IX_C where $X_C = 1/2\pi fC$) must be laid off in such a direction that the current will lead it by 90°. Then it must be drawn vertically downward. Since E_R and E_L are 90° out of phase, E must be equal to the vector sum of E_R and E_L. The two methods of performing this addition are shown in figs. 11-2b and 11-2c. In either case.

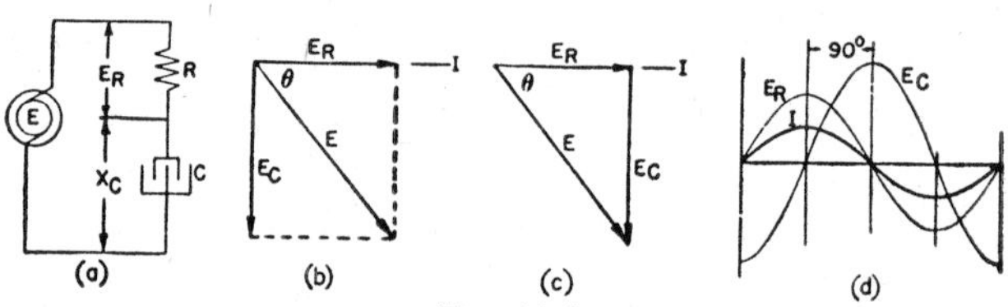

Fig. 11-2

$$E = \sqrt{E_R^2 + E_C^2} = \sqrt{(IR)^2 + (IX_C)^2} = I\sqrt{R^2 + X_C^2}$$

Then

$$Z = \sqrt{R^2 + X_C^2} \qquad (11\text{-}4)$$

Hence

$$E = IZ \qquad (11\text{-}5)$$

The voltage-current relations of fig. 11-2a are represented by sine curves in fig. 11-2d. The cosine of the angle between the line voltage and the current is the power factor, and may be expressed as

$$\text{Cos}\,\theta = \frac{E_R}{E} = \frac{IR}{IZ} = \frac{R}{Z} \qquad (11\text{-}6)$$

It will be noted that the vector diagrams of figs. 11-1 and 11-2 are similar, the only difference being that the current in the inductive circuit of fig. 11-1a lags θ degrees behind the impressed voltage (E), while the current in the capacity circuit of fig. 11-2a leads the impressed voltage by 90 degrees. It is obvious that the power taken by the circuit in fig. 11-2a can be determined by equation 10-41.

Resistance, Inductance, and Capacity in Series—Fig. 11-3a shows a resistance R, an inductance L, and a capacity C connected in series, with E volts impressed across the combination. Suppose we consider the relations of the quantities involved in the circuit. As it is a series circuit, the current vector is laid off horizontally. The voltage E_R across the resistance R (which equals IR) is in phase with the current, and must therefore be laid off along the current vector. The voltage E_L across the inductance must be laid off vertically upward, so that the current will lag this voltage by 90 degrees. The voltage across the capacity E_C must be drawn vertically downward, so that the current will lead this voltage by 90 degrees. In this example the voltage across the condenser is greater than the voltage across the inductance. The line voltage E must equal the vector sum of voltages E_R, E_L, and E_C. This vector addition is most readily performed by

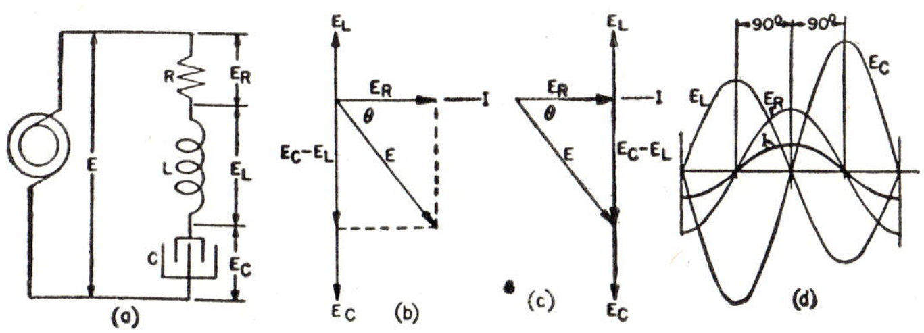

FIG. 11-3

first combining E_L and E_C. Since E_L is less than E_C, it is subtracted from E_C, and the vector $(E_C - E_L)$ is obtained (fig. 11-3b). The vector $(E_C - E_L)$ is then combined with the vector E_R to give the line voltage E. The two methods of performing the addition are shown in figs. 11-3b and 11-3c. In either case,

$$E = \sqrt{E_R^2 + (E_C - E_L)^2} = \sqrt{(IR)^2 + (IX_C - IX_L)^2}$$

or

$$E = I\sqrt{R^2 + (X_C - X_L)^2} \tag{11-7}$$

and

$$\sqrt{R^2 + (X_C - X_L)^2} = Z, \text{ the impedance} \tag{11-8}$$

Hence

$$E = IZ \tag{11-9}$$

The voltage-current relations of fig. 11-3a are represented by sine curves in fig. 11-3d. The cosine of the angle between the line voltage and the current (power factor) may be expressed as

$$\text{Cos}\,\theta = \frac{E_R}{E} = \frac{IR}{IZ} = \frac{R}{Z} \qquad (11\text{-}10)$$

In any series circuit, then, the power factor may be determined by equation 11-10.

To illustrate how the above formulas may be applied to a circuit containing resistance, inductance, and capacity, the following example will be considered.

EXAMPLE—A series circuit containing 10 ohms resistance, .2 henry inductance, and 20 microfarads capacity is connected to a 240 volt, 60 cycle line. Determine: (a) the impedance of the circuit, (b) the current that flows, (c) the voltage required to force the current through the resistance, (d) the voltage required to force the current through the inductance, (e) the voltage required to force the current through the capacity, (f) the power factor, (g) the power taken.

$$X_L = 2\pi fL = 2 \times 3.14 \times 60 \times .2 = 75.4 \text{ ohms}$$

$$X_C = \frac{1}{2\pi fC} = \frac{1}{2 \times 3.14 \times 60 \times .00002} = 132.63 \text{ ohms}$$

(a) $Z = \sqrt{R^2 + (X_C - X_L)^2} = \sqrt{10^2 + (132.63 - 75.4)^2} = 58.1$

(b) $I = \frac{E}{Z} = \frac{240}{58.1} = 4.13$ amperes

(c) $E_R = IR = 4.13 \times 10 = 41.3$ volts

(d) $E_L = IX_L = 4.13 \times 75.4 = 311.4$ volts

(e) $E_C = IX_C = 4.13 \times 132.63 = 547.76$ volts

(f) $\text{Cos}\,\theta = \frac{R}{Z} = \frac{10}{58.1} = .172$ or 17.2 per cent

(g) $W = EI \cos\theta = 240 \times 4.13 \times .172 = 170.49$ watts

Impedance Diagram of a Series Circuit—It is apparent from equation 11-8 that the impedance of a series circuit containing resistance, inductance, and capacity is found by combining, at right angles, the difference between the inductive and capacitive reactance with the resistance. To draw an impedance diagram for fig. 11-4a, the inductive reactance (X_L) should be laid off 90 degrees ahead of the resistance, and the capacitive reactance (X_C) must be laid off 90 degrees behind the resistance (fig. 11-4b). Their resultant is Z. A simpler method of making the impedance diagram is shown in fig. 11-4c.

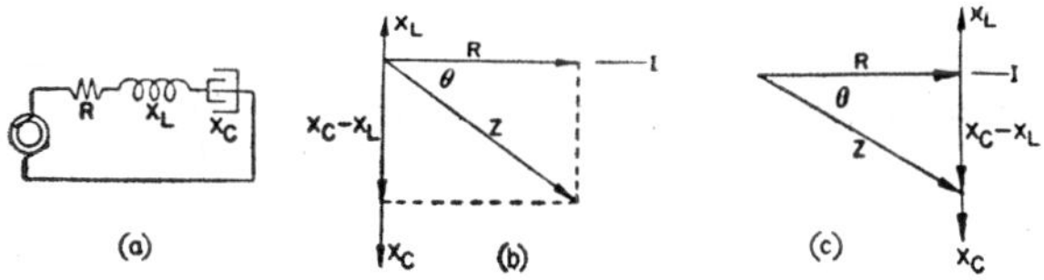

FIG. 11-4

Resistance and Inductance in Parallel—Across each branch of a parallel circuit, the voltage is the same. Then it is convenient to take the voltage vector as a reference by representing it along the horizontal. The current through each branch may then be represented with reference to the common voltage vector, and the total current may be determined by adding vectorially the individual currents. The method of determining the total current and other quantities will be illustrated by means of examples.

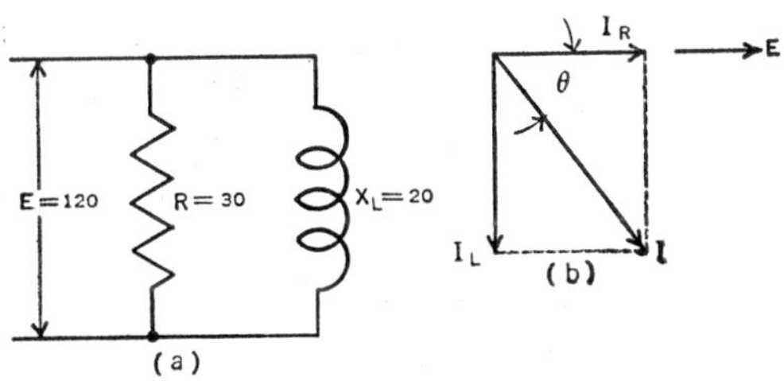

FIG. 11-5

EXAMPLE—In fig. 11-5a, determine:

(a) The angle that line current lags voltage; (b) the power factor of the circuit; (c) the line current.

The vector diagram is shown in fig. 11-5b.

$$I_R = \frac{120}{30} = 4 \text{ amps (current through R)}$$

$$I_L = \frac{120}{20} = 6 \text{ amps (current through } X_L\text{)}$$

$$\text{Tan}\ \theta = \frac{I_L}{I_R} = \frac{6}{4} = 1.5 \text{ or } \theta = 56°\ 19' \text{ (angle current lags voltage)}$$

(b) $\text{Cos}\ \theta = .555$ (or power factor = 55.5 per cent)

$$\text{(c)}\ I = \frac{I_R}{\text{Cos}\ \theta} = \frac{4}{.555} = 7.2 \text{ amps}$$

Resistance, Inductance, and Capacity in Parallel—A circuit containing resistance, inductance, and capacity in parallel will next be considered.

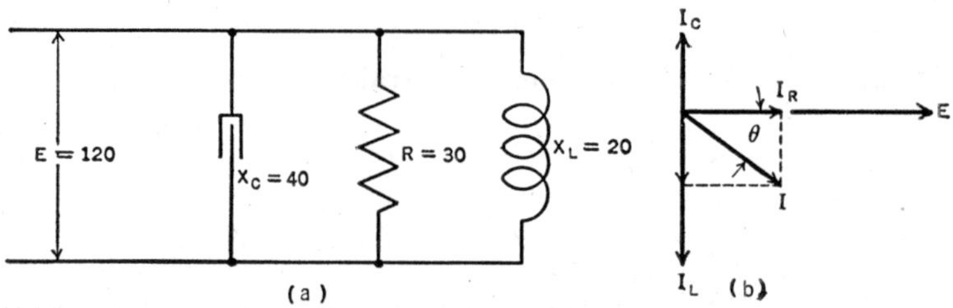

FIG. 11-6

EXAMPLE—In fig. 11-6a, find: (a) the power factor of the circuit; (b) the line current; (c) the total power taken.

$$\text{(a)}\ I_R = \frac{120}{30} = 4 \text{ amps,}$$

$$I_L = \frac{120}{20} = 6 \text{ amps, and}$$

$$I_C = \frac{120}{40} = 3 \text{ amps}$$

The vector diagram is given in fig. 11-6b, in which it is preferable to first combine I_L and I_C, since they are in opposition. This resultant is then combined with I_R to give the line current I.

Then,

$$I_L - I_C = 6 - 3 \text{ or } 3 \text{ amperes}$$

$$\text{Tan}\ \theta = \frac{I_L - I_C}{I_R} = \frac{3}{4} = .75 \text{ or Cos}\ \theta = .8$$

(b) The line current is

$$I = \frac{I_R}{\text{Cos}\ \theta} = \frac{4}{.8} = 5 \text{ amps}$$

(c) Since resistance only consumes power,

$$P = EI_R = 120 \times 4 = 480 \text{ watts.}$$

Impedance of a Parallel Circuit—This can be found merely by dividing the line voltage by the line current. For instance, in the previous example, the impedance

$$Z = \frac{E}{I} = \frac{120}{5} = 24 \text{ ohms}$$

To find the impedance of a parallel circuit when the voltage is not given, any convenient voltage may be assumed.

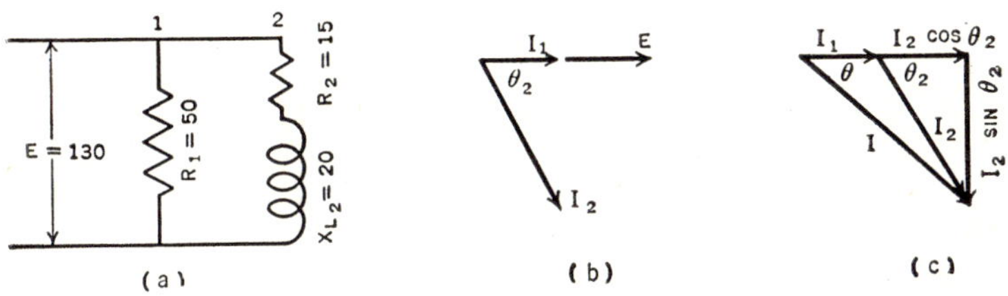

FIG. 11-7

Series Parallel Circuits—If the paths of a parallel circuit contain more than pure resistance, inductance, or capacity, each path should first be treated as a separate series circuit with the common voltage impressed across it. Then the current of each branch is determined, and a vector diagram drawn, in which the currents are indicated with reference to the common voltage vector. The method of determining quantities in such a parallel circuit will be considered by means of an example.

EXAMPLE—In fig. 11-7a calculate (a) the current through each branch, (b) the power factor of the circuit, and (c) the total current.

$$I_1 = \frac{130}{50} = 2.6 \text{ amps}$$

(a) In branch 2, $Z = \sqrt{15^2 + 20^2} = 25$ ohms

Then,

$$I_2 = \frac{130}{25} = 5.2 \text{ amps}$$

(b) Branch 2 contains resistance and reactance, which means that I_2 is neither in phase nor 90° out of phase with E. The power factor of branch 2 must next be determined, which can be done by using equation 11-10.

Then,

$$\text{Cos } \theta_2 = \frac{R}{Z} = \frac{15}{25} = .6 \text{ (or } \theta_2 = 53° \ 8')$$

The vector diagram, which may now be drawn, is shown in fig. 11-7b. The current vectors may be added more readily if the origin of I_2 is moved to the end of I_1, as indicated in fig. 11-7c. I_2 may be replaced by I_2 Cos θ_2 and I_2 sin θ_2, since the resultant of these two vectors equals I_2.

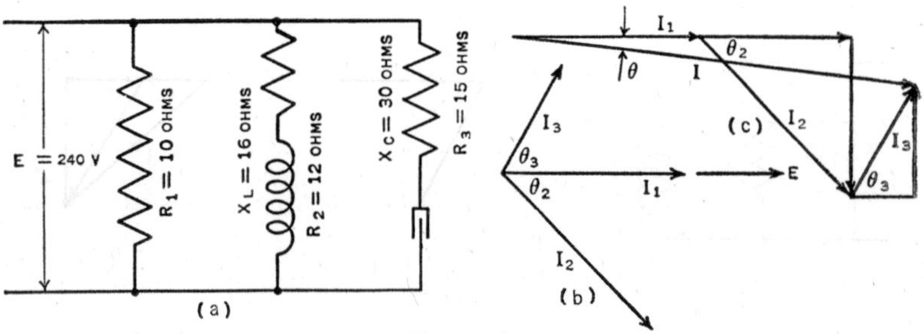

Fig. 11-8

The total current in phase with the voltage is

$$I_1 + I_2 \text{ Cos } \theta_2 = 2.6 + 5.2 \times .6 \text{ or } 5.72 \text{ amps}$$

The total current 90° displaced with the voltage is

$$I_2 \text{ Sin } \theta_2 \text{ or } 5.2 \times .8 = 4.16 \text{ amps}$$

$$\text{Tan } \theta = \frac{4.16}{5.72} = .7273, \text{ or Cos } \theta = .809$$

(c) The total current may be determined by dividing the current in phase with the voltage by the power factor. Hence,

$$I = \frac{5.72}{.809} = 7.07 \text{ amps}$$

Consideration will now be given to a more complex circuit.

Example—In fig. 11-8a, determine (a) the current of each path,

(b) the power factor of the circuit, (c) the line current, (d) the impedance of the circuit, and (e) the total power taken.

(a) $$I_1 = \frac{240}{10} = 24 \text{ amps}$$

$$I_2 = \frac{240}{\sqrt{12^2 + 16^2}} = \frac{240}{20} = 12 \text{ amps}$$

$$I_3 = \frac{240}{\sqrt{15^2 + 30^2}} = \frac{240}{33.54} = 7.16 \text{ amps}$$

(b) $$\text{Cos } \theta_2 = \frac{12}{20} = .6$$

$$\text{Cos } \theta_3 = \frac{15}{33.54} = .447$$

The vector diagram is shown in fig. 11-8b, and is given in a simpler form in fig. 11-8c.

The total current in phase with the voltage is

$$I_1 + I_2 \cos \theta_2 + I_3 \cos \theta_3 \text{ or } 24 + 12 \times .6 + 7.16 \times .447 \text{ or } 34.4 \text{ amps}$$

The total current 90° displaced with the voltage is

$$I_2 \sin \theta_2 - I_3 \sin \theta_3 \text{ or } 12 \times .8 - 7.16 \times .8945 \text{ or } 3.2 \text{ amps (lagging)}$$

$$\text{Tan } \theta = \frac{3.2}{34.4} = .093$$

$$\text{Cos } \theta = .996$$

(c) The line current is found by dividing the current in phase with the voltage by the power factor. Hence,

$$I = \frac{34.4}{.996} = 34.54 \text{ amps}$$

(d) $$Z = \frac{E}{I} = \frac{240}{34.54} = 6.95 \text{ ohms}$$

(e) $$W = EI \cos \theta = 240 \times 34.54 \times .996 = 8256 \text{ watts}$$

Power Factor Correction—In chapter 10 it was shown that the power delivered to a circuit may be computed by the formula

$$P = EI \cos \theta$$

This equation is represented vectorially in fig. 11-9a. The factor P is represented along the horizontal, designates the true power delivered to the load, and is called effective power (see fig. 11-9b). The product EI is represented θ degrees below the horizontal, to designate a lagging power factor, and is known as apparent power. Alternators, transformers, etc., are rated in kilovolt amperes ($EI \div 1000$), because their maximum voltage and current are fixed, depending upon the insulation and wire size used, while the number of watts delivered depends upon the power factor.

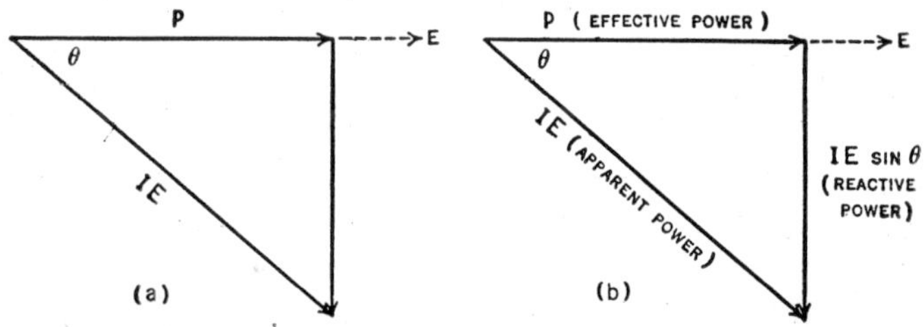

FIG. 11-9

In fig. 11-9, the side opposite θ equals the voltage times the current 90° lagging the voltage, or EI sin θ. This power, which is known as reactive power, circulates between alternators and loads, heating the winding through which it passes, and doing no useful work. Hence, it should be reduced as much as is practical. The method of doing this will be considered later.

EXAMPLE—In a power plant the reactive k v a meter reads 130 and the kw meter reads 150. Determine the power factor of the system.

$$\text{Tan}\,\theta = \frac{130}{150} = .8666$$

$$\text{Cos}\,\theta = .655$$

Causes of Low Power Factor or Large Reactive Power— In plants using alternating current, a large percentage of the loads generally consist of induction motors, because these machines are the simplest and most rugged known, and because they can be manufactured with operating characteristics suited for most practical requirements. However, such machines, and other equipment, such as magnets, welders, and furnaces, operate with a lagging power factor. In the average plant using alternating current, the average power factor rarely exceeds 85 per cent lagging, and may be as low as 40 per cent. A power factor less than unity renders the following detrimental effects.

(1) More current is required to deliver a given amount of power. A current increase causes a greater power loss ($= I^2R$) and a greater voltage drop ($= IZ$) in all the conductors through which the current flows.

(2) Loads taking a lagging current cause transformers and alternators to operate with a poor voltage regulation. In other words, the terminal voltage of alternators and transformers decreases with a greater lagging power factor or greater load current, even though the alternator is driven at a constant speed with a constant field excitation, or the transformer is supplied with a constant voltage. These conditions can be remedied by using equipment that stabilizes the voltage.

Apparently it is more economical for power producers to sell power, measured in kilowatt hours, at a high power factor; and it is reasonable to charge a consumer whose load has a low power factor more per kilowatt hour. This induces the consumer to employ various means to improve the power factor. This may be accomplished with large condensers, which are generally called capacitors, and synchronous motors.

As the current in a capacity leads the voltage by 90°, it takes a leading reactive power, which compensates for lagging reactive power taken by inductive loads. Individual capacitors may be connected directly across the terminals of induction motors or inductive equipment, but more generally they are connected to the junction points of feeders.

The reactive power taken by synchronous motors depends upon two factors, the D C field excitation and the mechanical load delivered by the machine. Maximum leading power is taken by synchronous motors with maximum field excitation and zero load.

Synchronous motors are more in practical use than capacitors, because such motors may be used to deliver a mechanical load in addition to operating with a leading power factor.

Power Factor Problems—In practice, the ratings in kva and the power factors of motors connected in parallel are usually known.

The total apparent power taken by these machines, which is the power supplied by the alternators, may be found by adding vectorially the apparent powers.

EXAMPLE—A certain load takes 50 kva at 50% power factor lagging, and another load connected to the same power line takes 100 kva at 86.6% power factor lagging. Find the total (a) effective power, (b) reactive power, (c) power factor, and (d) apparent power.

(a) The apparent power taken by each load, expressed with reference to the voltage, is represented in fig. 11-10a. The angle corresponding to a power factor of 50% is 60°, and to a power factor of 86.6% is 30°. To find the resultant, it is more convenient to place

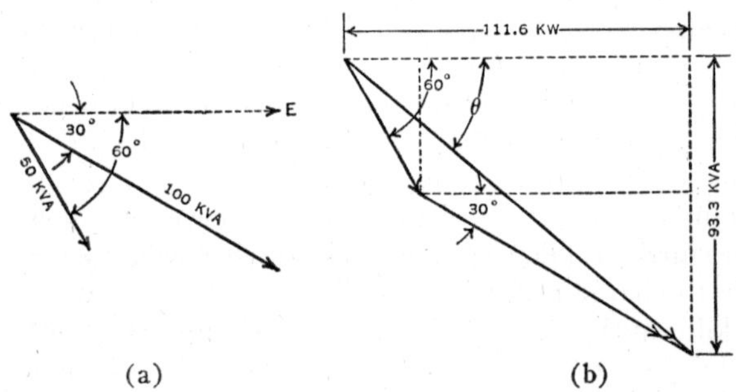

FIG. 11-10

one vector at the end of the other, as indicated in fig. 11-10b. The sum of the effective powers, which is indicated along the horizontal, is

$$50 \times .5 + 100 \times .866 \text{ or } 111.6 \text{ k w}$$

(b) The total reactive power, which is along the vertical, is

$$50 \times .866 + 100 \times .5 = 93.3 \text{ k v a}$$

(c)
$$\text{Tan } \theta = \frac{93.3}{111.6} = .836$$

$$\text{Cos } \theta = .767$$

(d)
$$\text{kva} = \frac{111.6}{.767} = 145.5$$

It was previously explained that if motors operating at a leading power factor are connected to the same power line as those operating at a lagging power factor, the power factor of the system is improved. This will be illustrated by means of an example.

EXAMPLE—A group of induction motors take 100 kva at 84% power factor lagging. A synchronous motor is connected to the same line and takes 60 k v a at 70.7% power factor leading. Determine the total (a) effective power, (b) reactive power, (c) power factor, and (d) apparent power.

(a) Lagging apparent power is conventionally represented below the reference voltage vector, and leading power is represented above. This is indicated in fig. 11-11a, in which the angles shown correspond to the power factors of each type of load. Fig. 10-11b shows the lead-

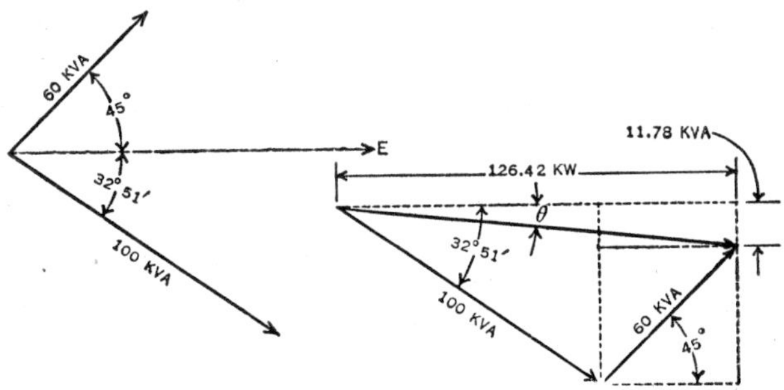

(a) (b)

FIG. 11-11

ing power vector represented from the end of the lagging power vector, which is done for the convenience of obtaining the vector sum.

From fig. 11-11b

$$k\ w = 100 \times .84 + 60 \times .707 = 126.42$$

(b) The total reactive power is the difference of the two, since one is leading and the other is lagging. Then the reactive power is

$$100 \times .542 - 60 \times .707 = 11.78 \text{ k v a lagging}$$

At the instant that the induction motors need 54.2 reactive k v a, the synchronous motor has 42.42 reactive k v a to return to the transformer, so it gives the 42.42 k v a to the induction motors instead. The induction motors then take only 11.78 reactive k v a lagging from the transformer. At another instant, the synchronous motor needs 42.42 k v a and the induction motors are just ready to return 54.2 k v a to the transformer. The induction motors supply 42.42 k v a to the synchronous motor and return the rest, 11.78 k v a, to the transformer. Hence, besides carrying the effective power, the transformers, etc., must carry only 11.78 k v a lagging reactive power.

$$\text{(c)} \qquad \text{Tan } \theta = \frac{11.78}{126.42} = .093$$

$$\text{Cos } \theta = .996$$

$$\text{(d)} \qquad \text{Apparent power} = \frac{126.42}{.996} = 126.93 \text{ k v a}$$

Cost of Power—It was pointed out that producers are justified in charging more for power sold at a low power factor. A method used by one power producing company to compute the charge on power consumed will be considered. For customers whose maximum demand exceeds 100 k w, charges are made as follows:

Power Factor			Discount Allowed
74.9 and lower			none
75	to	79.9	1%
80	to	84.9	2%
85	to	89.9	3%
90	to	94.9	4%
95	to	100	5%

$300 per month for first 100 k w of demand (or less)

$3.00 per month per k w of demand for the next 100 k w

$2.00 per month per k w of demand for the next 800 k w

$1.50 per month per k w of demand for the next 4000 k w

$1.00 per month per k w of demand for the excess

In addition .8 cent per k w h must be paid. A discount is allowed for a higher power factor, as specified in the table above, and a 5% discount is allowed for prompt payments.

Example—In one month a consumer obtains 40,000 k w h, with a maximum demand of 335 k w at a power factor of 90 per cent. Determine the power cost.

The first 100 k w cost \$300, the next 100 k w cost \$300, and the next 135 k w cost \$270.

40,000 k w h at .8 cents per k w h cost \$320. 300 + 300 + 270 + 320 = \$1190.

A power factor of 90 per cent permits a 4% discount. Then the cost is 1190 − 47.60 = \$1142.40. Considering that prompt payment is to be made, which allows 5% discount, the final cost becomes

$$1142.40 - 57.12 = \$1085.28$$

Exercises

1. A condenser with a capacity of 41 microfarads is connected to a 120 volt, 60 cycle line. Compute the current that flows in the condenser.
2. A resistance of 50 ohms and a condenser having a reactance of 35 ohms are connected in series across a 120 volt, 60 cycle supply. Find: (a) The amount of current that flows; (b) the angular difference between the current and the supplied voltage.
3. In problem 2, what is the power taken by each part of the circuit?
4. A condenser of 60 microfarad capacity, a coil with negligible resistance, and an inductance of .22 henry, are connected in series across a 120 volt, 60 cycle supply. Find: (a) The reactance of the combination; (b) the current that flows in the circuit.
5. What current will flow in the circuit of fig. 11-12 if it is connected to a 110 volt, 60 cycle source of supply?

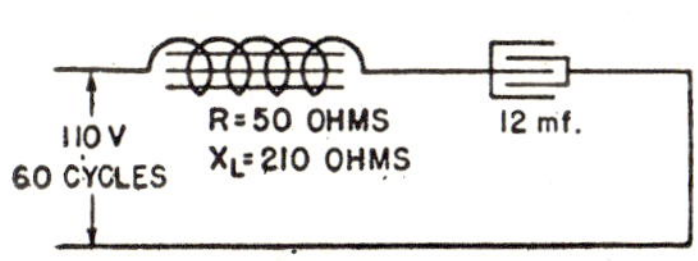

FIG. 11-12

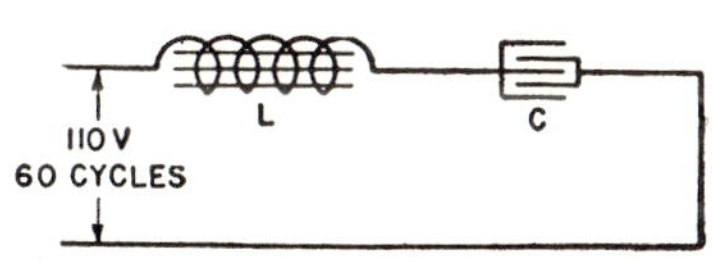

FIG. 11-13

6. In fig. 11-13 the impedance coil "L" has a resistance of 50 ohms and a reactance of 225 ohms. What must be the capacity of the condenser to make the power factor of the line equal unity?
7. A coil with a resistance of 15 ohms and an inductance of .1 henry is connected in series with a condenser of 50 microfarad capacity. What voltage at 60 cycles would force 3 amperes through this combination?
8. Compute the power factor of the circuit in problem 7.

9. Determine the current that will flow and the power taken by the series circuit of fig. 11-14.

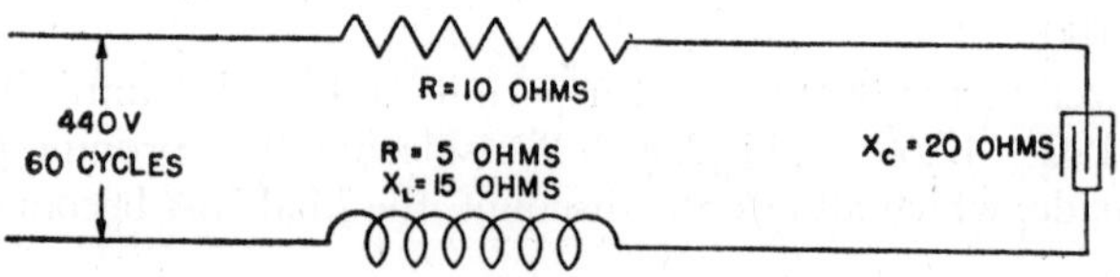

FIG. 11-14

10. Find the impedance of a coil taking 2 amperes when connected to a 440 volt A C main.
11. If the resistance of an appliance is 20 ohms and the reactance is 50 ohms, what current will flow through it when it is connected to a 110 volt supply?
12. A circuit has an inductance of .1 henry and a non-inductive resistance of 10 ohms. (a) If the circuit is connected across a 110 volt, 60 cycle main, compute the voltage across the resistance. (b) Across the inductance.
13. A resistance of 10 ohms and a reactance of 20 ohms are connected in parallel across 120 volts. Determine the power factor and the current of the line.
14. In problem 13, compute the impedance and the power taken by the circuit.
15. A 30 ohm inductance and a 50 ohm capacitance are connected in parallel across a 150 volt line. Determine the current taken.
16. In problem 15, (a) how many degrees is the line current out of phase with the voltage? (b) What is the power factor of the circuit? How much power is taken by the circuit?
17. A resistance of 20 ohms, an inductive reactance of 15 ohms, and a capacity reactance of 35 ohms are connected in parallel across a 120 volt supply. Determine (a) the number of degrees the line current is out of phase with the voltage, and (b) the total current taken by the circuit.
18. In problem 17, find the impedance and total power taken by the circuit.
19. In problem 17, to what capacity must the condenser be changed to make the power factor of the circuit unity?
20. Determine the impedance of a resistance of 32 ohms and a reactance of 10 ohms connected in parallel.
21. A circuit contains two paths connected across 110 volts. One path has a resistance of 25 ohms and the other path has a resistance of 10 ohms connected in series with a reactance of 20 ohms. Determine the power factor and total current of the circuit.

22. Determine the impedance and power taken by the circuit of problem 21.
23. When delivering 5 H P, a motor connected to a 120 volt, 60 cycle line has an efficiency of 80 per cent and a power factor of 60 per cent. What must be the capacity of a condenser connected across the terminals of the motor to raise the power factor of the line to unity?
24. In problem 23, determine (a) the current taken by the motor, and (b) the current taken by the combination.
25. In problem 23, what must be the capacity of the condenser to raise the power factor of the line to 85 per cent?
26. A circuit connected to a 120 volt line has a coil and a condenser connected in parallel. The coil has a resistance of 20 ohms and a reactance of 30 ohms, and the condenser has a reactance of 40 ohms. Determine (a) the power factor of the system, and (b) the line current.
27. Determine the impedance and total power taken by the circuit of problem 26.
28. A circuit has two paths, one containing a resistance of 8 ohms in series with an inductive reactance of 20 ohms, and the other containing a resistance of 8 ohms in series with a capacity of 16 ohms. Determine the impedance of the circuit.
29. Does a wattmeter read apparent, effective, or reactive power?
30. What is (a) effective power? (b) reactive power? (c) apparent power?
31. What are the advantages of delivering power at a high power factor?
32. How is the power factor of a load improved?
33. What are the advantages and disadvantages of each method of improving the power factor?
34. An alternator delivers 50 k w and 83.3 reactive k v a. Determine (a) the apparent power delivered, and (b) the power factor of the load.
35. A 40 k w load operating at a power factor of 70 per cent is connected to the end of a No. 6 gage line 10 miles long (a No. 6 gage conductor has a resistance of .37 ohm per 1000 feet). If the voltage at the load is 13,200 volts, determine (a) reactive power of the load, (b) apparent power of the load, (c) the line current, (d) the line loss, and (e) the transmission efficiency.
36. What is the power factor of the alternator in problem 35?
37. A load connected to a 60 cycle, 440 volt line takes 10 k w. The power factor is 60 per cent lagging. Determine the k v a rating of the capacitor required to raise the power factor to unity.
38. An alternator supplies power to a 50 k w lighting load and a 100 k v a inductive load operating at 86.6 per cent power factor. Determine (a) the total effective power, (b) the total reactive power, (c) the power factor of the alternator, and (d) the total apparent power.
39. An induction motor takes 30 k v a at a power factor of 70 per cent, and another induction motor takes 50 k v a at 85.4 per cent power

factor. Determine (a) the power factor of the line, and (b) the total apparent power.

40. An inductive load that takes 100 k v a at a power factor of 84.1 per cent, and a capacity load that takes 80 k v a at a power factor of 60 per cent are connected to an alternator. Determine (a) the total effective power, (b) resultant power factor, and (c) the total apparent power supplied by the alternator.

41. An induction motor taking 30 k w at 75 per cent power factor is in parallel with a synchronous motor taking 20 reactive k v a at 82.2 per cent power factor leading. Determine (a) the power factor, and (b) the total apparent power taken by the two motors.

42. A 25 H P induction motor that has a power factor of 80 per cent and an efficiency of 75 per cent is connected to the same line as a 50 H P induction motor that has a power factor of 90 per cent and an efficiency of 70 per cent. Determine the apparent power taken by the motors.

43. The power factor of the line in problem 42 is raised to unity by connecting to this line a synchronous motor operating at a power factor of 70.7 per cent. Compute the k v a taken by the synchronous motor.

44. An inductive load takes 100 k v a at a power factor of 50 per cent. A synchronous motor operating at a power factor of 20 per cent leading is connected in parallel with the inductive load, and raises the power factor of the system to 76.3 per cent. Determine the k v a rating of the synchronous motor.

45. An alternator supplies 50 k w at 80 per cent power factor to inductive loads by means of three feeders, A, B, and C. Feeder A takes 10 k w at 85 per cent power factor, and feeder B takes 20 k w at 60 per cent power factor. Determine the k v a taken and the power factor of feeder C.

46. In one month a consumer is supplied with 60,000 k w h at a power factor of 86 per cent. The maximum demand is 98 k w. Assuming that prompt payment is to be made, determine the cost of the power, if the consumer's bill is computed by the method given in this chapter.

CHAPTER 12

TRANSFORMERS

Alternating current may be transmitted to a region of consumption far more economically than direct current because of the existence of an A C voltage-changing device called a transformer. Transformers have no moving parts, and are simple, rugged, and durable in construction, thus requiring little attention. They also have a very high efficiency (as high as 99%).

How Transformers Increase the Efficiency of Transmission—Lamps, motors, etc., are designed to be used on a low voltage partly for the sake of safety. However, power is never transmitted any great distance at a low voltage because of large power (I^2R) losses in the line, or because of heavy investments in copper wire. To overcome these difficulties, one transformer is used to step up the voltage for the transmission line and another is used to reduce the voltage to the desired value at the region of consumption.

Power is usually generated at 14,000 volts and for moderate distances it is generally transmitted at this value. At the load, transformers are used to step the voltage down to 440, 220, and 110 volts for motors and lights. If the power is to be transmitted a long distance the voltage is usually stepped up at the generator station, then stepped down at the load.

Transformer Principle—A simple transformer may consist of two coils placed on an iron core, as shown in fig. 12-1a. When an alternating current is supplied to one coil (called the primary), a magnetic field is produced. This expands and contracts, cutting the turns of the second coil (called the secondary) and inducing a voltage in it. Since the same number of magnetic lines cuts both primary and secondary, the voltage induced per turn in each winding is the same. Further, since the induced voltage in a coil of negligible resistance equals the supplied voltage (see chapter 10), it follows that the voltage per turn on the primary equals the voltage per turn on the secondary. Then the voltage per turn on either winding is

$$\frac{E_p}{T_p} = \frac{E_s}{T_s}$$

or

$$\frac{E_p}{E_s} = \frac{T_p}{T_s} \qquad (12\text{-}1)$$

where E_p = primary voltage, E_s = secondary voltage, T_p = primary turns, and T_s = secondary turns.

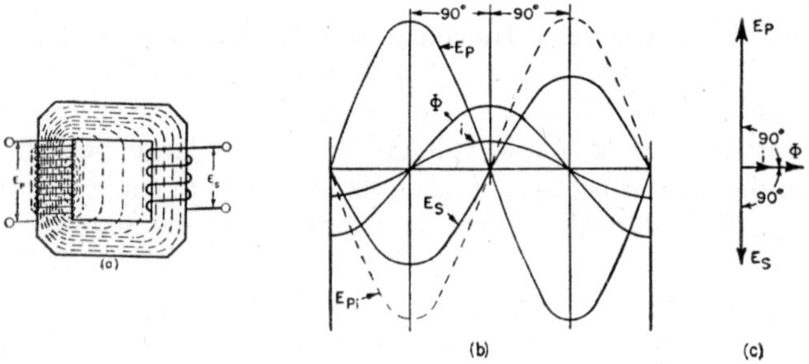

FIG. 12-1

This shows that the voltage of each winding is proportional to the number of turns. If the transformer is intended to raise the voltage, the winding having the lesser number of turns is used as the primary; if the transformer is intended to lower the voltage, the winding having the greater number of turns is used as the primary.

EXAMPLE—It is desired to build a transformer to step down the voltage from 120 to 6 volts. If 200 turns are used on the primary, determine the number of turns that should be used on the secondary.

Substituting in equation 12-1,

$$\frac{120}{6} = \frac{200}{T_s}, \text{ or } T_s = 10 \text{ turns.}$$

When considering coils of negligible resistance, we learned that the impressed voltage leads the current by 90° and the induced voltage lags the current by 90°, or the impressed and the induced voltage are 180° out of phase. Thus, in a transformer with an unloaded secondary, the impressed voltage E_p, the primary induced voltage E_{pi}, and the current i, have the phase relation shown in figs. 12-1b and 12-1c. The voltage induced in the secondary E_s is in phase with E_{pi}, because they are both produced by the same flux. Then E_p and E_s are opposite in direction, or 180° out of phase.

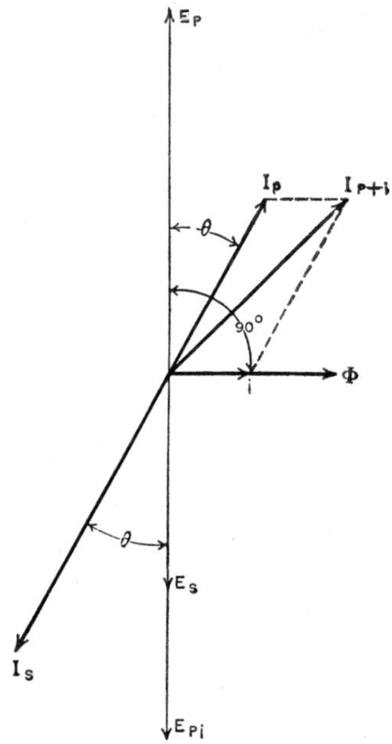

Fig. 12-2

A current I_s will flow when a load is connected to a secondary, due to the induced voltage E_s. The angle at which I_s leads or lags E_s depends upon the resistance and reactance of the load. In fig. 12-2 I_s is represented as lagging behind E_s by θ°. This current sets up a flux in the opposite direction to the flux originally set up in the primary due to i. But the slightest decrease in flux causes a corresponding decrease in the primary induced voltage. Then the current in the primary will increase until a magnetizing action is produced that is equal and opposite to that produced by I_s. This restores the flux to the zero or no-load value and the current remains steady again. In order to produce at all times a magnetizing action in the primary equal and opposite to that of the secondary, I_p must be exactly opposite I_s (see fig. 12-2) and the ampere-turns of the primary due to $I_p (T_p I_p)$ must equal the ampere-turns of the secondary $(T_s I_s)$.

At full load, i is very small compared to I_p and considerably out of phase with it. Then, if i is neglected, the ampere-turns of the primary equal the ampere-turns of the secondary. Thus

$$T_p I_p = T_s I_s$$

or

$$\frac{I_p}{I_s} = \frac{T_s}{T_p} \qquad (12\text{-}2)$$

Equation 12-2 shows that the primary and secondary currents vary inversely as the number of turns.

According to equation 12-1,

$\frac{T_s}{T_p} = \frac{E_s}{E_p}$. Substituting $\frac{E_s}{E_p}$ for $\frac{T_s}{T_p}$ in equation 12-2 gives

$$E_pI_p = E_sI_s, \tag{12-3}$$

which shows that the volt-amperes of the primary equals the volt-amperes of the secondary.

Construction of Transformers—The windings of commercial transformers are not placed on separate legs as indicated in fig. 12-1, because the primary winding would produce considerable flux that would not cut the secondary winding. This leakage flux would induce a back voltage in the primary, thus introducing a primary reactance drop. Similarly, the secondary current would set up a flux that would not cut the primary. Hence it would not be neutralized by the primary flux, and would produce a reactance drop in the secondary. The effect of these flux leakages would be the same as though a

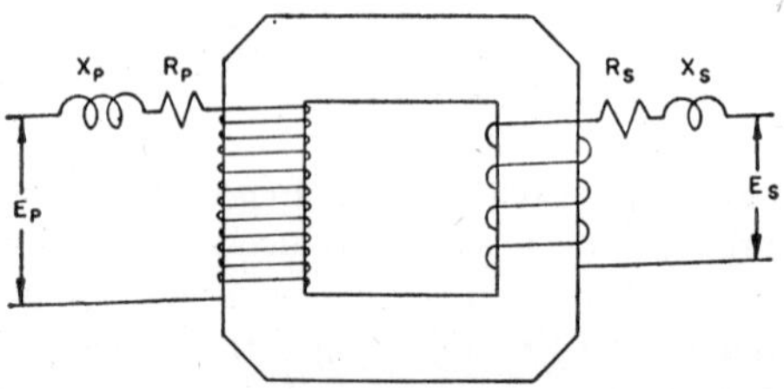

FIG. 12-3

reactance were connected in series with each winding of a transformer that had no flux leakage. In addition to reactance, each winding has resistance. These reactances and resistances may be represented externally as in fig. 12-3. Thus in each winding of a transformer there is an IX drop due to the reactance produced by the flux leakage and an IR drop due to resistance.

The IX drop may be made small by reducing the flux leakage to a minimum, which may be done by placing primary and secondary windings on the same leg of the core. The windings may be cylindrical in form and placed one inside the other, with the necessary insulation between them (fig. 12-4a), or they may be built up in thin flat sections called pancake coils. These sections are sandwiched together with the required insulation between them (fig. 12-4b). Note that in fig. 12-4a the low voltage winding is placed next to the core and the high voltage winding is placed on the outside. This arrangement requires only one layer of high voltage insulation, which is placed between the two windings. If the high voltage winding were placed next to the core, two layers of high voltage insulation would be required, one next to the core and another between the two windings.

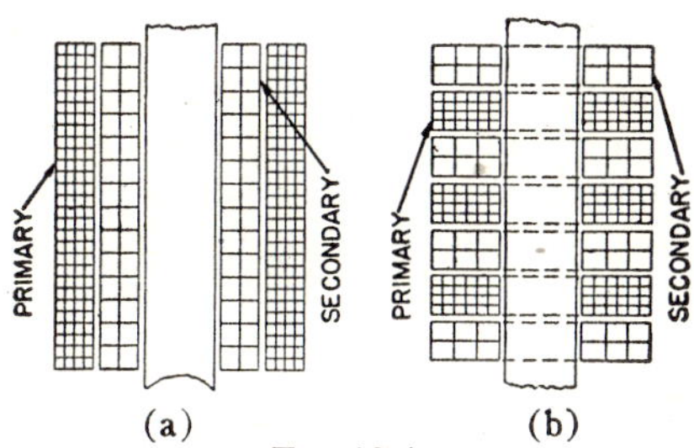

FIG. 12-4

Forms of Transformer Structures—The primary and secondary may be placed on the same leg in either of two types of core structure—the core type or the shell type. In the core type transformer a core of the shape shown in fig. 12-1a is used, but one-half of the primary and secondary is placed on each leg (fig. 12-5a). In the shell type transformer the core has a middle leg upon which both primary and secondary windings are placed (fig. 12-5b). This structure provides a much shorter magnetic path, but the average length per turn of wire is much greater.

The choice of type (whether core or shell) will not greatly affect the efficiency of the transformer. The core type is generally more suitable for high voltage and small output, while the shell type is generally more suitable for low voltage and high output.

The most modern transformers have a modified shell type structure, as shown in fig. 12-5c. This differs from the standard shell type in that the section of the core surrounding the winding is divided into four parts, symmetrically spaced around the center leg. In this arrangement the center leg has a higher flux density than the four outer ones because it is made comparatively small in cross section. This small cross section does not make the reluctance and losses high when the entire magnetic circuit is considered. However, it does make possible a shorter length per turn, which means that the winding will have a low resistance, and thus a low IR drop. Furthermore, this arrangement is compact and more easily cooled than the other types.

General Transformer Equation—The maximum flux density B_m in the core of a transformer is of considerable importance because it has a bearing on the operating characteristics. The value of B_m depends upon many factors. These are introduced in the general transformer equation 12-9 given below.

The average voltage induced by an alternating flux in a core may be determined by equation 10-16, which is

$$E_{av} = \frac{\Phi_m T}{10^8 t} \qquad (12\text{-}4)$$

where E_{av} = average generated e m f, Φ_m = maximum flux, T = number of turns, and t = time in seconds required for the flux to change from zero to maximum or vice versa.

Since $t = \frac{1}{4f}$, equation 12-4 becomes

$$E_{av} = \frac{\Phi_m T}{10^8 \left(\frac{1}{4f}\right)} = \frac{4f\Phi_m T}{10^8}$$

But

$$E_{av} = \frac{E_{eff}}{1.11}$$

Then

$$\frac{E_{eff}}{1.11} = \frac{4f\Phi_m T}{10^8}$$

Or

$$E_{eff} = \frac{4.44 f\Phi_m T}{10^8} \qquad (12\text{-}5)$$

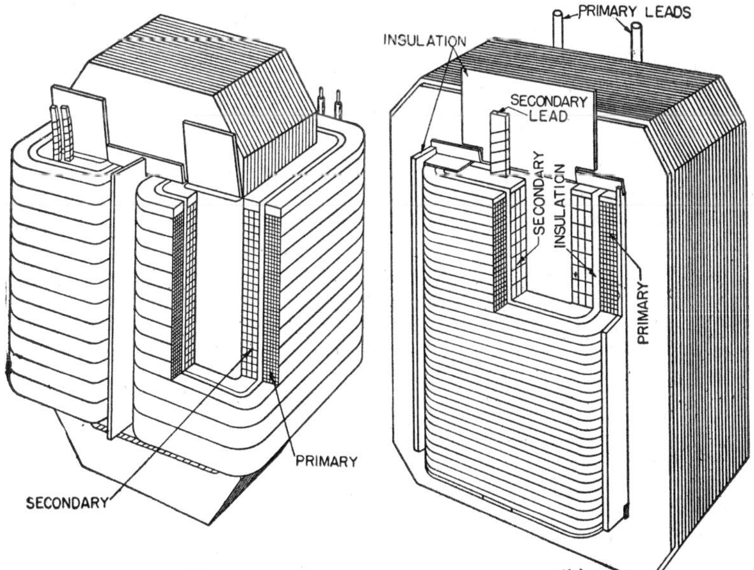

(a) (b)

When considering the primary, equation 12-5 is written

$$E_p = \frac{4.44 f \Phi_m T_p}{10^8} \quad (12\text{-}6)$$

and when considering the secondary it is written

$$E_s = \frac{4.44 f \Phi_m T_s}{10^8} \quad (12\text{-}7)$$

The above two equations are fundamental in the calculation and design of transformers and all kinds of A C machines and devices.

Since $\Phi_m = B_m A$, where B_m = maximum flux density and A = core cross section, equation 12-6 may be written

$$E_p = \frac{4.44 f B_m A T_p}{10^8} \quad (12\text{-}8)$$

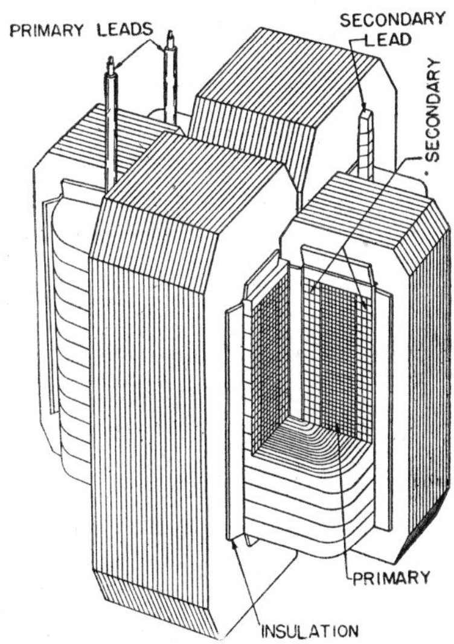

(c)

Fig. 12-5

The effect of changing the primary voltage of a transformer may be conveniently analyzed by equation 12-8 when it is expressed as

$$B_m = \frac{10^8 E_p}{4.44 f A T_p} \qquad (12\text{-}9)$$

If we increase the primary voltage E_p and keep the other factors constant, B_m will increase. If B_m is carried much beyond the knee of the saturation curve, the reluctance and the no-load current increase considerably (see fig. 12-6). Thus the iron and copper losses become large, causing the temperature rise to increase unless the output is considerably reduced. Hence, when the primary voltage is increased the efficiency and current output are reduced.

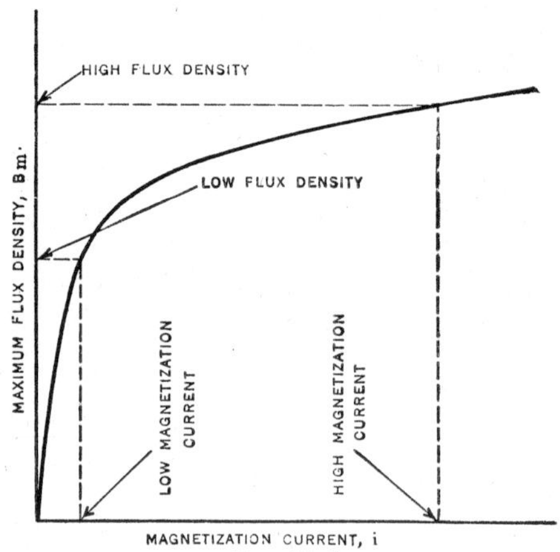

FIG. 12-6

It was mentioned above that formula 12-8 is also very useful in transformer design. In this case E_p, f, and B_m are known, and A and T_p are to be determined. If A is made large and T_p correspondingly small, the voltage regulation will be good, but the transformer will be rather large for a given k v a rating. If A is made small, a correspondingly larger number of turns will be required. This gives a smaller transformer but the voltage regulation will be poorer. The resistance of the winding will produce an appreciable IR drop as the load is applied. No definite equation can be given for determining the best value of A or T_p to use, but the curves in fig. 12-7 give the voltage per turn for transformers of different ratings. These curves were determined by experiment and represent good practical values.

EXAMPLE—Suppose it is desired to build a shell type transformer of 1 k v a, 440 to 110 volts, and 60 cycles. Using a flux density of 60,000 lines per square inch, determine (a) the number of turns required by the primary and secondary windings, and (b) the cross-sectional area of the iron.

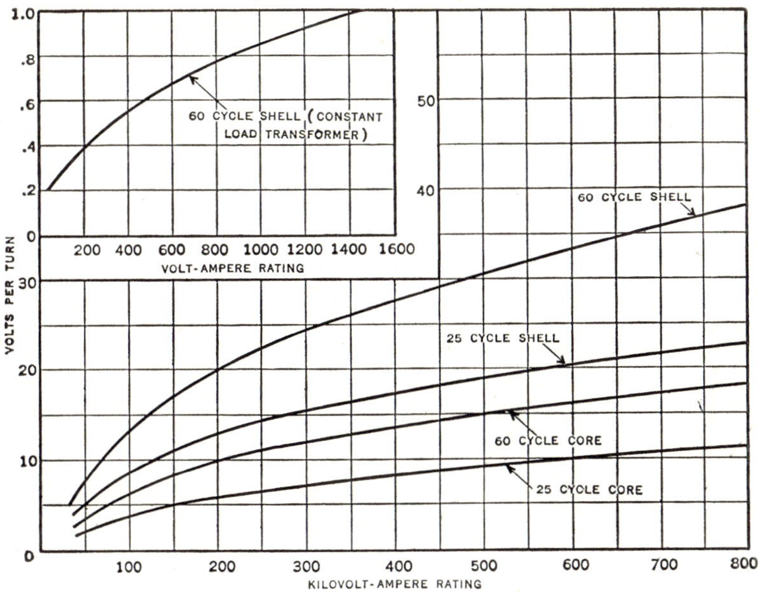

FIG. 12-7

(a) For a transformer of this capacity, fig. 12-7 recommends .85 volt per turn. Then $T_p = 440 \div .85$ or 517 turns, and $T_s = 110 \div .85$ or 129 turns

(b) In equation 12-8,

$$A = \frac{10^8 E_p}{4.44 f T_p B_m} = \frac{10^8 \times 440}{4.44 \times 60 \times 517 \times 60{,}000} = 5.32 \text{ square inches}$$

The middle leg therefore requires 5.32 square inches of sheet steel. The rest of the magnetic circuit carries half as much flux, and thus requires half the cross-sectional area, or 2.66 square inches.

Additional turns should be placed on the secondary to compensate for the IR drop in the two windings.

Losses and Efficiency—The power losses in a transformer are made up of copper losses and iron losses, though these amount to only a small portion of the total input (from 1% to 15%). The copper losses consist of the I^2R losses of both windings. The core losses are due to the hysteresis and eddy currents set up in the iron (see chapter 4). In that chapter it was shown that power is required to reverse a magnetic field, the amount of power required depending upon the kind of steel used. The alternating flux induces a voltage in the core as well as in the winding. This induced voltage causes eddy currents to flow in the laminations (see fig. 12-8), which are made thin to reduce these losses.

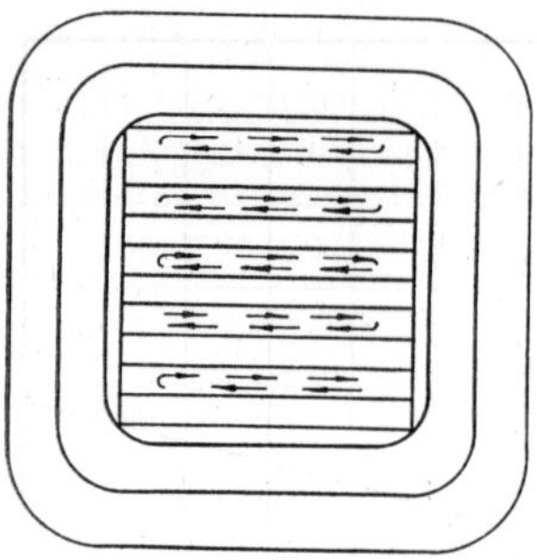

FIG. 12-8

Since losses in a transformer are very low compared to output, the efficiency is very high, varying from 85% to 99% (the larger the transformer the higher the efficiency). Fig. 12-9 shows how the efficiency varies in a 10 k v a, 60 cycle transformer. Note that the efficiency is practically constant from 1/8 load to 50% above full load.

Methods of Cooling Transformers—In all electrical machines the losses produce heat, and means must be provided to keep the temperature low. In generators and motors the rotating unit serves as a fan, causing air to circulate and carry away the heat. A transformer has no rotating parts, however, and some other method of cooling must be used. Transformers rated at less than 5 k v a are generally air cooled; that is, the heat produced is carried away by the surrounding air. Small and medium size power or distributing transformers are generally cooled by housing them in specially designed tanks filled with oil. The oil serves a double purpose, carrying the heat from the winding to the surface of a tank and insulating the primary from the secondary.

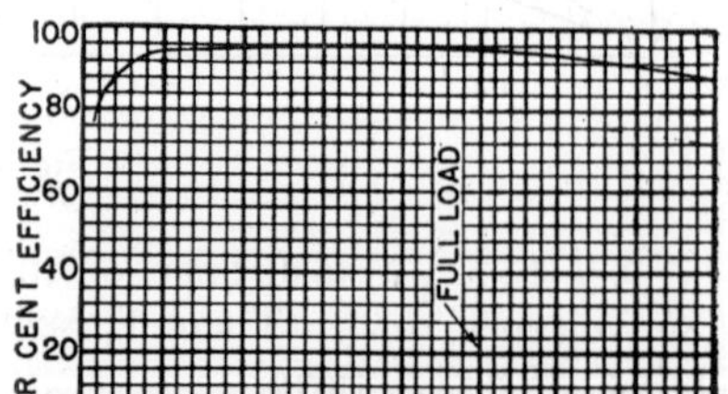

FIG. 12-9

A moderate size transformer is shown in fig. 12-10a (note that a corrugated cast iron tank is employed). The tubular steel tanks

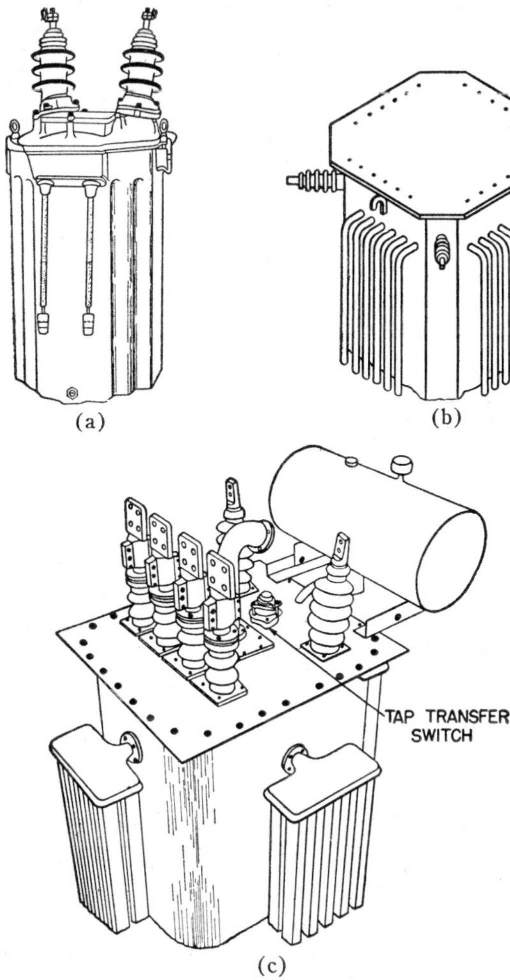

FIG. 12-10

of fig. 12-10b are sometimes used to increase the radiating surface. The external radiators of fig. 12-10c are in common use for large units (around 10,000 k v a capacity). In the largest sizes the oil cannot carry away the generated heat fast enough, so a coil of copper pipe is placed near the top where the oil is the hottest, and water is circulated through the coil (see fig. 12-11).

Transformers are sometimes cooled with an air blast. With this method the air is forced up from the bottom of the transformer against the core and coils, thus carrying away the excess heat. For the same k v a rating these transformers are lighter than the water

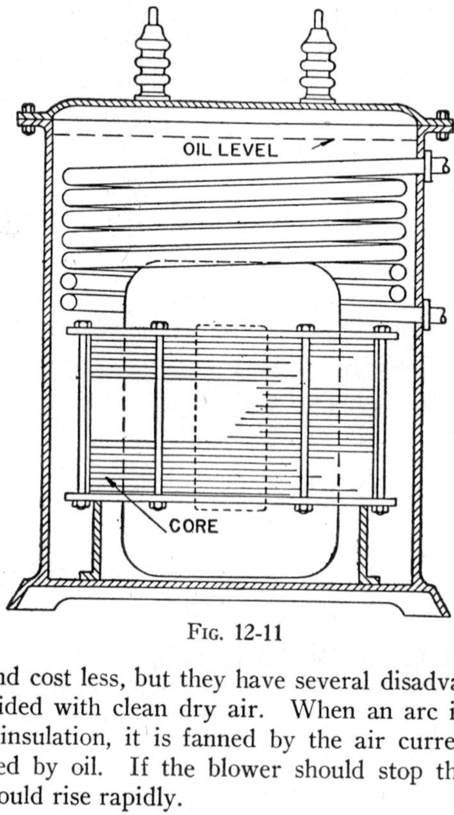

Fig. 12-11

cooled type and cost less, but they have several disadvantages. They must be provided with clean dry air. When an arc is formed by a break in the insulation, it is fanned by the air currents instead of being quenched by oil. If the blower should stop the temperature of the coils would rise rapidly.

Tap Changer—It is often desirable to vary slightly the turn ratio of transformers. Suppose we compare the primary voltage of a transformer near a generating station with that of one attached to the same power line at considerable distance from the station. The primary voltage of the second transformer will be lower than that of the first, due to the impedance drop in line. To obtain the same secondary voltage in both transformers the turn ratios must be different.

It is also essential to change the turn ratio slightly in keeping a constant voltage with a change in seasonal load.

Changes in transformer turn ratio are made on a deenergized transformer by a tap changing switch. The switch of fig. 12-12a, which is a common type, is mounted on a terminal board located above the transformer core. It is operated outside the tank by a shaft

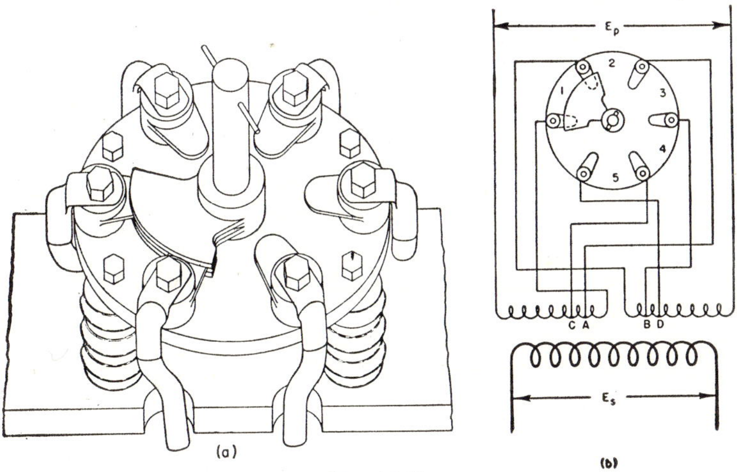

FIG. 12-12

which protrudes through the cover (see figs. 12-10c and 12-12a). A knob on the outside end of the shaft is marked to indicate the voltage ratio positions. Such switches make it possible to change taps without removing the transformer cover or making serious mistakes.

Fig. 12-12b shows how the tap changing switch is connected to a transformer winding. When the connector is in the position shown, all the primary winding is connected in the circuit. When the connector is shifted to position 2, section A is removed from the circuit, which causes the secondary voltage to be raised. When the connector is shifted consecutively to positions 3, 4, and 5, sections B, C, and D are removed from the transformer and the secondary voltage increases in steps.

Auto-Transformers—The auto-transformer differs from the standard transformer in having only one winding. Fig. 12-13 shows how this winding is connected. A portion of the winding serves for both the primary and secondary. If the transformer is used to step down the voltage, the turns between H_1 and H_2 constitute the primary winding and the turns between L_1 and L_2 constitute the secondary winding. The ratio of voltage transformation, as in a two-coil transformer, is equal to the ratio of primary to secondary turns.

If the exciting current is neglected, the primary and secondary ampere-turns in an auto-transformer are equal. Then equations 12-1, 12-2, and 12-3, which apply to two-coil transformers, also apply to auto-transformers.

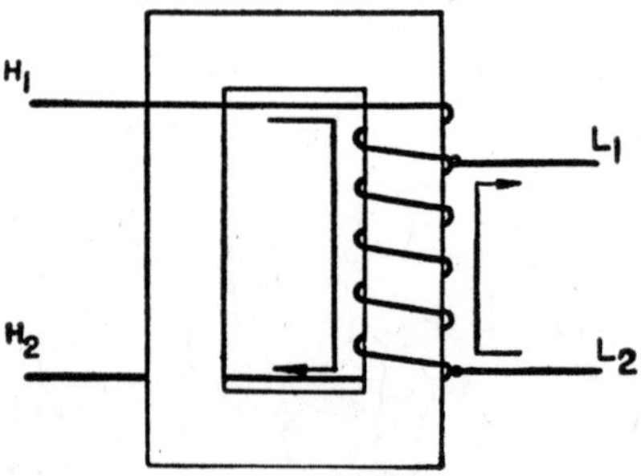

FIG. 12-13

The currents in the primary and secondary of a transformer flow in opposite directions. In an auto-transformer, the current in the portion of the winding that is common to both the primary and secondary equals the difference between the supplied and load currents. Due to this fact, auto-transformers require little copper, particularly if the voltage ratio is near unity.

Auto-transformers are less expensive, operate at a higher efficiency, and have a better voltage regulation than two-coil transformers of the same rating. They are not safe, however, for stepping down the voltage from a high to a low value, as from 13,200 to 110 volts for lamps and motors. There is a possibility that the section of the winding common to the primary and secondary may become open. In such a case, any one coming in contact with the secondary is subject to the high voltage, or the primary may become grounded and establish a high voltage between one of the low voltage line wires and the ground.

Auto-transformers are used for reducing the voltage supplied to alternating current motors during the starting period. They are also used for raising voltages, or to compensate for voltage drops in transmission lines, in which case they are known as booster transformers.

Instrument Transformers—Measuring instruments and relays in alternating current circuits should not be connected directly to high voltage circuits but should be linked to them by instrument transformers, commonly called voltage and current transformers. Such transformers will insulate the instruments from high voltage. Furthermore, by using instrument transformers of various ratios it has been possible to standardize electrical instruments to operate on 110 volts and 5 amperes.

Voltage Transformers—Voltage transformers are similar to the transformers already discussed except that their power rating is small (their load seldom exceeds 300 watts). A voltage transformer has the high voltage winding connected directly across the high voltage mains. The low voltage coil is generally wound to supply 110 volts and is connected to the instruments. This low voltage coil should always be grounded so as to eliminate static from the instruments and to insure the safety of anyone that may come in contact with them.

Current Transformers—Ammeters with separate shunts are employed in D C circuits when only a small fraction of the load current is to pass through the instrument. The use of shunts for A C ammeters is generally unsatisfactory because of the inductance effect and because of the fact that the instruments are not insulated from the high voltage line. Hence, a current transformer is used to eliminate these undesirable conditions.

The primary winding of a current transformer consists of one or more turns of heavy wire wound on an iron core and connected in series with the line. If the primary is to have a large current rating, it may consist of a straight conductor passing through a core, as shown in fig. 12-14. The secondary then consists of many turns of finer wire, with ammeters and other instruments and relays with current coils connected to it. Measuring instruments are usually designed for a full scale deflection of five amperes, and relay coils are rated at that value.

It is shown by equation 12-2 that the current ratio is inversely proportional to the number of turns. For example, if the primary has one turn and the secondary has twenty turns, the current ratio will be twenty to one. This means that the instrument will have one-

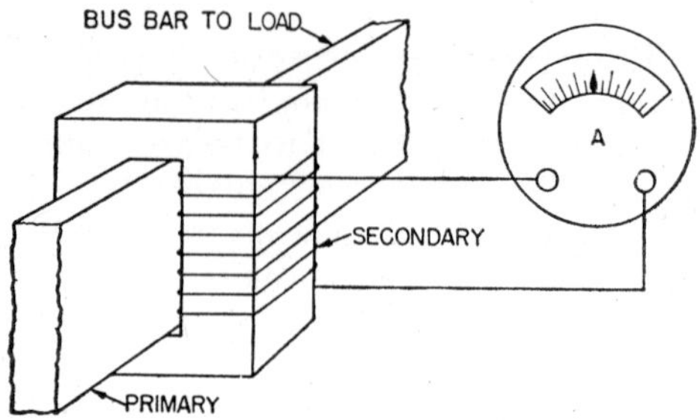

Fig. 12-14

twentieth as much current flowing through it as the load, and its reading must therefore be multiplied by twenty.

The current transformer differs from the ordinary constant potential transformer in that the primary current depends entirely upon

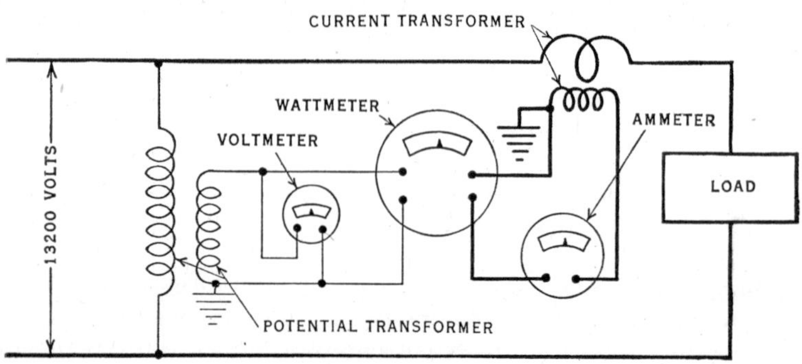

Fig. 12-15

the load on the system and not upon the load of the secondary. If the secondary is opened while the primary is carrying a heavy load, the demagnetizing effect of the secondary current will no longer exist and the flux in the core will increase. This increased flux will induce a high voltage in the secondary circuit, which may puncture the insulation or produce a dangerous shock to anyone coming in contact with it. *Therefore the secondary of a current transformer should never be opened while the primary is energized.*

Fig. 12-15 shows the manner of connecting a voltmeter, ammeter, and wattmeter to a 13,200 volt line. In permanent installations the instruments are calibrated to take care of the turn ratio of the transformers, so that current, voltage, and power are read directly.

Exercises

The maximum flux density in transformer cores should be kept below the saturation point. Densities between 50,000 and 90,000 lines per square inch are permissible. Values near 90,000 are generally used in artificially cooled power transformers. In the following problems allow 60,000 lines per square inch.

The least number of circular mils per ampere that may be used is limited by the temperature rise, for if this is excessive it will damage the insulation. For air cooled transformers, 1400 to 2000 circular mils per ampere are permissible, and for oil emersed transformers, 900 to 1400 circular mils per ampere are used. In the following problems, allow 1700 circular mils per ampere.

1. It is desired to build a transformer to step down the voltage from 440 to 110 volts. If 2,080 turns are used on the secondary, how many turns should be used on the primary?
2. The low voltage side of a welding transformer has one turn and delivers 400 amperes. If the high voltage side has 800 turns, how much current flows through it?
3. If a 13,200/440 volt transformer has 3 volts per turn, determine the number of turns in the primary and in the secondary windings.
4. If a 3 ohm impedance is connected to the secondary of a 440/6 volt transformer, find the primary current.
5. What are the different methods of arranging the primary and secondary windings in a transformer?
6. What are the detrimental effects of operating a transformer above its rated voltage? Below its rated voltage?
7. What are the different methods of cooling transformers?
8. What is the purpose of tap changers? When are they used?
9. For what purposes is an auto-transformer suitable? For what purpose should it never be used?
10. A 13,200/440 volt transformer is rated at 100 k v a. Determine the current ratings of the primary and secondary windings.
11. What is a suitable number of volts per turn for a 60 cycle, 440/110volt, 500 k v a, core type transformer?

12. For the transformer in problem 11, determine (a) the number of turns in each winding, (b) the cross section of the core, and (c) the circular mil area of the wire in each winding.
13. If the cross section of the core in problem 12 is made twice as great, how many turns are required in each winding to maintain the same flux density and the same voltage ratio?
14. For a 220/6 volt, .5 k v a, shell type transformer, determine (a) the number of turns in each winding, (b) the cross section of the middle leg, and (c) the circular mil area of the wire in each winding.
15. An auto-transformer is used to step up the voltage from 13,200 to 23,000 volts. If the transformer delivers 46 k v a, determine the currents in the line wires and in each portion of the transformer winding.
16. A 220/110 volt shell type auto-transformer is rated at 1 k v a. Determine (a) the number of turns in each section, (b) the cross section of the core, and (c) the size of wire in each section of the winding.

CHAPTER 13

POLYPHASE SYSTEMS

In general, polyphase systems are superior to single phase systems in a number of ways. Polyphase machines are smaller, simpler in structure, and have better operating characteristics than single phase units of the same capacity. Furthermore, a three-phase power line requires about three-quarters as much copper as a single phase line of the same capacity, voltage, and transmission efficiency.

To avoid errors when considering polyphase circuits, it is essential to employ some systematic method of combining voltages and currents of different phases. The method of doing this will be considered in the following discussion.

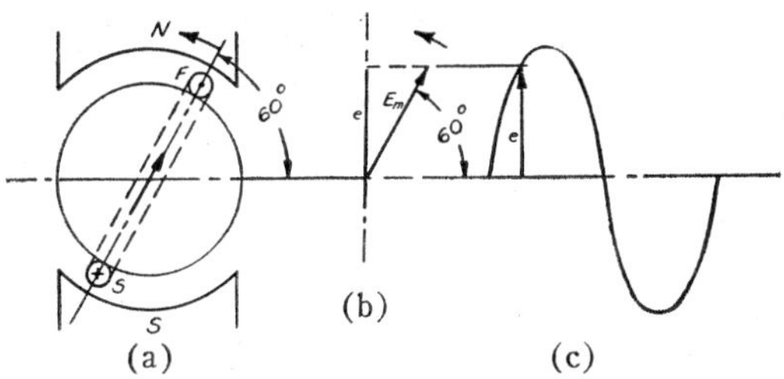

FIG. 13-1

In fig. 13-1a, assume that the coil has rotated 60 degrees in the counter-clockwise direction. By applying the right-hand rule, it will be found that the induced voltage is in the direction shown. This voltage may be conveniently indicated by means of a vector drawn as in fig. 13-1b, whose length represents the maximum value. The vertical projection e indicates the instantaneous value of voltage. The e m f induced in the coil is also indicated by e in fig. 13-1c. According to the conventions of trigonometry, e in figs. 13-1b and 13-1c is positive for the first half-cycle and negative for the second half. Then the voltage indicated in the coil of fig. 13-1a is positive. The coil end which this voltage enters will be called the start end (denoted by S), and the end it leaves will be called the finish end (denoted by F).

In fig. 13-1a, suppose that another coil Y is placed 30° behind the indicated coil, which we will call X. This is indicated in fig. 13-2a. Coil Y will generate a voltage E_Y which lags the voltage E_X of coil X by 30°. This is indicated in figs. 13-2b and 13-2c.

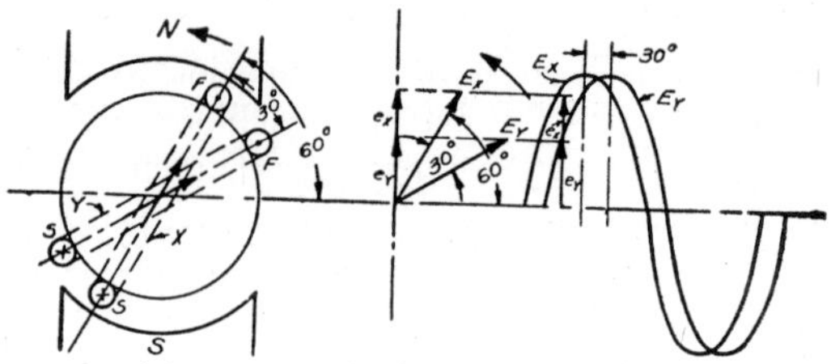

FIG. 13-2

Now suppose that these coils are to be connected in series. Four connections are possible, from X_F to Y_S, X_F to Y_F, X_S to Y_F, or X_S to Y_S. If we wish to connect the two coils so that a high voltage is produced between their open ends, which is usually the connection

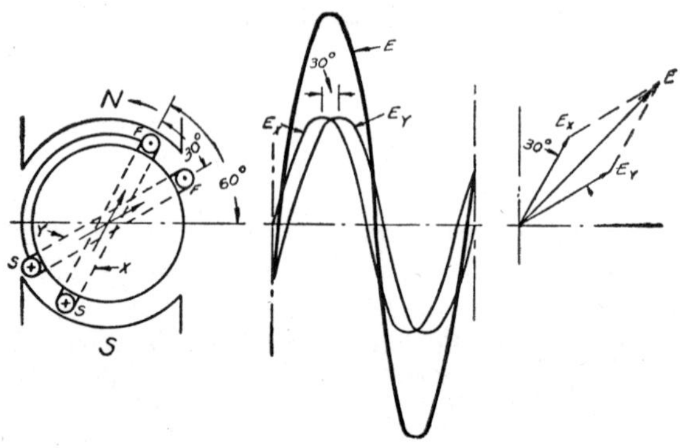

FIG. 13-3

desired, we must connect the finish end of coil X to the start end of coil Y. This connection, which is the first one suggested above, is shown in fig. 13-3a. The voltage produced at the terminals of the two coils may be obtained by adding the instantaneous voltages generated by each. This is shown in fig. 13-3b, where E is the resultant voltage. Instead of adding the voltage waves, we can more easily obtain the resultant by vector addition. This is shown in fig. 13-3c.

In fig. 13-2a, suppose that X_F and Y_F are connected, obtaining X_S and Y_S for the open ends (see fig. 13-4a). In this case, the voltage wave of coil X must be added to the voltage wave of coil Y in reverse, because coil Y is now connected to coil X in reverse. The voltage wave of coil Y in reverse is indicated by $-E_Y$ in fig. 13-4b. The voltage $-E_Y$ is added to the voltage E_X and gives the resultant voltage E. The corresponding vector diagram is shown in fig. 13-4c.

The remaining two possible connections give the same resultant voltages as those shown in figs. 13-3 and 13-4 respectively.

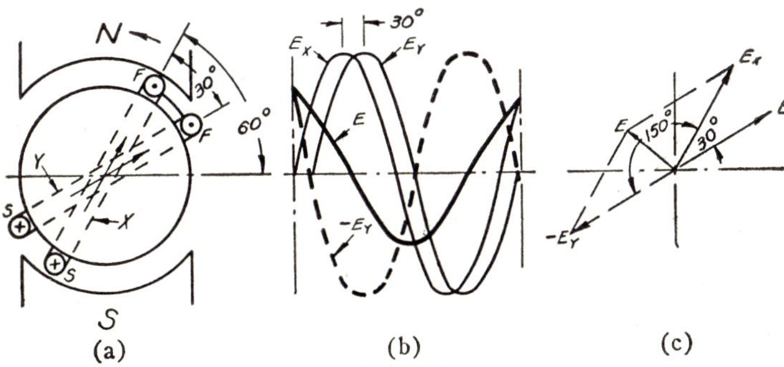

FIG. 13-4

The conclusion to be drawn from this consideration is as follows: *The total voltage supplied by two coils in series, formed by connecting a start and a finish end, may be determined by finding the resultant of their individual voltage vectors. If two start ends or two finish ends are connected, one vector must be reversed before the resultant is obtained.*

The correct resultant voltage of two coils connected in series will be obtained, regardless of which coil ends are called the start ends. For instance, suppose that in fig. 13-4a the start ends are selected as shown in fig. 13-5a. Then coil X lags coil Y by 150 degrees, and their voltage waves will also be 150 degrees displaced (fig. 13-5b).

Since the start end of coil X is connected to the finish end of coil Y, or since the arrows are in the same direction through the coils, the voltage waves must be added to obtain the resultant voltage. This is shown in fig. 13-5b. The corresponding vector diagram is shown in fig. 13-5c. In fig. 13-5 it will be noted that the resultant voltage is the same as that obtained in fig. 13-4.

When considering coils generating voltages out of phase, it is unnecessary to draw coils on a core. The idea may be fully conveyed in a simpler manner by means of a conventional diagram. Such a diagram is shown in fig. 13-6a, which conveys the same idea as fig.

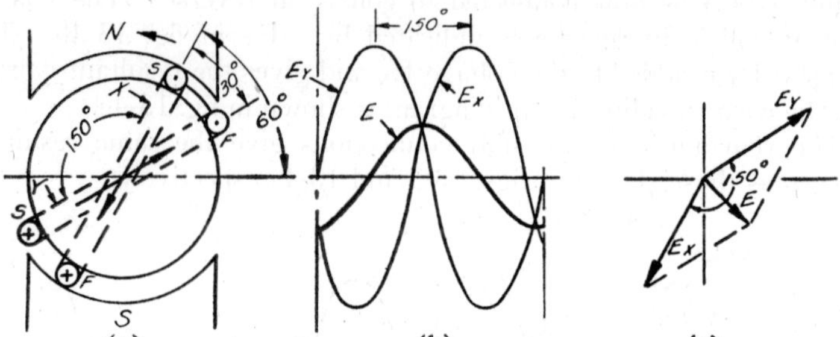

(a) (b) (c)

FIG. 13-5

13-2a—that the voltage of coil Y lags the voltage of coil X by 30 degrees. The arrows do not serve to indicate the direction of voltage at the same instant, but indicate the positive direction of voltage, which is always from the start to the finish end of a coil. The corresponding vector diagram is shown in fig. 13-6b, in which the vectors are drawn to represent the effective instead of the maximum values.

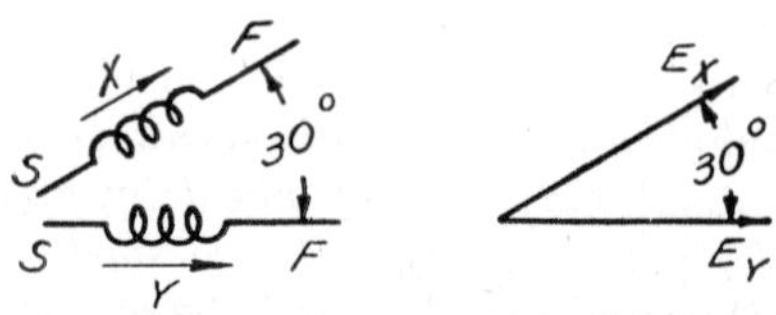

(a) (b)

FIG. 13-6

Suppose that the coils of fig. 13-6a are connected as shown in fig. 13-7a, which is the same connection as that of fig. 13-3a. In this case the resultant voltage is obtained by adding vectorially the voltages of the two coils, because their positive voltages are in the same direction, or because the start of phase Y is connected to the finish of phase X, as shown in fig. 13-7b.

Suppose that it is desired to determine the voltage between the open ends of the coils in fig. 13-6a when terminal X_S is connected to

Y_S. This connection, which is the same as that shown in fig. 13-4a, is shown in fig. 13-8a. In tracing through the coils from X_F to Y_F, we pass through coil X in the opposite direction to the arrow, which indi-

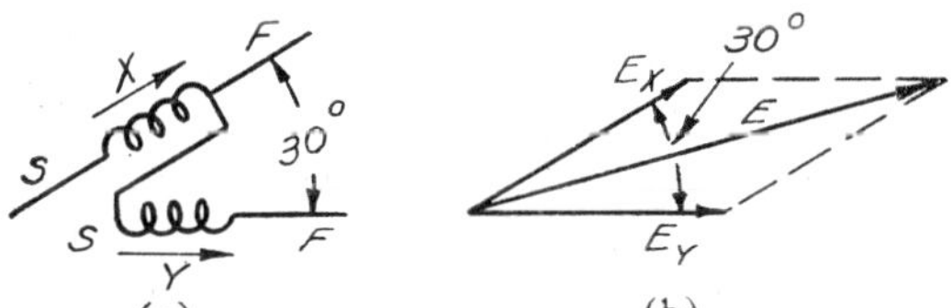

(a) (b)

FIG. 13-7

cates that coil X is connected to coil Y in reverse. Then the voltage vector of phase X must be reversed before the vectors are added. This is carried out in fig. 13-8b.

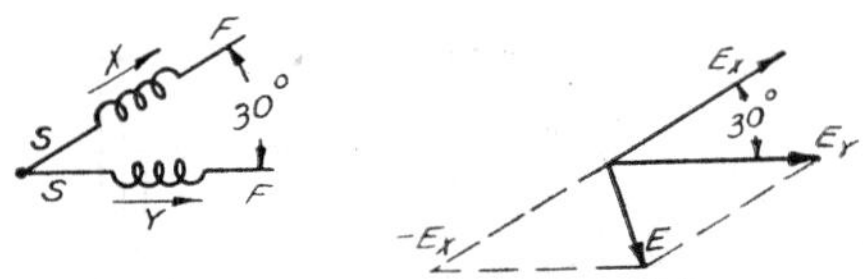

(a) (b)

FIG. 13-8

Two-Phase Alternator—If a two-pole alternator has an armature with two coils X and Y placed at right angles, or 90 degrees apart

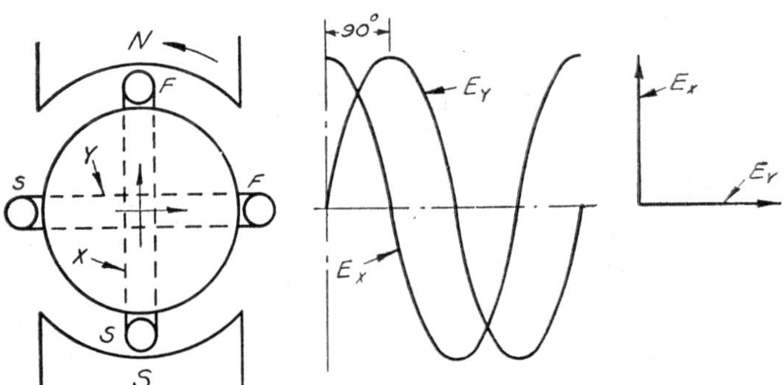

(a) (b) (c)

FIG. 13-9

(fig. 13-9a), the machine is known as a two-phase alternator. The voltages generated in these two coils are 90 degrees out of phase, as shown in figs. 13-9b and 13-9c. In some systems, the coils are not electrically connected, and four wires are required for supplying power to the external load. This arrangement is known as a two-phase, four wire system. To such a system, both single phase and polyphase loads may be connected. This is shown in fig. 13-10, in which there are two single phase lamp loads and a two phase motor load. Regardless of the type of loads, each phase may be treated as a single phase circuit.

EXAMPLE—How much power must each phase of the alternator in fig. 13-10 deliver if each lamp takes 500 watts and the two-phase motor takes 8000 watts?

Only the loads connected to phase X will be considered. The lamps take 5 × 500 or 2500 watts. Half the motor will take 8000 ÷ 2

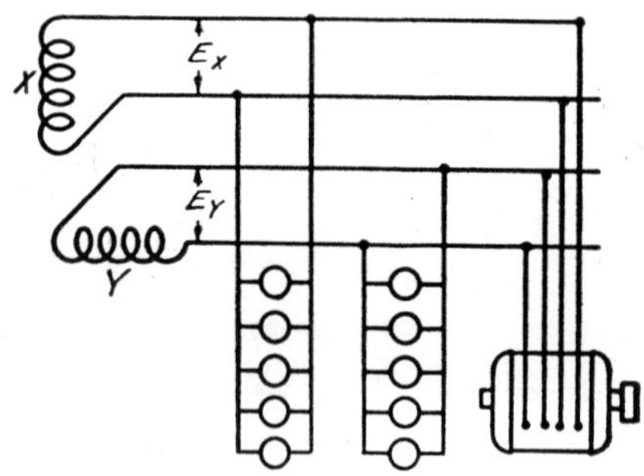

FIG. 13-10

or 4000 watts. Then the power delivered by each phase of the alternator is 2500 + 4000 or 6500 watts (6.5 kw).

If the two coils of the alternator shown in fig. 13-9a are connected in series, only three wires are required to supply the power to the external load (see fig. 13-11). This arrangement is known as a two-phase, three wire system. If the e m f generated in each coil is 120 volts, the voltage between line wires L_X and L_C, and between L_Y and L_C, will each be 120 volts. The voltage between L_X and L_Y may be determined from the vectors of fig. 13-9c, which are also shown in fig. 13-11b. When tracing through the coils from L_X to L_Y (fig. 13-11a), we pass through coil X in the opposite direction to the arrow, and through coil Y in the same direction as the arrow. This indicates that coil X is connected in reverse. Then, in fig. 13-11b, E_X must be reversed before it can be vectorially added to E_Y. Carrying this out gives the resultant E. Since the e m f of each coil is 120 volts, the voltage between L_X and L_Y is

$$\sqrt{120^2 + 120^2} = 169.7 \text{ volts}$$

If the connections to one coil in fig. 13-11a are reversed, the voltage between the open ends will remain the same. This is due to the fact that if one of two vectors 90 degrees displaced is reversed, the two vectors will still be 90 degrees displaced.

Due to the non-inductive resistances of 12 ohms each, connected as shown in fig. 13-11a, equal currents I_X and I_Y flow in each phase. The accompanying arrows show the positive direction of the currents, which are from the start to the finish end of the coils. Each resistance, outside line wire, and coil carries the same amount of current,

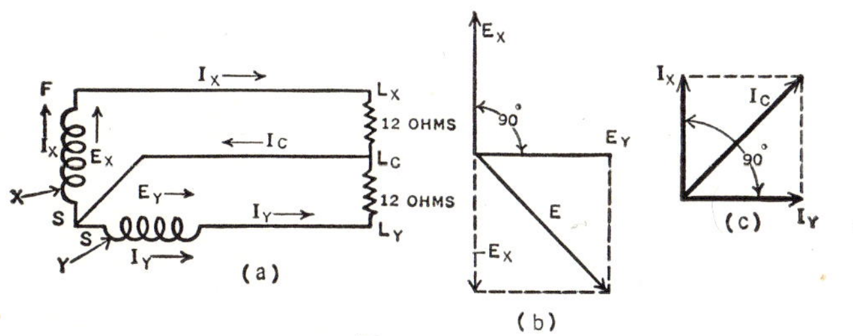

FIG. 13-11

which is 120 ÷ 12 or 10 amperes. The current I_C through the common wire equals the resultant of the currents through the two coils. The positive direction of I_X and I_Y is the same as that of I_C. Then, to obtain I_C, the currents I_X and I_Y must be vectorially added. This is shown in fig. 13-11c.

Since the current per phase is 10 amperes, then

$$I_C = \sqrt{10^2 + 10^2} = 14.14 \text{ amperes}$$

This shows that the common wire carries more current than either of the other two, consequently it should have a greater cross section.

Three-Phase Alternator—A three-phase alternator may be conveniently represented by a diagram such as that of fig. 13-12a. The coils are displaced 120 degrees, which is accomplished by placing their start ends (lettered S) 120 degrees apart. At the instant shown, coil X is in the neutral plane, and the e m f induced in it is zero. This is indicated in fig. 13-12b, in which E_X is shown starting at zero and increasing in the positive direction. As soon as the winding has rotated 120 degrees, the e m f in coil Y will be zero and starting in the positive direction, as shown by curve E_Y. When the winding has rotated another 120 degrees, the e m f in coil Z is zero and starting in the positive direction (see curve E_Z). An examination of fig. 13-12b will clearly show that the three voltages are displaced 120 degrees. This is shown vectorially in fig. 13-12c.

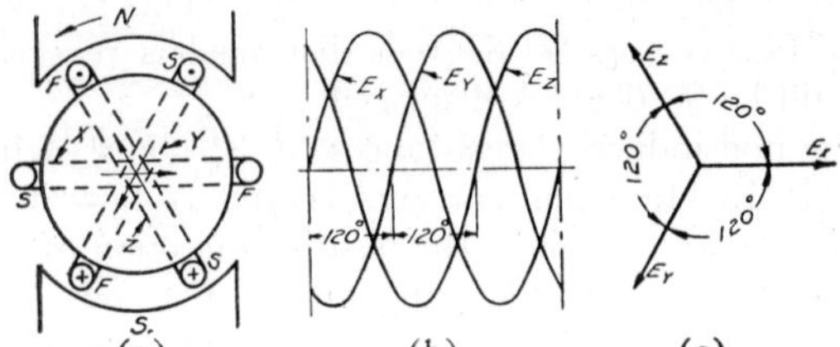

Fig. 13-12

The three coils shown in fig. 13-12a are generally connected together inside the machine in one of two schemes, either by a star connection or a delta connection.

Star Connection—In this scheme the start ends are connected together and the finish ends are used as terminals of the machine, to which the line wires are connected (fig. 13-13a). Represented

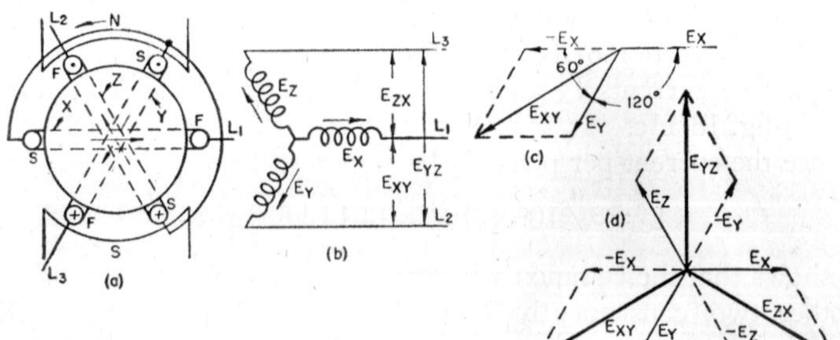

Fig. 13-13

diagrammatically, this method looks like a star (fig. 13-13b). It is therefore known as a star connection. To find the voltage between the line wires, which are indicated by E_{XY}, E_{YZ}, and E_{ZX}, it is necessary to find the resultant of the phase voltages. The method of finding the voltage between L_1 and L_2 will first be considered.

In fig. 13-13b, by tracing from L_1 to L_2, we pass through coil X in the direction opposite the arrow, and through coil Y in the same direction as the arrow. This indicates that coil X is connected in reverse to coil Y. Then, in fig. 13-13c, we must reverse E_X before adding the vectors, which gives E_{XY}. Fig. 13-13d shows the relation between the phase voltages and the line voltages, in which it will be noted that the line voltages are equal and displaced from one another by 120 degrees.

A numerical relation exists between the phase voltages and line voltages. Referring again to fig. 13-13c, the vector E_{XY} bisects the angle between $-E_X$ and E_Y, which is 60 degrees. Then $E_{XY} = 2E_X \cos 30° = 1.73E_X$. A similar relation exists between the other phase voltages and line voltages.

It then follows that in a star connected, three-phase system, the line voltage E_L equals 1.73 times the phase voltage E_f. Expressing this by an equation,

$$E_L = 1.73E_f \qquad (13\text{-}1)$$

The current I_L in each line wire is evidently the same as the current I_f in each phase. Then

$$I_L = I_f \qquad (13\text{-}2)$$

EXAMPLE—Each coil of a star connected, three-phase alternator generates an e m f of 254 volts. Determine the e m f between the line wires.

$$E_L = 1.73 \times 254 = 439.4 \text{ volts}$$

Incorrect Star Connection—The three coils of fig. 13-12a may be connected 60 degrees out of phase. This is accomplished by selecting three coils ends 60 degrees displaced, considering them as start ends, and connecting them together to form the star center. This is shown in fig. 13-14a.

At the instant shown, coil X is in the neutral plane and the e m f induced in it is zero. This is indicated in fig. 13-14b, in which E_X is shown as starting from zero and increasing in the positive direction. As soon as the winding has rotated 60 degrees, the e m f in coil Y is zero and starting in the positive direction, etc. (see curve E_Z). The three voltage waves in fig. 13-14b will then be 60 degrees displaced. The connection of fig. 13-14a is shown diagrammatically in fig. 13-14c, and its corresponding vector diagram is shown in fig. 13-14d. Fig. 13-14e shows the resultant line voltages, which, it will be noted, are neither equal nor 120 degrees displaced. Because of these reasons, a machine is never connected this way in practice.

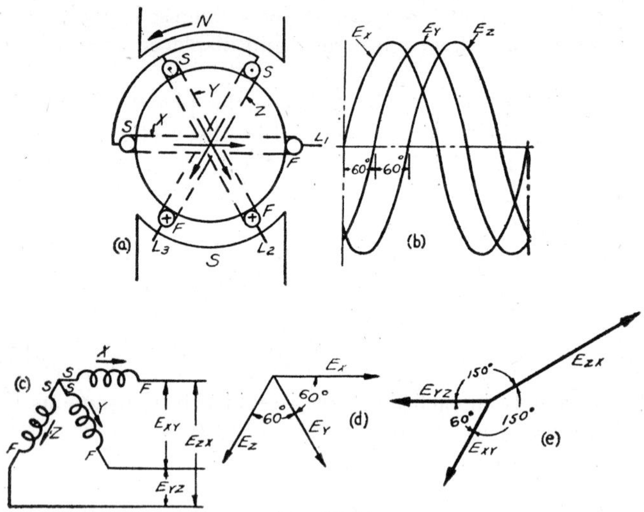

FIG. 13-14

Delta Connection—The other scheme for connecting a three-phase alternator is illustrated in fig. 13-15a. In this case the finish end of coil X is connected to the start end of coil Y, the finish end of coil Y is connected to the start end of coil Z, and the finish end of coil Z is connected to the start end of coil X. The line wires are brought out at the points where the coil connections are made. This method of connecting the winding is shown diagrammatically in fig. 13-15b. It is known as a delta connection because the diagram has the appearance of the Greek letter delta (Δ).

In fig. 13-15b it may appear that the three phases are short-circuited on themselves. The resultant voltage actually existing in the closed circuit may be found by combining the resultant voltages of the three phases. The same result may be obtained more easily by combining the voltage vectors, which are shown in fig. 13-15c. The vector resultant of E_X and E_Y is E_{XY}; and since E_X is displaced 120 degrees from E_Y, the resultant voltage $E_{XY} = E_X = E_Y$. But E_{XY} is equal and opposite to E_Z. Then, in a delta connection, the voltage of one coil is equal and opposite to the voltage of the other two, and the resultant voltage in the closed circuit is zero.

If one phase of the delta connection is reversed, the resultant voltage in the closed circuit will no longer be zero, but will be double the voltage of one phase. This voltage will establish a high current, even without an external load, and will damage the winding.

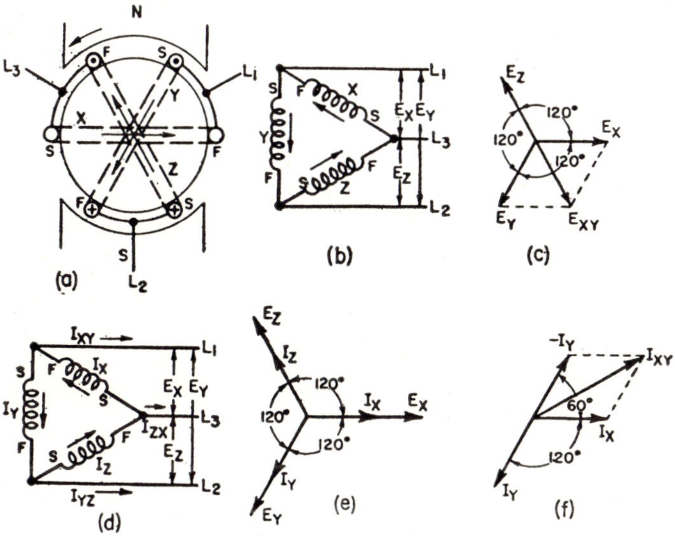

FIG. 13-15

In practice, the voltages of a delta connection are not exactly equal, nor are they an exact sine wave. Consequently, some current always circulates in the three phases, but it is so small that it may be neglected.

Evidently, in a delta connection the line voltages E_L and the phase voltages E_f are equal. Then in this case,

$$E_L = E_f \tag{13-3}$$

The current in any one of the line wires is the resultant of the currents in the two phases attached to that line wire. Suppose that the winding is loaded so that equal currents flow in the three phases, and that these currents are in phase with their respective voltages. This is assumed in fig. 13-15d, in which the arrows indicate the positive direction of the currents. The vector diagram for this condition is shown in fig. 13-15e. Since no current circulates in the delta, then the phase currents must flow in or out of the line wires. The positive direction of current in the lines is represented away from the alternator (fig. 13-15d). Since the current I_X is toward line L_1 and I_Y is away from L_1, to get I_{XY} we must reverse I_Y and add it to I_X. This is carried out in fig. 13-15f. This diagram shows that the line current I_{XY} is the vector sum of two equal currents I_X and $-I_Y$, 60 degrees apart, and that it bisects the angle formed by them. Then

$$I_{XY} = 2I_Y \cos 30° = 1.73I_Y$$

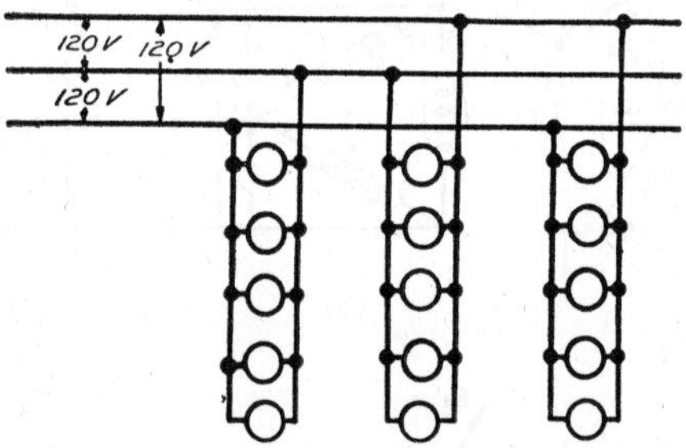

FIG. 13-16

Obviously, a similar relation exists for the currents in the other line wires. Then, for a balanced load, the line current I_L equals 1.73 times the phase current I_f. Expressed by an equation,

$$I_L = 1.73I_f \tag{13-4}$$

If the three currents (I_X, I_Y, and I_Z) are displaced from their corresponding voltages by the same angle, equation 13-4 still applies.

Not only are sources of energy, such as alternator coils, connected star or delta, but such loads as heating resistors, transformers, and lamps are also connected in this way. In the case of such units attached to a three-phase power line, the same relation exists between line and phase voltages, or line and phase currents, as exists between these quantities at the alternator.

EXAMPLE—To a three-phase, 120 volt power line, it is desired to connect a balanced load consisting of fifteen 120 volt lamps, each taking 2 amperes. Determine the current through each line wire.

For a balanced load, the lamps must be connected as shown in fig. 13-16. The current supplied to each group of lamps, which is the phase current, is 2 × 5 or 10 amperes. Since the lamps form a delta connection, the current through each line wire may be determined by equation 13-4. Hence, the line current is 1.73 × 10 or 17.3 amperes.

Comparison of Two- and Three-Phase Systems—It was pointed out that polyphase systems are superior to single phase systems. In comparing two and three phase systems, the latter is slightly superior, and consequently is the standard system used. The three phase system is superior to the two phase system for the following reasons:

(1) A three phase alternator utilizes the stator surface more efficiently, and for this reason is smaller. For the same rating, three phase machines cost about six per cent less.

(2) The efficiency of transmission in three phase systems is slightly greater.

(3) Three phase systems make use of star and delta connections, each of which has certain industrial applications.

Power in a Three-Phase System—The power taken (or delivered) by a balanced three-phase system is obviously three times the power per phase. Expressed by an equation, this is

$$P = 3E_f I_f \cos \theta_f \qquad (13\text{-}5)$$

where E_f = phase voltage, I_f = phase current, and $\cos \theta_f$ = phase power factor.

In a balanced star connected system,

$$E_f = \frac{E_L}{1.73}$$

and

$$I_f = I_L$$

Substituting in equation 13-5, we have

$$P = 3 \times \frac{E_L}{1.73} \times I_L \times \text{Cos } \theta_f = 1.73 E_L I_L \text{ Cos } \theta_f \qquad (13\text{-}6)$$

Then,

$$\text{Cos } \theta_f = \frac{P}{1.73 E_L I_L} = \frac{\text{total power}}{\text{total volt-amperes}} = \text{Cos } \theta_L$$

This shows that the power factor of a phase equals the power factor of the line. Then equation 13-6 may be written as

$$P = 1.73E_LI_L \cos \theta_L \qquad (13\text{-}7)$$

In a similar way it can be shown that equation 13-6 applies to a balanced delta connected three-phase circuit.

Hence, the power taken (or delivered) by any three-phase system equals the product of 1.73, the line voltage, the line current, and the power factor of the system.

Measurement of Power—The power taken by a three phase, star connected load, whether balanced or unbalanced, may be measured with three wattmeters connected as shown in fig. 13-17. In this

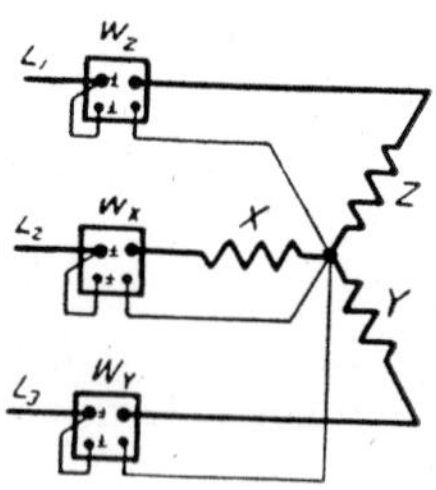

FIG. 13-17

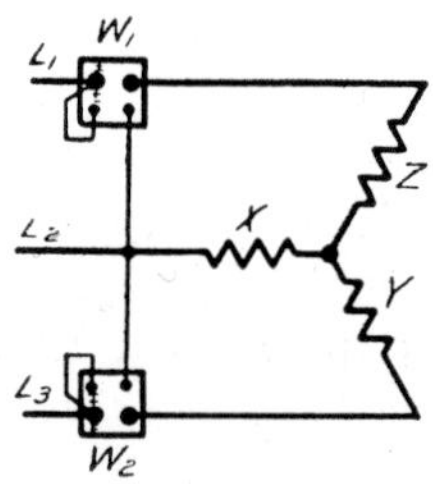

FIG. 13-18

case, wattmeter W_X indicates the power taken by phase X, wattmeter W_Y indicates the power taken by phase Y, etc. Of course the total power taken is the sum of the wattmeter readings. If the star center of a load is not available, which is often the case, or if the load is connected delta, correct wattmeter readings may be secured by connecting together the three ends of the voltage coils, which in fig. 13-17 connects to the star center. This may be done only if the impedances of all the potential coils are equal.

The power taken by a three phase load, however, can be measured by two wattmeters connected as shown in fig. 13-18. The current coil of each wattmeter is connected in a line wire, and the potential coil of each meter is connected to the line which includes its current coil and to the line which includes no current coil. The ± terminals should be connected to the section of the line extending from the alternator.

If the power factor of the load is unity, each wattmeter will read one-half the total power. If the power factor is 50 per cent, one wattmeter will read zero and the other will read the total power. If the power factor is between 50 and 100 per cent, one wattmeter will read more than the other, in which case the total power is found by adding the two wattmeter readings. If the power factor is below 50 per cent, one wattmeter will tend to indicate backwards. The connection of the current or voltage coil of this wattmeter must be reversed in order that a reading may be obtained. In this case the power taken by the circuit is determined by subtracting the reading of the wattmeter that reversed from the other wattmeter reading. Hence, when two wattmeters are used, the total power taken is obtained by adding or subtracting the readings, the operation performed depending upon whether the power factor is greater or less than 50 per cent.

It was stated that the terminals marked ± should be connected to line wires extending from the alternator. Suppose the wattmeter terminals are not marked, and it is not known whether the power factor of the load is greater or less than 50 per cent. In this case it is necessary to conduct a simple test, after the hookup is complete, to determine whether the wattmeter readings should be added or subtracted. The test is made as follows: A line wire that includes the current coil of one wattmeter is opened. This reduces the system to a single phase circuit with a wattmeter. If this wattmeter tends to indicate in reverse, its reading must be subtracted from the other, when the three phase circuit is complete. A similar test should also be made, leaving the wattmeter not previously checked in the circuit.

Polarity—In order to facilitate the making of proper transformer connections, the American Institute of Electrical Engineers has specified that the high voltage leads are to be marked H_1, H_2, etc., and the low voltage leads are to be marked X_1, X_2, etc. The order of the subscripts must be as follows: When H_1 and X_1 are connected together and a voltage is impressed on either the high voltage or low voltage winding, the e m f between the H lead with the highest subscript and the X lead with the highest subscript must be less than the e m f of the high voltage winding. If this is not the case, the order of the subscripts on the low voltage (or high voltage) leads must be reversed, bringing about the desired condition. After the desired result is obtained, the voltage between H_1 and H_2 will be in the same direction as that between X_1 and X_2.

With the leads marked as specified, the transformer has a subtractive polarity if H_1 and X_1 are adjacent to each other (see fig.

13-19a), and an additive polarity if H_1 and X_1 are diagonally opposite each other (fig. 13-19b).

Transformers with leads marked according to this rule may be connected, without difficulty, for parallel operation and for operation in three-phase circuits.

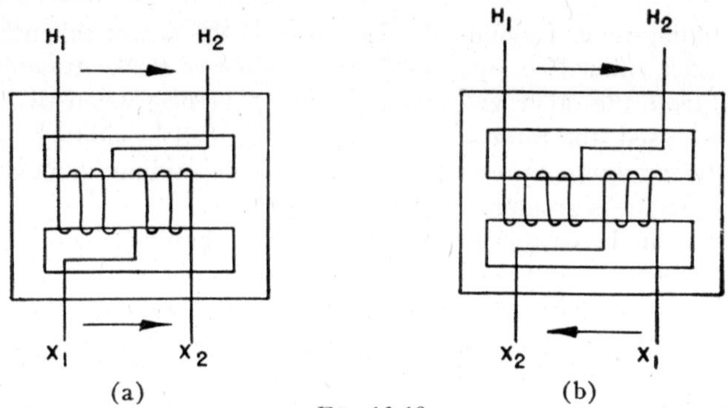

FIG. 13-19

Transformers in Three-Phase Systems—In a three-phase system, the voltage is usually lowered or raised by means of three single-phase transformers or by one three-phase transformer. In either case, the windings are connected star-star, delta-delta, star-delta, or delta-star. The connections for the first three are shown in

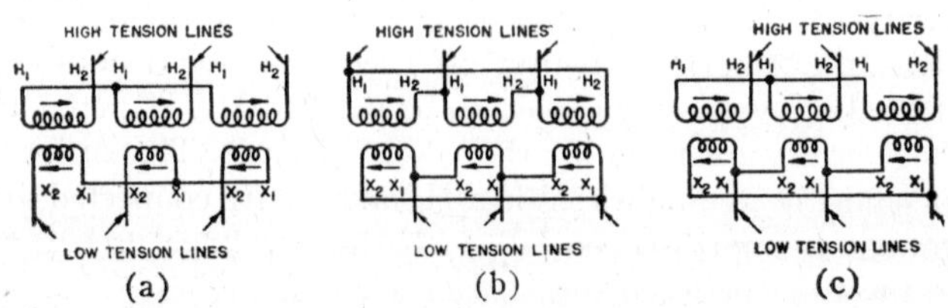

FIG. 13-20

fig. 13-20. The last connection is not shown because it is the same as the third, except that the power is supplied to the delta side of the transformer. In fig. 13-20 it will be noted that the coils are joined as suggested for alternators. For a star connection, all the arrows point away from (or all point toward) the star center. In a delta connection, all the arrows point clockwise or counterclockwise.

In three-phase systems, the voltages at the transformer terminals depend upon the turn ratios and the kind of connection. If transformer windings are star connected, the voltage across each winding equals the line voltage divided by 1.73. If the windings are delta-delta connected, the line and transformer voltages will be the same.

For high voltage transmission lines, transformers for a star-star connection are cheaper because they are designed for 57.7 per cent of the line voltage; whereas, for a delta connection, the transformers must be designed for line voltage.

If one transformer breaks down in a star-star connected system, three-phase power cannot be supplied until the defective transformer has been replaced or repaired. To eliminate this undesirable condition, single phase transformers are generally connected delta-delta. In this case, if one transformer breaks down, it is possible to continue

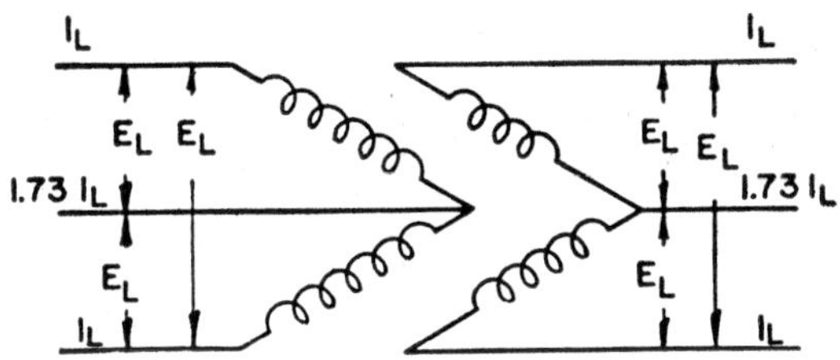

Fig. 13-21

supplying three-phase power with the other two (see fig. 13-21), because this arrangement maintains the correct voltage and phase relations on the secondary. With two transformers, however, only 57.7 per cent of the rated power can be supplied.

The star-delta connection is suitable for stepping down a high voltage, in which case the primaries are designed for 57.7 per cent of the high tension line voltages. The delta-star connection is obviously suitable for stepping up the voltage.

Three-Phase Transformers—A three-phase transformer weighs less, occupies less space, and costs about twenty per cent less than three single-phase transformers of equal capacity. Because of these advantages, three-phase transformers are in common use, especially for large power transformations.

Suppose that three single-phase, core type, transformers, each with windings on only one leg, have their unwound legs combined to provide a path for the returning flux (see fig. 13-22a). This arrangement gives a three-phase transformer in which the center leg carries the fluxes produced by the three primary windings. Since the sum of the three primary currents at any instant is zero, the sum of the three fluxes passing through the middle leg must also be zero. Hence, no flux exists in the middle leg, and it may therefore be eliminated. This modification gives a three leg transformer which in practice is generally constructed as shown in fig. 13-22b.

Fig. 13-22c shows the shell type of three-phase transformer. It is equivalent to three single-phase, shell type transformers, placed one on top of the other.

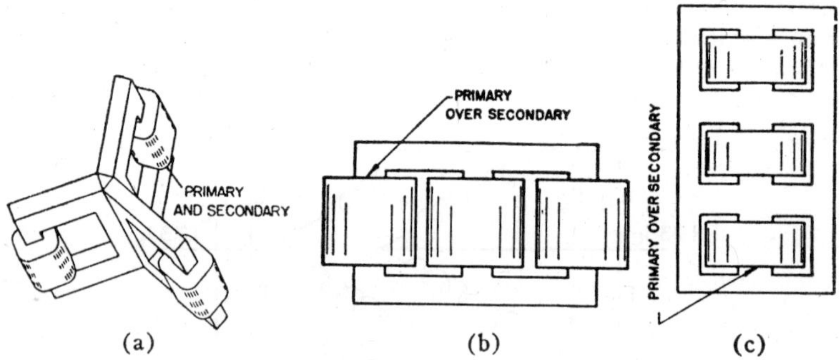

FIG. 13-22

All the connections of three-phase transformers are made inside the case, so that only three primary leads and three secondary leads are brought out of the case.

Exercises

1. Two coils 45 degrees apart each generate 120 volts. If they are connected in series, what two voltages may be obtained between their open ends?
2. In fig. 13-2a, suppose that the finish end of coil Y is connected to the start end of coil X. If the e m f of each coil is 140 volts, determine the voltage between the open ends.
3. In fig. 13-6a, suppose that the start ends of the two coils are connected together. If the e m f of each coil is 240 volts, determine the voltage between the finish ends.
4. A two-phase, four wire, 220 volt supply line carries a total load of 300 k w at a power factor of 80 per cent. What current flows through each line wire?

5. In fig. 13-11a, suppose that the 12 ohm resistance between L_X and L_C is replaced by a coil having a resistance of 4 ohms and a reactance of 6 ohms. If an e m f of 150 volts is generated in each coil, what current flows in each line wire?
6. A two-phase alternator connected like the one shown in fig. 13-11a has a current I_X of 50 amperes lagging E_X by 24 degrees, and a current I_Y of 50 amperes leading E_Y by 20 degrees. What current flows through each line wire?
7. Determine the line voltages of the alternator in fig. 13-13 if the e m f of each coil is 120 volts.
8. Solve problem 7 if the connections to coil Y are reversed.
9. Suppose that each coil in fig. 13-13b has an e m f of 6600 volts and a current-carrying capacity of 10 amperes. Determine (a) the line voltage and (b) the current in each line wire at full load.
10. Solve problem 9 assuming that the coil are connected delta.
11. A star connected, three-phase alternator supplies a line current of 100 amperes at 240 volts. What is (a) the phase voltage, and (b) the phase current?
12. If 4600 volts are generated in each phase of fig. 13-23, which is a star connection, determine the voltage (a) between O and A; (b) between A and B; (c) between A and C.

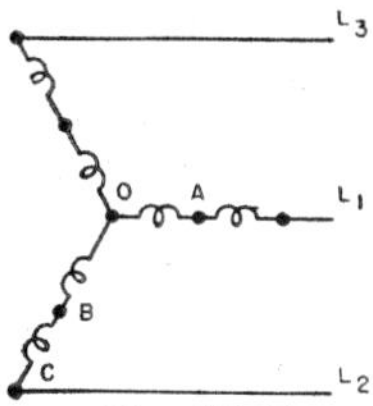

FIG. 13-23

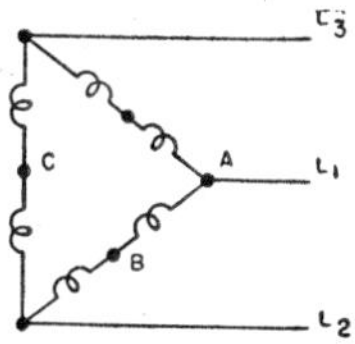

FIG. 13-24

13. The current through each phase of a delta connected alternator is 20 amperes. Determine the current in the line wires.
14. Three heating elements are connected star across a three-phase, 220 volt line. If 10 amperes flow through each line wire, determine the resistance of each heating element.
15. What current will flow through the line wires in problem 14 if the heating elements are reconnected delta?
16. If 440 volts are generated in each phase of fig. 13-24, which is a delta connection, determine the voltage (a) between A and B; (b) between B and C; (c) between A and C.
17. If the connection to one phase in problem 16 is reversed, what is the resultant voltage of the three phases?

18. Calculate the current that would circulate in problem 17 if each phase has an impedance of 2 ohms.
19. Suppose the winding in fig. 13-24, which is a delta connection, is open at C. (a) If each phase generates 440 volts, what are the voltages between each pair of line wires? (b) If each phase (excluding the open one) carries 50 amperes, what are the currents in the line wires?
20. A 120 volt, three-phase feeder delivers power to thirty 120 volt, 40 watt lamps. If the system is balanced, determine the current through each line wire.
21. A 240 volt, three-phase line delivers power to three resistances of 10 ohms each. Determine the power delivered (a) when they are connected star, and (b) when they are connected delta.
22. A three-phase line supplies 100 k w at a power factor of 60 per cent. If the line voltage is 6600, determine the current in each conductor.
23. A three-phase alternator is rated at 4600 volts and 300 k v a. Compute its full load current.
24. A 6600 volt, 500 k v a, star connected alternator is reconnected delta. What is its new voltage and k v a rating?
25. Each phase of a three-phase alternator has a current carrying capacity of 100 amperes and a voltage rating of 6600. Determine (a) the k v a rating of the alternator when it is connected star, and (b) when it is connected delta.
26. Each of two wattmeters connected as in fig. 13-18 indicates 20 k w. (a) What is the total power supplied to the load? (b) What is the power factor of the load?
27. Each phase of a three-phase, 440 volt line carries a load of 30 k w. How much power is delivered?
28. Suppose the power in problem 27 is measured by two wattmeters. What would each indicate (a) if the power factor is unity, and (b) if the power factor is 50 per cent?

CHAPTER 14

ALTERNATORS

It was shown that in D C motors a commutator is used to change the alternating e m f induced in the coils to a unidirectional e m f for the external circuit. An alternating current would be supplied to the external load if slip rings were used instead of a commutator. In this case the machine would be called an alternator.

Revolving Field Alternator—Since the armature of an alternator has no commutator, it need not be the rotating member. In fact it is more desirable to have a stationary armature with the field poles rotating inside it (fig. 14-1a), as this structure has several advantages. For polyphase power a rotating armature requires three or more slip rings to deliver power to an external load. These rings, being exposed, are hard to insulate, particularly for the high voltages that alternators are required to supply. Arc-overs and short circuits are apt to occur. Furthermore, it is considerably harder to properly insulate a high voltage rotating winding than a stationary one, because the heavy insulation must withstand the mechanical forces due to centrifugal force. No slip rings are required in a stationary armature. The leads from the winding may be insulated entirely to the switchboard.

Field Structure—An alternator with a stationary armature requires a rotating field which is excited through slip rings. The current required for the field is supplied at a pressure that need not exceed 250 volts. Thus slip rings for the field winding offer no serious difficulty. A stationary armature winding is called a stator and a rotating field structure is usually called a rotor. A type of rotor used in low speed machines, called the salient pole type, is shown in fig. 14-1b. It consists of a number of laminated poles dovetailed or bolted to a cast iron spider. The slip rings are also attached to the spider, and the terminals of the field winding are connected to the slip rings. The direct current is supplied through brushes resting on the slip rings. The field coils for small machines are wound with round wire, while rectangular copper strips wound on edge are used for large machines.

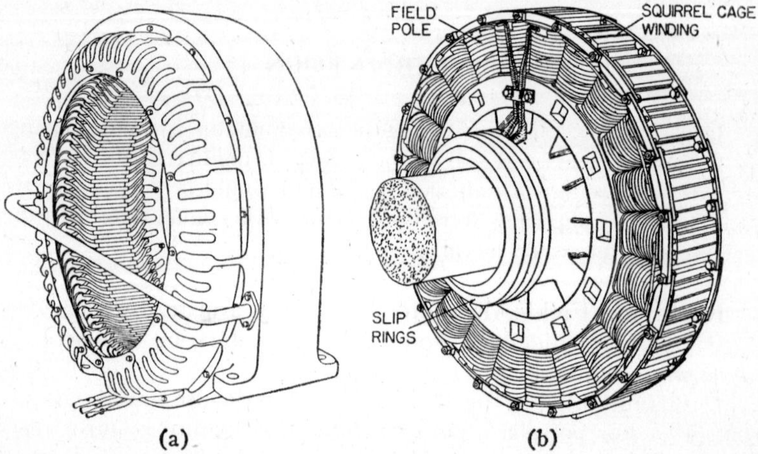

FIG. 14-1

Note that copper bars are embedded in the face of the field poles and connected at their ends with brass rings. Such a winding is used particularly on the field poles of alternators driven by internal combustion engines, to improve the parallel operation. They are not usually required in alternators driven by reciprocating steam engines, water wheels, or steam turbines.

Turbo-Type Alternators—When the steam engine is operated at a high speed it has a high efficiency, and for this reason it is often used to drive alternators. But for high speeds it is difficult to build a rotating field with projecting poles strong enough to withstand the centrifugal force. Projecting poles also cause excessive wind losses and make the alternator noisy. To overcome these undesirable features, alternators intended for steam turbine drive have their field structure made cylindrical in form and small in diameter, with two or four poles (fig. 14-2a). The stator for this type of rotor is shown in fig. 14-2b. A machine of this type is called a turbo-type alternator. In this alternator considerable heat is liberated in small spaces, and this heat must be carried away by air currents forced through passages in the heated parts. Hence forced ventilation is required. The machine must be enclosed to control the direction of the air currents, as well as to reduce noise.

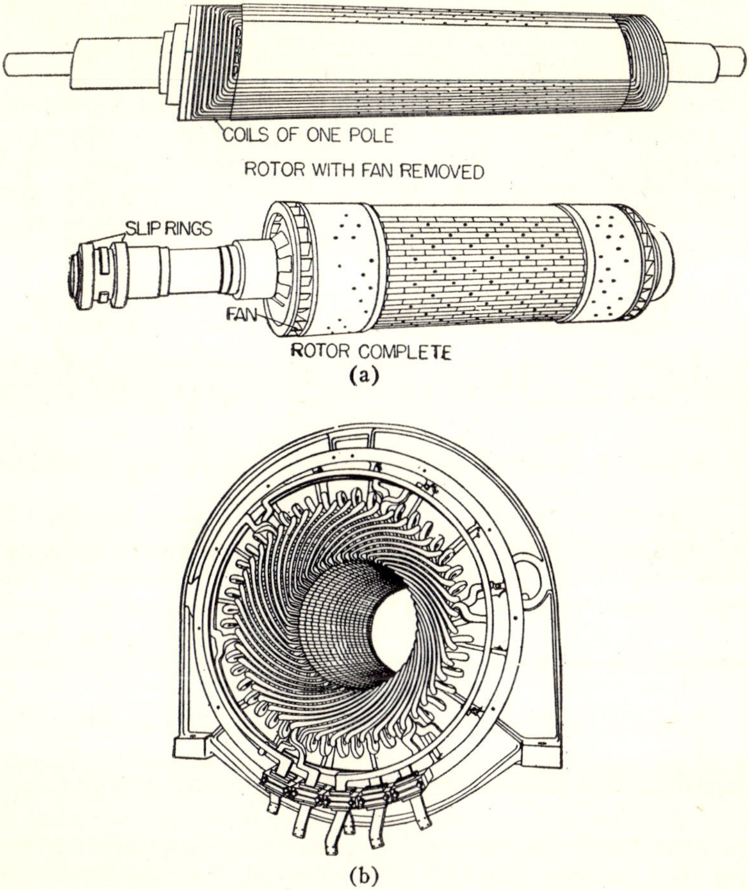

Fig. 14-2

Frequency—In the United States alternators are designed to supply power at a frequency of 60 or 25 cycles per second. Sixty cycle alternators are generally used to supply power to loads situated a short distance from the generating station. Twenty-five cycle alternators are generally used to supply power at a long distance, because this lower frequency gives a correspondingly lower IX drop in the transmission line. In European countries frequencies of 50 and 25 cycles are commonly used.

The frequency of the e m f generated by an alternator depends upon the number of poles and the speed. It was shown in fig. 10-1 that a complete cycle of voltage is generated when a conductor has passed one pair of poles. Then in an alternator with P poles, P/2 cycles of voltage will be generated for each revolution of the field poles. But the number of cycles per second, or the frequency (denoted by f), equals the number of cycles per revolution (P/2) times the number of revolutions per second, or times the number of revolutions per minute (N) divided by 60.

$$f = \frac{P}{2} \times \frac{N}{60}$$

or

$$f = \frac{PN}{120} \qquad (14\text{-}1)$$

EXAMPLE—What is the frequency of the e m f generated in an eight pole alternator operating at 900 r p m?

$$f = \frac{8 \times 900}{120} = 60 \text{ cycles per second}$$

Forms of Alternator Windings—Even though windings of many different forms may be used on the stator of an alternator, only two types are very common—the lap winding and the chain or basket winding (the basket winding is a further development of the chain winding). These windings are much simpler than those used on D C machines, because no commutator is used.

Lap Winding—Lap winding is the most common type of winding used in alternators, synchronous motors, and induction motors. This winding is double layer, made up of form wound coils with a twist between the coil sides so that the coils can be lapped (see fig. 14-3). All the coils are identical; hence only one form of coil is necessary.

Thus lap winding used on alternators is similar to that used on D C machines. In fact a D C machine may be converted into a rotating armature alternator by simply placing slip rings on the shaft and properly tapping the winding to the slip rings. However, the so-called open circuit winding is more suitable in alternators and is therefore universally used. An open circuit winding is one that does not close upon itself, but the open ends form the terminals of the winding.

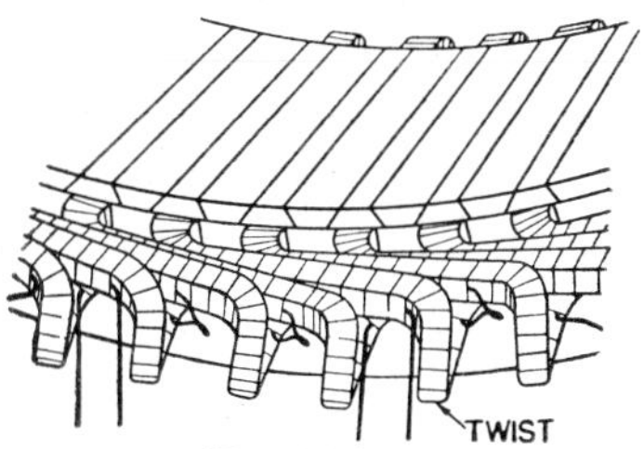

FIG. 14-3

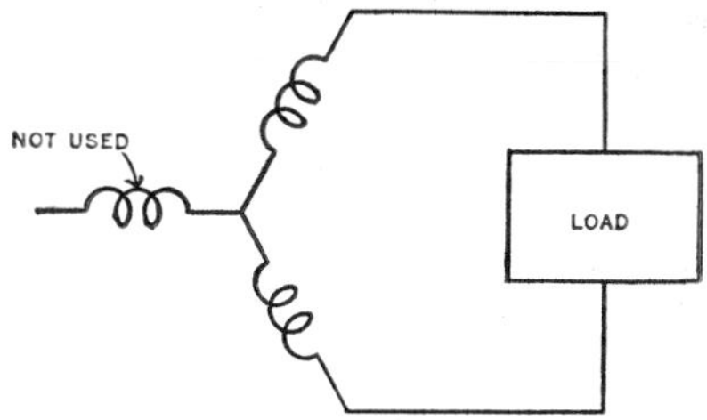

FIG. 14-4

Single-Phase Winding—Single-phase alternators are seldom manufactured. However, if only a single phase voltage is desired (such as for electric railways), a standard three phase, star connected alternator is used, but line wires are extended from the terminals of only two phases, as shown in fig. 14-4. The unused phase serves as a spare for the other two. Single-phase windings will be discussed

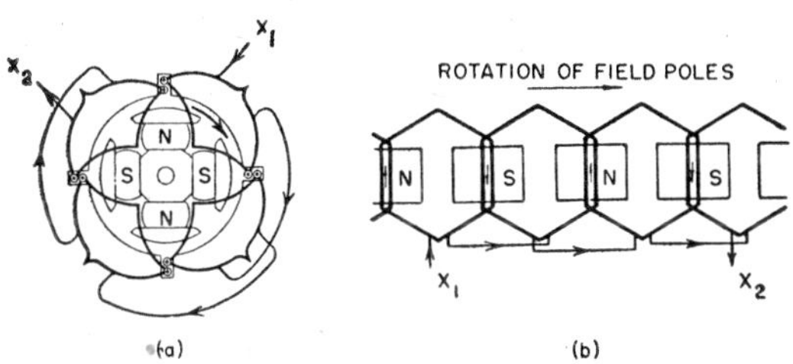

FIG. 14-5

briefly, however, because a knowledge of such windings enables one to more easily understand polyphase windings.

Fig. 14-5a shows a single-phase winding for a 4 pole alternator. In this circuit the connections to the coils are reversed. Inspection of the developed diagram in fig. 14-5b discloses that reversals are necessary to make the e m f's additive.

Now if another winding similar to the first is placed on the stator, but displaced 90 electrical degrees from the first, a two phase winding results (remember that the distance between pole centers is 180 electrical degrees). Such a winding is shown in fig. 14-6. Phase Y will generate a voltage exactly as phase X but displaced from X by 90 degrees.

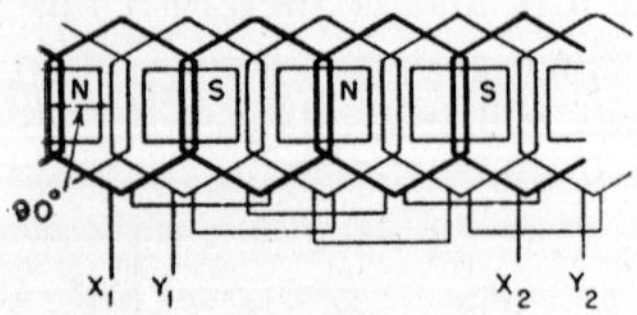

FIG. 14-6

If three single phase windings spaced 120 degrees apart are used, a three phase, six lead winding results. Such a winding is shown in fig. 14-7. In this case note that the adjacent coils are 60 degrees apart. The three phases may be connected either star or delta.

Distributed Windings—A winding with only one slot per pole per phase, as the one shown in fig. 14-7, is called a concentrated winding. In this type of winding the e m f generated per phase is the greatest that can be obtained for a given number of turns, field flux, and speed, because the e m f's of the coils are in phase.

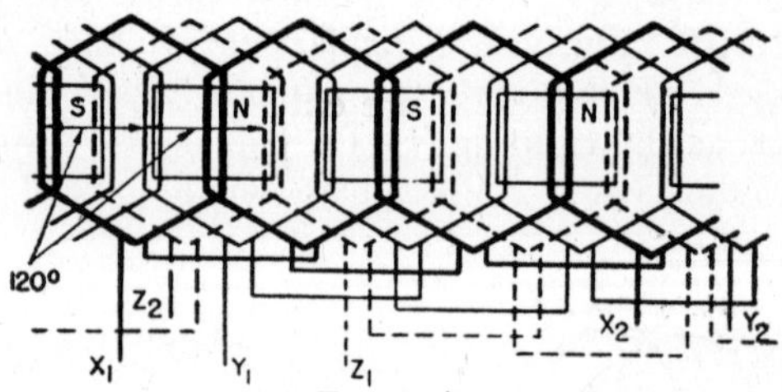

FIG. 14-7

The voltage induced per phase in each of the above windings may be computed by the fundamental transformer equation (12-6). This may be expressed for alternators as:

$$E_p = \frac{4.44\,\Phi\, f T_p}{10^8} \qquad (14\text{-}2)$$

where E_p = voltage induced per phase, Φ = flux per pole, f = frequency in cycles per second, and T_p = number of turns in each phase.

The concentrated winding has these two undesirable features: (1) In large machines only a small portion of the stator surface is utilized, and (2) there is serious difficulty in producing a sine wave form of e m f, since in this case the wave form can be altered only by shaping the pole faces.

A sine wave of e m f is very desirable. If machines with unlike wave forms are paralleled, current will circulate between them because the terminal voltages at certain instances will be different. Thus the output of the system will be reduced. If the form of the voltage wave is peaked, the machines will require better insulation than if it is a sine curve. If the wave form has a flat top, the hysteresis loss is greater than for a sine curve. This loss reduces the efficiency of the alternator and of the electrical machinery to which the alternator supplies power.

To overcome the undesirable features of concentrated windings, commercial machines use two or more coils per pole per phase instead of one. Fig. 14-8 shows how the coils in one phase of a three-phase winding are connected. The three coils of each group are first connected in series and then the groups are joined in series, as shown in fig. 14-8. Note that the alternate groups are reversed.

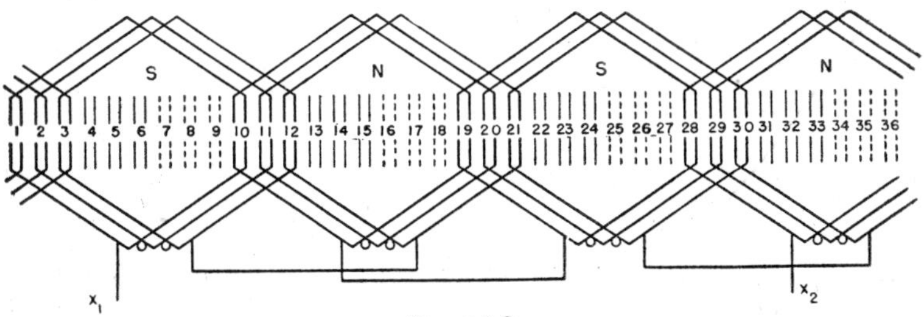

FIG. 14-8

To show why a distributed winding generates a voltage wave that is nearly a sine curve, one group from fig. 14-8 will be considered in fig. 14-9a. Between pole centers there are 180 electrical degrees and 9 slots. Then there must be 180 ÷ 9 or 20 degrees between slots or coils. Because the field poles move from left to right, the e m f induced in coil 2 lags that of coil 1 by 20 degrees. Likewise, the e m f induced in coil 3 lags that of coil 2 by 20 degrees. This 20 degree displacement of induced e m f is shown in fig. 14-9b. The resultant e m f is obtained by adding the instantaneous values of the individual e m f's. Note that it is more nearly a sine wave than the individual e m f's. This also shows that in a distributed winding the voltages induced in the coils of a group do not reach their maximum value at the same time. Thus the generated e m f is less than it would be if the windings were concentrated.

Aside from generating a voltage wave that is more nearly a sine curve, a distributed winding has other advantages. Since the copper is more evenly distributed on the rotor surface, the heating is more uniform and this winding is more easily cooled. Thus a higher current density is permissible than in a concentrated winding, and the machine will have a greater capacity for a given size.

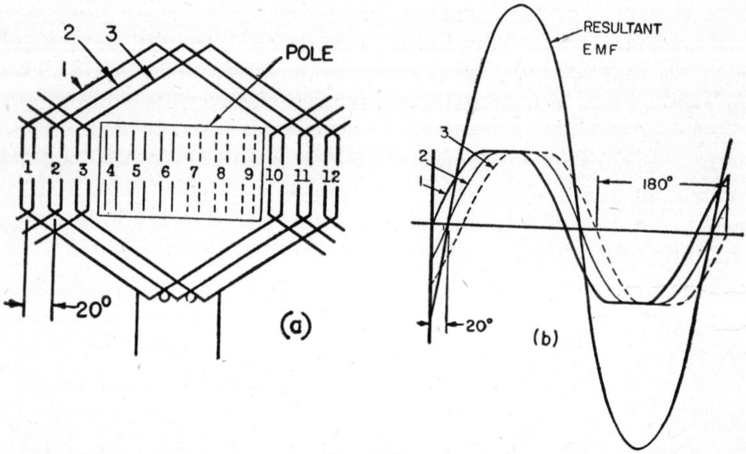

FIG. 14-9

Pitch—The winding in fig. 14-8 is full pitch because the coils span centers of adjacent poles. This span of nine slots (or 1 to 10), is

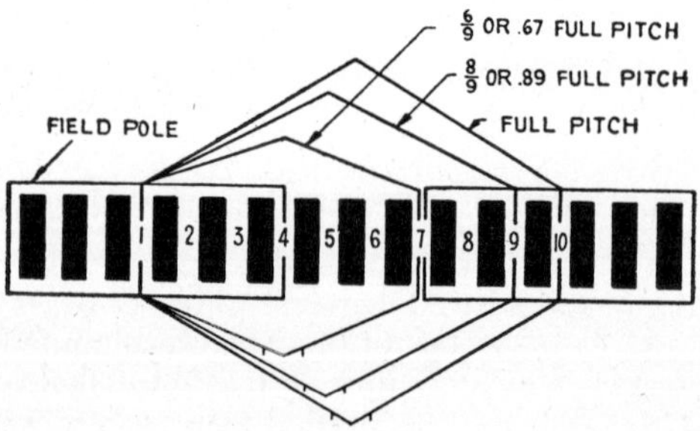

FIG. 14-10

also shown in fig. 14-10. This full pitch was determined by the following equation developed in armature winding.

$$Y_s = \frac{S}{P} \tag{14-3}$$

Y_s = slot span, S = number of slots, and P = number of poles.

If the coil span is reduced to eight (or 1 to 9) as in fig. 14-10, the result is a fractional pitch winding which is 8/9 or .89 full pitch. If the coil span is six the winding will be 6/9 or .67 full pitch.

Fractional Pitch Windings—A winding identical to that of fig. 14-8 except that a .89 full pitch is used as shown in fig. 14-11. This figure shows that in fractional pitch windings some slots contain coils of different phases.

In a fractional pitch winding the two sides of a given coil are not at corresponding points under adjacent poles. Hence the e m f induced in the two sides is out of phase and the coil e m f is less than if full pitch were used with the same number of turns.

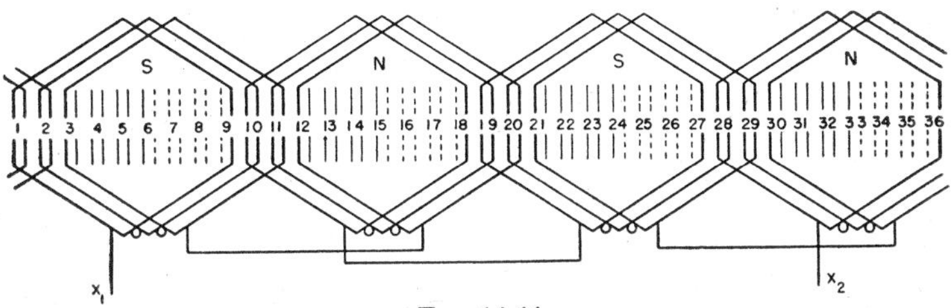

FIG. 14-11

A fractional pitch is desirable, however, because less copper is required per coil and because the sine wave in alternators is improved.

In order to take into account the winding distribution and the pitch, the right member of equation 14-2 must be multiplied by two factors. Equation 14-2 then becomes

$$E_p = \frac{4.44\Phi f T_p k_d k_c}{10^8} \qquad (14\text{-}4)$$

where k_d is a constant which accounts for the distribution of the coils in a group, and it is called a distribution factor. The value of k_d for the most common types of windings is given in Table 14-I. The constant k_c takes account of the pitch of the coil, and is called the chord factor. The value of k_c for different pitches may be obtained in fig. 14-12.

EXAMPLE—The stator of a 60 cycle, 4 pole, 3 phase alternator is wound as shown in fig. 14-11. Each coil has 20 turns and the flux per pole is 13×10^6 lines. Determine the e m f generated per phase.

Since there are 12 coils per phase and 20 turns per coil, each phase must have 12×20 or 240 turns. For three coils per group, $k_d = .96$, and for .89 pitch, $k_c = .984$. Substituting in equation 14-4,

$$E_p = \frac{4.44 \times 13 \times 10^6 \times 60 \times 240 \times .96 \times .984}{10^8} = 7851.5 \text{ volts}$$

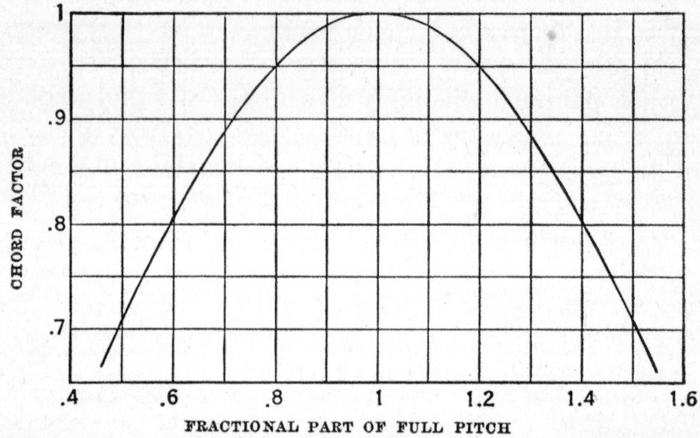

FIG. 14-12

Chain Winding—Chain windings have only one coil per pole per phase and one coil side per slot (see figs. 14-13 and 14-14). Hence such windings are single layer and concentrated. In order that the coil ends may pass each other, coils of two shapes and sizes are used (fig. 14-15). For this reason, such windings are known as two range windings. In figs. 14-13 and 14-14 the two ranges are distinguished by long and short coils. In fig. 14-15 note that the ends of half the coils are straight while the ends of the other half are bent so as to pass around the coil sides with straight ends.

TABLE 14-I

Phase	Coils per Group	Value of k_d
1	3	.667
2	2	.924
2	3	.911
3	2	.966
3	3	.960
3	4	.958
3	5	.957

In the two-phase winding of fig. 14-13, the top and bottom range coils are displaced 90 degrees. The phases are formed by connecting the coils of each range in series.

In the three-phase winding of fig. 14-14, three line wires extend from the start end of three adjacent coils which are displaced 120

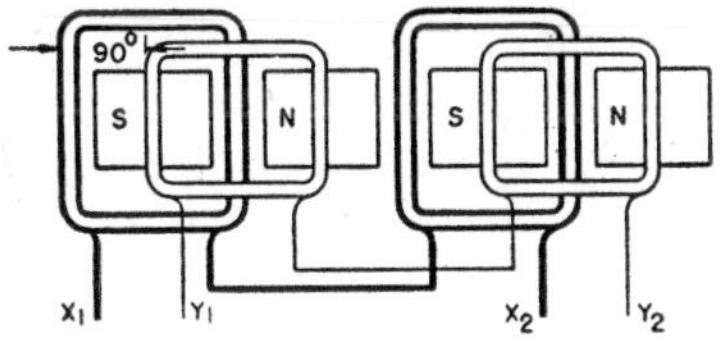

FIG. 14-13

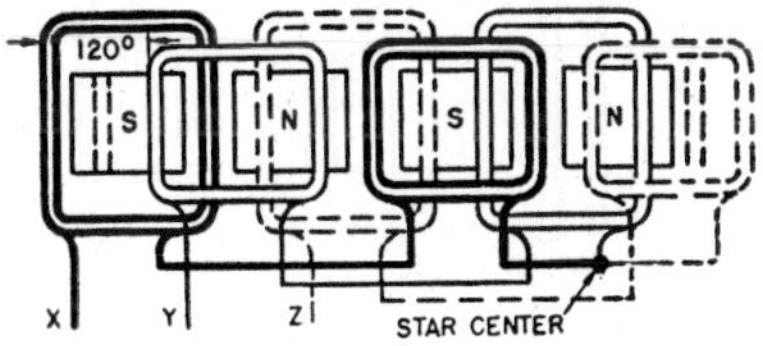

FIG. 14-14

degrees. Note that each phase has a long coil and a short one; that is, a coil in the top range and one in the bottom. The average lengths of the ranges per coil turn are different, but the three phases are identical, with conductors of the same length and therefore of the same resistance.

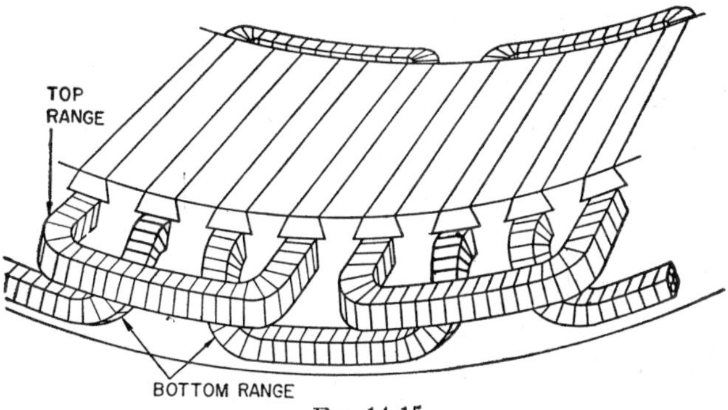

FIG. 14-15

Basket Winding—As it is undesirable to have only one coil per pole per phase, each coil in the chain winding may be replaced by two or more coils with different spans, one placed within the other (fig. 14-16). The concentric coils are then connected in series to form groups and the groups are connected the same as the coils in the chain winding.

The advantage of the chain or basket winding is that the number of cross-overs at the coil ends is considerably less than in the lap winding. Because of the wide air space that may be easily provided between the coil ends, this winding is very suitable for high voltage

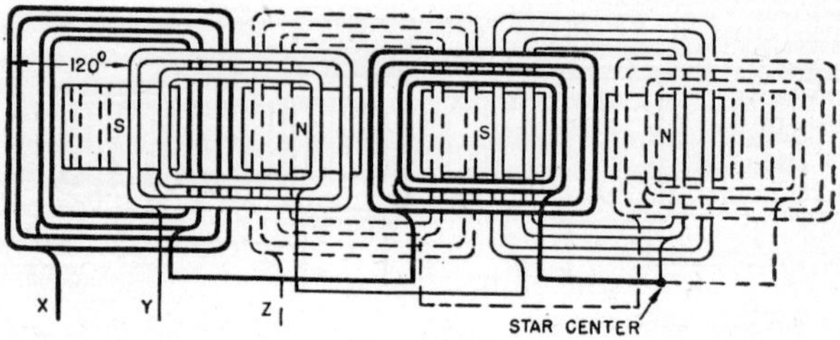

Fig. 14-16

machines, and is extensively used in small alternators of high voltage and low speed. However, the use of high speed steam turbine units is increasing, and the double layer lap winding is more suitable in these.

Alternator Characteristics—In an alternator the three following factors cause the voltage to change with an increase in load.

(1) Since the stator winding has some resistance, there will be an IR drop when a current flows through it. Thus if the generated voltage is constant the terminal voltage will decrease as the load is applied, the amount of decrease being equal to the IR drop.

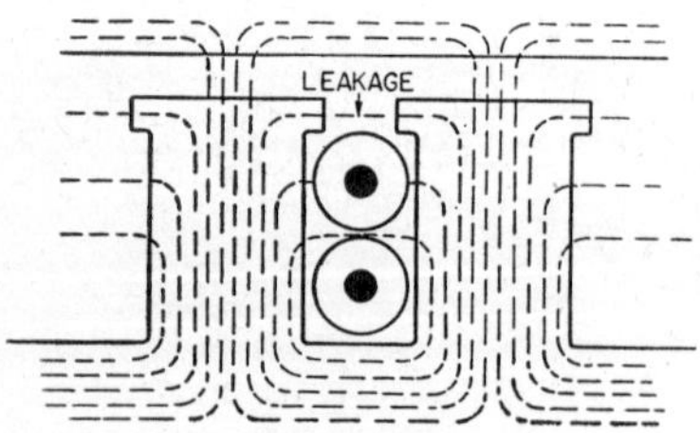

Fig. 14-17

(2) The current through the winding sets up a flux, a portion of which links the coil sides as shown in fig. 14-17. This magnetic flux leakage alternates with the current and gives the winding self-inductance. Then there will be an IX drop which is also effective in reducing the terminal voltage.

(3) The voltage variation in an alternator is produced mainly by the magnetizing action of the armature. In a direct current generator the armature produces a magnetizing force which distorts the flux distribution in the air gap. The extent of flux distortion depends

upon the amount of current flowing through the armature and the position of the brushes.

The flux distribution in the air gap of an alternator depends on the amount of stator current and on the phase relation existing between the current and voltage; that is, the power factor. The reason for this will be considered in the following:

Fig. 14-18 shows a portion of a rotor and stator of a three-phase alternator having three coils per group. The three phases have the same magnetizing action in the air gap; then, for the sake of simplicity, the magnetizing effect of the phase drawn with heavy lines will be the only one considered.

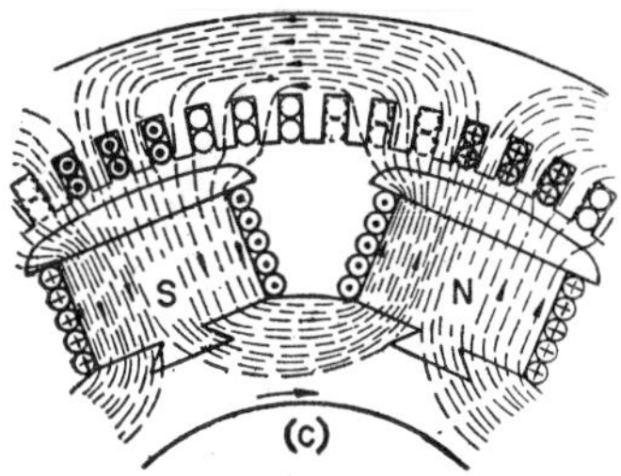

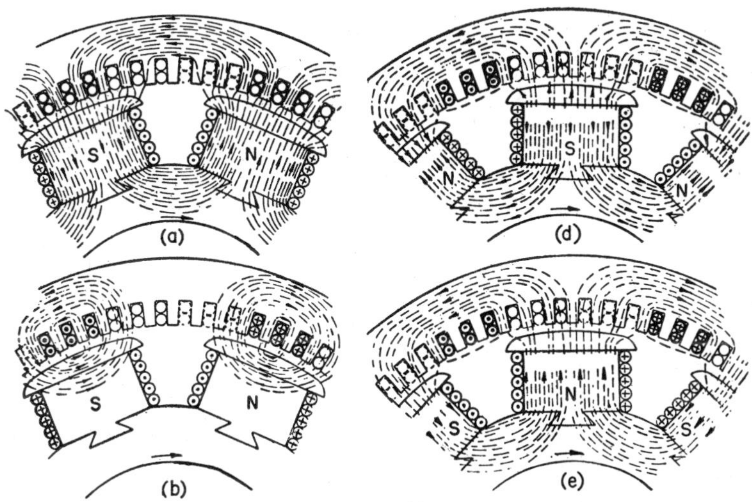

FIG. 14-18

At the instant shown in fig. 14-18a, this phase is cutting maximum flux or generating maximum voltage. In this case the stator carries no current; therefore its magnetizing action is zero and the field flux is not affected.

In fig. 14-18b this phase carries current in phase with the induced voltage (the field flux is not shown). At the instant shown, the e m f is at its maximum value; and since the current is in phase with the induced voltage, it is also at its maximum value. This is the case when the alternator is supplying power at unity power factor. According to the right-hand rule, the direction of the induced e m f and of the current is "out" under the south pole and "in" under the north pole. The flux produced is in a direction that will strengthen the field flux at the left side of each pole and weaken it at the right side. The resultant flux distribution is shown in fig. 14-18c. In this case the flux in the air gap is distorted but not weakened.

A stator current which lags the induced e m f by 90 degrees will now be considered. In this case the current does not reach its maximum until the south pole has advanced 90 electrical degrees, as indicated in fig. 14-18d. All the flux produced by the stator opposes the field flux, and therefore weakens it. This causes a reduction in the generated e m f.

A stator current which leads the induced e m f by 90 degrees will next be considered. In this case the current in the heavily drawn phase reaches its maximum value 90 electrical degrees before the south pole moves under the conductors of this phase (see fig. 14-18e). Observe that the flux produced is now in the same direction as the field flux, and therefore strengthens it. This causes an increase in the generated e m f.

Thus we conclude that the terminal voltage of an alternator varies with the IR and the IX drops in the stator winding and the power factor of the load. This is illustrated by the characteristic curves of fig. 14-19.

Paralleling Alternators—Alternating current power systems generally consist of several generating units connected in parallel to a common bus line. Fig. 14-20a represents a station bus line to which alternators (not shown) are connected. Suppose it is desired to connect to this bus line the alternator shown. Before the switch may be closed safely the following conditions should be fulfilled:

(1) Alternator must generate a voltage wave of approximately the same shape as that of the line;

(2) Terminal voltage of the alternator should equal that of the bus line;

(3) Frequency of alternator must equal approximately that of the bus line;

(4) With reference to the load, the voltage of the alternator and that of the bus line must be in phase;

(5) Phase rotations of the alternator and bus line must be the same.

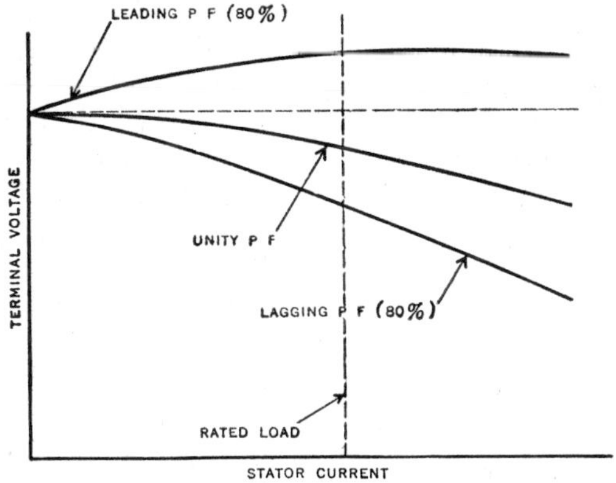

FIG. 14-19

For reasons previously explained, modern alternators are designed to generate a voltage wave of practically a sine curve shape. Slight differences in wave shapes among alternators are usually not enough to produce unsatisfactory parallel operation.

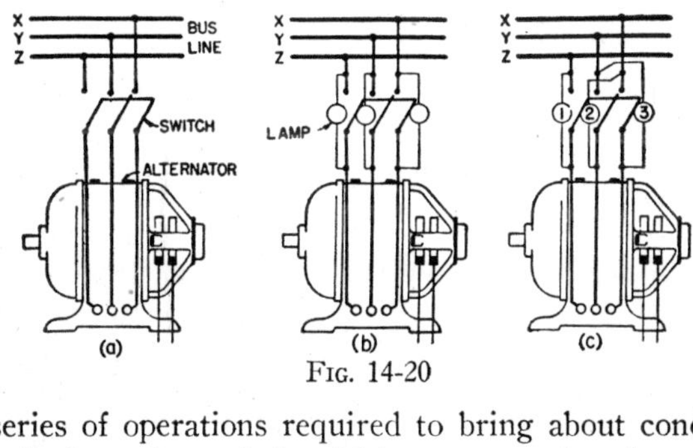

FIG. 14-20

The series of operations required to bring about conditions 2, 3, and 4, and close the switch which connects the alternator to the bus line, is called synchronizing. Requirements 2 and 3 are satisfied as follows: The alternator voltage is made equal to the bus line voltage by manipulating the field rheostat; and the frequency, which depends directly on the speed, is altered by changing the speed of the prime mover (driving machine). A synchronizing device must be used to satisfy requirement 4. Either incandescent lamps or a synchroscope may be used, but a synchroscope must always be used on large machines, because indication by lamps is not accurate enough. Condition 5 is satisfied by means of lamps.

Fig. 14-20b shows one method in which lamps are used for synchronizing. One lamp is connected across each pole of the three pole switch, which when closed connects the alternator to the bus line. The voltage rating of the lamps should not be less than the line voltage of the machine. For instance, if 240 volts are generated by the alternator, three 240 volt lamps should be used, one across each pole (of course, two 120 volt lamps in series across each pole might be used instead). If the alternator is properly connected to the knife switch, the three lamps become bright and then dim together. If the lamps become bright and then dim, one after the other, the phase rotation of the alternator and that of the bus line are in opposition. Reversing the connection of any pair of leads from the alternator will correct this error.

When the lamps are connected correctly and are bright, the alternator and bus line voltages add relative to the lamps, and when the lamps are dark the alternator and bus line voltages are in opposition, and thus neutralize. This varying phase displacement of the two sets of voltages causes the lamps to flicker at a rate equal to the difference in frequency of the alternator and bus line. Then as the voltages of the alternator and bus line become nearer in step, the flickering becomes slower. The knife switch may be closed when the lamps are dark.

With this method the lamps are dark even when there is appreciable voltage between their terminals. Hence the operator is likely to close the switch when considerable voltage exists across each pole. This will do no harm to a small capacity or low speed unit, but will cause considerable disturbance in a high speed turbo unit.

This difficulty may be eliminated partly by cross-connecting two of the lamps, as shown in fig. 14-20c (be sure that the bus lines are not cross-connected). With this connection, when the alternator is near synchronism, lamp 1 is dark while lamps 2 and 3 are bright. One of the bright lamps will be increasing in brilliancy and the other will be decreasing. At the instant of synchronism lamps 2 and 3 will be equally brilliant, thus indicating definitely when the switch should be closed.

When machine voltages are exceedingly high, potential transformers should be used to reduce the voltage for the lamps.

Synchroscope—In practice alternators are generally synchronized by means of an indicating instrument called a synchroscope (see fig. 14-21). This instrument (which is discussed in chapter 19) has

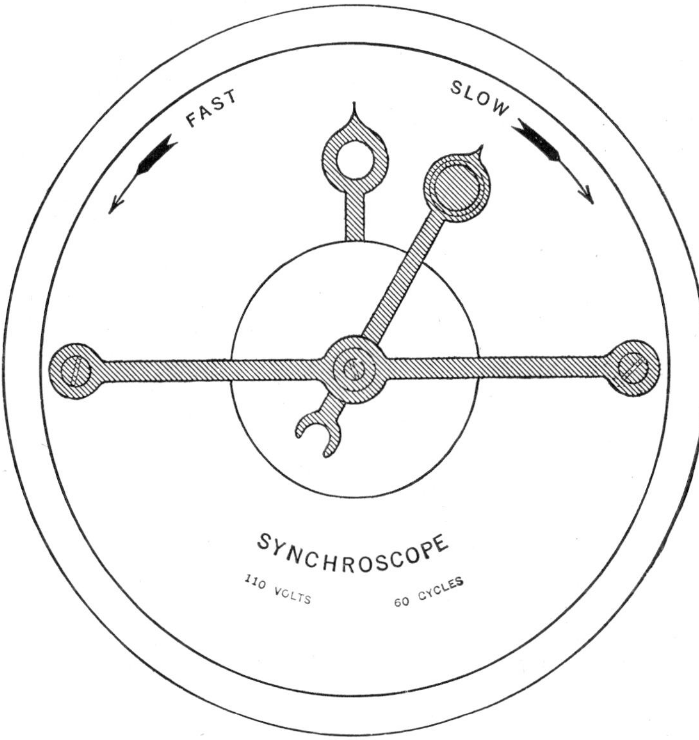

Fig. 14-21

two independent circuits, one being connected to the alternator and the other to the bus line. The magnetic fields set up by these two circuits cause the hand to rotate. When the hand comes to a standstill at the mark on the dial indicating synchronism, the switch connecting the alternator to the bus line may be closed.

Alternator Panel—Fig. 14-22 shows a typical wiring diagram of an alternator panel. Since it is unsafe to connect the instruments directly to the alternator leads, current and potential (instrument) transformers must be used. The voltmeter and ammeter are connected to the instrument transformers by inserting their respective plugs in the provided receptacles. When the synchronizing plug A

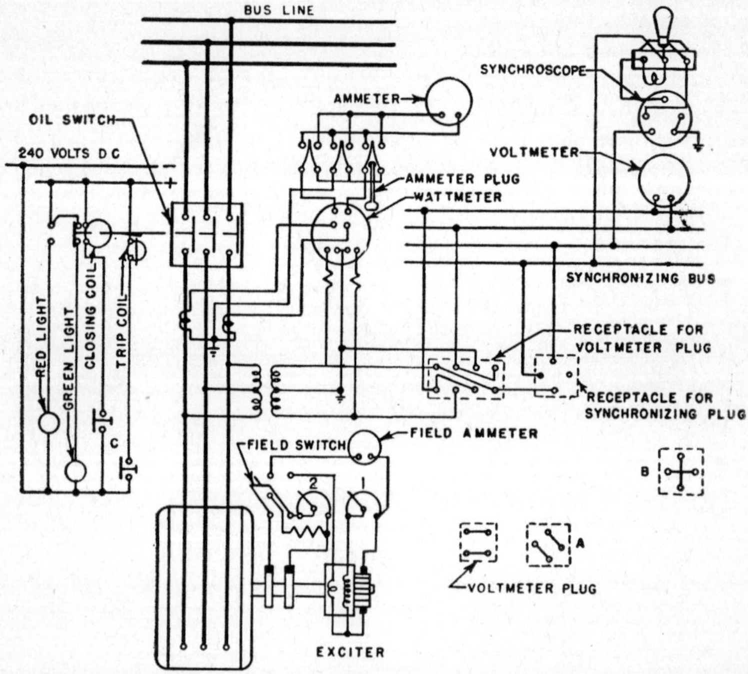

Fig. 14-22

is inserted in the receptacle provided for it (see fig. 14-22), and plug B is inserted in a corresponding receptacle of another alternator panel, which is connected to the common synchronizing bus, the connections to the synchroscope are complete.

After the alternator is properly synchronized, the oil switch may be closed from control C.

The exciter, which is a small D C generator usually installed on the shaft of the alternator, provides the field excitation for the machine. The alternator voltage may be controlled from two sources—by rheostat 1, which controls the terminal voltage of the exciter, or by rheostat 2, which varies the resistance of the alternator field circuit.

Division of Load—No current circulates between the alternator and the bus line immediately after they are synchronized. Thus the alternator does not deliver power to the bus or receive power from it. Relative to the load attached to the bus line, the alternator voltage E_a and the bus line voltage E_b are in phase. This is represented vectorially in fig. 14-23a. In order to make the alternator carry some load, its driving torque must be increased. This is accomplished by supplying more steam to the prime mover (assuming that the prime mover is a steam turbine). The prime mover will then try to speed

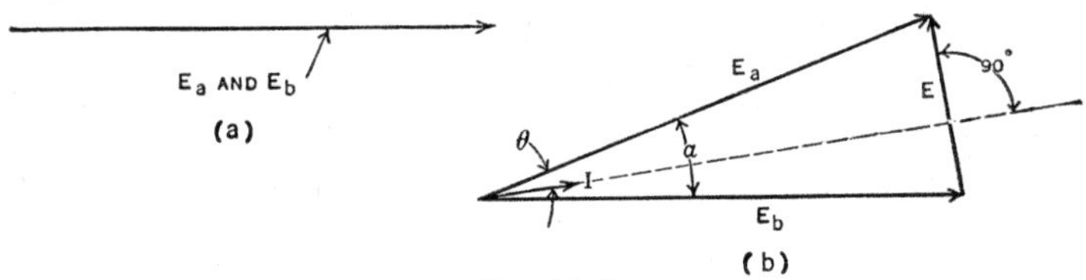

FIG. 14-23

up the alternator, but this speed increase can occur for only a fraction of a revolution, until the alternator voltage has pulled slightly ahead of the bus line voltage in its phase relation. While this is taking place the load on the alternator increases to such an extent that it causes the driving unit to slow down to its former speed. The alternator voltage is then no longer in phase with the bus line voltage, but is ahead by the amount represented by angle a in fig. 14-23b. The voltages of the alternator and bus line therefore no longer neutralize, for a voltage difference E exists. This voltage is effective in sending a current I through the alternator and bus line. The numerical value of I may be determined by dividing the voltage difference E by the sum of the impedances of the alternator and bus line. As the resistance of the alternator is very small compared to its reactance, the current I will lag the voltage E by nearly 90°, as indicated in fig. 14-23b. The power delivered from the alternator will then be

$$W = 1.73 E_a I \cos \theta$$

The operator in a power plant generally has control of the steam supply to the prime movers. *By varying this supply to the various machines, the operator can shift the load from one alternator to another as he desires.*

Effect of Varying the Field Excitation—In fig. 14-23b the field excitation is such that the voltage generated in the alternator E_a equals the bus line voltage E_b. In this case the voltage difference E causes the current which flows to be nearly in phase with the bus line voltage. Thus power is delivered by the alternator at nearly unity power factor.

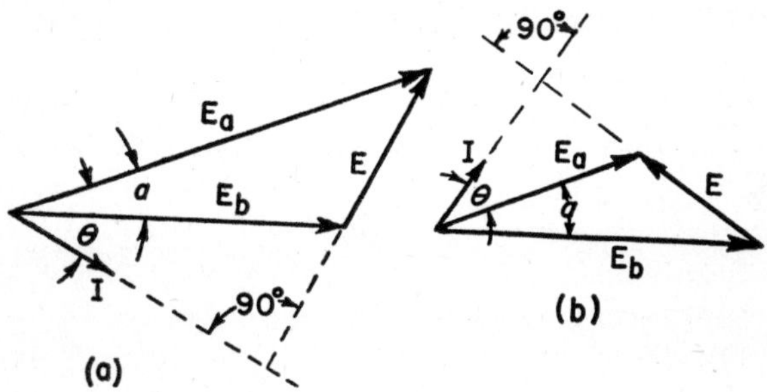

FIG. 14-24

If the field excitation is increased, the generated voltage E_a increases correspondingly. This establishes a greater voltage difference E (fig. 14-24a), which causes a greater current I to flow. But since I always lags E by nearly 90 degrees (due to the fact that the stator winding is highly inductive), this current will lag the alternator voltage E_a a greater amount. A greater field excitation thus causes the current I to increase and the power factor to become more lagging, thus becoming poorer. This change in current and power factor is such as to cause the power delivered to remain constant.

If the field excitation (and therefore the generated voltage E_a) is reduced below the values considered in fig. 14-23b, the voltage difference E will again increase (fig. 14-24b). As the current I always lags E by nearly 90 degrees, in this case I will lead the alternator voltage E_a. Thus a reduction in field excitation causes the current I to again increase and the power factor to become more leading (and therefore poorer). The power delivered will again remain the same, as in the two cases previously considered.

Thus a change in the field excitation of an alternator (operating in parallel with others) changes the power factor and kilovolt-amperes that the machine supplies, but does not change the kilowatts it delivers.

It should also be realized that the generated voltage of an alternator (in parallel with others) does not change a great deal with large changes in field current. When the field current is strengthened, the lagging stator current sets up a magnetizing action that is effective in reducing the field flux; and when the field current is weakened, the leading stator current sets up a magnetizing action that increases the field flux. Thus an effect takes place in the machine which tends to neutralize the changes we attempt to produce.

Hunting—Sometimes an alternator will not operate satisfactorily in parallel with others. The load shifts from the alternator in question to the other alternators, and then back again. This is due to the fact that there are periodic fluctuations of angle θ (fig. 14-23b). This action is called hunting, and under certain conditions it is accumulative. The successive oscillations increase in magnitude, causing the alternator to pull out of synchronism in a short time, or the loads at certain instances may be sufficient to open the circuit breakers.

Hunting generally occurs in alternators driven by engines, because the driving torque of engines is not uniform in a revolution of the flywheel. The methods generally used to reduce hunting are: (a) dampening of the oscillations by the use of a squirrel cage winding, placed on the field poles as shown in fig. 14-1b; (b) changing the natural period of vibration of the machine by changing the flywheel (a heavier flywheel usually gives a more dampening effect); (c) dampening the governor if the oscillations are started by the action of the governor. Turbo-alternators seldom hunt because the prime mover supplies a uniform driving torque.

Exercises

1. What factors cause the voltage delivered by an alternator to change as the load is applied?
2. On what kind of machines are chain windings generally used?
3. Explain the difference between a chain and a basket winding.

4. The winding of a 60 cycle, three-phase, eight pole alternator is full pitch and has three coils per group (or a total of 72 coils). The flux per pole is 2,000,000 magnetic lines. Determine the voltage generated in each phase if each coil has six turns.
5. Determine the voltage per phase in problem 4 if the coil span is 8.
6. A 60 cycle, three-phase, star connected, 4 pole alternator has a basket winding with two coils per group, each coil having 10 turns. If the flux per pole is 2,500,000 magnetic lines, determine the terminal voltage delivered.

CHAPTER 15

INDUCTION MOTORS

THE induction motor is the most common type of alternating current machine because it is simple, rugged, relatively cheap, and can be manufactured with characteristics to suit most industrial requirements. It consists essentially of two units, called stator and rotor. The stator does not differ in any respects from that used in alternators. The rotor consists of a laminated cylinder with slots in its surface. The winding placed in these slots may be one of two types. One type, called a squirrel cage winding (see fig. 15-1a), usually consists of heavy copper bars, connected together at each end by a conducting end ring made of copper or brass. The joints between the bars and the end rings are commonly made by an electric weld. In some cases the rotor, including the bars, is placed in a mold and the bar ends are cast together with cast copper. In small squirrel cage rotors, the bars, end rings, and blowers are of aluminum cast in one piece (fig. 15-1b).

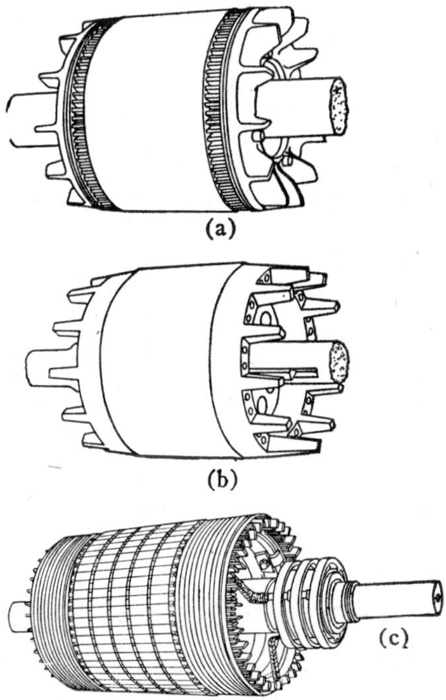

FIG. 15-1

Some industrial applications require a motor with a wound rotor instead of the squirrel cage type. In this case the rotor is provided with a winding similar to that used on the stator. The winding is generally connected star, and its open ends are fastened to slip rings mounted on the shaft (fig. 15-1c). Brushes mounted on the rings are attached to a star-connected rheostat. The purpose of this arrangement is to provide means of varying the rotor resistance.

Rotating Magnetic Field in a Two-Phase Motor—The stator of the induction motor, like that of the alternator, has no projecting

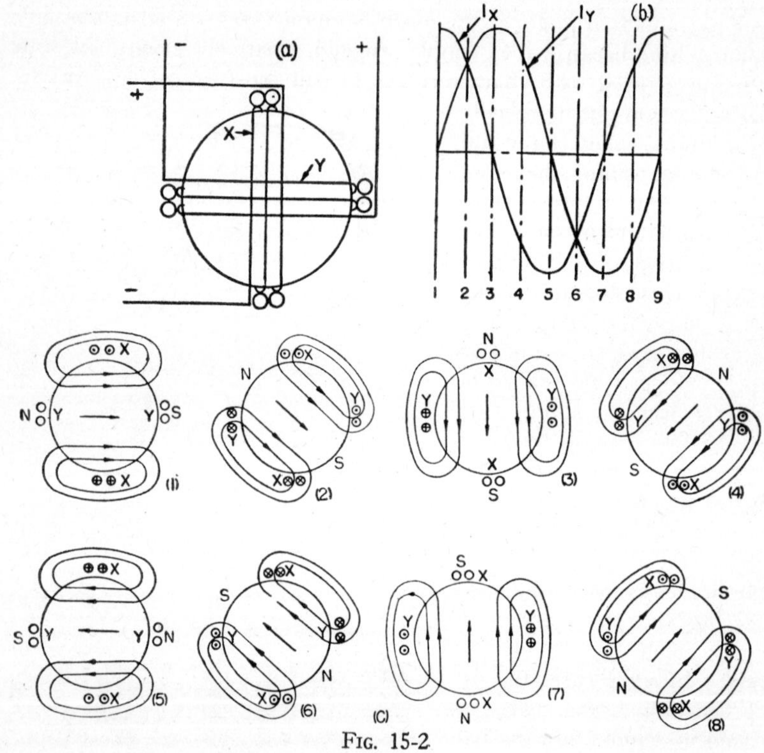

Fig. 15-2

poles. The winding is embedded in slots. In fig. 15-2a is shown a two pole, two phase stator winding. When this winding is energized from a two phase source, the currents vary as indicated by I_x and I_y in fig. 15-2b.

At instant 1 (figs. 15-2b and 15-2c), the current is zero in phase Y and maximum in phase X. With the current in the direction shown, a flux is established toward the right as indicated by the arrow. At

2 the current is still in the same direction in phase X and an equal current flows through phase Y. This establishes a resultant flux of the same strength 45 degrees clockwise from position 1. At 3 the current in phase X has decreased to zero and the current in phase Y has increased to maximum. This establishes a flux downward. At 4 the current in phase X has reversed and has the same value as that of phase Y. This establishes a flux 45 degrees clockwise from position 3. At 5 the current in phase X is maximum and in phase Y is zero, establishing a flux to the left. Diagrams 6, 7, and 8 indicate the direction of the flux during the remaining successive instances, resulting from the combined magnetizing effect of the two phases. At 9 the flux is in the same direction as at 1.

The successive instants considered in fig. 15-2 *indicate that a rotating magnetic field of constant strength is produced, which makes one complete revolution per cycle.* Thus the speed in revolutions per second (r p s) equals the frequency. That is, if the frequency is 60 cycles per second, the rotating magnetic field makes 60 r p s or 3600 r p m.

Rotating Magnetic Field in a Three-Phase Stator—In fig. 15-3a is shown a two-pole, 3-phase stator winding. When this winding is energized from a three-phase source, its current varies as indicated in fig. 15-3b. In fig. 15-3c is shown four successive positions of the rotating magnetic field. At instant 1 (figs. 15-3b and 15-3c), the current in phase X is zero and the currents in phases Y and Z are equal and opposite. These currents flow outward in the top conductors and inward in the bottom conductors. This establishes a flux toward the right. At 2 the current is maximum in phase Y

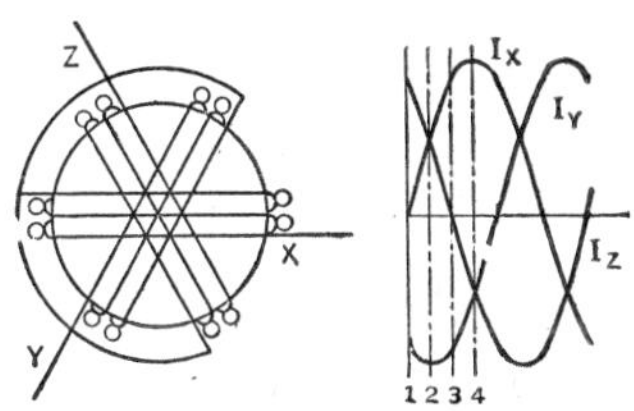

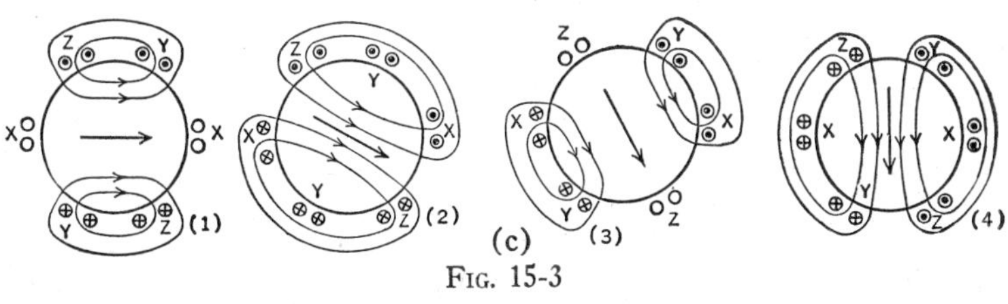

FIG. 15-3

and .5 maximum in phases X and Z. This establishes a flux 30° clockwise from position 1. At 3 the current in phase Z is zero and the currents in phases X and Y are .866 maximum, establishing a flux 30° further clockwise. At 4 the current in phase X is maximum and the currents in phases Y and Z are .5 maximum, the current in phase Z having reversed. This establishes a flux downward.

Apparently a three-phase winding gives a rotating magnetic field exactly the same as a two-phase winding.

Speed of Rotating Magnetic Field—The magnetic field of a two-pole stator winding makes 60 r p s or 3600 r p m when supplied with 60 cycle frequency. To obtain lower speeds it is necessary to increase the number of poles produced by the stator winding. In fig. 15-4a is shown one phase of a four-pole, three-phase stator winding. In fig. 15-4b is shown three successive positions of the rotating magnetic field. The current in the phase varies as indicated in fig. 15-4c. In this case the rotating magnetic field makes one revolution during two cycles. For a frequency of 60 cycles, the rotating magnetic field has a speed of

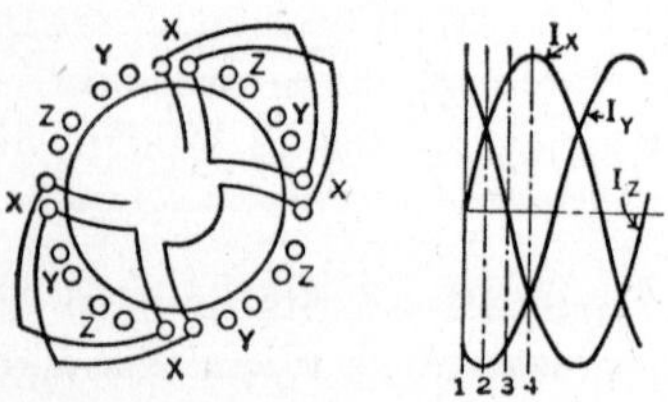

(a) (c)

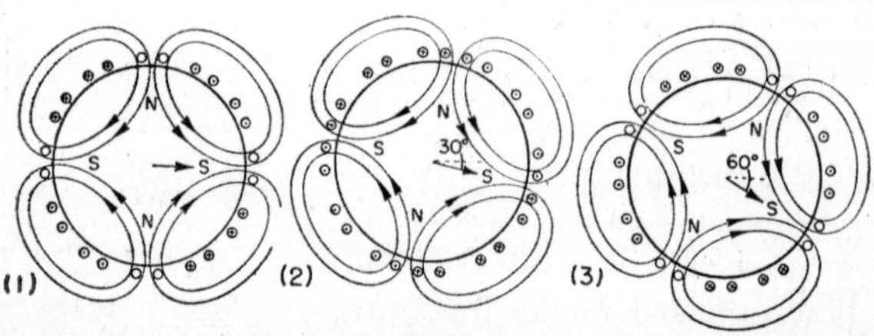

(b)

Fig. 15-4

30 r p s (1800 r p m) in a four pole winding and a speed of 20 r p s (1200 r p m) in a six pole winding. For any number of poles (P), the magnetic field makes one complete revolution during P/2 cycles. The number of r p s of the magnetic field times the number of cycles during one revolution (P/2) gives the number of cycles per second or the frequency (f); that is,

$$\text{r p s} \times \frac{P}{2} = f$$

Since revolutions per second equals revolutions per minute (N) divided by 60, then

$$\frac{N}{60} \times \frac{P}{2} = f, \text{ or } N = \frac{120f}{P} \qquad (15\text{-}1)$$

The speed of the rotating magnetic field is the same as the speed of the alternator that is supplying power to the motor if the two have the same number of poles. Equation 15-1 given above is the same as equation 14-1, which applies to alternators.

EXAMPLE—What is the speed of the rotating magnetic field in a 12 pole, 60 cycle stator?

$$N = \frac{120 \times 60}{12} = 600 \text{ r p m}$$

Direction of Rotation of Magnetic Field—The direction of rotation of a magnetic field is reversed by interchanging the connection of the leads of one phase in a two-phase motor and interchanging any two line leads in a three-phase motor.

Torque—In fig. 15-5a is shown a cross section of an induction motor with a squirrel cage rotor. Suppose the magnetic field is rotating in the clockwise direction and the rotor is at a standstill. On

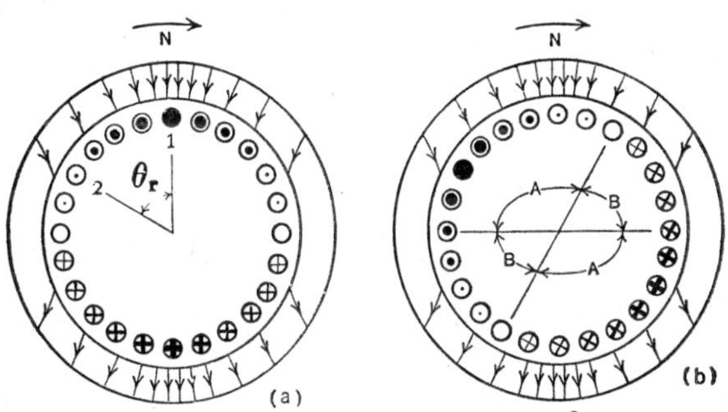

FIG. 15-5

applying the right-hand rule, it will be found that an e m f is induced outward under the north pole and inward under the south pole. The flux in the air gap is not uniform, but varies as a sine curve; it is more dense near the center of each pole and decreases toward the edges. Then the e m f induced in the rotor bars varies as indicated in fig. 15-5a. Since the bars are short-circuited at both ends, this e m f will produce a rotor current whose frequency is the same as that of the stator and whose effective value may be expressed by the following equation:

$$I_r = \frac{E_r}{\sqrt{R_r^2 + X_r^2}} \qquad (15\text{-}2)$$

At starting, the reactance of the rotor is large compared to the resistance, and therefore the rotor current lags the voltage by a large angle. At the instant that maximum voltage is induced in conductors 1, maximum current flows in conductors θ_r degrees behind 1, which are designated by 2. The flux and rotor current relation are shown in fig. 15-5b. In this figure the rotor conductors within angles A produce forces tending to turn the rotor clockwise, while conductors within angles B tend to turn the rotor counter-clockwise. Obviously, the conductors in angles A produce a greater torque than those in angles B, and so the resultant torque tends to turn the rotor in the same direction as that of the rotating magnetic field. The torque developed in the rotor depends upon the flux, the rotor current, and the phase relation of these two, that is, the cosine of the angle between them. Expressing this by means of an equation, we have

$$T = K\Phi I_r \operatorname{Cos} \theta_r \qquad (15\text{-}3)$$

where $K =$ a constant which is fixed for a given motor, $\Phi =$ flux, $I_r =$ rotor current, and $\cos \theta_r =$ rotor power factor.

Starting Condition—If the resisting torque is less than the torque developed at standstill, the rotor will revolve in the direction of the rotating magnetic field. As the rotor speed increases, the rate at which the field sweeps past the rotor conductors decreases and therefore the rotor voltage becomes less. If the rotor speed could increase until it turned at exactly the same speed as that of the rotating magnetic field, there would be no induced voltage, no rotor current, and therefore no torque. Apparently, even at no load, the final rotor speed is less than the speed of the rotating magnetic field which is called the synchronous speed. The rotor will run just enough below synchronous speed to establish sufficient rotor current to produce a torque equal to the resisting torque.

If the external load is increased the motor speed will decrease until the e m f produced supplies sufficient current to develop a torque equal to the new resisting torque. After this the motor will continue to run at a constant speed until the load is again changed.

Slip—The difference between the synchronous speed N_s and the rotor speed N_r is called the slip S. The slip is generally expressed in terms of synchronous speed, and is calculated as follows:

$$S = \frac{N_s - N_r}{N_s} \tag{15-4}$$

The rotor frequency is directly proportional to the slip, hence

$$f_r = Sf_s \tag{15-5}$$

Example—Determine the slip and rotor frequency of a 60 cycle, eight pole motor running at 840 r p m.

From equation 15-1, the synchronous speed is 900 r p m. Substituting in equation 15-4,

$$S = \frac{900 - 840}{900} = .067 \text{ or } 6.7 \text{ per cent}$$

Substituting in equation 15-5,

$$f_r = .067 \times 60 = 4.02 \text{ cycles per second}$$

Efficiency—Neglecting the stator copper losses, the power input to the stator is transferred to the rotor through the medium of the magnetic field. Then the power transferred to the rotor of a three-phase motor is

$$W_r = 1.73E_sI_s \text{ Cos } \theta_s$$

where E_s, I_s, and Cos θ_s represent respectively the voltage, current, and power factor of the stator.

If r represents the ratio per phase of the turns on the stator to the turns on the rotor, then the rotor induced voltage at standstill is E_s/r and at slip S is SE_s/r. The rotor current is rI_s. The stator and rotor power factors are approximately the same. The rotor losses are therefore

$$1.73\frac{SE_s}{r} \times rI_s \times \text{Cos } \theta = 1.73SE_sI_s \text{ Cos } \theta$$

$$\text{motor efficiency} = \frac{\text{output}}{\text{input}} = \frac{\text{input} - \text{losses}}{\text{input}}$$

$$= \frac{1.73E_sI_s \cos\theta - 1.73SE_sI_s \cos\theta}{1.73E_sI_s \cos\theta}$$

Then

$$\text{Efficiency} = 1 - S \tag{15-6}$$

Since the stator losses are neglected, the actual efficiency is less than that given by this equation. *The actual efficiency of any induction motor is less than the rotor speed expressed in per cent of synchronous speed.*

EXAMPLE—What is the approximate efficiency of a 60 cycle, 6 pole induction motor running at 1050 r p m?

Since the synchronous speed is 1200 r p m,

$$S = \frac{1200 - 1050}{1200} = .125 \text{ or } 12.5 \text{ per cent}$$

$$\text{Efficiency} = 1 - S = 1 - .125 = .875 \text{ or } 87.5 \text{ per cent}$$

Stator Current—The induction motor is fundamentally a transformer, in which the stator is the primary and the rotor is a short-circuited secondary. This is evident, particularly when the rotor is stationary. The rotor current establishes a flux which opposes and therefore tends to weaken the stator flux. This causes more current to flow in the stator winding, just as an increase in secondary current in a transformer causes a corresponding increase in the primary current. Because an induction motor has an air gap between the stator and the rotor, sufficient flux leakage occurs to limit the starting current to from 4 to 6 times the full load current.

Loading a Squirrel Cage Motor—In the case of a low resistance rotor motor, full load current occurs at a low slip. Then even at full load, f_r and therefore X_r (which equals $2\pi f_r L$) is low. Between zero and full load, $\cos\theta_r$ and Z_r remain practically constant. Then I_r (which equals E_r/Z_r), and therefore T in equation 15-3, increases directly with the slip. I_s increases in proportion to I_r. This is shown in fig. 15-6, where T and I_s are indicated as straight lines from no load to full load. Above full load, X_r becomes appreciable, causing Z_r to increase faster and I_r to increase slower, while $\cos\theta_r$ decreases faster. As a result, T and I_s increase at a lower rate, as indicated by the curves. For a standard motor, maximum torque is reached at about 25 per cent slip. Beyond this point, $\cos\theta_r$ decreases faster than I_r increases, and therefore T decreases (see fig. 15-6). Then if a motor is loaded beyond maximum torque, it will quickly slow down and stop. In fig. 15-6 the value of T at starting, which is at 100 per cent slip, is 1.5 full load torque. The starting current is about five times full load current. This motor is essentially a constant speed machine, having speed characteristics about the same as a D C shunt motor. Its

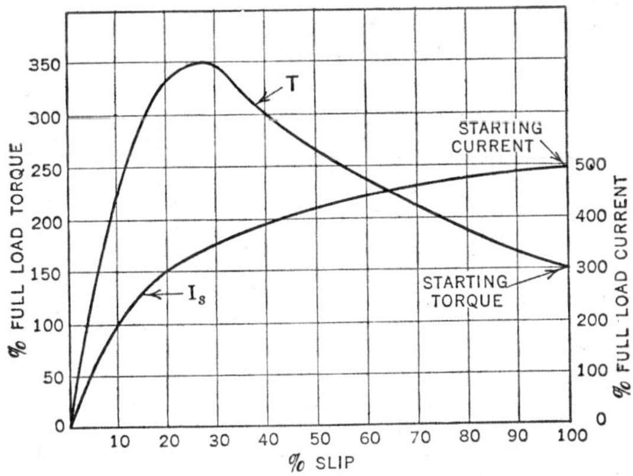

FIG. 15-6

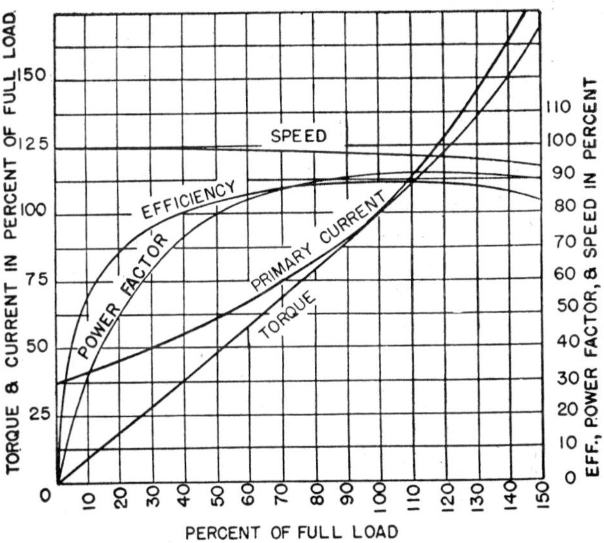

FIG. 15-7

performance curves are given in fig. 15-7. This motor should be used where average starting torque is necessary and constant speed operation is desired.

Torque of Induction Motor with Change in Line Voltage—If the stator voltage is doubled, the flux doubles. Hence, twice as much rotor voltage is induced, doubling the rotor current. Then, according to equation 15-3, the torque becomes four times as great. *Hence the torque of an induction motor varies as the square of the voltage.*

Effect of a Higher Rotor Resistance—Suppose that a motor with torque characteristics as indicated in fig. 15-6 and also by curve

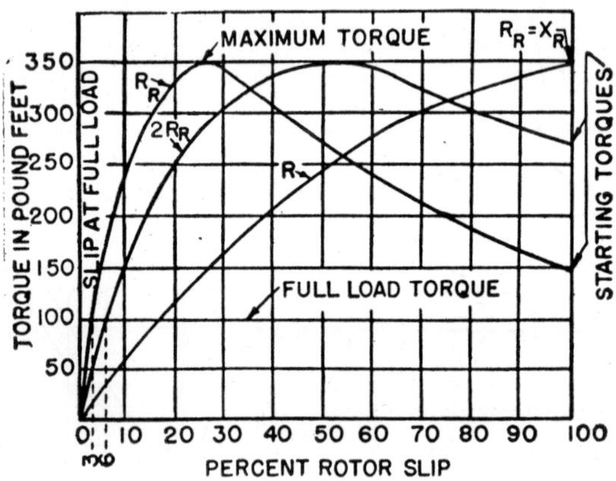

FIG. 15-8

R_r in fig. 15-8 has its rotor resistance doubled. Now if this motor is loaded until twice as much slip occurs, X_r, Z_r, and E_r double (see fig. 15-9). I_r (which equals $2E_r/2Z_r$) and cos θ_r remain the same as in the previous case. Then, according to equation 15-3, T remains as before. Hence for a constant torque T, doubling R_r doubles S. This is indicated by curve $2R_r$ in fig. 15-8. This figure also shows that an increase in rotor resistance does not change the maximum value of T, but merely shifts it toward the standstill point.

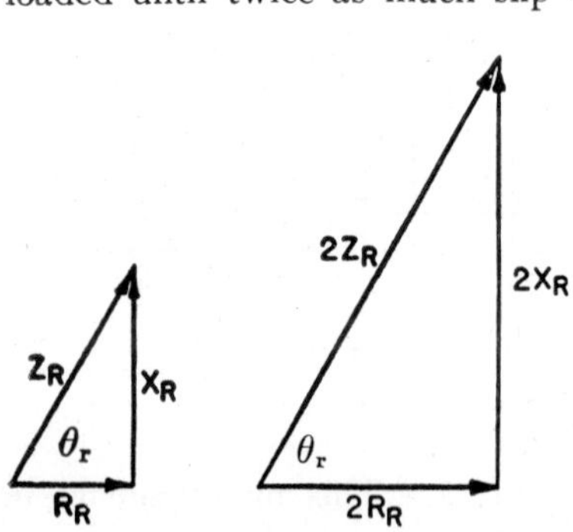

FIG. 15-9

There is a value of rotor resistance that gives the maximum starting torque as indicated by R (fig. 15-8). In equation 15-3 consider that the two following values are substituted for I_r and $\cos \theta_r$:

$$\frac{E_r}{\sqrt{R_r^2 + X_r^2}}$$

$$\frac{R_r}{\sqrt{R_r^2 + X_r^2}}$$

Then equation 15-3 becomes

$$T = \frac{K\Phi E_r R_r}{R_r^2 + X_r^2}$$

In this equation T is maximum when $R_r = X_r$, and when this occurs, $\theta_r = 45°$, and the power factor is 70.7 per cent.

In industry, induction motors of a given rating and different operating characteristics are required. This is attained merely by using motors having rotors with different resistance to reactance ratio at standstill.

Squirrel cage motors are manufactured in which the rotor resistance equals the rotor reactance at standstill. In these the starting current is less than in the standard type, being about four times full load current. These motors have wide speed variations with changes in load (see curve R in fig. 15-8) somewhat as an overcompounded D C motor. The high rotor resistance increases the copper losses, hence the efficiency is less than in the standard type. These motors are employed where high starting torques at safe starting currents are required, and where it is desired that the motors slow down without drawing large increases in currents. They are used for driving cranes and elevators, and large presses with heavy flywheels.

Wound Rotor Motors—The undesirable features of a high resistance rotor are eliminated by using a wound rotor (see fig. 15-1c), in which the high resistance is used only at starting and is cut out as the motor gains speed. In this case maximum torque is secured at starting and maintained as the motor approaches normal speed, where the running characteristics are about the same as the standard squirrel cage type. The wound rotor induction motor has the following advantages over the squirrel cage type:

(1) High starting torque with low starting current;
(2) Smooth acceleration under heavy loads;
(3) No abnormal heating during starting;
(4) Good running characteristics;
(5) Adjustable speed.

Its disadvantage is that its initial and maintaining costs are greater than that of the squirrel cage motor.

Double Squirrel Cage Rotor—One type of induction motor with excellent operating characteristics has a rotor with two squirrel cage windings (fig. 15-10). The bars of the inner winding are made of low resistance metal surrounded by iron except between the two windings. Then the resistance of this winding is low and its induc-

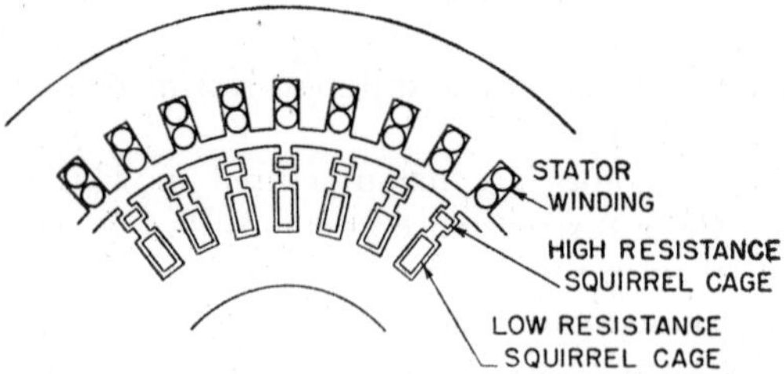

Fig. 15-10

tance is high. The bars of the outer winding are made of high resistance metal with iron at their sides only. Then the resistance of this winding is high and its inductance is low.

When a rotating magnetic field sweeps across the two windings, equal e m f's are induced in each. At starting, the rotor frequency is the same as that of the line, making the reactance of the lower winding much higher than that of the upper. Then more current flows through the higher resistance winding at a good power factor and a high torque is produced. As the rotor nears synchronous speed, the rotor frequency becomes low and the current division is governed largely by the resistance. As a result the lower winding carries practically all the current, and the machine operates as a standard squirrel cage induction motor. Of course, while the motor is accelerating, the current decreases in the high resistance winding and increases in the low resistance winding, thus supplying a high torque at all times. The operating characteristics of this type of motor are shown in fig. 15-11.

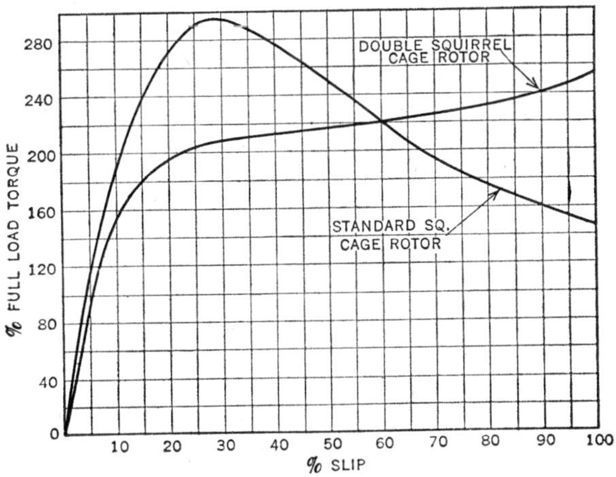

FIG. 15-11

Adjustable Speed Motors—The speed of a standard squirrel cage motor is inherently constant. However, special squirrel cage motors are manufactured with one or two stator windings, whose poles may be changed by changing the external connections, giving a limited number of different speeds. For instance, a motor with a single special winding may be connected for either four or eight poles, giving synchronous speeds of 1800 and 900 r p m. The motor may have two special windings, each good for two speeds. For instance one winding may be connected for either four or eight poles and the other for either six or twelve poles. The synchronous speeds for such a motor are 1800, 1200, 900, and 600 r p m. A detailed explanation of these motors is given later.

The speed of a wound rotor motor may be adjusted. If such a motor is carrying a given load and resistance is inserted in the rotor circuit, the rotor current and therefore the developed torque will become less and the motor will slow down. As the speed decreases, more e m f is induced in the rotor and the rotor current increases. At some lower speed, sufficient rotor current is developed to supply the required torque. With this method of varying the speed, the efficiency decreases with an increase in resistance and the speed changes considerably with a change in load.

A polyphase motor that operates reasonably well as an adjustable speed motor is shown in fig. 15-12. The input or primary winding is on the rotor and is energized through slip rings. In the slots containing the primary winding, there is also another winding, known as a regulating winding, which connects to a commutator the same as those in D C machines. The secondary winding is stationary with the two ends of each phase brought to a pair of brushes (A and B) resting on the commutator. Brushes A are fastened to a yoke at one end of the commutator, and brushes B are fastened to another yoke

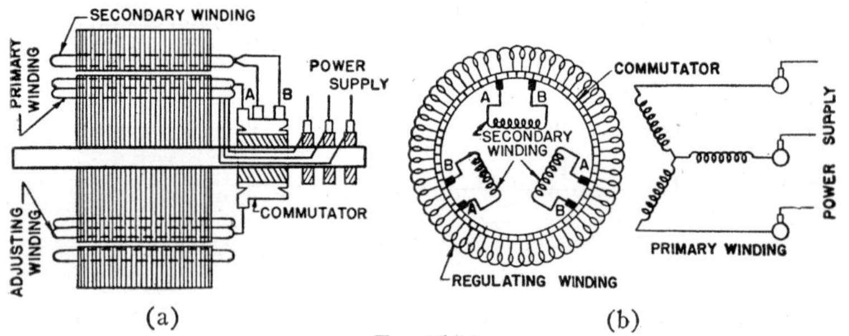

FIG. 15-12

at the other end. The brushes are shifted by a brush shifting mechanism, which on moving brushes A in one direction moves brushes B the same distance in the opposite direction.

Suppose that brushes A and B in each phase are placed in line as indicated in fig. 15-12a and the primary is excited. This winding establishes a rotating magnetic field, as is developed in the standard induction motor. This field induces a voltage in the regulating and in the secondary winding. The regulating winding is not effective at this position of the brushes. Current flows through the secondary winding, which acts with the flux established by the primary winding current, causing the primary to rotate. In this case the machine operates as a standard induction motor.

If brushes A and B in each phase are moved apart, the voltage induced in the regulating winding is impressed on the secondary winding and has the same frequency. With the brushes in each phase moved apart one way, the voltage induced in the regulating winding is in the opposite direction to that induced in the secondary winding, and therefore the secondary current decreases. This decreases the developed torque and the rotor slows down. In so doing, the secondary is cut by the magnetic field at an increasing rate until enough current is established in this winding to supply the required torque. If brushes A and B in each phase are shifted past each other, the voltage induced in the regulating winding is impressed in the secondary winding in the same direction as the induced secondary voltage. In this case the secondary current increases, causing a higher torque to be developed. Consequently, the motor speed will increase, in which case the secondary will be cut at a decreasing rate and therefore the secondary current decreases. As soon as the secondary current reaches a value that gives a torque equal to the load torque, the motor will again run at a constant speed.

High Frequency Motors—For certain classes of work, motors are required that operate at higher speeds than those obtainable with the normal 60 cycle supply. With a 60 cycle frequency, a two pole motor gives a maximum synchronous speed of 3600 r p m. Higher speeds may be obtained by increasing the frequency, and in practice three phase power at a frequency of 180 cycles is employed. Motors operating at this higher frequency are generally used where it is desired to avoid the use or reduce the size and number of transmission gears. These motors are small for their rated horsepower, which is desirable in portable units.

The most common uses for high frequency motors are for portable grinders, nut runners, drills, screw drivers, and for wood working machines, whose spindle speed must be high.

Frequency Converter—A frequency of 180 cycles is generally obtained from an induction generator, which has the same structure as a wound rotor induction motor, and is generally driven by a squirrel cage motor (fig. 15-13).

An induction generator is fundamentally a transformer. If its rotor is held stationary, the output frequency is the same as the supplied line frequency. The output voltage, compared to the input voltage, is proportional to the ratio of the stator to the rotor turns. If the generator is turned in a direction opposite to that of its rotating magnetic field, the frequency and voltage will be greater in proportion to the speed. The output then depends upon transformer action and the speed of the rotor.

The frequency output may be expressed by the following equation:

$$f = f_L + \frac{PN}{120}$$

where f = frequency output, f_L = line frequency, P = number of poles, and N = rotor speed.

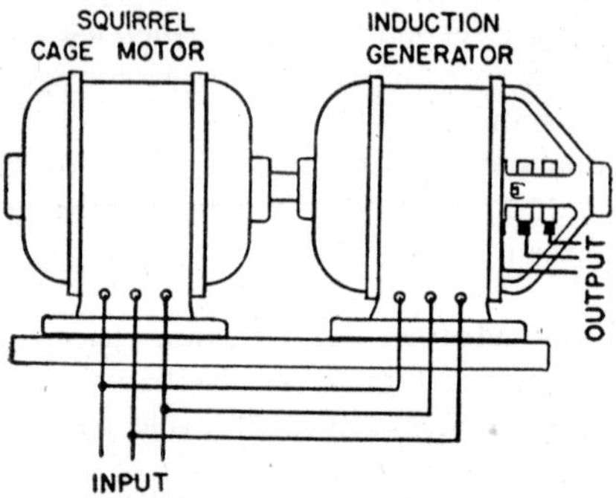

Fig. 15-13

Example—A frequency converter commonly used consists of an 8 pole induction generator driven by a 4 pole squirrel cage motor. Determine the output frequency.

$$f = 60 + \frac{8 \times 1800}{120} = 180 \text{ cycles per second}$$

Example—The voltage supplied to the above induction generator is 440 and the voltage output is 220. Determine the turn ratio.

If the rotor is turned at synchronous speed, the output voltage is twice as much as when the rotor is at standstill. But in this example the rotor turns twice synchronous speed, therefore the output voltage is three times as great as when the rotor is at standstill. Then at standstill the output voltage is 220 ÷ 3 or 73.3, and the turn ratio is 440 ÷ 73.3 or 6 to 1.

Synchronous Motors—It will be recalled that a D C generator runs as a motor. In like manner an alternator will run as a motor, and when so used it is called a synchronous motor. Synchronous motors are generally of the salient pole type (fig. 15-14) while alternators are either of the turbo or salient pole type.

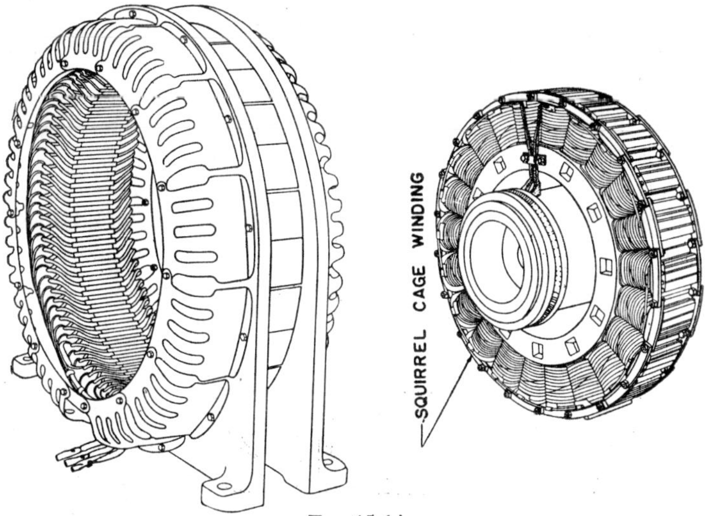

FIG. 15-14

Operating Principle—In a synchronous motor, a polyphase current is supplied to the stator winding, and it produces a rotating magnetic field as in an induction motor. A direct current is supplied to the rotor winding, and it produces a fixed polarity at each pole.

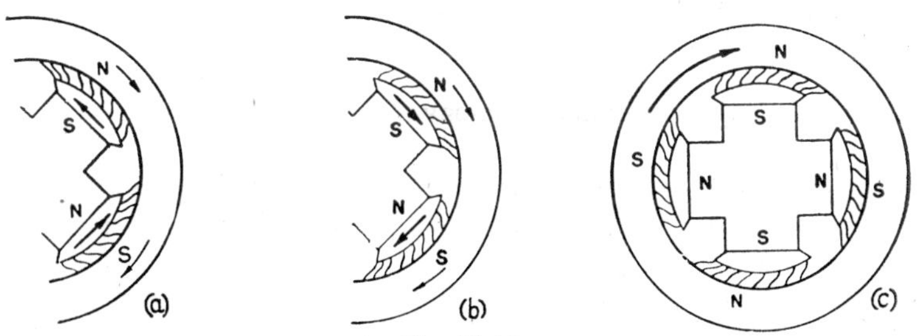

FIG. 15-15

Suppose that the stator and rotor winding of a synchronous motor are energized. As the poles of the rotating magnetic field approach rotor poles of opposite polarity (fig. 15-15a), the attractive force tends to turn the rotor in the direction opposite to that of the rotating field.

Just as the rotor gets started in this direction, the rotating field poles are leaving the rotor poles (fig. 15-15b), and this tends to pull the rotor poles in the opposite direction. Hence, the rotating field tends to pull the rotor poles first in one direction and then in the other, and as a result the starting torque is zero.

Starting Synchronous Motors—A squirrel cage winding is generally placed on the rotor of a synchronous motor to make the machine self-starting (see fig. 15-14). To start the motor, the rotor is left deenergized and a polyphase voltage is supplied to the stator. After the motor comes up to speed, which is slightly less than synchronous speed, the rotor is energized. Now opposite polarity poles

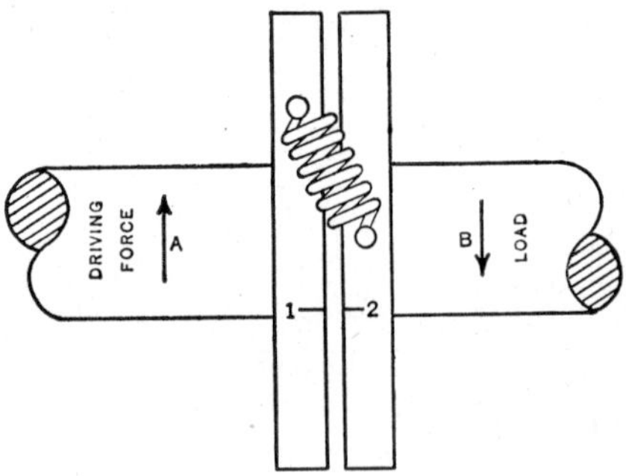

FIG. 15-16

on the stator and rotor will attract. The rotor will lock in step with the rotating field and will be pulled around at synchronous speed (fig. 15-15c).

The magnetic coupling in a synchronous motor can be illustrated by means of the mechanical analogy shown in fig. 15-16. Shaft A is coupled to shaft B by the spring. If A turns in the direction indicated and there is no load on B, then 1 and 2 remain in line as shown. This corresponds to an unloaded synchronous motor. Now if a load is applied on shaft B, the spring will stretch and 2 will drop behind 1 by a small angle, but after this, shaft A and B will continue to run at exactly the same speed. This corresponds to a loaded synchronous motor.

When the load is applied in a synchronous motor, the motor poles are pulled behind the stator poles through a small angle (about 20° in a fully loaded, two pole machine). Then the magnetic coupling keeps the rotor turning at exactly the same speed as that of the rotating field.

Suppose that a synchronous motor is started with its rotor winding open. At the instant that its stator is energized, a rotating magnetic field is established, which sweeps past the rotor winding at a rapid rate, since it is at standstill. This winding has a large number of turns and therefore a high voltage is induced in it, which will appear between its terminals. A high peak terminal voltage also appears when the rotor circuit of a machine in operation is opened. These high voltages will puncture the insulation unless the winding is highly insulated.

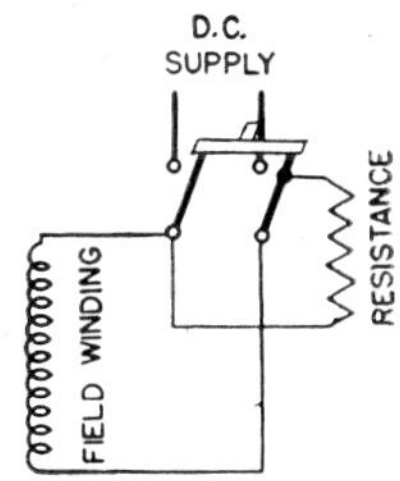

FIG. 15-17

In practice a low resistance is connected to the switch of the rotor circuit as shown in fig. 15-17. During the starting period, the switch is in the open position, in which case the resistance is connected across the field terminals. This allows alternating current to flow in the field winding. Since the impedance of this winding is high, compared to that of the external resistor, a high voltage drop is produced in the winding, reducing the terminal voltage to a safe value. The voltage induced when the field switch is open is reduced by the back voltage set up in the winding, as is evident from Lenz's law. This, together with the IX drop in the winding, gives a low terminal voltage.

Starting Torque—The current established in the rotor circuit during the starting period and the current in the squirrel cage are effective in producing the starting torque. In fig. 15-18, curve T_r indicates the variation in torque due to the rotor winding and curve T_s that due to the squirrel cage winding. Curve T is the sum of T_r and T_s and indicates the variation in torque during the starting period. Note that curve T_r shows that the rotor winding is very effective in producing torque after the rotor nears synchronous speed.

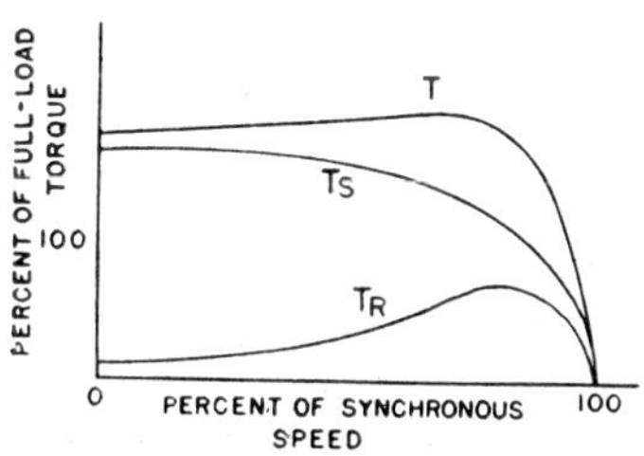

FIG. 15-18

Varying Load and Field Excitation of Synchronous Motor—In fig. 15-19a is shown a section of a synchronous motor in which only one coil C will be considered for simplicity. The e m f induced in this coil is maximum when its sides are opposite the pole centers and minimum when its sides are midway between pole tips. Thus the induced voltage varies as indicated by curve E'_i. When the motor carries no load, the induced voltage E'_i is practically 180 degrees out of phase with the applied voltage E_a. If the rheostat in the rotor circuit is adjusted so that E'_i equals E_a (fig. 15-19b), the stator current is small and may be neglected.

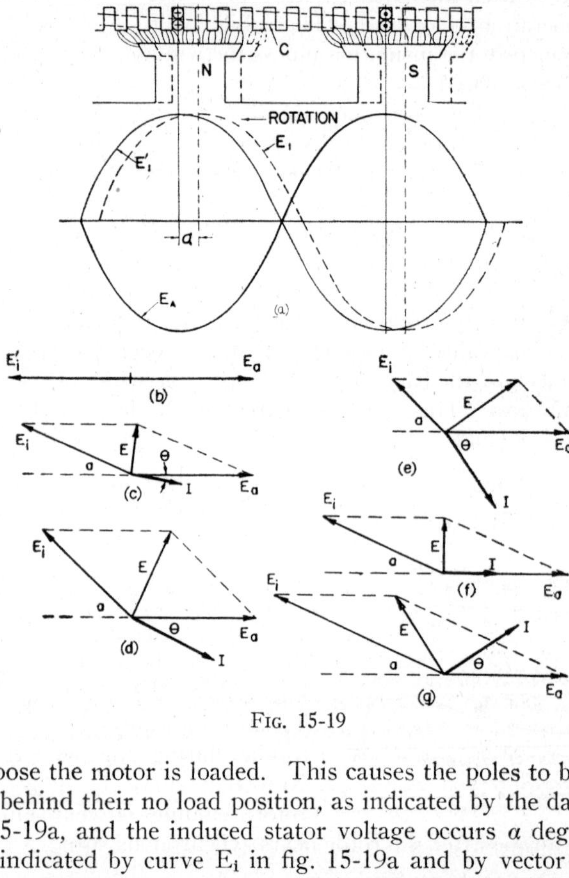

FIG. 15-19

Suppose the motor is loaded. This causes the poles to be pulled α degrees behind their no load position, as indicated by the dashed lines in fig. 15-19a, and the induced stator voltage occurs α degrees later. This is indicated by curve E_i in fig. 15-19a and by vector E_i in fig. 15-19c.

In a D C motor, the armature current I is determined by dividing the difference between E_a and E_i by the armature resistance R. Similarly, in a synchronous motor the stator current is determined by dividing the voltage vector difference between E_a and E_i by the stator impedance Z. This vector difference is indicated by E in fig. 15-19c. The stator reactance is large compared to its resistance, and therefore the current I lags the resultant voltage E by nearly 90 degrees, and hence the current I lags the applied voltage E_a by θ degrees. In this case the synchronous motor is operating with a lagging power factor. If the load is increased, the rotor poles are pulled further behind the stator poles, and this causes E_i to lag further or α to increase (fig. 15-

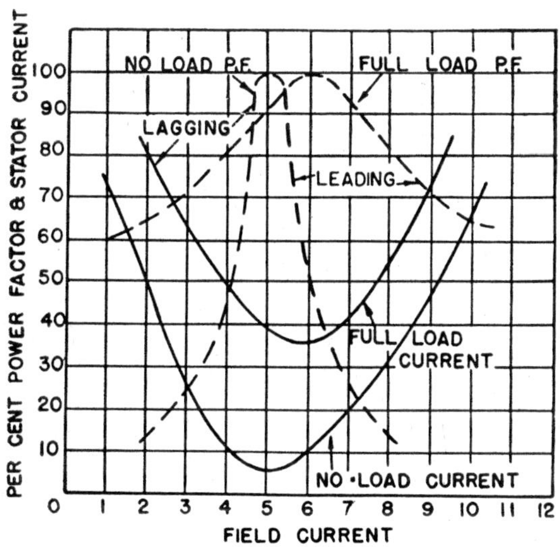

FIG. 15-20

19d). As a result, the stator current I lags the applied voltage E_a through a greater angle.

Suppose the field excitation is decreased. Since the speed is constant, the induced voltage E_i decreases (fig. 15-19e). Then the resultant voltage becomes greater. As a result the current becomes greater and lags the applied voltage a greater amount. Suppose that the field excitation is increased until the current I is in phase with the applied voltage E_a, making the power factor of the synchronous motor unity (fig. 15-19f). For a given load, at unity power factor the resultant E and therefore the stator current I are minima. In fig. 15-19g the field excitation is still further increased. In this case the current I has increased and leads the applied voltage E_a. *Then for a given load, the power factor is governed by the field excitation; a weak field produces a lagging current and a strong field produces a leading current.* The current is minimum at unity power factor and increases as the power factor becomes poor, either leading or lagging. The variation of power factor and stator current at no load and at full load with a varying field excitation is shown in fig. 15-20. If the motor is heavily overloaded, the field poles are pulled out of step with the rotating field and a heavy current will flow through the armature. This causes the circuit breakers to open and the motor to come to a standstill.

Uses of Synchronous Motors—In plants in which the load consists chiefly of induction motors, the power factor is usually lower than 70 per cent. It is desirable that this power factor be near unity for reasons given in chapter 11. To improve the power factor, synchronous motors are generally connected in the system. If these motors are operated without load, their power factor may be adjusted to a value as low as 10 per cent leading. Motors so operated are generally referred to as synchronous condensers, because they take a leading current, as do condensers. In this case the motors take little effective power, which supplied its losses, and high leading reactive power, which reduced the lagging reactive power taken by the inductive loads.

Synchronous motors cannot be used where there are sudden applications of heavy loads, because such loads will pull the rotor out of step with the rotating magnetic field. They are generally used for driving loads requiring constant speed operation and infrequent starting and stopping, such as D C generators, blowers, and compressors.

Single-Phase Induction Motor—Fig. 15-21a shows diagrammatically a single-phase motor that has a squirrel cage rotor and a single winding on the stator, which is energized by an alternating e m f. During the time current flows from L_1 to L_2 and is increasing, flux is established which induces an e m f in the rotor, causing current to flow as indicated. This rotor current will set up poles at N_r and S_r in direct line with the stator poles N_s and S_s. It then follows that no torque is developed, due to the relative position of the stator and rotor poles. If other instances are considered, it will be found that torque is developed at no time. This shows that a single phase motor, such as is shown in fig. 15-21, is not self starting, but requires some starting means.

Now suppose that by some means that will be described later, the rotor is turned in the clockwise direction. Due to this rotating action, the conductors will cut the flux established by the stator winding, and as a result an e m f will be induced in the rotor, which will cause current to flow as indicated in fig. 15-21b. This current is such as to set up poles on the rotor at N_r and S_r, which are midway between the stator poles, and since the rotor is highly inductive, the rotor current will lag the induced e m f by nearly 90 degrees. Hence, the flux produced in the rotor pulsates nearly 90 degrees out of phase with that of the stator. Since these two fields are at right angles, a rotating magnetic field is established, as in a two-phase winding.

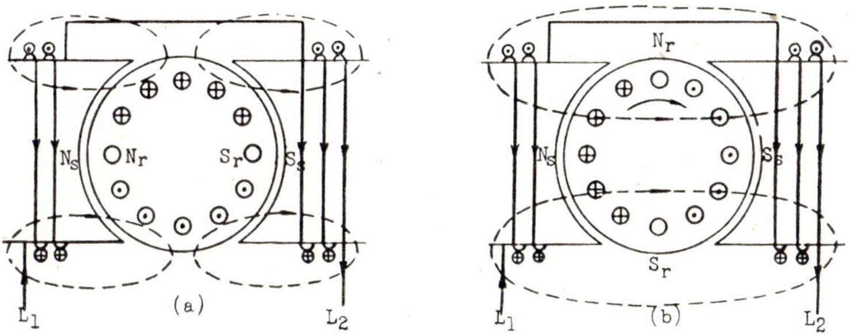

Fig. 15-21

If the rotor turns at synchronous speed, the rotor field becomes nearly as strong as the stator field, and the resultant rotating field remains at nearly constant strength. When the rotor turns at a slow speed, the resultant rotating field has a much greater strength in the direction N_sS_s than in the direction N_rS_r. This produces a varying torque.

The performance of a single-phase motor, after it is up to speed, is nearly the same as that of a polyphase motor. As in a polyphase motor, the rotor turns slightly below synchronous speed at a rate depending upon the nature of the load. For the same rating, however, single-phase motors (including auxiliary equipment) are larger and generally operate at a lower power factor and efficiency than polyphase motors.

Single-Phase Operation of Polyphase Motors—If one phase of a polyphase motor is opened, the machine becomes virtually a single-phase motor. It will not be self starting, but if started it will continue to run at a considerably reduced capacity. It will overheat if rated load is applied.

Split Phase Motor—Even though the polyphase motor has a number of distinct advantages over the single phase machine, it is occasionally necessary to install small single phase motors, particularly where polyphase power is not accessible. In such a case, the motors are energized from lighting circuits, which are invariably single phase, and are used for driving fans, sewing machines, contact-making drums, relays, etc.

To be most satisfactory, the single phase motor must be self-starting. A number of methods are used to accomplish this, some of which will now be considered.

Split Phase Starting—One method of making a motor self-starting is to employ the phase splitting principle. Fig. 15-22a shows an

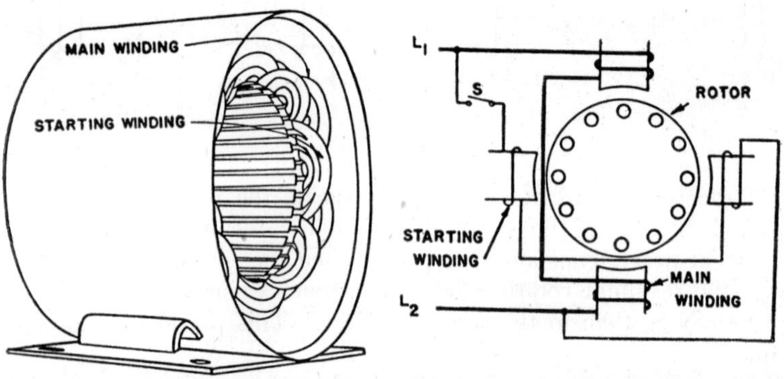

(a) (b)

Fig. 15-22

arrangement in which the stator has two windings—a main winding and a starting winding. The main winding, which is connected across the line in the usual manner (fig. 15-22b), has a low resistance and a high inductance, because it consists of heavy wire placed near the bottom of the slots. The starting winding, which is displaced 90 degrees from the main winding, has a high resistance and a low inductance, because it consists of fine wire placed near the top of the slots. The currents in the two windings are then out of phase with each other. For the best condition, the two currents should be 90 degrees out of phase, but with this scheme, this is not readily obtainable nor necessary. The two windings will produce a weak rotating magnetic field, which is sufficient to provide a low starting torque. When the motor reaches full speed, the centrifugal switch S (fig. 15-22b) opens and cuts out the starting winding. The motor then operates as a single phase induction motor, as previously explained.

In the last few years, another type of split phase motor, known as the condenser motor, has become considerably popular because it has better efficiency and power factor than the one previously discussed, and can be constructed to have excellent starting and running characteristics.

As in fig. 15-22, the condenser motor contains two windings 90 degrees displaced. In this case both windings are connected across the

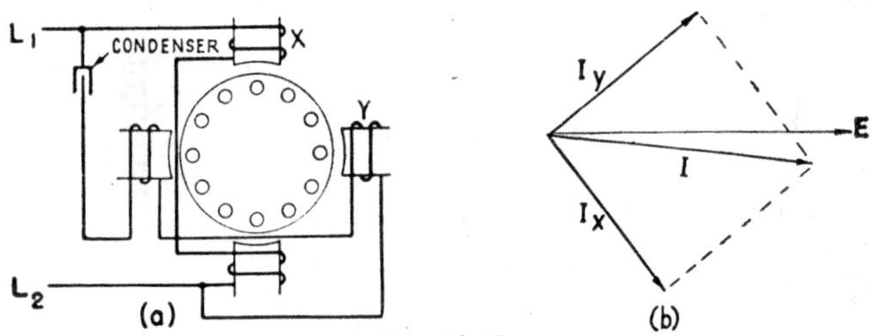

Fig. 15-23

line, but one is in series with a condenser (see fig. 15-23a), and generally contains more turns of smaller wire than the other.

The vector diagram for the starting condition is shown in fig. 15-23b. In phase X, the current I_x lags the voltage by approximately the same angle that the current lags in the main winding of fig. 15-22. In phase Y, the current I_y leads the current of phase X by 90 degrees if the condenser is of the proper capacity. The total current I, being the vector sum of the other two, is nearly in phase with the voltage, hence the power factor is nearly unity. The starting torque will be large because I_x and I_y are nearly 90 degrees out of phase. This motor will supply the same torque as a three-phase motor, at about 70 per cent as much current, because I_x and I_y are at right angles.

Condenser motors supplying as high a starting torque as the one just considered require a large condenser, which is too large during the normal load. Hence, in some motors, means are provided to reduce the capacity as soon as the motor nears full speed. Two common methods of doing this are illustrated in fig. 15-24. In fig. 15-24a, at the instant of starting, the two condensers are in parallel, providing a high capacity. When the motor reaches a certain speed, the centrifugal switch disconnects the large condenser, and the motor operates with only the small condenser in the circuit.

The scheme shown in fig. 15-24b permits circuit Y to draw a large leading current with the use of a comparatively small condenser. When the motor is started, the centrifugal switch connects phase Y to point A on the auto-transformer, thereby supplying about 600 volts to the condenser. The high transformer ratio provides a current through phase Y that is about 20 times as large as would flow if the condenser were connected directly in phase Y. This heavy current provides a high starting torque.

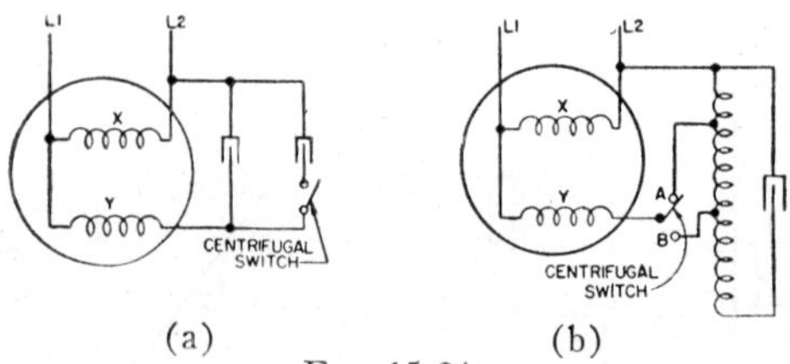

Fig. 15-24

When the motor reaches nearly full speed, the centrifugal switch connects phase Y to point B on the transformer. Now the transformer ratio is about one to two. The current through phase Y will now be about twice as much as would flow if the condenser were connected directly in this phase. This motor has operating characteristics about the same as those of a three-phase motor.

Shading Coil Starting—Single phase motors are sometimes started by means of a low-resistance, short-circuited coil placed around one tip of each pole (see fig. 15-25). When the current, and therefore the field flux, are increasing, a portion of the flux cuts the shading coil. This establishes a current in it, which sets up a flux opposing the main field flux. Hence, lines pass only through the unshaded sections of the poles.

During the time the main flux is at its maximum value, the shading coil is not being cut. Then no opposing flux is established, and as a result the main field flux also passes through the shading coil.

During the time the main flux is decreasing, an e m f is induced in the shading coil, causing a current to flow, which sets up a flux in the same direction as the main field flux. Hence, a high flux passes through the shading coil.

The effect of the shading coil is then to cause a flux to sweep across the pole faces, from the unshaded to the shaded section of the pole, producing a weak rotating magnetic field.

Single phase motors with shading coils produce an exceedingly weak starting torque, and consequently are suitable only for driving small devices such as small fans and relays.

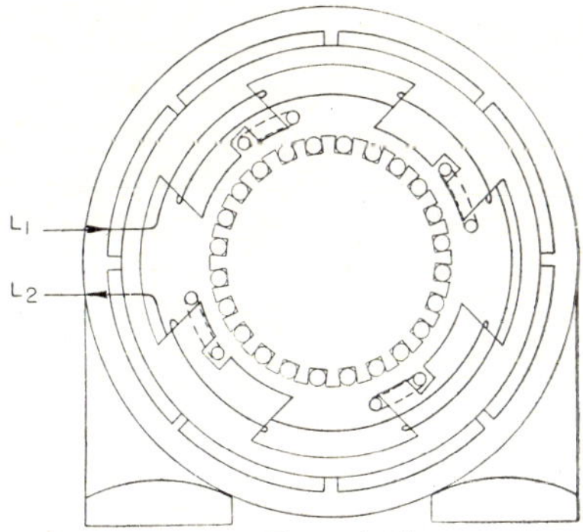

FIG. 15-25

Repulsion Motors—A single phase induction motor may be repulsion starting, in which case the rotating member consists of a wound

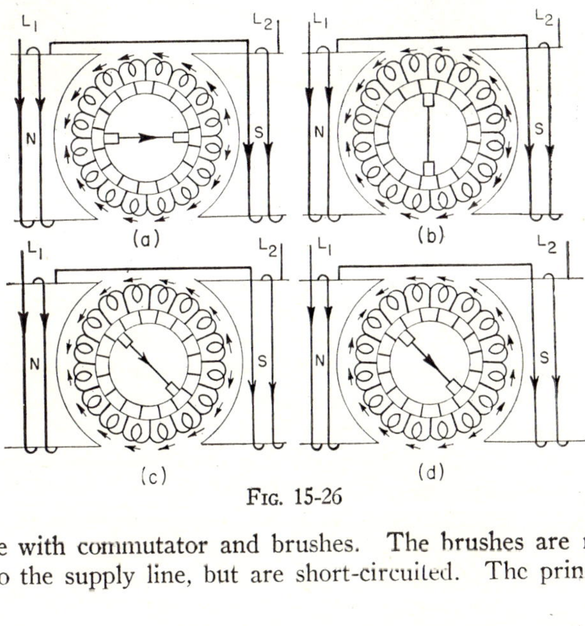

FIG. 15-26

armature with commutator and brushes. The brushes are not connected to the supply line, but are short-circuited. The principle involved in this method of starting may be understood from a consideration of fig. 15-26. Although the current through the stator is alternating, the arrows indicate the direction of the current during a half-cycle. In fig. 15-26a, suppose the stator current is increasing and is in the direction indicated. The flux produced will induce voltages in the armature conductors as indicated. These voltages are additive on each side of the brushes, and therefore send a high current through the armature and short-circuited brushes. No torque will be developed, however, because half of the conductors under each

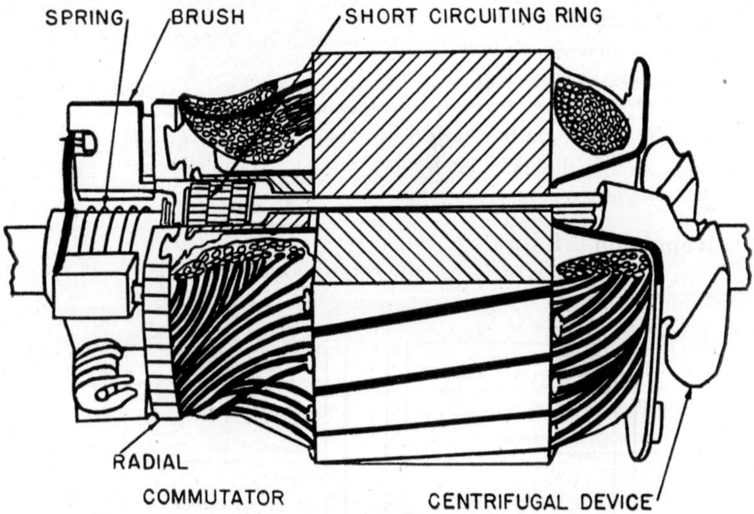

Fig. 15-27

pole carry current in one direction and the other half carry current in the opposite direction.

If the brushes are shifted 90 degrees, as shown in fig. 15-26b, the e m f in each path will be neutralized. Then no voltage exists at the brushes, consequently no current flows through the armature, and of course no torque is produced.

If the brushes are shifted to the position shown in fig. 15-26c, a resultant voltage will exist in each path, sending current through the armature as indicated in fig. 15-26d. Now all conductors under one pole carry current in one direction, and all conductors under the opposite pole carry current in the opposite direction. Then a torque is developed. If the brushes were shifted to the opposite direction from the position shown in fig. 15-26c, the armature would turn in the opposite direction.

This method is employed for starting single phase motors, in which case the armature is provided with a commutator that has either a radial structure as shown in fig. 15-27 or a cyclindrical structure as in D C machines.

At starting, the machine functions as a repulsion motor, developing a high starting torque. As soon as the armature, assumed to have a structure as shown in fig. 15-27, reaches nearly full speed, the centrifugal device forces the short-circuiting ring in contact with the inner surface of the commutator. This short circuits the segments, and therefore converts the machine into an induction motor. At the same time, the centrifugal device raises the brushes, which reduces wear of the brushes and commutator.

Since this machine starts as a repulsion motor and runs as an induction motor, it is generally called a repulsion induction motor. It is used principally where a large starting torque is desired.

Series A C Motor—If the direction of the current through a D C motor is reversed without changing the relative connections of the field and armature, the direction of rotation will remain unchanged, because the polarity of the magnetic poles and the direction of the current through the armature are both reversed. It is to be expected that even when the polarity is reversed rapidly, torque will continually be developed in only one direction. Then if D C motors are supplied with A C, a unidirectional torque will be developed.

In a shunt motor the high reactance of the field winding will permit only a small field current to flow, which will produce a weak field. Furthermore, due to the high reactance, the field flux will be considerably out of phase with the armature current, and therefore the torque developed will be small. These undesirable features are reducible, however, so that A C shunt motors are built in extremely small sizes.

In a series motor, the armature current and field current are in phase, and consequently, in such a machine, considerable torque will be developed. The ordinary D C series motor, however, does not function satisfactorily on A C, for the following reasons:

(a) The alternating flux sets up large eddy currents in the unlaminated portion of the magnetic circuit, causing excessive heat and a lower efficiency.

(b) The high field reactance establishes a large voltage drop across the field winding, which reduces the current input and power factor to such an extent as to make the motor impractical.

(c) The alternating flux establishes high currents in the coils short-circuited by the brushes; and the rupture of these currents, which takes place when the short-circuited coils break contact from the brushes, causes excessive sparking.

The method of reducing these difficulties to a minimum will now be considered.

The eddy currents are reduced to a negligible value by making the magnetic circuit entirely laminated. The laminated field poles and yoke must then be supported in a cast steel housing. Hence, A C motors require a more expensive construction than D C motors.

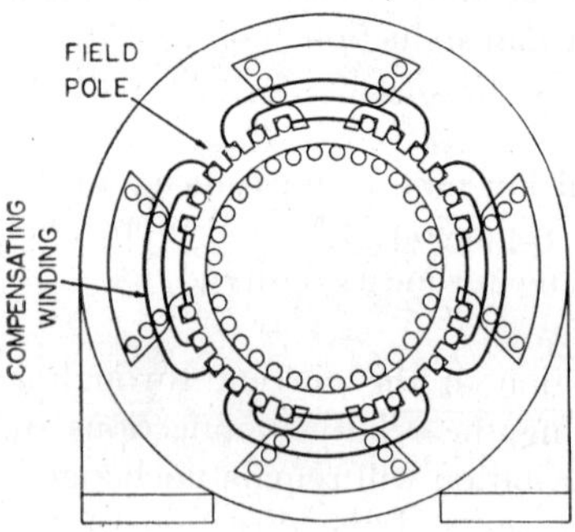

FIG. 15-28

To reduce the field voltage drop and improve the power factor, it is necessary to decrease the reactance of the field and armature winding. This may be done by using a low frequency. The usual frequency of 60 cycles per second is too high, except for small motors. Medium size series motors are generally supplied with a 25 cycle voltage, and in the case of large railway series motors, a 15 cycle voltage is sometimes used.

Regardless of the frequency used, the field reactance is reduced by using as few turns on the field as possible. A high field flux is obtained, however, by reducing the reluctance of the magnetic circuit as low as possible, which is accomplished by making the iron cross section large and the air gaps short (see fig. 15-28).

A common method of reducing armature reactance is to make use of a compensating winding embedded in the field pole faces, as indicated in fig. 15-28. This winding is arranged to supply a magnetizing action that is equal and opposite at all loads to that of the armature coils. This is accomplished by connecting the compensating winding in series with the armature or short-circuiting it upon itself. In the latter case, the magnetizing action of the compensating winding is obtained by transformer action. If the motor is designed for both

D C and A C, the compensating winding should be connected in series with the armature.

The excessive sparking at the brushes, caused by the flux cutting short-circuited coils, can be eliminated by using high resistance leads to connect the coils to the commutator segments (see fig. 15-29). These leads are generally wound at the bottom of the slots, and each has a resistance of about twice that of a coil.

Each coil short-circuited by a brush has two resistance leads in series with it, hence the current through it is small; but for the main armature current there are two leads in parallel at each brush, so that in this case the effective resistance is small. Moreover, only those resistance leads connected to the commutator segments in contact with the brushes carry current at any instant. Hence, resistance leads do not affect the resistance of the armature as a whole.

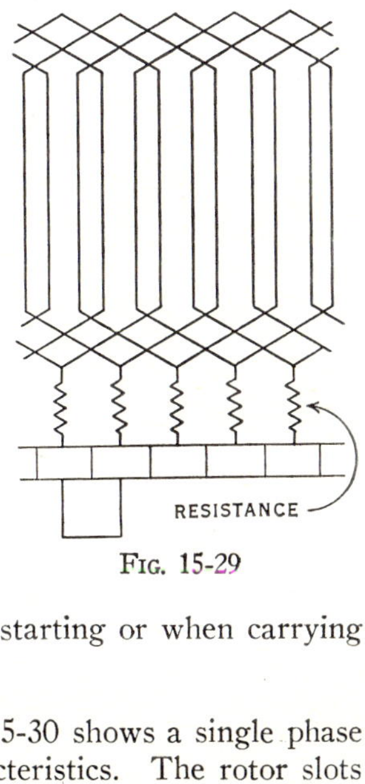

FIG. 15-29

Owing to modifications that must be made to reduce eddy currents and reactance, series motors are more complex in structure, heavier per horsepower, and therefore more expensive than D C motors of the same rating.

The operating characteristics of A C series motors are much the same as those of the D C series motors. At light loads, the speed is high and the power factor is near unity. As the load is applied the speed decreases, and at full load the power factor drops to about 90 per cent. When starting or when carrying an overload, the power factor is lower.

Unity Power Factor Motor—Fig. 15-30 shows a single phase motor which has excellent operating characteristics. The rotor slots contain a winding at the top, which is connected to the commutator the same as a winding in D C machines, and a squirrel cage winding at the bottom (see fig. 15-30a). These windings have a magnetic separator between them.

Referring to fig. 15-30b, the brushes d and e are in line with the stator poles and are short-circuited together. Another pair of brushes are midway between the poles, and these brushes are connected in series with the main field. The compensating winding is shunted across brushes a and b, and this circuit includes a centrifugal switch S.

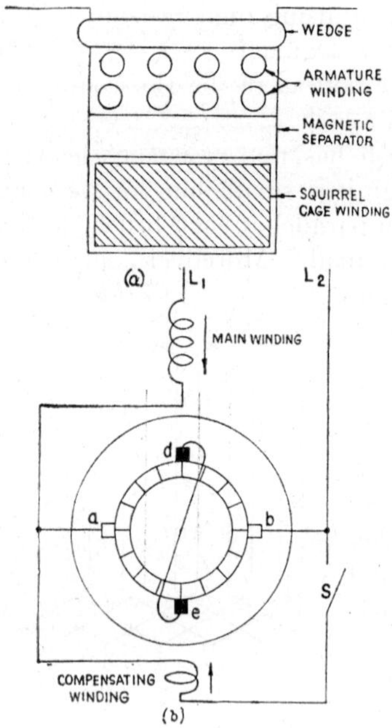

FIG. 15-30

On starting, the main winding is connected through brushes a and b, while the compensating winding is open at switch S. The current at this instant is not large, because the rotor has considerable reactance, produced by the high frequency and high inductance. The inductance is high because of the magnetic separator. As the rotor approaches synchronous speed, the rotor frequency, and therefore the rotor reactance, decreases, consequently the current increases to the full load value.

At starting, the torque produced is due primarily to the shorted brushes (e and d), causing the machine to behave as a repulsion motor. As the speed increases, the rotor frequency becomes less, and the squirrel cage rotor current increases until the machine operates as a squirrel cage motor. When the rotor nears full load speed, the centrifugal switch S closes, and the compensating winding serves to keep the power factor near unity.

Single-Phase Stator Winding—The most common types of stator winding for single phase motors are hand, form, and skein windings.

Hand Winding—A hand winding is one in which the wire is wound from a reel directly into the slots. One pole of such a winding is usually wound as shown in fig. 15-31a. In this case a coil is wound in slots 2 and 4, a second coil is wound in slots 1 and 5, and then the wire is cut. These two coils, called a group, establish a pole when current flows through them. The other groups are wound in exactly the same manner. All the groups are connected together as those of each phase in three phase machines. In both cases the alternate groups are connected in reverse to give the correct polarities (fig. 15-31b). It is possible to wind all the coils without cutting the wire, by winding alternate groups in reverse as shown in fig. 15-31c.

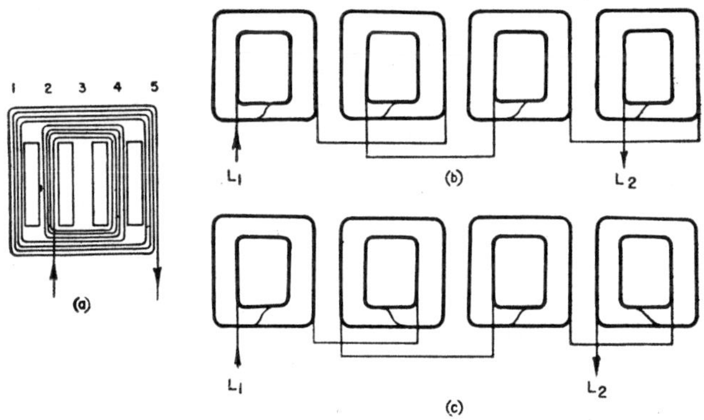

FIG. 15-31

If a stator has semi-closed slots or requires large wire, it is generally hand wound. More time is usually required to place a hand winding on a stator than to wind any other type.

Form Winding—The coils for a stator with open slots are generally form wound; that is, they are wound on a form, removed, taped, and then inserted in the slots (fig. 15-32). In most cases, all the coils of one group are wound without cutting the wire, so that it will be unnecessary to make a large number of connections after the coils are in the slots.

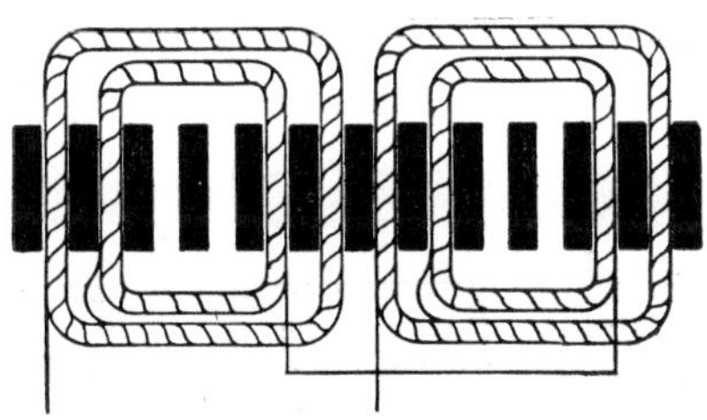
FIG. 15-32

Form wound coils are frequently inserted in stators with semi-closed slots. In this case, however, the coils must be taped at the ends only, as shown in fig. 15-33a. With coils so taped it is possible to insert the coil sides in the slots, in which the insulation is placed as shown in fig. 15-33b, by inserting one or two conductors at a time. After all the coils are in the slots, the insulation extending out is cut off and the ends are folded around the coil sides. Then the wedge is inserted (fig. 15-33c).

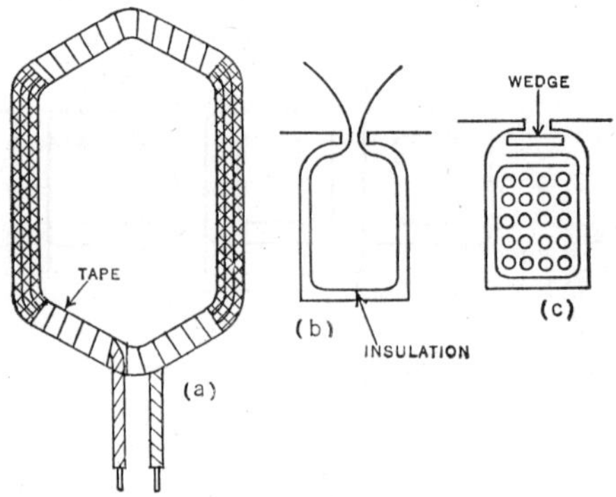

FIG. 15-33

Skein Winding—One type of winding frequently placed on single phase motors is known as a skein winding. The coils for this winding

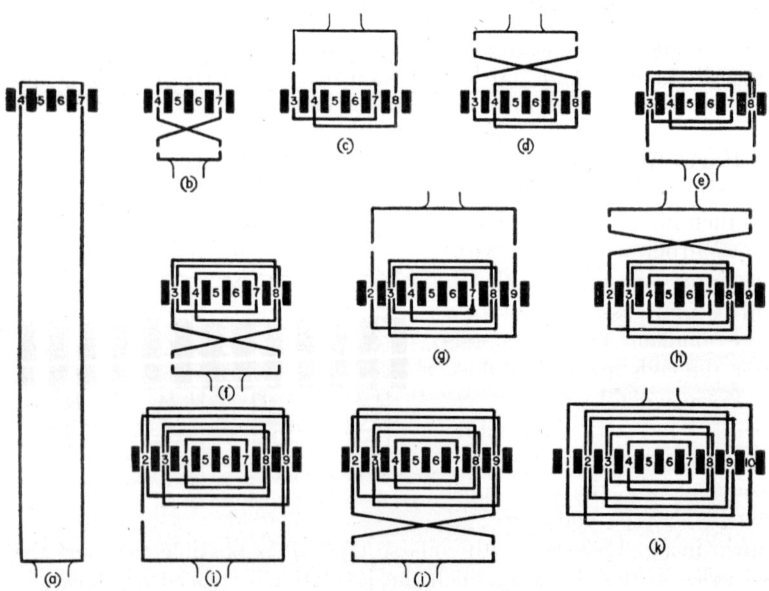

FIG. 15-34

are first wound in large loops, called skeins, of the proper length and with the proper number of turns (the length and the number of turns is best determined from an old skein, which is obtained by carefully removing it from the stator).

To wind one pole of a 4 pole, 36 slot stator, the procedure is as follows: A skein is placed in slots 4 and 7 (fig. 15-34a) and its end is then forced against the core. Next, a half twist is made (fig. 15-34b)

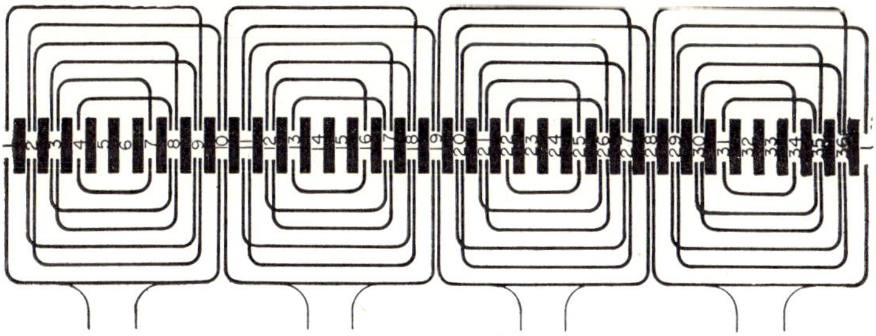

FIG. 15-35

and the free end is drawn into slots 3 and 8 (fig. 15-34c). A half twist is again made, but in the opposite direction (fig. 15-34d), and the free end is again drawn into slots 3 and 8 (fig. 15-34e). Now a half twist is again made in the same direction as the first (fig. 15-34f), and the free end is drawn into slots 2 and 9 (fig. 15-34g). This pro-

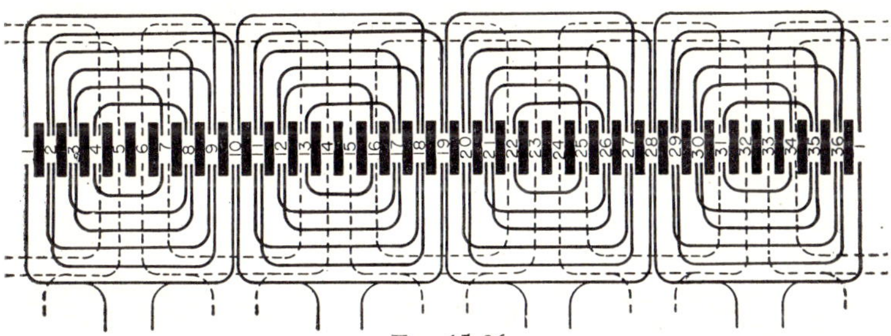

FIG. 15-36

cedure is continued as shown in the figure, and the winding of one pole will end as shown in fig. 15-34k.

The twists are made first in one direction and then in the other until the winding is complete. If this is not done, the skein will twist and will be difficult to wind.

The other poles are wound in the same manner, giving the complete winding illustrated in fig. 15-35. This type of winding can generally be placed on the core more quickly than the other types.

The stators of some single phase motors have one winding, while split phase motors have two, one known as the main winding and the other as the starting winding. These two windings are placed 90 electrical degrees apart. This is illustrated in fig. 15-36, which shows a main winding and a starting winding of the skein type. The main winding is shown solid, and is the same as the one shown in

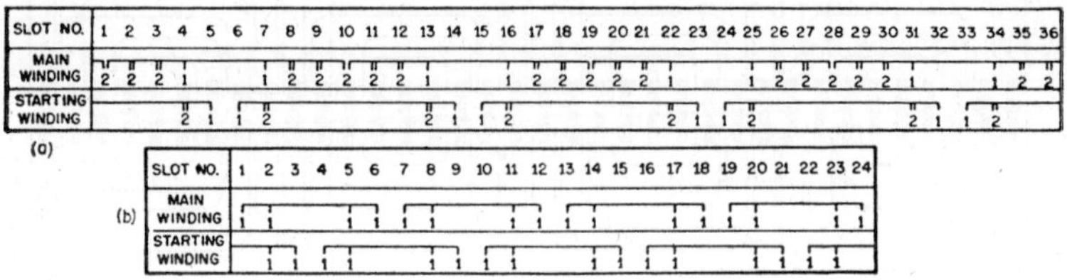

Fig. 15-37

fig. 15-35, and the starting winding is shown dotted. The coil sides are distributed as shown in fig. 15-37a. Each number indicates the number of times a skein side is drawn into a slot. For 4 poles and 36 slots, this is considered a very convenient arrangement, because both windings are distributed. For 24 slots, the windings may be distributed as shown in fig. 15-37b. Two-hand or two-form windings may be placed on a stator and may also be distributed as in fig. 15-37.

Exercises

1. During one cycle, what fraction of a revolution does the rotating magnetic field make in (a) four pole motors? (b) Six pole motors? (c) Ten pole motors?
2. (a) What is the speed of the rotating magnetic field of a 12 pole, 60 cycle induction motor? (b) Of a 2 pole, 180 cycle induction motor?
3. A 6 pole motor connected to a 60 cycle line has a no load speed of 1,194 r p m and a full load speed of 1,140 r p m. Determine: (a) the synchronous speed; (b) per cent slip at no load; (c) per cent slip at full load.
4. What is the rotor frequency of the motor in problem 3 (a) at no load? (b) At full load? (c) When the rotor is at standstill?
5. Why does an induction motor with a low resistance rotor develop little torque at starting, even though it takes a very large current?
6. Upon what three factors does the torque developed by an induction motor depend?
7. What is the approximate efficiency of a 4 pole motor running at 1,650 r p m?

8. Explain the industrial applications of (a) low resistance rotor motors, (b) high resistance rotor motors, and (c) wound rotor motors.
9. At starting, the rotor of an induction motor has an impedence of 5 ohms and a power factor of 30 per cent. Determine the power factor at 10 per cent slip.
10. Each phase of a rotor winding has a resistance of .5 ohm and a reactance of 2.5 ohms at standstill. What must be the resistance of each branch of a star connected rheostat to provide maximum torque at starting?
11. An induction motor develops a starting torque of 15 lb.-ft. when it is connected across a 440 volt line. Determine the torque developed when it is connected across a 220 volt line.
12. In what ways is a double squirrel cage induction motor superior to a standard squirrel cage motor?
13. What is the disadvantage of using a wound rotor motor for work that requires a variable speed?
14. The speed of an induction motor depends upon what three factors?
15. (a) State the different methods of speed control of induction motors. (b) What are the advantages and disadvantages of each?
16. A 4 pole, 60 cycle, 220 volt wound rotor induction motor, without external resistance is driven at a speed of 1800 r p m in a direction opposite to that of the rotating magnetic field. If the stator and the rotor are connected the same way and have the same number of turns, determine (a) the frequency and (b) the terminal voltage of the rotor.
17. Solve problem 16 assuming that the rotor is driven at twice the speed of the rotating field and in the opposite direction.
18. An 8 pole, 60 cycle synchronous motor will run at what speed?
19. How many poles must a synchronous motor have to operate at 450 r p m on a 60 cycle system?
20. Why is the rotor of a synchronous motor equipped with a squirrel cage winding?
21. How is a synchronous motor started?
22. Why is a resistance placed across the field terminals of a synchronous motor when the field is not energized?
23. Under what condition does a synchronous motor take the least current for a given load?
24. Explain what happens when the field excitation of a synchronous motor is (a) weakened, (b) strengthened.
25. A line is loaded with induction motors taking 100 k w at a power factor of 60 per cent. What size synchronous motor running at a power factor of 3 per cent leading is necessary to raise the power factor of the line to unity?
26. What is a synchronous condenser?
27. For what classes of work are synchronous motors suitable?

28. Why is the motor in fig. 15-21 not self-starting?
29. Give the different methods of making the motor referred to in Question 28 self-starting.
30. If one of the three line wires delivering power to a three-phase motor is opened, will the motor continue running? Explain your answer.
31. The rotor of a motor connected as shown in fig. 15-23a was locked, and (without the condenser) like readings were obtained for each circuit, as follows: 90 volts at a frequency of 60 cycles per second, 2 amperes, and 100 watts. Determine the size of the condenser required to produce maximum starting torque. (HINT: Maximum starting torque occurs when the currents in the two circuits are displaced 90 degrees.)
32. What is the power factor of the motor of problem 31 at starting?
33. Why is torque developed in a repulsion motor only when the brush axis makes some angle greater than zero and less than 90 degrees with the pole axis?
34. What operation converts a repulsion motor into an induction motor?
35. What are the advantages and disadvantages of single-phase induction motors compared with polyphase induction motors?
36. Why is a series motor more suitable for alternating current than a shunt motor?
37. What commutation difficulty exists in an A C commutator motor and does not exist in a D C motor?
38. In what ways does an A C series motor differ from a D C series motor?
39. Why is low frequency desirable for A C series motors?
40. Compare the starting torques of the different types of single-phase motors.
41. How is the direction of rotation of the different types of single-phase motors reversed?

CHAPTER 16

STATOR WINDING

IN ANY double layer winding (see chapter 14) each phase must have as many groups as there are poles. Hence, the total number of groups (G) must equal the number of poles (P) times the number of phases (n). Expressing this by an equation, we have

$$G = Pn \qquad (16\text{-}1)$$

In fig. 16-1a, as well as in previous figures pertaining to double layer windings, all the groups have the same number of coils. In these cases the number of coils per group (C_g) was found by dividing the total number of coils (C) by the number of groups (G). This may be expressed by an equation as

$$C_g = \frac{C}{G} \qquad (16\text{-}2)$$

Single Line Diagrams—After the number of coils per group has been determined, it is not necessary to draw the winding in detail as suggested in fig. 16-1a. Windings of this nature are tedious to draw, especially if they contain a large number of coils. Hence, single line diagrams should be drawn instead, as these diagrams are usually satisfactory for all practical purposes.

A single line diagram is developed in fig. 16-1 as follows: A heavy dashed line is drawn between the terminals of each group (fig. 16-1a), each dash representing a coil in the group. For instance, coils 1, 2, and 3 are represented by the three dashes indicated as group 1; coils 4, 5, and 6 are represented by the three dashes indicated as group 2; etc. These dotted lines are placed at an angle to distinguish between the top and the bottom lead of each group. The direction from the start to the finish end of each phase (positive direction) is indicated by arrows.

Now suppose that the diagram within the dashed line is removed. Then only the single line diagram remains, as indicated in fig. 16-1b. Single line diagrams are obviously easy to make, and are easy to follow when actually connecting stator windings. Furthermore, by changing the number of dashes in each group the same diagram may be used to represent any stator having a given number of poles.

Two-Phase Diagrams—Fig. 16-2 shows the steps that should be followed in constructing a diagram for a two-phase series winding when short jumpers (top to top connection) are used. A 4 pole 32 slot core will be selected for convenience. Equation 16-1 shows that this winding must contain eight groups, and equation 16-2 shows that each group must contain four coils. Then the first step is to represent groups containing four coils each, as suggested in fig. 16-2a.

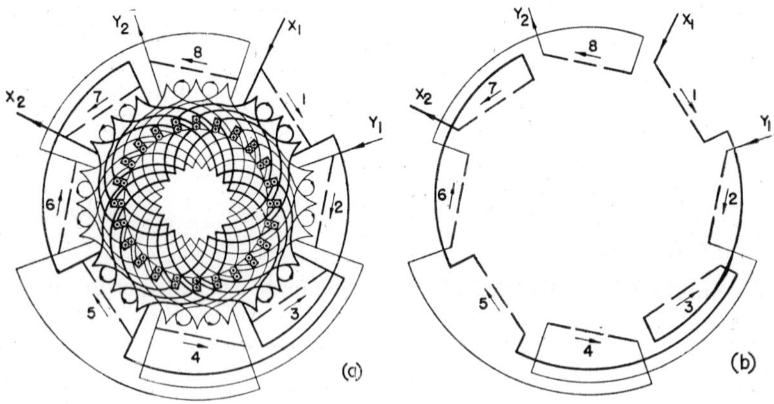

FIG. 16-1

In fig. 16-1 it was shown that the arrows are clockwise for groups 1 and 2, counterclockwise for 3 and 4, again clockwise for 5 and 6, etc., reversing in this way throughout the winding. Hence, the next step is to put arrows near the dotted lines that alternate in this way. Fig. 16-2b shows how the groups of phase X should be connected. Note that every other group belongs to the same phase. The connections from group to group are made so that in following through the winding from the start to the finish end we always trace in the direction of the arrows. We proceed in exactly the same way to connect phase Y (see fig. 16-2c).

Since short jumpers are used in figs. 16-2b and 16-2c, top to top connections result. Fig. 16-2d shows the groups of fig. 16-2a connected with long jumpers. To make these connections proceed as follows: A long jumper is extended from group 1 to group 5, the two groups that belong to the same phase and have current passing through them in the same direction. Next, a long jumper is extended from group 3 to group 7, the remaining two groups of the same phase. These likewise have current passing through them in the same direction. Lastly, a short jumper is extended from group 3 to group 5. When tracing through each phase of this winding we follow the direction of the arrows, as before.

Long and Short Jumpers—When short jumpers are employed, the leads from the groups may be left long enough to serve as jumpers. Thus, less soldered joints will occur. Furthermore, short jumpers

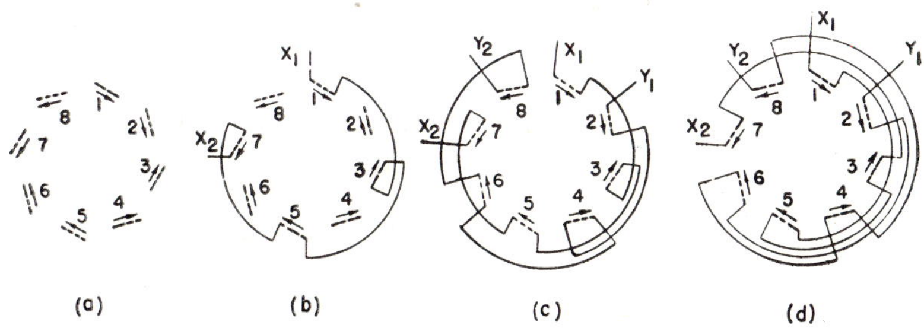

Fig. 16-2

give a top to top connection, which eliminates practically all cross-overs and results in a much neater job.

Long jumpers must be used in multispeed motors. This type of machine will be discussed later in detail. In large machines or in machines with a large number of coils in each group, long jumpers are too long and therefore should not be used.

Parallel Connections of Two-Phase Winding—A parallel connection is one in which there are more than one path per phase. Such a connection is easily made after a series connection has been formed. Suppose it is desired to make two parallel connections of the winding in fig. 16-2c. For the sake of clearness, the method of reconnecting only one phase will be shown. The other phase should be reconnected in exactly the same way. To make two parallel paths, the series winding is opened at its midpoint, which is between groups 3 and 5 (fig. 16-3a). Then the top lead of group 5 is connected to the top lead of group 1. Next the top lead of group 3 is connected to the top lead of group 7. The completed parallel connection is shown in fig. 16-3b.

Fig. 16-4 shows the method of paralleling a series winding when long jumpers are used. The series connection is cut at the short jumper, which is between groups 3 and 5 (fig. 16-4a). Then the bottom lead of group 3 is connected to the top lead of group 1 and the top lead of group 7 is connected to the bottom lead of group 5. This is shown in fig. 16-4b.

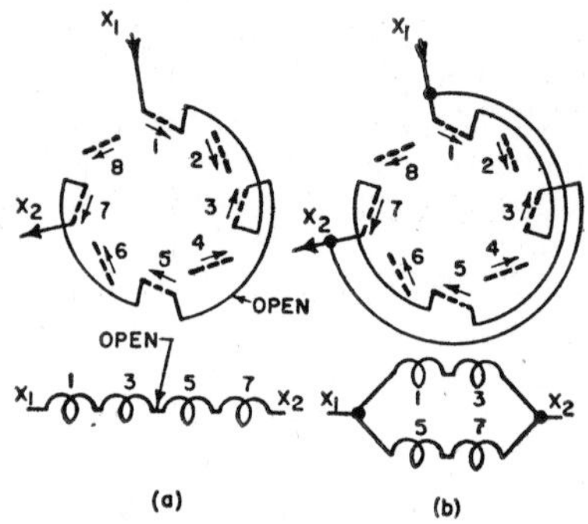

Fig. 16-3

Note again in figs. 16-3 and 16-4 that when following through the parallel paths we trace in the direction of the arrows.

One should never attempt to parallel the coils in a group. The voltages induced in two adjacent coils are out of phase. Hence the voltage of one coil is at times considerably higher than that of some other coil connected in parallel with it. This would cause circulating current.

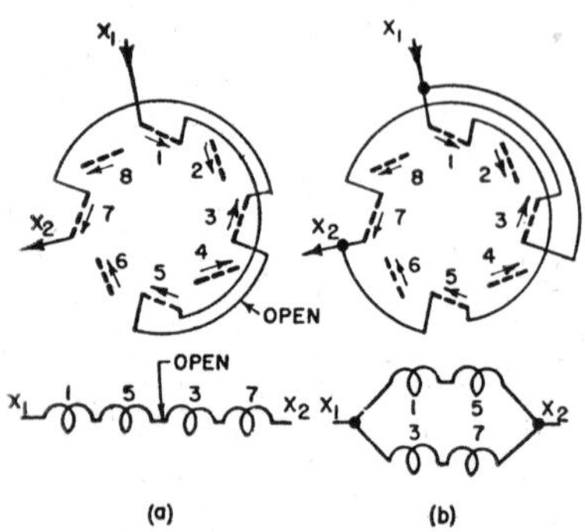

Fig. 16-4

Three-Phase Winding with Short Jumpers—It was shown in Chapter 14 that a three-phase winding is fundamentally three single-phase windings displaced 120 degrees, and connected star or delta. To show how a one line diagram may be developed for a star connected winding, a 36 slot, four pole stator will be considered. Equation 16-1 shows that this winding has 4×3 or 12 groups, and equation 16-2 shows that there will be $36 \div 12$ or three coils per group. A detail drawing of this winding is shown in fig. 16-5a. A single line diagram may be developed in somewhat the same manner as in a two-phase winding.

First, heavy dotted lines are drawn to represent the groups. Suppose we follow through each phase of the winding, beginning at the line wires and tracing to the center of the star. The direction in which we trace through each group will be indicated by arrows. Now suppose the diagram is removed from within the dotted lines so that only the single line diagram remains (this is indicated in fig. 16-5b). Note in this figure that the arrows near the dotted lines representing the groups alternate in direction; that is, the arrow near group 1 is clockwise, the one near group 2 is counter-clockwise, etc.

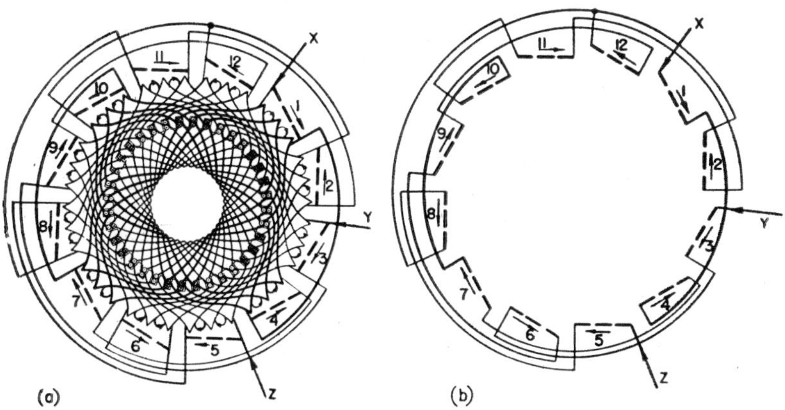

FIG. 16-5

To study the steps that may be followed in constructing a single line diagram for a three-phase, series star winding, when short jumpers are used, a 48 slot, four pole stator will be considered.

The winding for this core will have 12 groups, and each group will contain four coils. The first step is to represent this condition by dashed lines, as suggested in fig. 16-6a. Next arrows are placed on this diagram—arrows that alternate in direction. Fig. 16-6b shows how the groups of one phase should be connected. Note that every third group is connected in the same phase. This is true in all three-phase windings. Note further that the connections from group to group are made so that in following through the winding we always trace in the direction of the arrows.

The line wires for phases Y and Z are located by counting three and six groups from group 1 (it should be realized that adjacent groups are 60 electrical degrees apart). The connections for the other two phases are then made as explained for phase X (fig. 16-6c). The leads coming out of the three phases are then connected together to form a star connection (fig. 16-6d).

To make a delta connection of fig. 16-6c it is necessary to join the phases 120 degrees displaced, so that they form a closed circuit, and then extend line wires to their junctions. In the case of fig. 16-6c

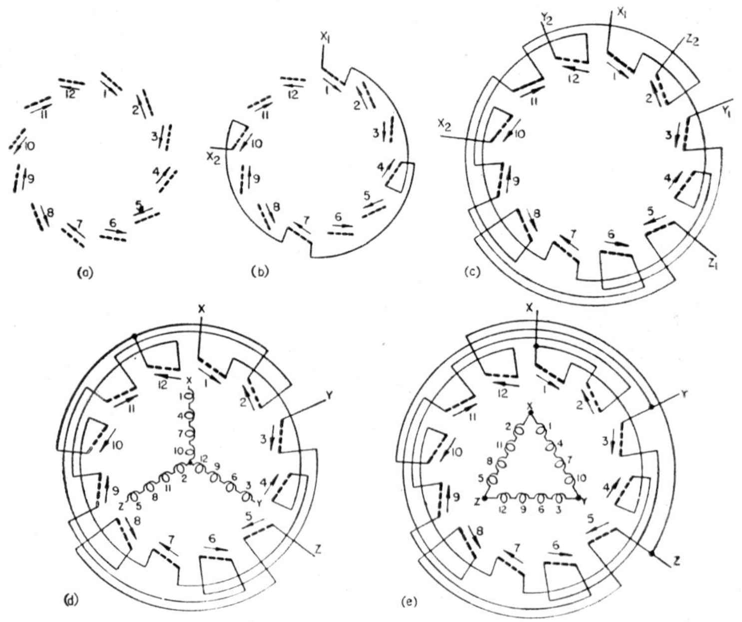

FIG. 16-6

X_2 and Y_1, Y_2 and Z_1, and Z_2 and X_1 are connected as shown in fig. 16-6e.

Three-Phase Winding with Long Jumpers—As in a two-phase winding, the groups that carry current in one direction and belong to the same phase are connected by means of long jumpers. Next, the groups of the same phase that carry current in the opposite directions are likewise connected by long jumpers. Then the two series connections are joined by a short jumper. Fig. 16-7 shows windings connected with long jumpers, one being series star and the other series delta.

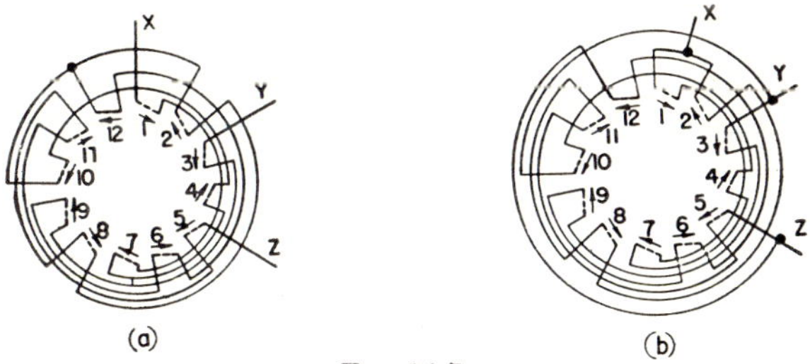

FIG. 16-7

Parallel Connections of Three-Phase Windings—The method of changing from a series to a parallel connection in a three-phase winding is shown in fig. 16-8. As in the two-phase winding, phase X (fig. 16-8a) is opened at its midpoint, which is between groups 4 and 7. Next the top lead from group 7 is connected to the top lead of group 1, and the top lead of group 4 is connected to the top lead of group 10 (fig. 16-8b). This phase should then be checked by tracing through each path. The other two phases should then be paralleled in the same way. Finally the three phases should be connected either star or delta.

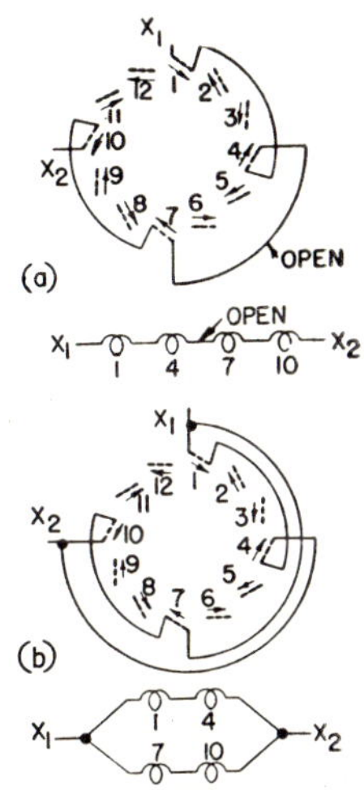

FIG. 16-8

Double Voltage Winding—It is the common practice of manufacturers to build machines that may be connected externally for two voltages in the ratio of 2 to 1. This may be accomplished, with either a star or delta connection, by dividing each phase into two sections and bringing out nine terminals from these sections, as suggested in fig. 16-9.

In fig. 16-9a the sections may be connected either series or parallel star. If the series connection is suitable for 440 volts, the parallel connection will be suitable for 220 volts.

The series connection is formed by joining T_4 to T_7, T_5 to T_8, and T_6 to T_9. Line wires are then extended to T_1, T_2, and T_3.

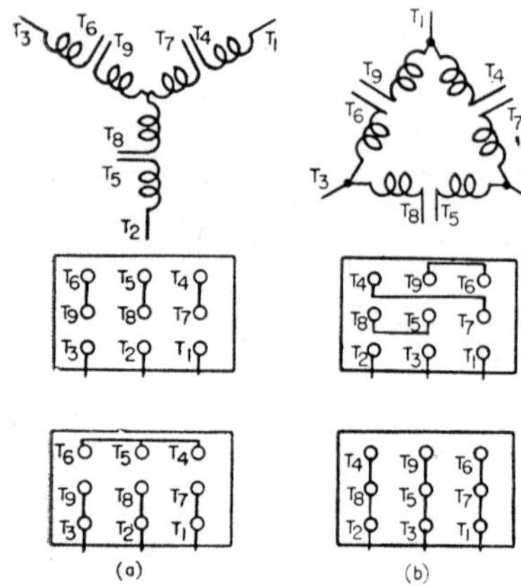

FIG. 16-9

The parallel connection is obtained by joining T_4, T_5, and T_6 to form a second star. The two stars are next joined in parallel by connecting T_1 to T_7, T_2 to T_8, and T_3 to T_9. Then the line wires are extended to these junctions.

Note that the method of making the above connections is also given in the terminal block diagrams under the figures.

Fig. 16-9b shows how the nine terminals are usually brought out of a winding so that it may be connected for two voltages by making either a series or parallel delta connection.

Unequal Coil Grouping—In a given winding all the paths must contain the same number of coils. This means that in some stators all the coils cannot be used. To determine the number of unused coils (usually called dummy coils), proceed as follows: First, find the total number of paths (b) by multiplying the paths per phase by the number of phases. Next, find the number of coils per path (C_b) by dividing the total number of coils by the number of paths. Expressing this by an equation,

$$C_b = \frac{C}{b} \tag{16-3}$$

If the above division leaves a remainder, the remainder gives the number of dummy coils. Dummy coils are placed as far apart as possible in the stator core, and are not connected into the winding. Their terminals are taped and arranged to give a neat appearance. *If the number of dummy coils exceeds eight per cent of the total number, the winding will be unsatisfactory for the rated horsepower.*

After the number of dummy coils has been determined, it is necessary to find the number of coils per group. This number may be found by using equation 16-2. If the division in equation 16-2 leaves a remainder, the winding must contain groups with unequal numbers of coils. This is known as unequal coil grouping.

Stators with dummy coils and unequal coil grouping are common in practice. They occur when windings are reconnected for different conditions; and are also frequently found because manufacturers, in order to reduce the number of dies, use the same stator stampings in machines having different numbers of poles, phases, or parallel paths.

To figure out a winding which will give the best operating results, it is good practice to determine (a) the number of dummy coils, (b) the number of coils per group, (c) the positions of the large groups, and (d) the positions of the dummy coils. Such a procedure is used in working out the following examples.

EXAMPLE—How should the coils in a six pole, 56 coil, three-phase, two-parallel star winding be grouped to give satisfactory operating results?

Substituting in equation 16-3, $C_b = 56 \div 6$ or 9, with a remainder of 2. This means that the winding must possess two dummy coils. This is permissible since it is less than eight per cent of the total number.

The winding will contain 6×3 or 18 groups (equation 16-1), and it will have $54 \div 18$ or 3 coils per group (equation 16-2). All the groups will contain the same number of coils, since the division does not give a remainder.

To distribute the dummy coils as far apart as possible, we will place one coil near group 1 (a group in phase X) and the other about half-way around the armature, or near group 11 (a group in phase Y).

EXAMPLE—Select a suitable coil grouping for a six pole, 96 coil, three-phase series winding.

Substituting in equation 16-3, $C_b = 96 \div 3$ or 32, without a remainder. Thus this winding will contain no dummy coils.

The winding will have 6×3 or 18 groups, and there will be $96 \div 18$ or 5 coils per group, with a remainder of 6. This indicates that the winding must contain 6 large groups with $5 + 1$ or 6 coils per group, and the remaining groups ($18 - 6$ or 12) must contain 5 coils each.

A convenient arrangement of the unequal groups may be determined by developing Table 16-I, as follows:

Six out of 18 groups are large. If the large groups are equidistant, the distance between any two of them will be 1/6 of 18 or 3 groups.

TABLE 16-I

Group number	1	2	3	4	5	6	7	8	9	10	11	12	13	14	15	16	17	18
Phase in which the group number falls	X	Y	Z	X	Y	Z	X	Y	Z	X	Y	Z	X	Y	Z	X	Y	Z
First step in selecting position of large groups	L			L			L			L			L			L		
Final position of large groups	L					L		L		L					L		L	
Coils in each group	6	5	5	5	5	6	5	6	5	6	5	5	5	5	6	5	6	5

TABLE 16-II

Group number	1	2	3	4	5	6	7	8	9	10	11	12	13	14	15	16	17	18	19	20	21	22	23	24	25	26	27	28	29	30
Phase in which the group number falls	X	Y	Z	X	Y	Z	X	Y	Z	X	Y	Z	X	Y	Z	X	Y	Z	X	Y	Z	X	Y	Z	X	Y	Z	X	Y	Z
Position of small groups	S					S					S					S					S					S				
Coils in each group	1	2	2	2	2	1	2	2	2	2	1	2	2	2	2	1	2	2	2	2	1	2	2	2	2	1	2	2	2	2

With this spacing, groups 1, 4, 7, etc., will be large (see line 3 of Table 16-I). This will not be satisfactory, of course, because all the large groups fall in phase X. The large groups should be distributed equally in the three phases and not adjacent to each other. Suppose the first large group is placed in phase X, the second in phase Z, and the third in phase Y, repeating this order throughout the winding, as indicated in line 4. The number of coils in each group will then be in the order indicated in line 5.

EXAMPLE—Suppose it is desired to obtain a suitable coil grouping for a winding containing 57 coils, which is to be connected for three phases, ten poles, and two parallel paths.

Substituting in equation 16-3 gives 57 ÷ 6 or 9, with a remainder of 3. Then the winding must contain three dummy coils, which is permissible since this is less than eight per cent of the total number.

This winding will contain 30 groups, with 54 ÷ 30 or 1 coil per group, and a remainder of 24. Then the winding must contain 24 large groups with 2 coils in each and six small groups with one coil in each. Since there are fewer small groups than large ones, it is more convenient to determine the position of the small groups.

Table 16-II shows the coil arrangement for this winding.

The distance between the small groups is 30 ÷ 6 or 5. This spacing, which is shown in line 3, need not be altered because the small groups are equally distributed in the three phases. Then the coils per group will be in the order indicated in line four.

Since this winding contains two parallel paths, it is now necessary to discover if this coil grouping will provide the same number of coils in all the paths when short jumpers are used. The number of coils in one path of phase X may be found by adding the coils in the first half of the groups belonging to this phase. This addition gives 9. The number of coils in the other path of phase X may be found by adding the coils in the last half of the groups. This also gives 9. On considering the other phases similarly it will be found that each path contains nine coils, which fulfils the requirement.

The three dummy coils will be spaced equal distances apart if the first is placed near group 1, the second one-third the distance around the armature, which is near group 11, and the third one-third the distance further around the armature, which is near group 21. This gives the best possible distribution because the dead coils are equidistant and are placed near small groups.

Sometimes an existing stator must be rewound for a different voltage, a different speed, or a different frequency. When such changes are made, the ampere-conductors per slot (amperes per conductor times the number of conductors per slot) and the air-gap flux density should not be changed.

Change in Voltage—If a stator is to be rewound for a new voltage, the flux per pole does not change, since the number of poles remains unaltered. Considering equation 14-4, expressed as

$$\Phi = \frac{10^8 E}{4.44 f T_p K_d K_c} \qquad (16\text{-}4)$$

in which Φ, f, K_d, and K_c remain constant, then

$$\frac{10^8 E_2}{4.44 f T_2 K_d K_c} = \frac{10^8 E_1}{4.44 f T_1 K_d K_c} \text{ or } \frac{E_2}{T_2} = \frac{E_1}{T_1}$$

Inverting,

$$\frac{T_2}{E_2} = \frac{T_1}{E_1} \qquad (16\text{-}5)$$

This equation shows that an increase in the voltage requires a corresponding increase in the number of turns. To get more turns in the same slot, the wire size must be correspondingly decreased. In other words, the wire size must vary inversely with the number of turns. This may be expressed as follows:

$$\frac{cm_2}{cm_1} = \frac{T_1}{T_2} \qquad (16\text{-}6)$$

Example—Each phase of a 440 volt, series star connected motor contains 480 turns of No. 18 gage wire. If the motor is to be rewound for 110 volts, determine the number of turns per phase and the size of the wire.

$$\frac{T_2}{110} = \frac{480}{440} \text{ or } T_2 = \frac{110 \times 480}{440} = 120 \text{ turns per phase}$$

As a No. 18 gage wire has 1624.3 circular mils, then

$$\frac{cm_2}{1624.3} = \frac{480}{120} \text{ or } cm_2 = \frac{480 \times 1624.3}{120} = 6497.2 \text{ circular mils}$$

This indicates that a No. 12 gage wire should be used, but since a 12 gage wire is too difficult to form into a coil, two 15 gage wires in parallel are recommended.

Change in Speed—To rewind a motor for a change in speed, it is necessary to change the number of poles. In order to keep the same flux in the air-gap, $\Phi_2 P_2$ must equal $\Phi_1 P_1$.

When the per cent pitch is not changed, $K_d K_c$ remains practically constant. Then

$$\frac{10^8 E P_2}{4.44 f T_2 K_d K_c} = \frac{10^8 E P_1}{4.44 f T_1 K_d K_c} \text{ or } \frac{P_2}{T_2} = \frac{P_1}{T_1}$$

Inverting,

$$\frac{T_2}{P_2} = \frac{T_1}{P_1} \qquad (16\text{-}7)$$

When making a change in the number of poles, the flux density in the core should be determined to make sure that it does not exceed the maximum allowable value, which is 90,000 lines per square inch. If the core flux density exceeds this value, the flux per pole must be reduced. The method of doing this will be considered later.

EXAMPLE—A six pole, series star, 440 volt winding has 480 turns of No. 18 gage wire per phase. Determine the number of turns per phase and the size of the wire required if the stator is to be rewound for four poles.

$$\frac{T_2}{4} = \frac{480}{6} \text{ or } T_2 = 320 \text{ turns}$$

$$\frac{cm_2}{1624.3} = \frac{480}{320} \text{ or } cm_2 = 2436.45 \text{ circular mils}$$

This indicates that No. 16 gage wire should be used.

Change in Frequency—When rewinding a stator for a change in frequency, the speed changes if the number of poles is not changed. To keep the speed constant, it is necessary to change the number of poles.

If the number of poles is not changed, equation 16-4 gives

$$\frac{T_2}{T_1} = \frac{f_1}{f_2} \qquad (16\text{-}8)$$

which shows that the number of turns required varies inversely with the frequency.

If the speed is kept constant by changing the number of poles, and a constant flux is maintained, which is essential, then $\Phi_2 P_2$ must equal $\Phi_1 P_1$. Considering the same fractional pitch maintained, then

$$\frac{10^8 EP_2}{4.44 f_2 T_2 K_d K_c} = \frac{10^8 EP_1}{4.44 f_1 T_1 K_d K_c} \tag{16-9}$$

or

$$\frac{P_2}{f_2 T_2} = \frac{P_1}{f_1 T_1} \tag{16-10}$$

The synchronous speed of a motor is

$$N = \frac{120f}{P}$$

In this equation it is apparent that if the number of poles is varied to the same extent as the frequency, the speed will remain constant. Then for a constant speed,

$$\frac{P_2}{f_2} = \frac{P_1}{f_1} \tag{16-11}$$

Substituting in equation 16-10,

$$T_2 = T_1 \tag{16-12}$$

In this case it is not necessary to change the number of turns or the size of the wire. Of course the stator must be either reconnected or rewound. If it is reconnected the coil pitch generally changes and therefore K_c changes. For this condition equation 16-9 must be expressed as

$$\frac{T_2}{K_{c_2}} = \frac{T_1}{K_{c_1}} \tag{16-13}$$

To determine the new size of wire, equation 16-6 applies.

The stator winding of an induction motor may overheat when the motor is normally loaded, indicating that the winding is incorrect, the data taken while stripping the stator may be lost or taken incorrectly, or that the stator is to be rewound for a lower number of poles, in which case it will be found that for the same air-gap flux, the flux density in the core is too high. In such cases it usually becomes necessary to determine, from the dimensions of the stator core, the number of turns and the size of the wire required. The method of doing this will be illustrated in the following example.

EXAMPLE—The core shown in fig. 16-10, which has 39 slots, is to be wound series star for 60 cycles, six poles, and 440 volts.

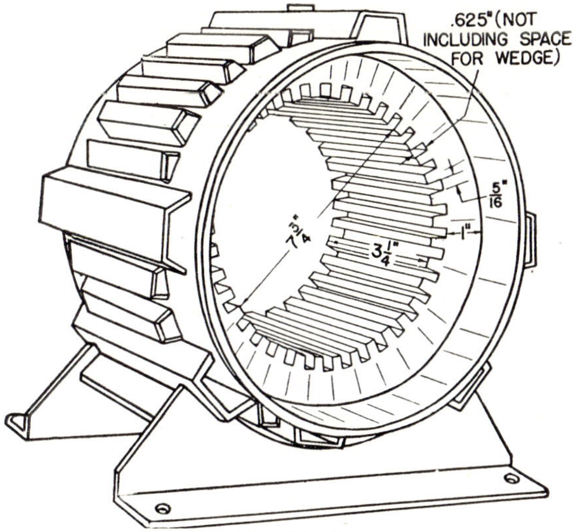

FIG. 16-10

Table 16-III shows that the flux density in the air-gap should not exceed 35,000 lines per square inch. The flux per pole in a stator is

$$\Phi = \frac{\pi D l B}{P}$$

TABLE 16-III

Parts	Lines Per square Inch
Air Gap	35000
Stator Core	90000
Stator Teeth	95000

where D = inside diameter of the core, l = net length of core, B = flux density in the air-gap, and P = the number of poles. Then, in this example,

$$\Phi = \frac{3.1416 \times 7.75 \times 3.25 \times 35000}{6} = 461{,}586 \text{ lines}$$

The flux determined is suitable for the air-gap, but it may establish too high a flux density in the core. Then the core flux density must be determined, and if it exceeds the value permitted in Table 16-III, the flux per pole should be reduced.

The stator core cross section is 3.25×1 or 3.25 square inches, and carries half the flux per pole, which is $461{,}586 \div 2$ or 230,793 lines. Then the flux density is $230{,}793 \div 3.25$ or 71,013 lines per square inch, which does not exceed the allowable value (see Table 16-III).

The voltage per phase is $E_p = 440 \div 1.73 = 254.34$. For full pitch, the coil span is $Y_s = 39 \div 6$ or 6.5. A pitch of six will be used, making K_c practically equal to one. In 15 groups there will be two coils per group and in three groups there will be three coils per group. Then $K_d = .966$.

Computing the number of turns per phase by means of equation 16-4,

$$T_p = \frac{10^8 \times 254.34}{4.44 \times 461586 \times 60 \times .966} = 214 \text{ turns}$$

Since there are two conductors per turn and three phases, then the total number of conductors is $214 \times 2 \times 3$ or 1284, and the number of conductors per slot is $1284 \div 39$ or 32.9. The nearest higher even number, which is 34, will be selected, giving an air-gap flux density slightly less than the maximum allowable value. An even number must be used so a top and a bottom coil side will contain an equal number of conductors, which in this case is 17. Hence, each coil will contain 17 turns.

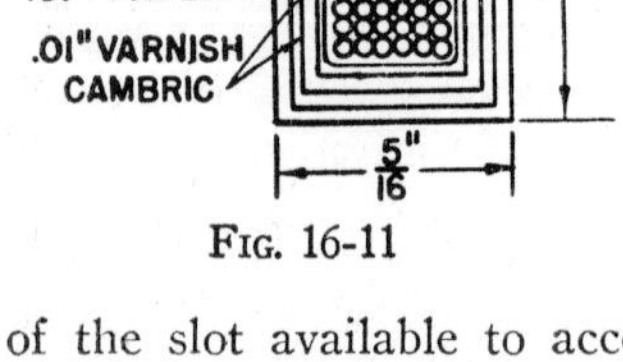

FIG. 16-11

Before the wire size can be computed, it is necessary to estimate the amount of space taken up by the coil and slot insulation. A typical insulation for a machine of this size and voltage is shown in fig. 16-11. The width of the slot available to accommodate the wire is $.3125 - 6 \times .01 - 4 \times .007 = .2245$ inch, and the depth is $.625 - 9 \times .01 - 8 \times .007 = .479$ inch. Then there is $.2245 \times .479$ or .1075 square inch of space available for 34 conductors, or the equivalent of 316.26 conductors per square inch. Table 3-IV shows that 374 turns of No. 17 gage enameled single cotton covered wire occupy one square inch. Consequently, 34 conductors of No. 17 gage wire can be placed in the available space with ease. Thus the number of turns of No. 17 enameled single cotton covered wire in each phase is

$$17 \times \frac{39}{3} = 221$$

Exercises

1. Why are the dashed lines in the single line diagrams, representing groups, placed at an angle?
2. What is meant by short jumpers? Long jumpers?
3. What is meant by a top to top connection? Top to bottom connection? In what type of winding is each found?
4. Determine the number of groups in an 8 pole, three-phase winding.
5. Compute the number of coils in a 4 pole, three-phase winding with five coils per group.
6. Sketch a single line diagram for a 6 pole, 72 coil, series star connected winding.
7. Show by another sketch how the winding of problem 6 is connected series delta.
8. Show by a single line diagram how to parallel one phase of an 8 pole, 48 coil, two-phase series winding.
9. Fill out the table given below.

No of poles	No of phase	Type of winding	No of coils	No of dead coils	No of large groups and coils per group	No of small groups and coils per group
(a) 2	2	series	13			
(b 2	3	series	16			
(c) 4	3	series	106			
(d) 6	3	2 parallel	136			
(e) 8	3	2 parallel	54			

10. What is the arrangement of the groups in problem 9?
11. A 440 volt, 4 pole motor has 480 turns of No. 18 wire per phase. If the motor is to be rewound for 220 volts, determine the number of turns per phase and the size of the wire.

12. Solve problem 11 considering that the voltage is not changed, but that the motor is rewound for six poles.
13. A 220 volt, 180 cycle, 4 pole motor has 12 turns per coil of No. 14 wire. If the motor is rewound for 110 volts, 60 cycles, and the same number of poles, determine the number of turns per coil and the size of the wire.
14. Solve problem 13 considering that the motor is rewound for 440 volts, 180 cycles, and six poles.
15. Determine the number of turns per coil and the size of the wire required for the stator core shown in fig. 16-10, if it is to be wound series delta for 60 cycles, 4 poles, and 440 volts.

CHAPTER 17

MULTI-SPEED INDUCTION MOTORS

SINCE the synchronous speed of an induction motor depends upon the supplied frequency and the number of poles (see equation 15-1), the only method of varying the speed is to change either or both of these factors. To change the speed by frequency control requires the provision of frequency changers, whereas altering the number of poles requires no additional equipment except the control panel. The former method will provide a fine speed variation similar in effect to that obtained by direct current motors, whereas the pole changing method will provide definite speeds corresponding to the number of poles selected. Squirrel cage motors with windings that may be connected for different numbers of poles offer the cheapest and simplest means of obtaining very definite speeds with little added equipment, and for this reason they are becoming very popular. These motors are manufactured for two, three, and four speeds, with constant horsepower, constant torque, and variable horsepower variable torque characteristics. Externally the appearance of these motors is the same as the standard three-phase induction motors; electrically they differ only in the stator winding.

Two speed motors with a single winding are in more general use because they require few leads and a very simple control. The control serves to change the connections of the stator winding, which causes the motor to run at two speeds, the high speed being approximately twice the low speed. That is, if the high speed is 1720 r p m, the low speed will be approximately 860 r p m. The two speeds are obtained by producing twice as many poles for the low speed operation as for the high speed operation. To understand clearly how this is accomplished, consider a single-phase motor with only two coils, connected as shown in fig. 17-1. In fig. 17-1a, terminal T_3 is left open and current enters T_1 and passes through the coils to T_2, thus producing one north pole and one south pole. With this connection the flux will pass from the north pole, downward through the rotor, through the south pole, and return to the north pole through the stator core. If this connection were used in a 60 cycle, single-phase motor, two poles would be produced and the synchronous speed would be 3600 r p m.

Now suppose that T_1 and T_2 are connected together and that current enters T_3 and comes out at T_1 and T_2. This gives a parallel connection with current flowing through the lower coil in the opposite direction, and as a result two north poles are produced. With this condition, the flux from the first north pole cannot return to the stator at the same place as before, because of the opposing magnetizing action of the second pole. Thus the fluxes emanating from the two north poles must return to the stator through the two regions between the coils; that is, at right angles to the axis of the coils, as indicated in

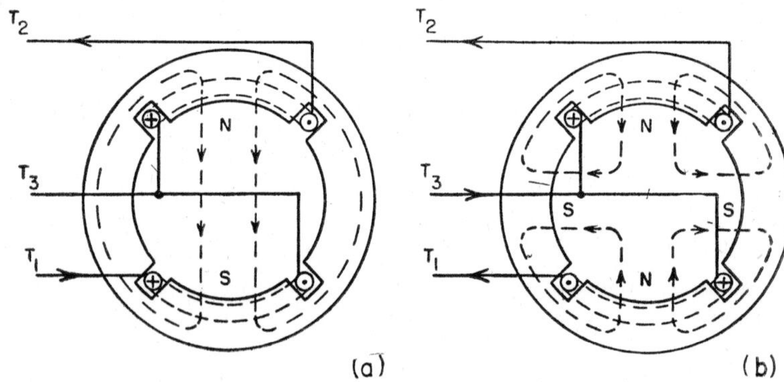

FIG. 17-1

fig. 17-1b. Since the region where the flux passes from the stator to the rotor is a north pole, and where the flux passes from the rotor to the stator is a south pole, there are two north poles produced by the coils and two south poles, called consequent poles, produced between the north poles, or a total of four poles. *From this discussion it should be concluded that the number of poles is doubled by reversing the current through half a phase.*

A more precise explanation of this action is given in connection with fig. 17-2. Since induction motors are usually wound with a distributed winding, the flux density of the magnetic poles varies as a sine wave, as suggested in fig. 17-2a. The north poles are indicated inside of the air gap and the south poles are indicated outside the air gap. These poles advance around the inner-periphery of the stator, forming a rotating magnetic field. For a 60 cycle frequency this field would rotate at 1200 r p m.

In fig. 17-2b the north poles are the same as before but the south poles are reversed, thus forming north poles. Since the magnetic paths formed must be closed circuits, the magnetic flux that leaves the north poles must return to the stator, and in so doing will form the south poles. In other words, the same number of lines that leave the stator must return to it, or in fig. 17-2b the areas outside the air gap, which indicate the south poles, must be equal to the areas inside the air gap, which indicate the north poles. Thus fig. 17-2b contains six salient poles and fig. 17-2b contains 12 poles, six salient and six consequent.

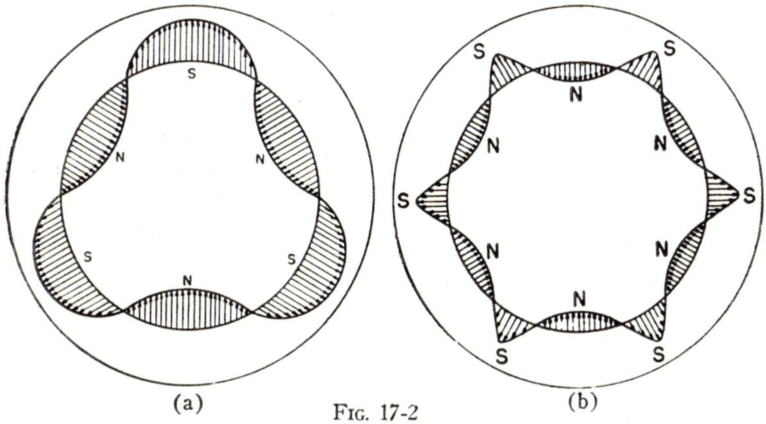

FIG. 17-2

For a 60 cycle frequency the magnetic field would rotate at 600 r p m, or half as fast as the field of fig. 17-2a.

Multi-speed motors are designed to give different operating characteristics. Some are designed to give the same maximum horsepower at all speeds. For this reason they are known as constant horsepower motors. In these the torque varies inversely as the speed, which is evident from equation 2-25 expressed as follows:

$$T = \frac{5252 \text{ HP}}{N}. \qquad (17\text{-}1)$$

Others are designed to give the same maximum torque at all speeds. These are known as constant torque motors. Their horsepower varies directly as the speed as is evident from equation 2-25. Still others are designed to give an increase in both horsepower and torque with an increase in speed. These are known as variable torque motors.

Constant Horsepower Connections—Fig. 17-3a shows the method of connecting one phase of a three-phase, four pole, constant horsepower induction motor. In this winding, as well as in all other windings for multi-speed motors, long jumpers are used. The four groups are connected so that when T_3 is left open, the current enters T_1 and passes through the groups to T_2, two north and two south poles being produced. This gives four salient poles, which will combine with an equal number of poles in each of the other phases to produce a four pole rotating magnetic field, which rotates at 1800 r p m. For the lower speed, the current enters T_3 and comes out at T_1 and T_2, which are connected together to give a parallel connection and a reversal of current in groups 7 and 1 (see fig. 17-3b). Four north poles are then produced, which establish four consequent south poles, as explained above, and give a total of eight poles. These combine with an equal

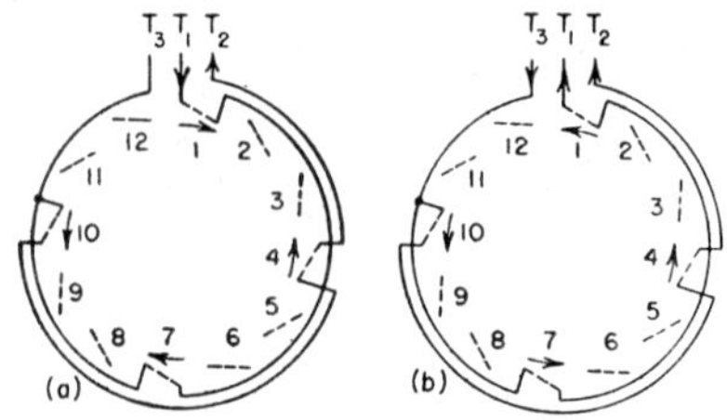

FIG. 17-3

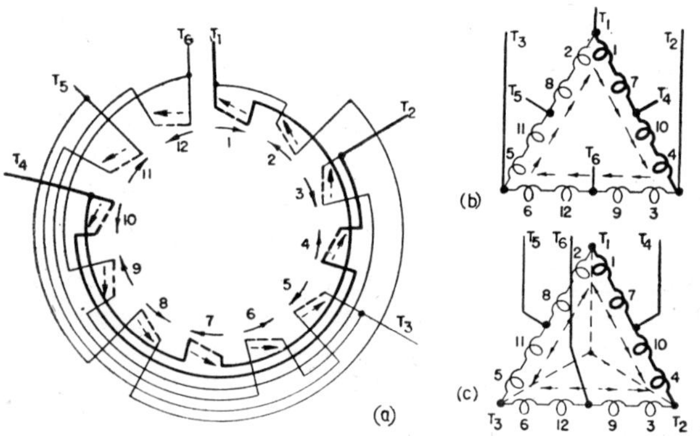

FIG. 17-4

number of poles in each of the other phases, producing an eight pole rotating magnetic field which rotates at 900 r p m.

Fig. 17-4a shows the connections of the phases in a constant horsepower two speed motor. Six leads are brought out. When the line wires are connected to leads T_1, T_2, and T_3 (see figs. 17-4a and 17-4b), a series delta connection is formed, and the current through it produces a four pole rotating field. When the line wires are connected to leads T_4, T_5, and T_6, and leads T_1, T_2 and T_3 are connected together, a parallel star connection is formed. This gives a reversal of current in half of each phase, and an eight pole rotating field is produced.

Motors having constant horsepower characteristics are used to drive certain types of machine tools and some classes of woodworking machines. Since most machine tools are to be operated at wide speed ranges, heavy cuts are taken at low speeds and light cuts are taken at

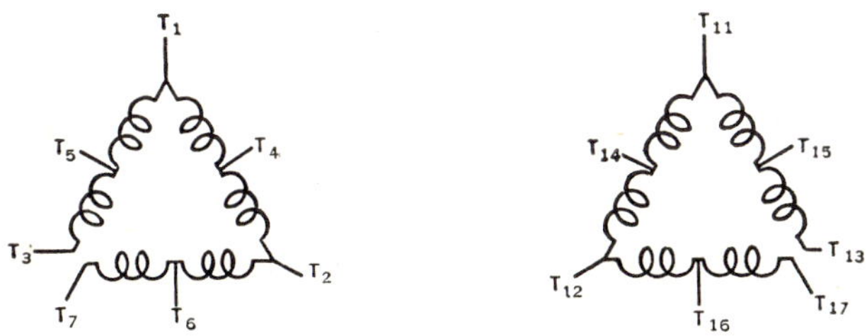

Speed	To line wires L_1-L_2-L_3 Connect respectively	Connect together
Low	T_4-T_5-T_6	T_1-T_2-T_3T_7
Second	T_{14}-T_{15}-T_{16}	T_{11}-T_{12}-$T_{13}T_{17}$
Third	T_1-T_2-T_3T_7	
High	T_{11}-T_{12}-$T_{13}T_{17}$	

FIG. 17-5

high speeds. In the case of rotating machinery, such as lathes and milling machines, the same cutting speed is desired for different diameters of work. For stock with a small diameter, the speed should be high and the torque required will be low. For stock with a large diameter, the speed should be low and the torque required will be high. In each case the horsepower required will be practically the same.

Fig. 17-5 shows the connection diagrams for a four speed, double winding motor. The windings have approximately the same horsepower capacity at all speeds; one winding gives four and eight poles (1800 and 900 r p m) and the other winding gives six and twelve poles (1200 and 600 r p m). It will be noted that the delta connections are open at one point. Since only one winding is used at a time, the other winding is left open to prevent circulating current in it.

Chording—When the connection of a stator is changed from salient to consequent poles, the pole width changes and therefore the chord factor also changes. That is, a stator connected for full pitch

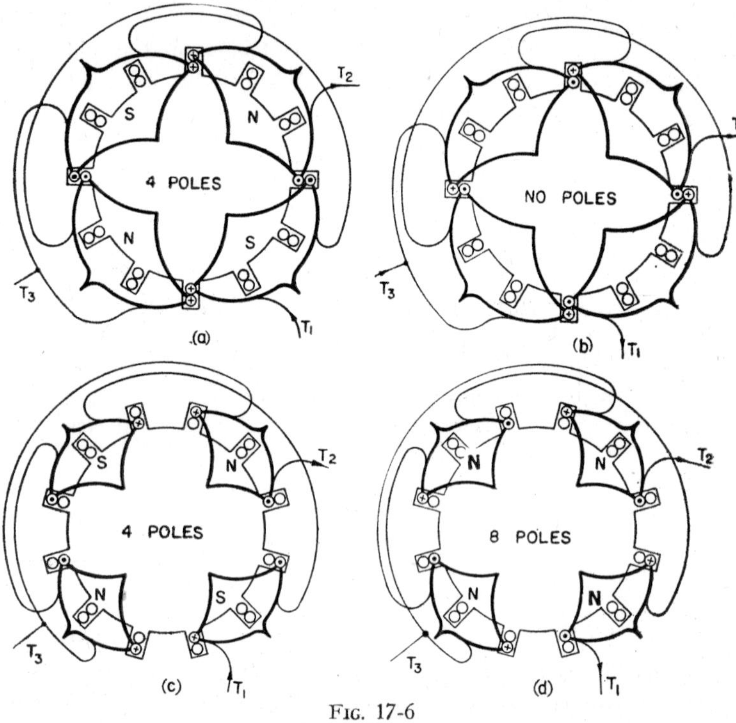

FIG. 17-6

for the high number of poles will be half pitch for the low number of poles. This pitch usually produces the best torque and flux conditions. When the pitch is greater than 80% for the salient pole connection, little or no power is developed for the consequent pole connection. Fig. 17-6, which shows each stator core containing 12 slots, makes this clear. For four salient poles, the full pitch is $12 \div 4 = 3$ slots (or 1 to 4). Figs. 17-6a and 17-6b show the connection of one phase. Note that the bottom half of coil 1 and the top half of coil 2 are in the same slot. Likewise, one side of coil 2 and one side of coil 3 fall in the same slot, etc. When the current flows from T_1 to T_2 (fig. 17-6a), the currents through the two coil sides in each slot flow in the same direction. Thus their magnetizing action is additive and four poles are produced. When the current is sent from T_3 to T_1 and T_2 (fig. 17-6b), the currents through the two coil sides in each slot are in opposite directions, and therefore no magnetizing action is produced.

Figs. 17-6c and 17-6d show the winding with a pitch of 2, which gives 2/3 of full pitch for four poles and 4/3 of full pitch for eight poles. In both cases the chord factor is .866. When current passes from T_1 to T_2 (fig. 17-6c), four poles are produced. Unlike the condition in fig. 17-6b, when current passes from T_3 to T_1 and T_2, no slots house conductors that carry current in opposite directions, and as a result the magnetizing action produces eight poles.

Constant Torque Connections—The manner in which groups are connected in a constant torque induction motor is shown in fig.

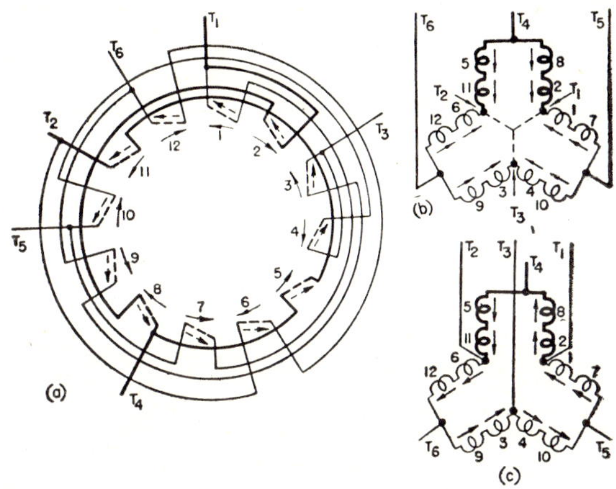

FIG. 17-7

17-7a. *These connections differ from those shown in fig.* 17-4 *in that half of each phase is reversed.* Hence, when the winding is connected as shown in fig. 17-7b, four poles are produced, giving a synchronous speed of 1800 r p m. When connected as shown in fig. 17-7c, eight poles are produced, giving a synchronous speed of 900 r p m. In a constant torque induction motor, then, the phase connections are made in the reverse order to the connections for a constant horsepower motor. *That is, in a constant horsepower motor the phases are connected series delta for high speed and parallel star for low speed, but in a constant torque motor they are connected parallel star for high speed and series delta for low speed.*

Constant torque motors are suitable for driving loads such as in stokers, conveyors, tumblers, reciprocating pumps, air compressors, constant pressure blowers, etc. To clarify the reason for utilizing a constant torque motor in these applications, consider a coal stoker driven by a two speed motor (say with synchronous speeds of 1800 and 900 r p m).

When considerable steam is required, the motor feeds coal into the furnace at high speed, running at a synchronous speed of 1800 r p m. When less steam is required, the motor runs at its lower speed, a synchronous speed of 900 r p m. Suppose the motor requires 3 H P to drive it at its high speed. What horsepower will then be required to drive the load at half speed? To answer this it is necessary to determine how the motor load varies with a change in speed. Obviously the volume and weight of coal handled per revolution of the motor is exactly the same regardless of whether the motor is running at high or low speed. In other words, the torque required to drive the motor is the same, irrespective of speed, and the motor should therefore have constant torque characteristics. From equation 17-1 the horsepower that the motor must supply varies directly as the speed. At half speed, then, 1.5 horsepower will be required.

It has been suggested that conveyors should also be driven with constant torque motors. This is due to the fact that for a given load the torque required to move a conveyor is the same, regardless of the speed, and the required horsepower therefore varies directly as the speed.

In the two above industrial applications, as well as in many similar ones, multi-speed motors are usually more suitable than slip ring motors, because they have constant speed characteristics or good speed regulation at each speed. The speed falls off very slightly with increasing torque requirements, because multi-speed motors behave like standard single speed squirrel cage motors.

Fig. 17-8 shows the connection diagram for a four speed, constant torque motor. The windings have approximately the same torque

Speed	To line wires L_1-L_2-L_3 Connect respectively	Connect together
Low	T_1-T_2-T_3T_7	
Second	T_{11}-T_{12}-$T_{13}T_{17}$	
Third	T_4-T_5-T_6	T_1-T_2-T_3T_7
High	T_{14}-T_{15}-T_{16}	T_{11}-T_{12}-$T_{13}T_{17}$

FIG. 17-8

capacity at all speeds. One winding gives four and eight poles and the other gives six and twelve poles.

Variable Torque Motors—Fans and blowers that must be operated at different speeds may be driven by slip ring induction motors. When so driven, the reduced speeds are obtained by increasing the external resistance, as explained in chapter 15. Such a method of lowering the speed is undesirable, however, since a portion of the electrical energy input to the rotor is converted into heat at the external resistance connected to the rotor. This electrical energy which is converted into heat supplies no mechanical power, but decreases the efficiency of the motor as the speed is reduced. On fans, blowers, and similar installations, reduced speeds are required for long periods. Thus the operating cost will be high if slip ring motors are used. One should consider whether the number of speeds available on a multi-speed motor will suffice, and if so, multi-speed motors should be used instead of slip ring motors.

Fans and blowers require an increase in both torque and horse-power when the speed is increased. The connections employed on a motor to give such characteristics are shown in fig. 17-9a. The two speeds are obtained by changing from parallel star to series star. Fig. 17-9b shows the winding connected for salient poles, giving four poles and a synchronous speed of 1800 r p m. Fig. 17-9c shows the winding connected for consequent poles, giving a synchronous speed of 900 r p m.

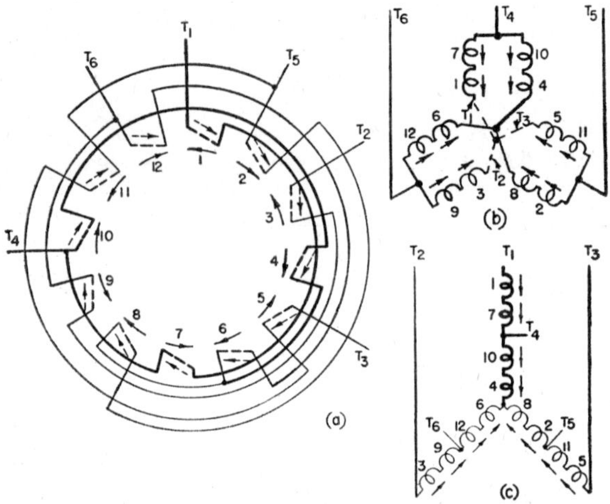

FIG. 17-9

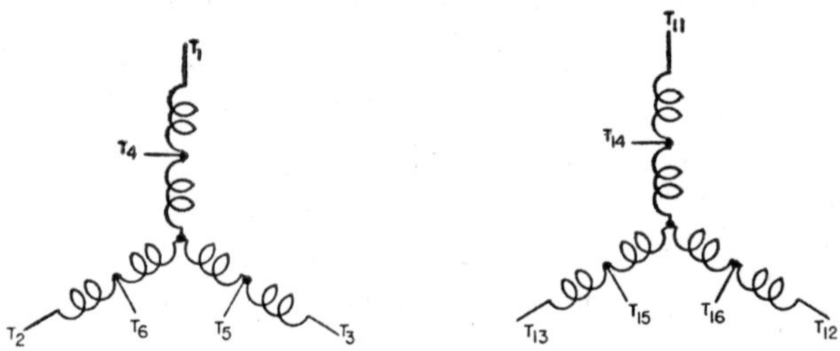

Speed	To line wires L_1-L_2-L_3 Connect respectively	Connect together
Low	T_1-T_2-T_3	
Second	T_{11}-T_{12}-T_{13}	
Third	T_4-T_5-T_6	T_1-T_2-T_3
High	T_{14}-T_{15}-T_{16}	T_{11}-T_{12}-T_{13}

FIG. 17-10

Fig. 17-10 shows the connection diagram for a four-speed, variable torque motor. As in the four-speed motors mentioned, each winding gives two speeds in a 2 to 1 ratio.

Multi-Speed Synchronous Motors—Multi-speed synchronous motors also have several important industrial applications, as in driving water pumps, blowers, air compressors, and similar machines requiring infrequent starting and stopping. Besides having constant speed characteristics for each of the selected speeds, these motors provide means of adjusting the power factor, as do single speed synchronous motors. Multi-speed synchronous motors are manufactured for only two-speed operation. Machines made for a higher number of speeds would be too complicated, as will be evident later.

Motors designed for two speeds, with a ratio of 2 to 1, have a single stator and a single rotor, each with a winding that can be connected for either speed. The stator winding is generally the same as that on a variable torque, two-speed induction motor. The field has the same number of poles as the winding for the lower speed, and the coils are connected as shown in fig. 17-11. Two sets of slip rings are required.

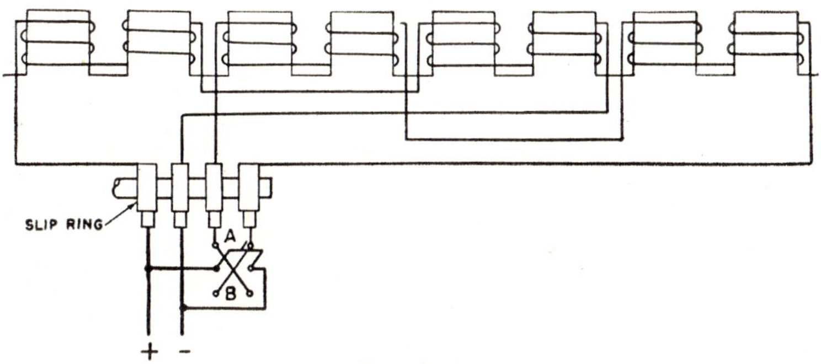

FIG. 17-11

When it is desired to run a motor at low speed, the stator winding is connected for the high number of poles (say eight). The field switch is closed at A (fig. 17-11), which establishes poles of alternative polarities, the same number as on the stator.

For high speed, the stator is reconnected for four poles. The field switch is closed at B, reversing the polarities of half the poles. This establishes two adjacent poles of like polarity, the next two of opposite polarity, and so on. In effect, this gives the same number of poles as there are on the stator.

When two speeds are desired in a ratio other than 2 to 1, two single speed motors are usually mounted on the same shaft. They are placed in a single frame or in separate frames mounted on a common bed plate. Only one motor is energized at a time.

Exercises

1. Distinguish between salient and consequent poles.
2. Are long or short jumpers used in two speed motors? Can either be used?
3. What is done to change the speed in multi-speed motors?
4. Why cannot a full pitch salient pole winding be connected for consequent poles?
5. Suppose that the winding of a two speed induction motor is full pitch at low speed. What is the pitch at high speed? What is the chord factor?
6. Discuss the operating characteristics of a constant horsepower induction motor.
7. Give two industrial applications for which each of the following multi-speed motors are suited: (a) constant torque induction motor; (b) variable torque induction motor; (c) synchronous motor.
8. What changes are made in the connections of the motors in problem 7 in order to change from the low to the high speed?
9. Describe the two types of synchronous motors.

CHAPTER 18

A C CONTROLLERS

Across-the-Line Starters—If an induction motor is started by connecting it directly across the line, it will have a higher starting torque than if started at a reduced voltage, and consequently the machine will come up to speed rapidly. The current drawn at starting will range from 4 to 6 times the full load current. Such a high current will not injure the motor because modern machines are strongly constructed, and unlike direct current motors, have neither

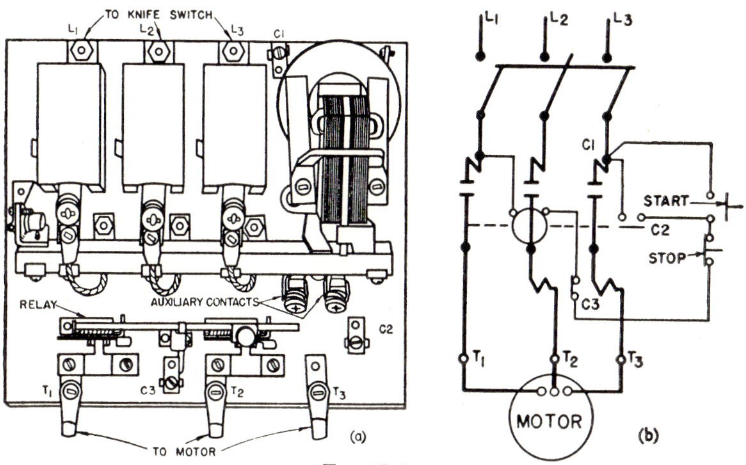

Fig. 18-1

commutator nor brushes to spark excessively when the current is high. If an induction motor is exceedingly large, however, the high starting current may cause too great a voltage fluctuation in the power line or may impose too great a stress on the driven machinery. Under these conditions, the starting voltage should be reduced during the starting period.

A motor is connected directly to the line by means of a so-called across-the-line starter, which is the simplest and therefore the most economical type. This starter is shown in fig. 18-1a, and consists of a three pole knife switch (not shown), a three pole magnetically operated contactor, and two overload relays. Its wiring diagram is shown in fig. 18-1b, in which a three wire push button control is used.

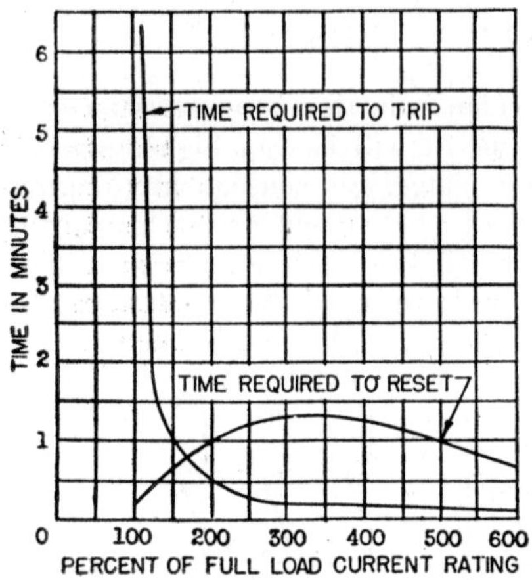

Fig. 18-2

Many other control systems, as explained in Chapter 8, may of course be substituted for the push button switch shown in fig. 18-1b.

In fig. 18-1, the auxiliary contacts serve to maintain the coil circuit after the start button is released. With this hookup, the contacts open with voltage failure. The relays protect the motor against excessive momentary overloads, normal overloads existing for long periods, or high currents due to an open phase, by opening the coil circuit and thus allowing the contactor to drop open and interrupt the power. Such relays have approximately the characteristics shown in fig. 18-2. The curve shows that for about 10 seconds the relay allows the starter to carry 500 per cent of the full load rating, but that for 150 per cent of the full load rating the relay will interrupt the power to the motor in one minute.

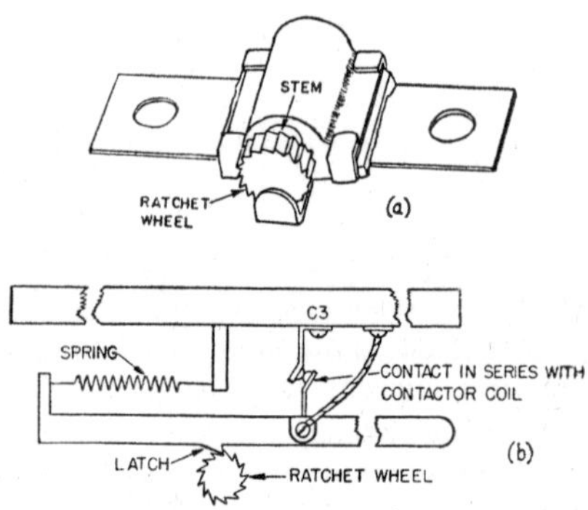

Fig. 18-3

The relays on the starter shown in fig. 18-1a each consist of a mica-insulated nichrome heating element placed adjacent to a cylinder containing a low-melting alloy and the stem of a ratchet wheel (fig. 18-3a). The relay is set by placing its spring under tension, and it is held this way by a latch engaging in the teeth of the wheel (fig. 18-3b). When an overload persists, the heat developed in the relay melts the alloy. The spring then turns the ratchet wheel and in contracting opens the contact in series with the contactor coil. The relay cannot be reset until the alloy hardens, and the time required for it to harden, which is

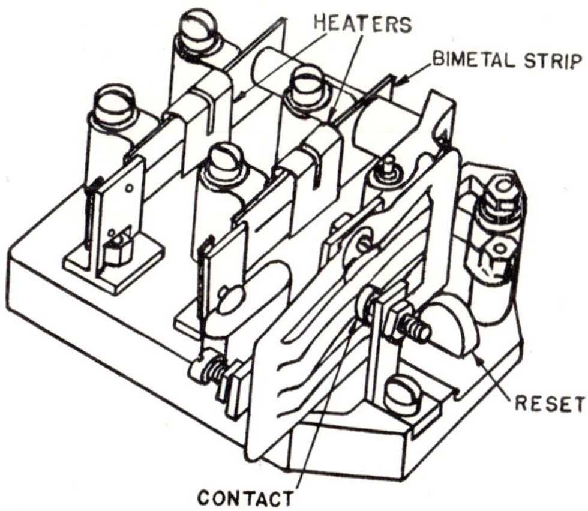

FIG. 18-4

indicated by a curve in fig. 18-2, depends upon the amount of overload which caused it to melt.

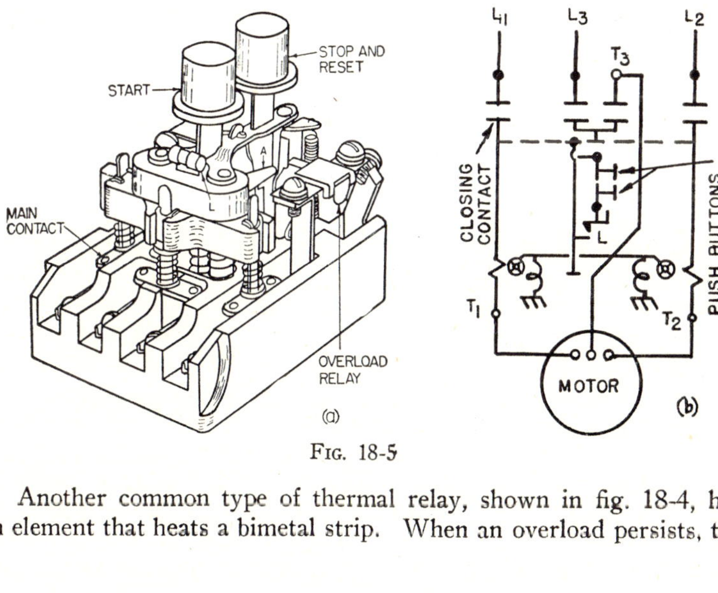

FIG. 18-5

Another common type of thermal relay, shown in fig. 18-4, has an element that heats a bimetal strip. When an overload persists, the

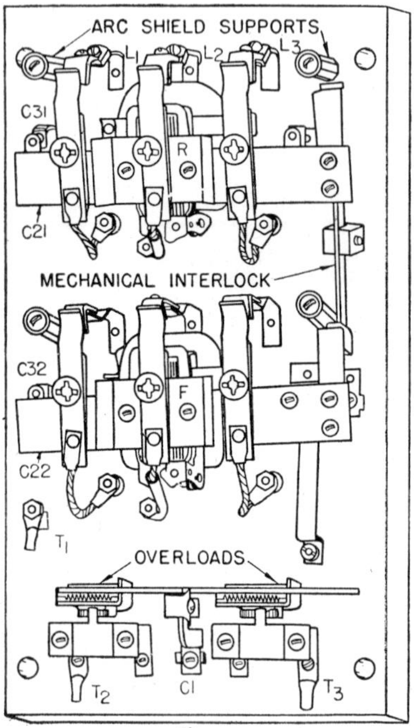

(a)

heat developed in the relay causes the bimetal to bend and open the contact in series with the contactor coil.

Fuses cannot be used instead of thermal overload relays, because if their capacity is low enough to protect the motor during the running period, they will blow when the motor is started. This will happen because fuses blow at about 50 per cent overload, and the starting current is greater than this. If fuses are large enough to carry the starting current, they will render no overload protection.

Another type of across-the-line starter, for motors not exceeding three horsepower, is shown in fig. 18-5a. Pressing the start button closes the contacts, which are held closed by

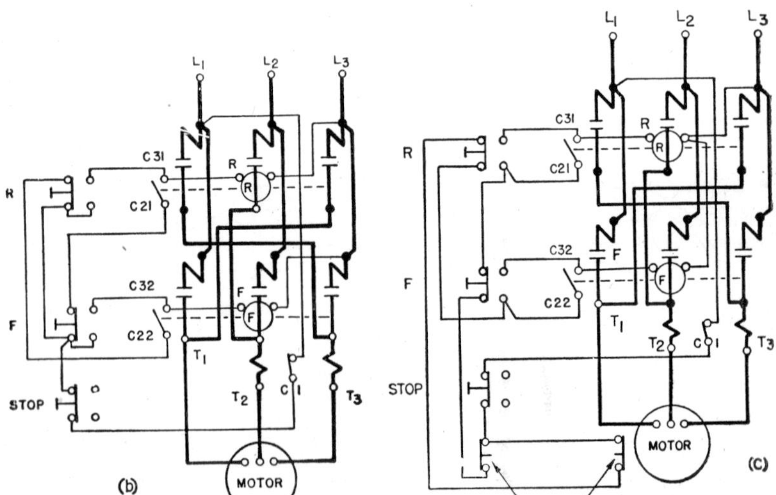

Fig. 18-6

latch L. This connects the motor to the power line as indicated in fig. 18-5b. The contacts may be opened by pressing the stop button, which trips latch L through arm A, or they may be opened by an overload actuating a relay, which likewise trips latch L through arm A. As soon as the alloy in the relay cools, it may be reset by pressing the stop button. Hence the stop button serves to stop the motor and also to reset the overloads.

Reversing Controllers—In some cases it is essential to repeatedly reverse the direction of rotation of a motor. This is accomplished, in a three-phase induction motor, by a controller which interchanges the connections of two of the three line leads to the motor.

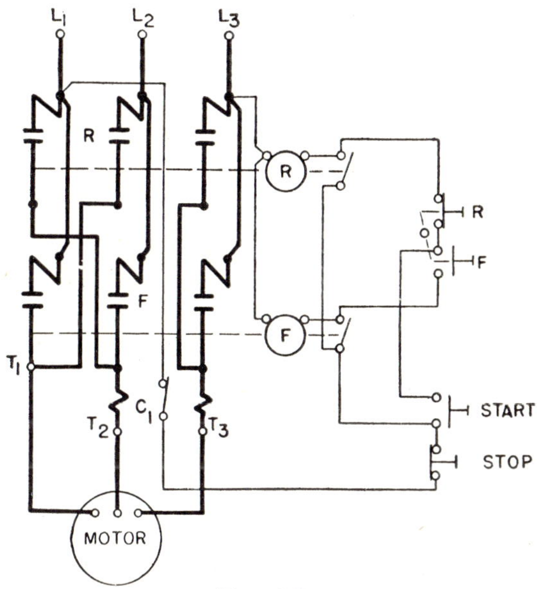

Fig. 18-7

Such a controller is shown in fig. 18-6a, and its wiring diagram is shown in fig. 18-6b. One button is provided for each direction of rotation, and only one stop button is necessary.

In fig. 18-6b, when the forward button (F) is pressed, coil F is connected between L_1 and L_3. Contacts F thus close, connecting the motor leads to corresponding line leads, and the motor runs in the forward direction. After the start button has been released, the circuit of coil F is maintained through its auxiliary contact.

If the reverse button (R) is pressed, the circuit of coil F is opened and the circuit of coil R is closed. Contacts F therefore open and contacts R close. This interchanges the connections of line wires L_1 and L_3 to the motor terminals, and the motor runs in the reverse direction. The circuit of coil R is also maintained through its auxiliary contact.

The common method of connecting the push buttons when two limit switches are included is shown in fig. 18-6c. Another common hookup for a reversing controller is shown in fig. 18-7. This hookup is used where it is desired to frequently start and stop a motor which normally runs in one direction.

Drum Controllers—Drum controllers are manufactured for across-the-line starting and reversing of squirrel cage motors. A simplified diagram of such a controller is shown in fig. 18-8. The motor is started by moving the handle from the "off" to the "forward" position, which connects T_1 to L_1, T_2 to L_2, and T_3 to L_3. To reverse the motor, the handle is thrown in the reverse direction. This connects T_1 to L_1 as before, but interchanges the connection of T_2 and T_3 to L_2 and L_3. When the handle is brought to the "off" position, the motor is disconnected from the line and therefore stops.

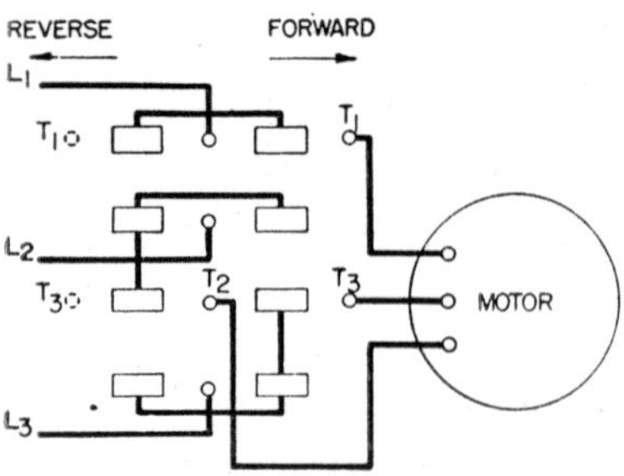

FIG. 18-8

Controllers for High Torque Starting—If a motor, equipped with a starter of the type shown in fig. 18-1a, has an extremely heavy load at starting, the overload relays may operate before the motor reaches maximum speed. This will not occur, however, if the starter shown in fig. 18-9 is used.

In this starter, coils M and N are energized when the start button is pressed, closing both sets of contactors. Most of the current will pass through contacts M because they are in paths of no resistance, since these paths contain no heating elements. A low current will therefore pass through the relays and they will not operate.

When the speed of the motor becomes constant, the start button is released. This deenergizes coil M, allowing contacts M to open.

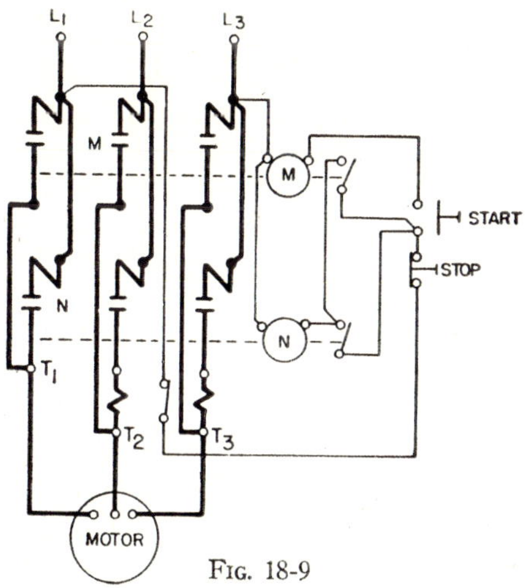

FIG. 18-9

After these open, all the current will pass through the relays and they will protect the motor against overloads.

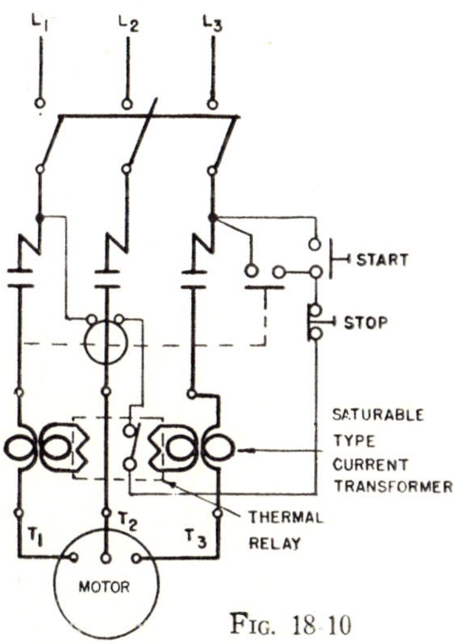

FIG. 18-10

Fig. 18-10 shows a controller which has the same industrial application as the controller in fig. 18-9. It is similar to the controller shown in fig. 18-1a except that the relays are connected to the line wires through two saturable core type current transformers. When the current to the motor in fig. 18-10 does not exceed 150 per cent of the full load, the current in the secondary, and therefore the current through the relays, is about the same as that through the motor. When the motor current exceeds 150 per cent of the full load current, which it does during the starting period, the transformer core becomes saturated, and, as a result, the relay current increases only slightly above 150 per cent. With this value of current through the relays, they will not operate unless the motor current remains above 150 per cent for more than about one minute (see fig. 18-2). This will not take place with a normal heavy load at starting, and therefore the motor will accelerate to full speed without the relays operating. After this, the current in the relays will be low and they will not operate unless an overload occurs and persists for some time.

Synchronous Motor Controller—Fig. 18-11 shows a controller for starting a synchronous motor directly across the line. The circuit of coil M is completed when the start button is pressed, closing the main contactor M, which connects the motor terminals to the line. After the start button is released, the circuit is maintained by the auxiliary contact 1. Current is then supplied to the stator, producing a rotating magnetic field. This field establishes a current in the squirrel cage bars and the field winding, which is closed through the discharge resistance (R) and coil C_1. The current in these two windings is effective in producing a starting torque.

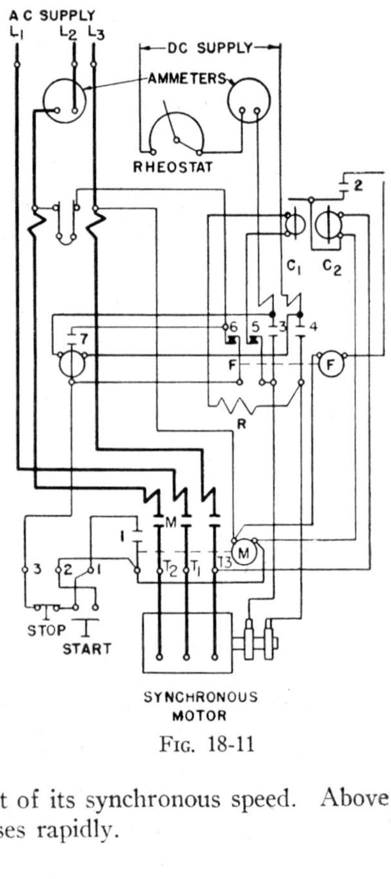

Fig. 18-11

When the motor is just starting, a high voltage is induced in the field at a frequency which is just the same as that in the stator. As the motor gains speed, the induced voltage and frequency decrease. The reactance of the circuit decreases with the frequency, and, as a result, the field current remains practically constant until the rotor reaches about 75 per cent of its synchronous speed. Above this point the field current decreases rapidly.

The field current flows through C_1, and while the current in this coil is constant, it produces sufficient force to hold contact 2 open. The current through coil C_2, which is connected across the line when contact 1 closes, establishes a constant force which tends to close contact 2. When the current through coil C_1 starts to decrease, which takes place only after the rotor reaches about 75 per cent of synchronous speed, coil C_2 closes contact 2, energizing coil F. This coil closes contacts 3 and 4 and opens contacts 5 and 6. The closing of contacts 3 and 4 energizes the field with direct current, completing the starting period of the motor. On opening, contact 5 disconnects the discharge resistance. With contact 6 open, contact 7 will disconnect the motor from the line if the D C voltage fails.

The synchronous motor is stopped merely by pressing the stop button, which allows the contactors to return to their normal position.

Another type of field relay has its two coils connected across the field terminals, one coil having a high resistance and a low inductance and the other a low resistance and a high inductance. When the motor is just starting, a high current flows through the high resistance coil and a low current flows through the high inductance coil, because of the high frequency. As the motor gains speed, the current through the resistance coil decreases, because the voltage decreases, and the current through the high inductance coil remains nearly constant, because both the voltage and the frequency decrease. Then, just before the speed of the rotor becomes constant, the relay contact closes and the coil circuit of the field contactor is completed. On closing, this contactor energizes the field.

Friction Brake—To bring a motor to a standstill quickly, a friction brake which is operated by a shunt coil, may be used. The coil, connected between two motor leads, and the motor are energized simultaneously, the current through the coil spreading the brake shoes so that the rotor can turn. When the power to the motor is interrupted, the brake coil is also deenergized. The brake spring then brings the shoes against the pulley, quickly stopping the rotor.

Dynamic Braking—During the study of direct current controllers, we learned that if a rotating armature is disconnected from the line and short-circuited through a resistance, while the field remains energized, the machine will become a generator and establish a current in the armature. This current and field flux develop a torque which opposes rotation, and can be used to stop a motor. Stopping a motor in this way, known as dynamic braking, is the most common method of quickly stopping a direct current motor.

Dynamic braking can also be used to stop an induction motor, but in this case the circuit is as shown in fig. 18-12. When the stop button is pressed, the motor is disconnected from the power line and the stator is excited with direct current, which establishes a stationary magnetic field. As the rotor with short-circuited bars revolves

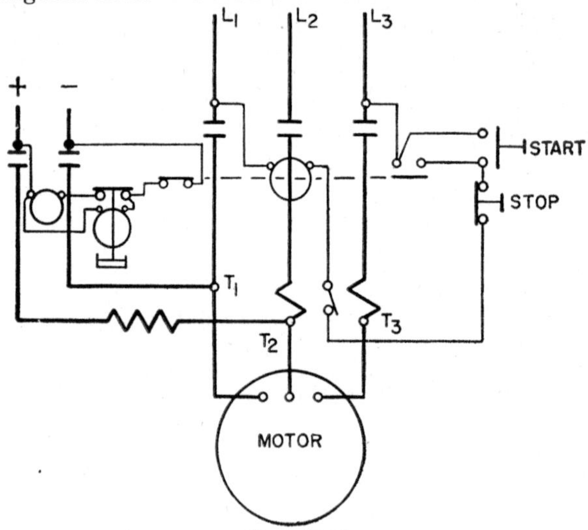

Fig. 18-12

through this field, a high rotor current is established. This acts with the stationary field to produce a counter torque that quickly stops the motor.

Plugging—If the line connections of any two leads from an induction motor are reversed, a counter torque will be developed which will cause the motor to quickly come to a standstill and then start in the opposite direction. If the power is interrupted at the instant the motor starts in the opposite direction, the rotor will turn slightly and then stop. The number of revolutions that the motor will make in the reverse direction varies inversely with the load, but is generally less than one. This method of quickly bringing a machine to a standstill, known as plugging, is much used in practice.

Plugging is accomplished with the use of a reversing controller (fig. 18-6a) and a plugging relay (fig. 18-13a). The plugging relay is generally belt driven from an auxiliary pulley on the shaft of the motor. The hookup generally used is shown in fig. 18-13b.

When the start button is pressed (fig. 18-13b), coil F is energized, closing contacts F and auxiliary contact 1, and opening contact 2. The motor comes up to speed, supplies mechanical power, and drives disk D (fig. 18-13a), upon which the two brushes B rest. Friction between the disk and the brushes causes the relay to make contact at C (figs. 18-13a and 18-13b). After the start button is

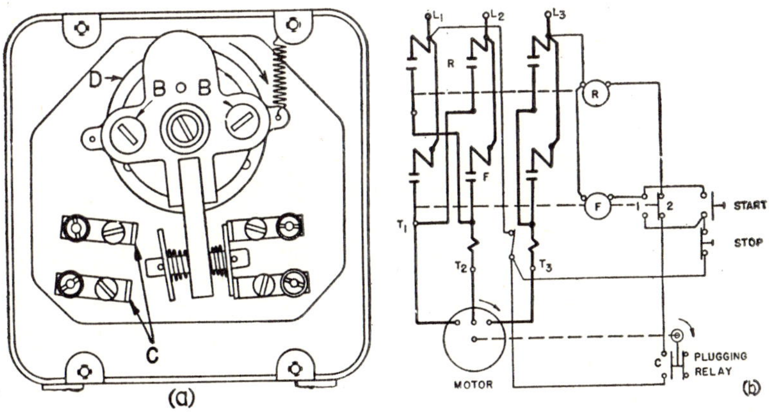

Fig. 18-13

released, the circuit of coil F is retained through contact 1 and the stop button, so that the motor continues to drive the load.

When the stop button is pressed, coil F is deenergized and contacts F open and auxiliary contact 2 closes. Since the relay is closed at C, the circuit of coil R is complete and contacts R close. This reverses the lead connections to the line and causes a counter torque to be developed, which brings the motor to a quick stop and starts it in the opposite direction. At the instant that the motor reverses contact C opens, coil R is deenergized and contacts R therefore open and interrupt the power supply. Having just started in the opposite direction, the motor barely turns before coming to a standstill.

Fig. 18-14 shows the method of making the connections if the motor is to be run in both directions. A limit switch (L) is shown, but a push button can be used instead. Coil F is energized when button F is pressed, closing contacts F and opening auxiliary contact 1. The motor then starts, supplying mechanical power and closing contact f in the plugging relay. The motor runs until the limit switch operates, causing contacts F to open and contact 1 to close. This completes the circuit of coil R and contacts R close. The motor stops quickly and then starts in the opposite direction. Starting in this direction, the motor opens contact f and coil R is deenergized. Contacts R then open, interrupting the power to the motor.

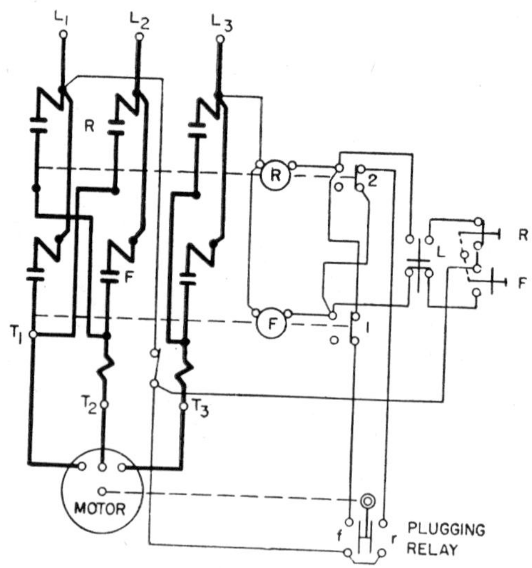

Fig. 18-14

Coil R becomes energized when button R is pressed, closing contacts R and opening auxiliary contact 2. The motor will then run in the reverse direction, supplying power and closing contact r. When the limit switch again operates, contacts R open and contact 2 closes. Coil F is therefore energized, closing contacts F. Then a counter torque is again developed, which quickly stops the motor. On starting in the opposite direction, the motor opens contact r, and thus the power is interrupted.

In the process of retooling a machine driven by a motor that is plugged, the motor shaft is sometimes turned slightly so as to close contact C (fig. 18-13b). This completes the circuit of coil R, closing contacts R (fig. 18-13b). As a result, the motor receives power, but only for an instant, because just as soon as the motor starts, contact C is opened. However, the sudden starting of the motor may injure the machinist. Then, too, during the plugging period, the motor turns in the reverse direction after coming to a standstill. This is sometimes

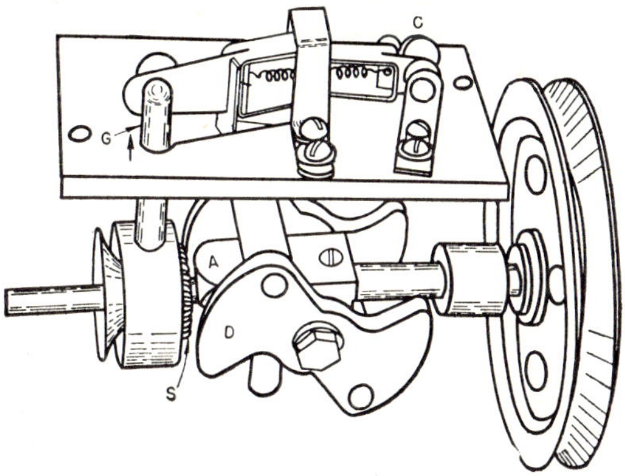

FIG. 18-15

undesirable. These actions may be eliminated by using a centrifugal type of relay (shown in fig. 18-15). When the relay is running above a certain low speed, the centrifugal device D forces A against spring S. Friction is established between S and its contacting surface, which causes the rotating shaft to move finger G in the direction indicated. This finger closes contacts C (which are connected to the controller the same as the contacts of the relay shown in fig. 18-13b) and holds them closed while the motor is running. After the stop button is pressed, the motor decelerates rapidly because contact C remains closed until the motor is nearly at a standstill. As soon as the speed becomes nearly zero, the centrifugal device removes the pressure on spring S and contacts C open, interrupting the power to the motor.

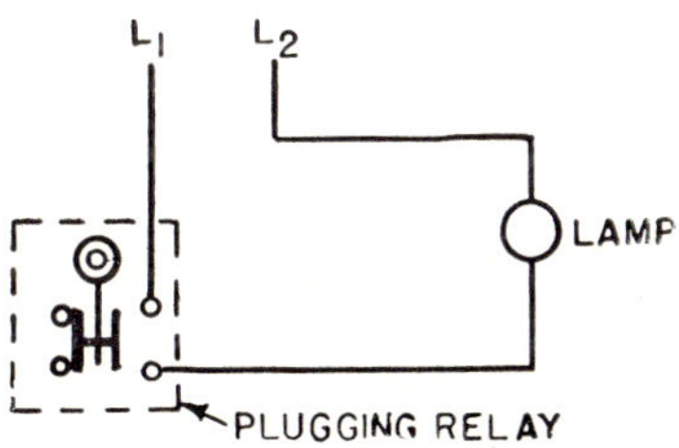

FIG. 18-16

Plugging relays have many other important applications. Fig. 18 16 shows a relay used to operate a pilot light. The relay is driven by a roller upon which a conveyor rides As long as the conveyor moves in the normal direction, the contact is held closed, completing the pilot light circuit. If the belt stops, the contact opens and the

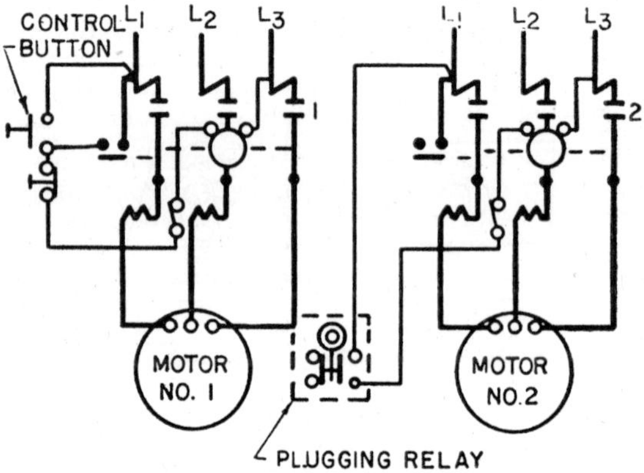

Fig. 18-17

light goes out. Hence, the light indicates whether the conveyor is moving or not.

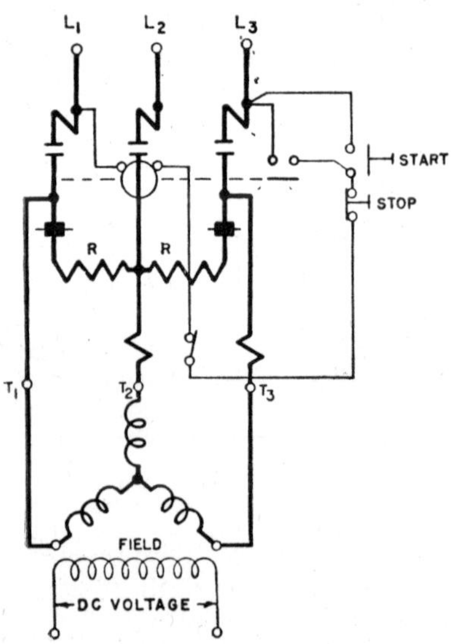

Fig. 18-18

Plugging relays are also used to operate motors in sequence. The hookup generally used is shown in fig. 18-17 (only two motors are considered).

Synchronous motor controls may also be equipped for dynamic braking, this generally being accomplished by means of a circuit similar to the one shown in fig. 18-18. When the stop button is pressed, the contactors disconnect the motor leads from the supply line and then short-circuit the motor through resistances (R), while the field remains energized.

Synchronous motors may also be stopped quickly by plugging, which is accomplished after the field is deenergized.

Star-Delta Starting—A reduced voltage may be supplied to a motor and then raised to normal merely by changing the motor terminal connections. This is accomplished by a so-called star-delta starter, one type of which is shown in fig. 18-19. When the start button is pressed, the main contacts (M) close and coil S and relay R become energized. Coil S closes contacts S, which connect the motor star, and as a result, 58 per cent of the line voltage is impressed on each phase. After a definite time, relay R opens the circuit of coil S and completes the circuit of coil D. Then contacts S open and contacts D close, changing the motor connections from star to delta. Full voltage is then impressed on each phase. Motors started in this way must have the leads of each phase brought out.

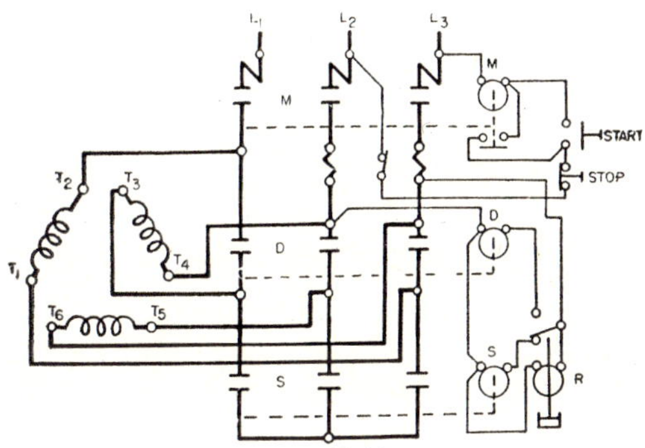

FIG. 18-19

Definite Time Relays—Definite time relays, of which there are numerous types, have many industrial applications. For instance, they may be used in conjunction with reversing controllers (fig. 18-6a) to allow a motor to run for a definite length of time in one direction and then reverse. Such relays are used on control systems in which a definite time must elapse before one operation stops and another takes place. Two similar types of such relays are shown in figs. 18-20a and 18-21a.

The relay shown in fig. 18-20a has a small single-phase motor (not visible) which drives disk D through a train of gears. The disk is revolved at a definite speed and therefore requires a definite time interval to rotate pin P through a given angle. This pin trips latch L, which controls the relay contacts. Fig. 18-20b shows the position of the contacts before and after the latch is tripped. Any time interval not exceeding 38 minutes can be obtained by changing the speed ratio of the train of gears and the starting position of the

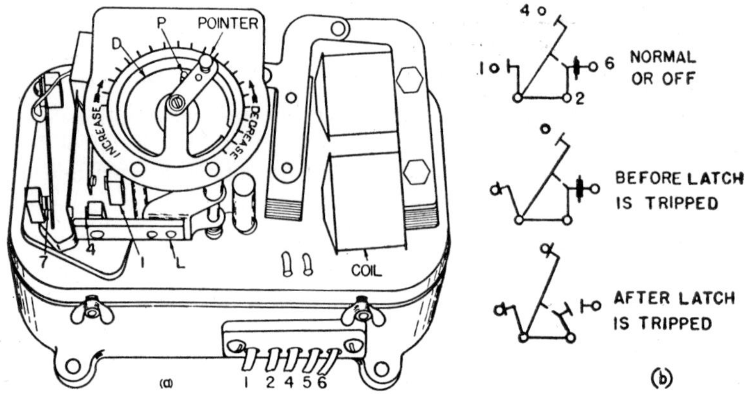

Fig. 18-20

tripping finger. The coarse adjustment is made by placing the pawl (which is located under the calibrating plate) in the proper notch

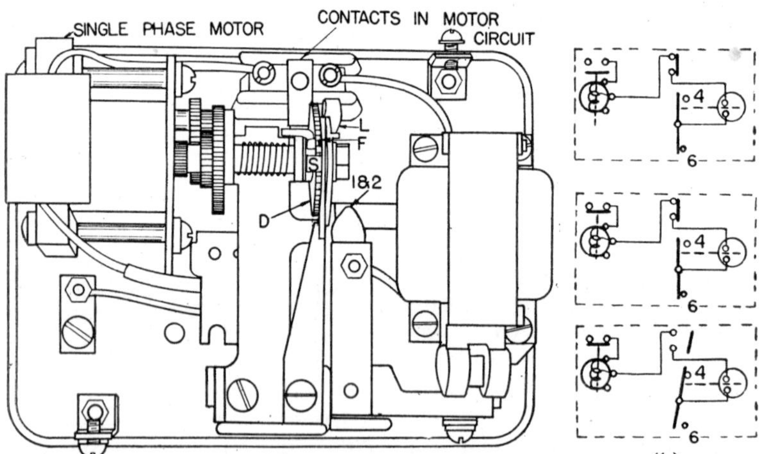

(a) (b)

Fig. 18-21

(three notches are provided). The fine adjustment is made by shifting the pointer, which changes the starting position of pin P.

The relay shown in fig. 18-21a also has a small single-phase motor, which drives finger F through a train of gears. After turning through a given angle, finger F strikes stop S and turns disk D sufficiently to trip latch L. This causes the contacts to change positions, as shown in fig. 18-21b. The timing can be changed in one-minute steps up to 18 minutes by shifting the position of stop S relative to finger F. The method of connecting this relay will be considered later.

Fig. 18-22 shows how the relay in fig. 18-20a is connected to stop one motor after it has run for a definite period of time and then start another. When the start button is pressed, the relay coil, the relay motor, and the coil of starter A are energized. On becoming energized, the relay coil closes contact 1, which maintains the circuit of the relay coil. This coil engages a pawl (not shown) with one of the gears that governs the speed of disk D (fig. 18-20). After running for a definite period of time (depending on the setting of the relay), the relay motor trips the latch, which allows contact 7 to open and contact 4 to close. Contact 7, on opening, deenergizes the coil of starter A and the relay motor, which stops. Contact 4, on closing, energizes the coil of starter B. Then the motor connected to starter A stops and the one connected to starter B starts.

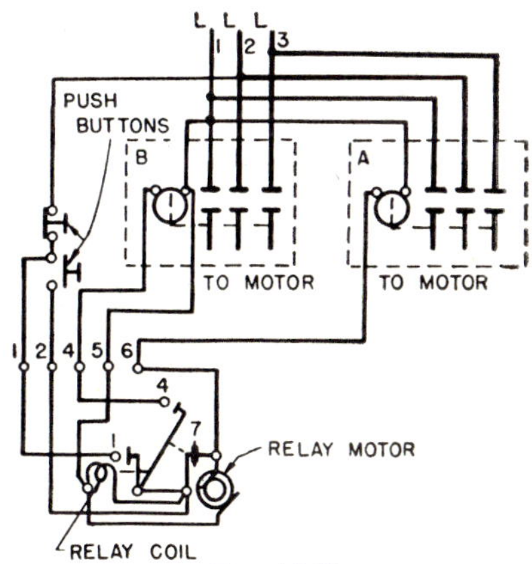

Fig. 18-22

The relay coil is deenergized when the stop button is pressed, allowing a spring to return the mechanism and contact fingers to their former positions.

Resistance Starter—The starting current taken by a squirrel cage motor may be reduced by using a controller that inserts equal resistances in each line wire and simultaneously cuts out these resistances as soon as the speed of the motor becomes constant. Fig. 18-23 shows the wiring diagram of such a controller. When the start button is pressed, coil M is energized, closing the main contacts and connecting the motor to the line through resistances R. The circuit of coil M is maintained through the auxiliary contact. The main contacts, on closing, start the dash-pot timer, whose contacts close after a definite time has elapsed, completing the circuit of coil N. This closes the contacts that short out resistances R, connecting the motor directly across the line.

When the stop button is pressed, the circuit of coil M is opened and the main contacts therefore open and interrupt the power to the motor.

The starting current in a resistance starter causes a relatively high voltage drop in the resistors. Therefore, a rather low voltage will be

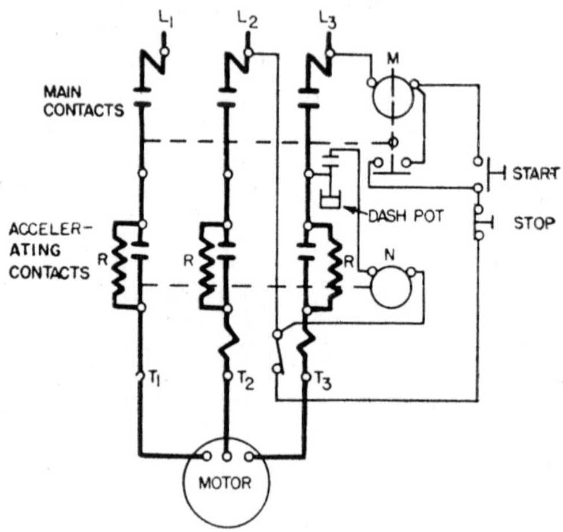

FIG. 18-23

available at the motor, and the heavier the load to be started, the lower the terminal voltage will be. As the motor gains speed, the current, and therefore the voltage drop, across the resistances decreases. The terminal voltage increases gradually, producing a smooth acceleration.

A feature of this starter is that it is simple in structure, and hence has a low initial cost and requires little maintenance. It is undesirable for heavy loads because it provides a low starting torque and low starting economy.

Controllers for Wound Rotor Induction Motors—Resistance starters are also employed for wound rotor induction motors, but these have a different wiring scheme from the one shown in fig. 18-23. Starters for wound rotor motors are manufactured in face plate, drum, and magnetic types, and serve for both starting and speed control. Where speed control is desired, however, the starter is generally of

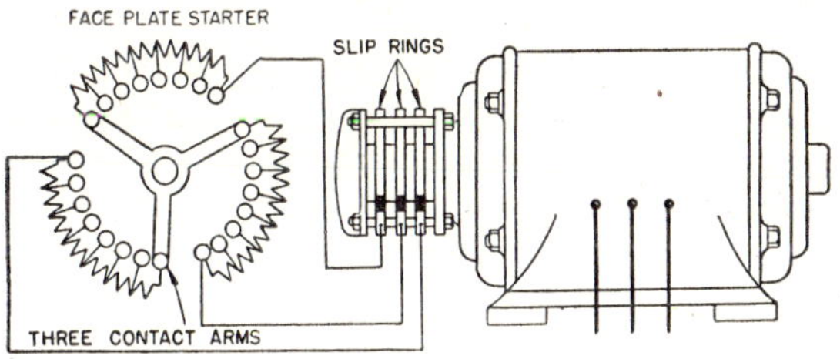

Fig. 18-24

the face plate or drum type. The magnetic type is inherently unsuitable for this, but may be used where a few special speeds are to be supplied. A starter selected for speed-varying service must be capable of carrying the rotor current continuously, while for constant speed it must be able to carry it only during the starting period.

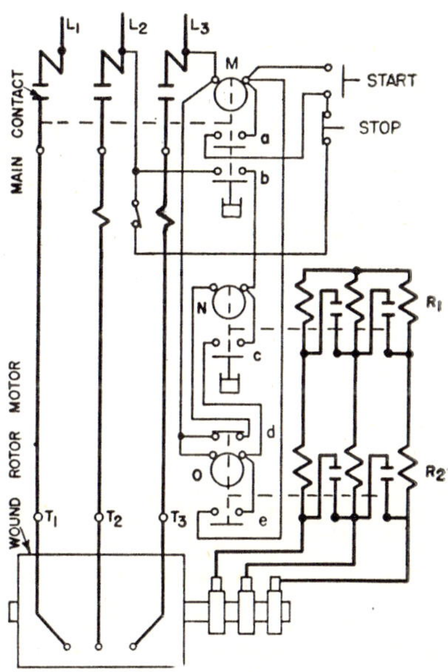

Fig. 18-25

The face plate starter consists of three resistances, each of which is connected to a phase of the rotor. They are varied equally by three contact arms (fig. 18-24). Since this starter serves only to vary the rotor resistance, an across-the-line starter must also be provided to connect the motor to the power line.

To bring a motor up to speed, all the resistance is first placed in the rotor circuit and the stator is energized. As the motor speeds up, the rotor resistance is gradually cut out. To operate the motor at a reduced speed, resistance is left in the rotor circuit.

The drum type controller serves the same purpose as the face plate type; that is, it provides a means of varying the resistance of the rotor circuit.

If the rotor resistance is to be used only for starting, the magnetic type of starter is generally used, especially for large motors. In this starter, the rotor resistance is automatically cut out of the rotor circuit by means of multi-finger contactors arranged to operate in sequence at definite time intervals. Fig. 18-25 shows a starter of this type.

When the start button is pressed, coil M is energized, closing the main contacts, which connect the motor to the power line. After the push button is released, the circuit of coil M is maintained by contact a. Contact b is held open by means of a dash-pot timer or some other type of retarding mechanism, and closes after a definite time. When b closes, coil N becomes energized and closes its contacts, shorting out resistances R_1. After a definite time, contact c closes and energizes coil 0, which closes the contacts to short out all the resistances. At the same time, contact d opens and deenergizes coil N, so that its contacts open. Contact e maintains the circuit of coil 0.

Compensators—Compensators are generally used for starting large squirrel cage and synchronous motors. They use an auto-transformer for reducing the impressed voltage at the time of starting. Fig. 18-26a shows a General Electric manually operated compensator. It consists essentially of a star-connected auto-transformer (see fig. 18-26b), which is disconnected from the line and motor when the handle is in the off position. When the handle is thrown in the start position, contacts A close, forming the connections shown diagrammatically in fig. 18-26c. The auto-transformer is connected across the line and supplies a reduced voltage to the motor. After the motor has accelerated as much as possible, the handle is thrown in the run position. This opens contacts A and closes contacts B (fig. 18-26b), connecting the motor directly across the line, with the thermal overload relays in the circuit. Undervoltage protection is provided by a retaining magnet which holds contacts B closed. The coil of this magnet is connected across the line, through the overload contacts and stop button. The stop button serves to stop the motor and reset the overload contacts.

A mechanism is provided which permits the handle to be operated only if it is first thrown in the start position and then quickly moved

(a) (b) (c) (d)

Fig. 18-26

to the run position. It is customary to provide each phase of the auto-transformer with several taps, so that various starting voltages may be secured.

Instead of being equipped with thermal overload relays, manually operated compensators are sometimes equipped with magnetic overload relays of the dash-pot type. This type will operate almost instantly on heavy overloads, but on light overloads it will not operate until a certain time interval has elapsed, the length of the time interval being inversely proportional to the amount of overload. Because of

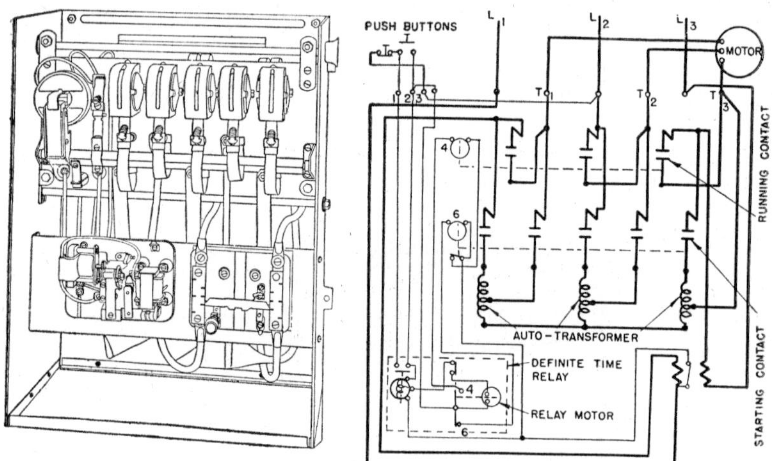

(a) (b)

Fig. 18-27

such operating characteristics, these overload relays are connected as shown in fig. 18-26d, in which it will be noted that they are not in the circuit when the motor is started, but are inserted when the controller is in the running position. This permits the motor to take a large starting current, which for a short period of time is not detrimental, and protects the motor during normal operation.

Automatic compensators are used where remote control is desired. One common type, which is manufactured by the General Electric Company, is shown in fig. 18-27a, and its wiring diagram is given in fig. 18-27b (note that this compensator contains the definite time relay considered in fig. 18-21a). When the start button is pressed (see fig. 18-27b), coil 6 is energized, which closes the starting contacts and connects the auto-transformer to the line and the motor leads to taps on the transformer. This provides the reduced voltage for starting. A circuit which energizes the relay coil is also completed, closing the contact which connects terminals 1 and 2, and thus completing the holding circuit for the relay coil. After a definite time, depending upon the relay setting, the escapement releases and allows the contact arm to break contact at 6 and make contact at 4. This opens the circuit of coil 6 and completes the circuit of coil 4, causing the starting contacts to open and the running contacts to close. As a result the motor is connected directly to the line.

In addition to equipping the compensator with an electrical interlock, which prevents the running contacts from closing until the starting contacts have opened, it is also safeguarded by a mechanical interlock between the two sets of contacts.

If the stop button is pressed, the relay coil is deenergized. The contact connecting 1 and 2 then returns to its normal position, and contact 4 opens. This allows the running contacts to open, and the motor is disconnected from the power line.

Oil Switches—For high voltage controllers, contacts operating in air cannot be used, and oil switches are used instead. Fig. 18-28a shows the construction of an electrically operated high voltage oil switch. In this case, each pair of poles (fig. 18-28b) is mounted in a separate tank of substantial construction, so as to make the system safe in case of an explosion, which may be produced by a short circuit. For moderate voltages, however, all the poles are placed in one tank, with wood separators between them. The auxiliary equipment, which is shown in fig. 18-28c, consists of a powerful closing coil, a relatively small trip coil, and auxiliary contacts. A common method of connecting this equipment is shown in fig. 18-28d. For the sake of simplicity, the protecting relays and indicating instruments have been omitted. When the oil switch is open, the green lamp circuit is completed by contact B. When the on button is pulled, the control relay coil is energized. Its contact closes, completing the circuit of the main coil, which closes the oil switch, opens contact B, and closes contacts A. Contact B deenergizes the green light, one contact A energizes the red light, to indicate that the oil switch is closed, and the other makes possible the energizing of the trip coil. After the on button is released, the oil switch and contacts A are held closed by the latch (see fig. 18-28c). As a result, the oil switch connects the load to the alternator.

When the off button is pulled, the trip coil is energized and trips the latch. The oil switch then opens, interrupting the power to the load.

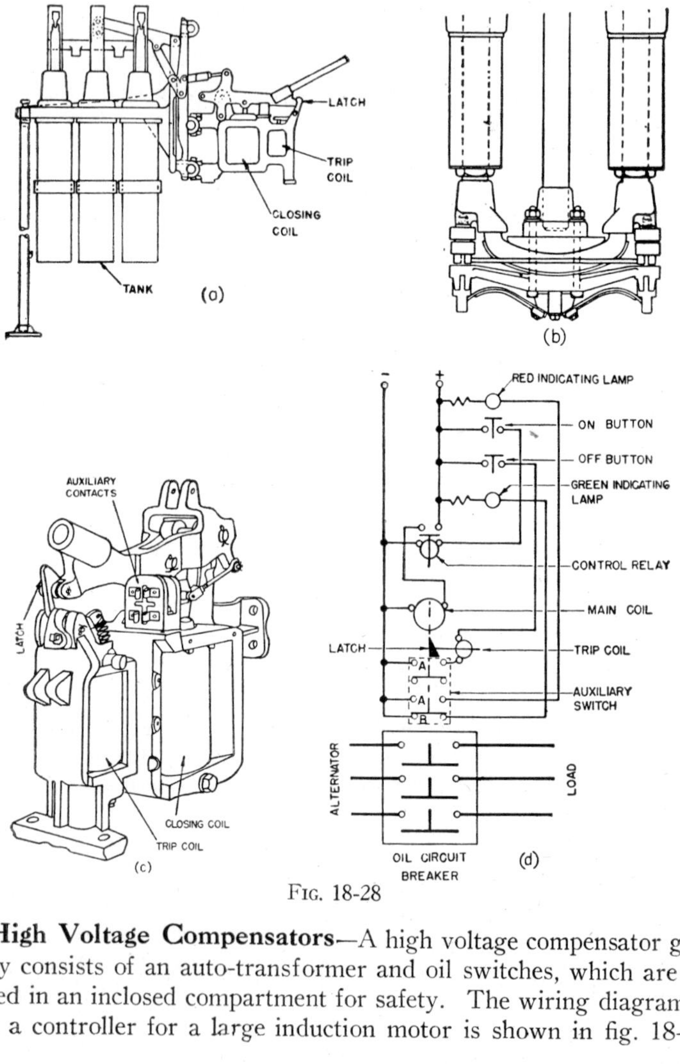

FIG. 18-28

High Voltage Compensators—A high voltage compensator generally consists of an auto-transformer and oil switches, which are installed in an inclosed compartment for safety. The wiring diagram of such a controller for a large induction motor is shown in fig. 18-29.

The same control is also used for synchronous motors, but when used for this purpose, the field should be energized manually after the motor has reached constant speed.

In fig. 18-29, the main coils S are energized when the start button is pulled. These coils close the starting oil switches and operate the

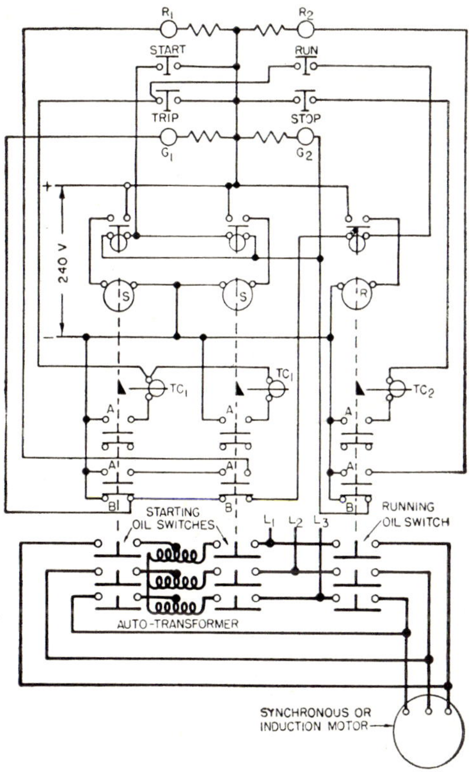

FIG. 18-29

auxiliary contacts. The oil switches connect the motor to the line through the auto-transformer. As a result, the motor receives a low voltage and will therefore take a low starting current. The auxiliary contacts deenergize light G_1, and energize light R_1, and make it possible to energize the two trip coils TC_1. After the speed of the motor becomes constant, which is detected by observing the indicating instruments, the run and the trip button are pulled simultaneously. This energizes the trip coils TC_1, which trip the latches of the starting oil switches, so that they open (contacts A open and contacts B close). Contacts B and the run button complete the circuit of the relay, that closes to energize coil R. This coil closes the running oil switch, which connects the motor directly to the power line. Contact B on the running switch opens and lamp G_2 is deenergized.

When contacts A close, one energizes the red lamp R_2, which indicates that the motor is across the line, and the other makes possible the energizing of the trip coil TC_2. After the machine reaches maximum speed, the field switch is closed.

The stop button is pulled to stop the motor. This energizes coil TC_2, tripping the running oil switch, which interrupts the power. The field switch is then opened.

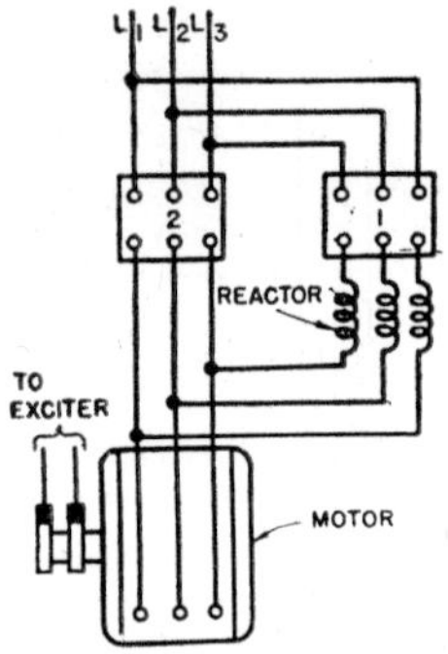

FIG. 18-30

Another starting method employed in control systems is illustrated in fig. 18-30. Oil switch 1 is first closed, connecting the motor to the power line through reactors. This will provide a low terminal voltage because the large starting current will establish a large voltage drop in the reactors. The current decreases as the motor gains speed, causing the voltage drop in the reactors to become less and the motor voltage to become greater. With a light load, the motor voltage will increase to nearly the line voltage. After the speed becomes constant, switch 2 is closed, connecting the motor directly to the power line. Switch 1 is then opened.

Unlike the controller in fig. 18-29, the controller in fig. 18-30 provides a voltage which increases gradually, at a rate which depends upon the nature of the load. This provides smooth acceleration.

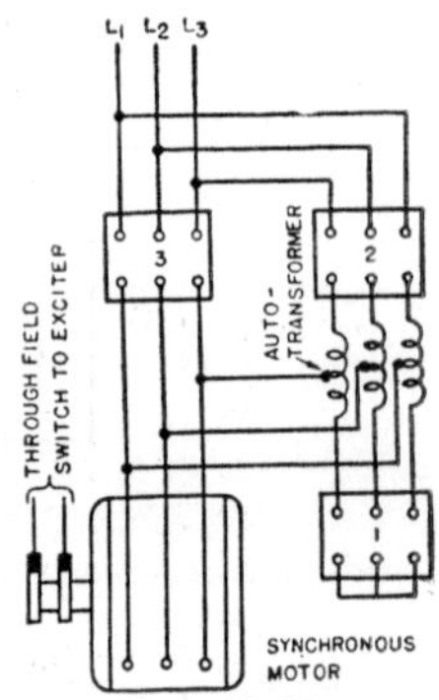

FIG. 18-31

Still another control system is illustrated in fig. 18-31, which is fundamentally a combination of the last two discussed. Switches 1 and 2 are first closed simultaneously, connecting the motor to the power line through the star-connected auto-transformer. Power is then supplied to the motor at a reduced voltage. As soon as the speed becomes constant, switch 1 is opened and switch 3 is closed, connecting the motor directly across the line. In this case, the terminal voltage increases to normal without dropping to zero, as it does in the control shown in fig. 18-29. Switch 2 opens immediately after switch 3 closes.

Comparison of Resistance Starter and Compensator— A motor starter with a compensator takes less current from the power line than one started with a resistance starter. The reason for this can be illustrated best by an example.

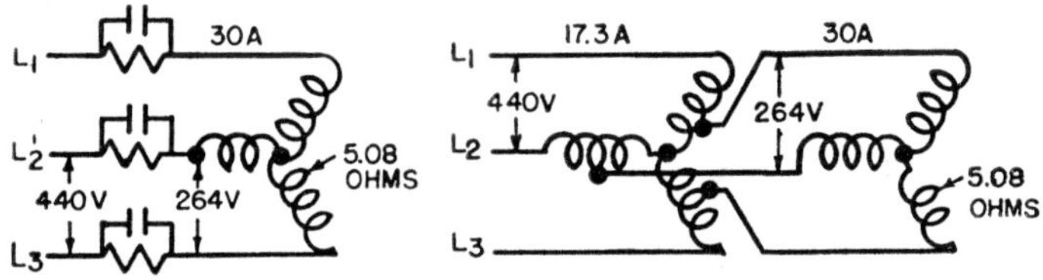

FIG. 18-32

EXAMPLE—A 10 H P, 440-volt, star-connected squirrel cage induction motor takes a current of 50 amperes when it is started directly across the line. What will be the starting current taken from the line (a) when a resistance starter is used, and (b) when a compensator is used, if both starters provide a starting voltage of 60 per cent normal value?

If the motor is started across the line, the phase voltage will be $440 \div \sqrt{3}$, or 254.

The impedance per phase at starting is $254 \div 50 = 5.08$ ohms (see fig. 18-32a).

Whether the resistance starter or the compensator is used, the terminal voltage at starting is $440 \times .6 = 264$, and the phase voltage is $264 \div \sqrt{3} = 152.4$.

If a resistance starter is used, the motor or line current is $152.4 \div 5.08 = 30$ amperes.

If a compensator is used, the motor current is 30 amperes, but due to the auto-transformer ratio, the line current is $(152.4 \div 264) 30 = 17.3$ amperes (fig. 18-32b).

In starting a motor with a compensator, therefore, the line current is less by 30 − 17.3 or 12.7 amperes (42.3%). This shows that the starting current will be less when compensators are used than when resistance starters are used.

Starting Several Motors from a Single Compensator—If it is necessary to start large motors one at a time, which is usually the case in substations, only one compensator is required, though it must be of sufficient capacity to carry the starting current of the largest motor. A typical wiring diagram of such a controller is shown in fig. 18-33. When a starting oil switch of one motor is closed, a reduced voltage is supplied to this motor. After the speed of the motor becomes

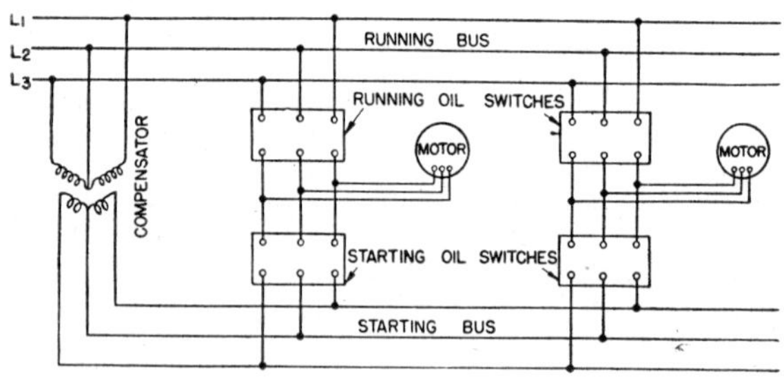

FIG. 18-33

constant, its starting oil switch is opened and its running oil switch is closed, connecting the motor directly to the line. The other motors are started in the same way.

Controllers for Two-Speed Motors—The main purpose of a controller for a multi-speed motor is to connect the motor externally for the various speeds. Fig. 18-34a shows a two-speed controller manufactured by the Square D Company, which may be used with constant horsepower, constant torque, or variable torque motors. The winding of a constant horsepower motor is connected to the controller as shown in fig. 18-34b. These connections are shown in a simplified manner in figs. 18-34c and 18-34d.

Pressing the low speed button (fig. 18-34b) energizes coil L, and its circuit is maintained through contact C22. This coil closes the five contacts L (two contacts connect the winding parallel star and the

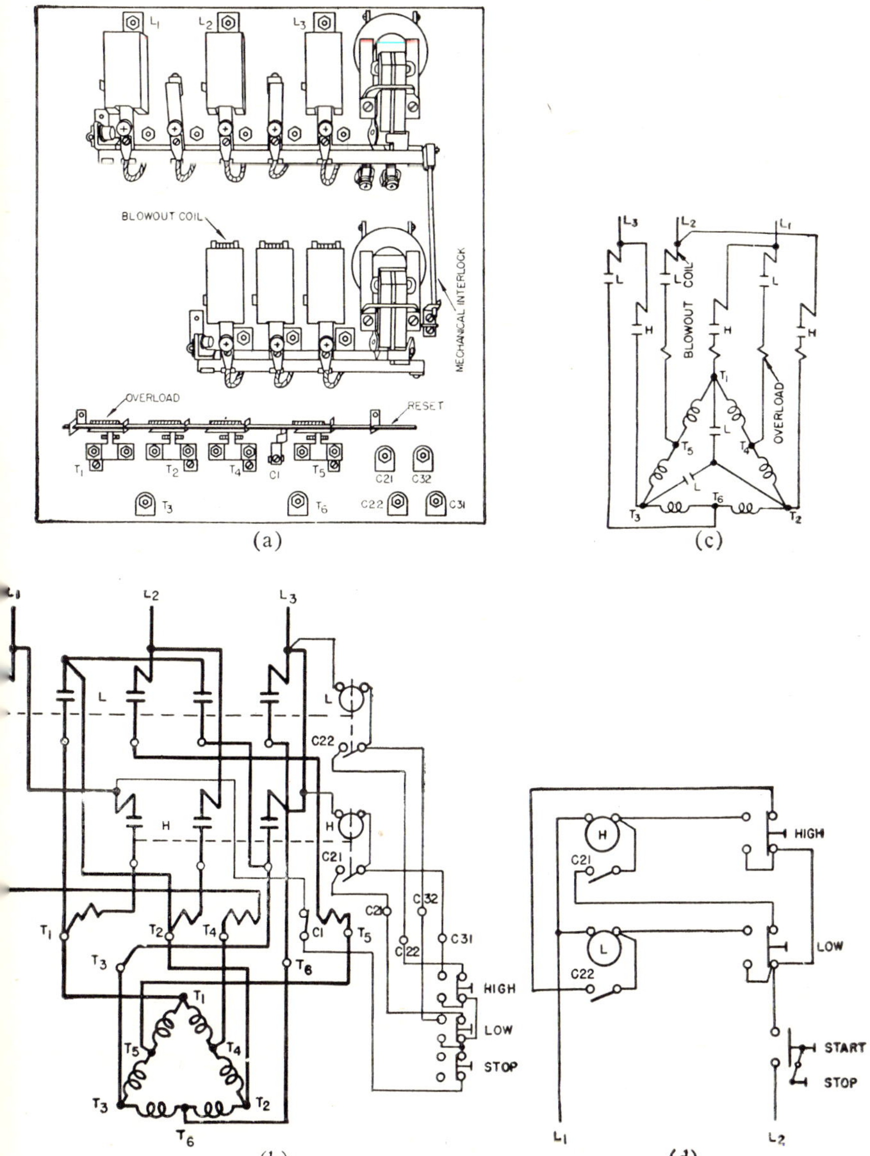
L1
L2
L3
BLOWOUT COIL
MECHANICAL INTERLOCK
OVERLOAD
RESET
T1
T2
T4
C1
T5
C21
C32
T3
T6
C22
C31
(a)
BLOWOUT COIL
OVERLOAD
L
H
(c)
HIGH
LOW
STOP
(b)
START
(d)

FIG. 18-34

other three connect it to the power line). As a result, the motor runs at a low speed.

When the high speed button is pressed, coil L is de-energized, contacts L open, coil H is energized, and contacts H close. The circuit of coil H is then maintained through contact C21. Contacts H form a series delta connection when they close, and the motor runs at high speed.

In figs. 18-34b and 18-34d, one button is used for each speed and only one stop button is necessary. The motor may be started at either speed. To change from one speed to another, it is merely necessary to press the button for the new speed. When the stop button is pressed, the circuit of the coil last energized is opened. The contacts of this coil then open, interrupting the power to the motor.

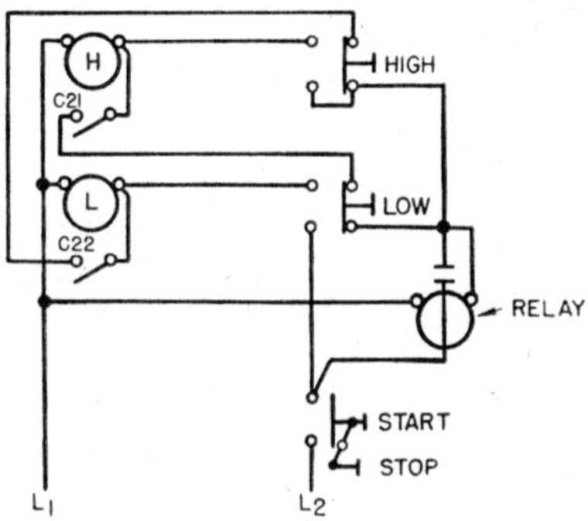

FIG. 18-35

A mechanical interlock is provided between the two sets of contacts (fig. 18-34a), so that they cannot be closed at the same time. If this took place, a short circuit would exist across the three line wires.

In fig. 18-34a, note that two contacts of the five-pole contactor have no blowout coils. These two contacts, which serve to form the parallel star connection (see fig. 18-34c), have a slight lead in closing and a slight lag in opening. Therefore they do not arc and require no blow-out coils.

In certain installations, excessive stress is imposed on the driven equipment if the motor is started at the high speed. Then, too, the motor may draw a high current long enough to trip the overloads. Such conditions may be avoided by starting the motor at the low speed. To make certain that the operator will do this, the auxiliary control must be connected so that the motor cannot be started at high speed. The method of doing this is shown in fig. 18-35 (the main circuits are similar to those shown in fig. 18-34b). The hookup in fig. 18-35 includes a single contact relay, but the two-contact type shown in fig. 18-36, which is manufactured by the Allen-Bradley Company, may also be used. The two contacts may be connected in parallel or one may be used to operate a pilot light.

With the hookup shown in fig. 18-35, coil H cannot be energized first because the relay contact is open. Coil L is energized if the low speed button is pressed, closing the low speed contactor and contact C22. This contact completes the circuit of the relay, which closes and maintains its circuit, and that of coil L, after the low speed button is released.

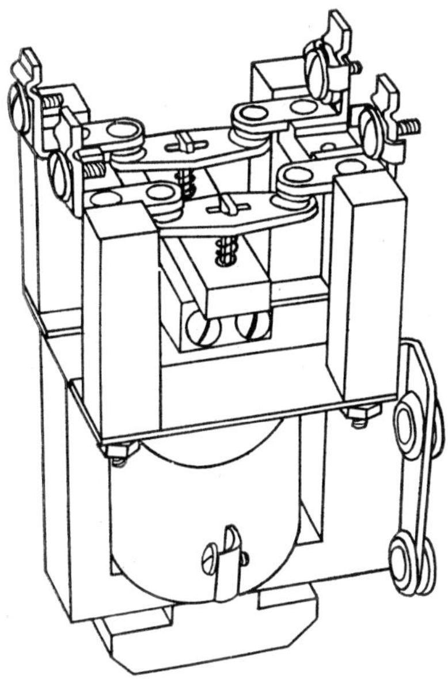

Fig. 18-36

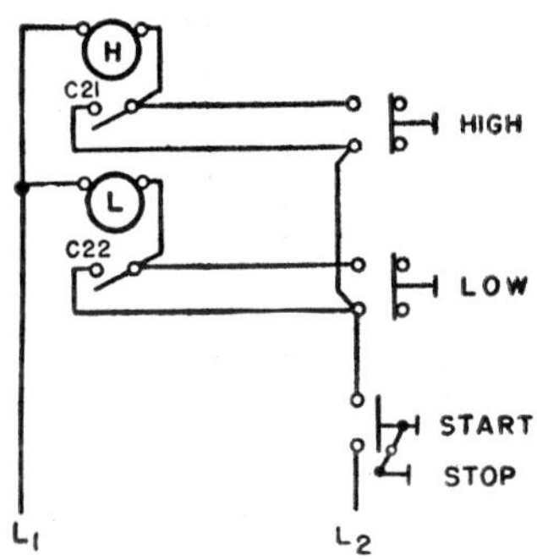

Fig. 18-37

If the high speed button is now pressed, the circuit of coil L is opened, the low speed contactor opens, and the circuit of coil H is completed through the relay contact. On becoming energized, coil H closes the high speed contacts, which connect the motor for high speed, and contacts C21, maintaining the coil circuit.

The push buttons are sometimes connected as shown in fig. 18-37. With such a hook-up, the motor may be started at either speed. To change from one speed to the other, the stop button must first be pressed, and then the button for the other speed.

Drum type multi-speed controllers are also manufactured. Such a controller for a two-speed motor is shown in fig. 18-38a, and its developed view is shown in fig. 18-38b. When the knob is turned to the left, contacts are made as shown in fig. 18-38c. This connects the motor series delta to give the low speed (assuming that the motor is a constant torque machine). When the knob is in the off position, the fingers are not in contact with the segments, and consequently the

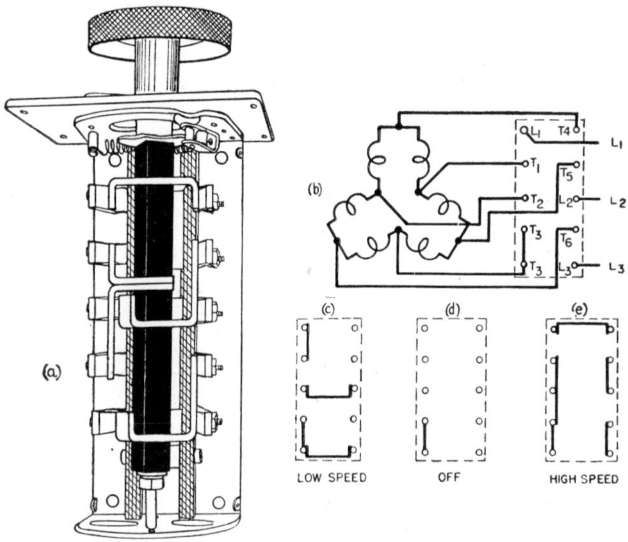

FIG. 18-38

motor is disconnected from the power line (fig. 18-38d). When the knob is turned to the right, contacts are made as shown in fig. 18-38e, and the motor is connected parallel star, which gives the high speed.

Controller for Two Speeds Not in the Ratio of Two to One—If two speeds not in the ratio of two to one are desired, the controller shown in fig. 18-39 may be used. In this figure a constant horsepower motor is considered, and is connected for speeds of 900 and 1200 r p m. For this connection, jumpers b and c must be removed. The push buttons may also be connected as shown in figs. 18-35 or 18-37.

Referring to fig. 18-39, when the L contacts are closed, three contacts connect winding B parallel star and the other three connect this winding to the line. This gives the low speed (900 r p m). When the L contacts are opened and the H contacts are closed, winding B is disconnected and winding A is connected series delta. This gives the high speed (1200 r p m).

Table 18-I gives the various methods of connecting a four-speed motor for different combinations of two speeds.

TABLE 18-I—CONNECTIONS OF A FOUR-SPEED CONSTANT H P MOTOR CONNECTED FOR TWO SPEEDS

Speeds	Omit Jumpers	Connect Terminals as Follows
600/900	a and c	T_1–T_{37}, T_2–T_{40}, T_3–T_{41}, T_4–T_{30}, T_5–T_{31}, T_6–T_{32}, T_7–T_{42}, T_{11}–T_{36}, T_{12}–T_{38}, T_{13}–T_{39}, T_{14}–T_{33}, T_{15}–T_{34}, T_{16}–T_{35}, T_{17}–T_{43}
600/1200	b and c	T_1–T_{33}–T_{37}, T_2–T_{34}–T_{40}, T_3–T_{35}–T_{41}, T_4–T_{30}, T_5–T_{31}, T_6–T_{32}, T_7–T_{36}–T_{42}
600/1800	b and c	T_1–T_{37}, T_2–T_{40}, T_3–T_{41}, T_4–T_{30}, T_5–T_{31}, T_6–T_{32}, T_7–T_{42}, T_{11}–T_{33}, T_{12}–T_{34}, T_{13}–T_{35}, T_{17}–T_{36}
900/1200	b and c	T_1–T_{33}, T_2–T_{34}, T_3–T_{35}, T_7–T_{36}, T_{11}–T_{37} T_{12}–T_{40}, T_{13}–T_{41}, T_{14}–T_{30}, T_{15}–T_{31}, T_{16}–T_{32}, T_{17}–T_{42}
900/1800	b and c	T_{11}–T_{33}–T_{37}, T_{12}–T_{34}–T_{40}, T_{13}–T_{35}–T_{41}, T_{14}–T_{30}, T_{15}–T_{31}, T_{16}–T_{32}, T_{17}–T_{36}–T_{42}
1200/1800	b and d	T_1–T_{30}, T_2–T_{31}, T_3–T_{32}, T_7–T_{37}, T_{11}–T_{33}, T_{12}–T_{34}, T_{13}–T_{35}, T_{17}–T_{36}

Controller for a Three-Speed Motor—Fig. 18-40a shows a wiring diagram of a three-speed controller connected to a constant horsepower, three-speed motor. One winding gives two speeds in the ratio of two to one and the other winding gives the third speed. The push buttons are connected so that the motor may be started at any one of the speeds, and a change in speed is made merely by pressing the button for the new speed.

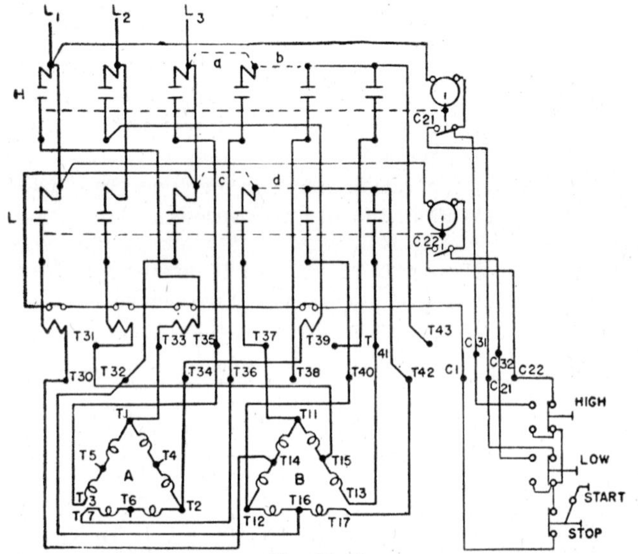

Fig. 18-39

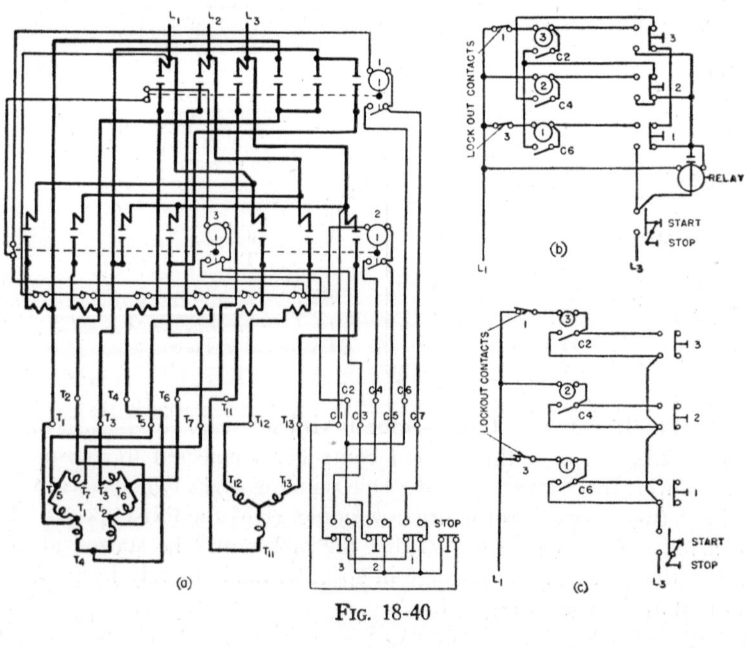

Fig. 18-40

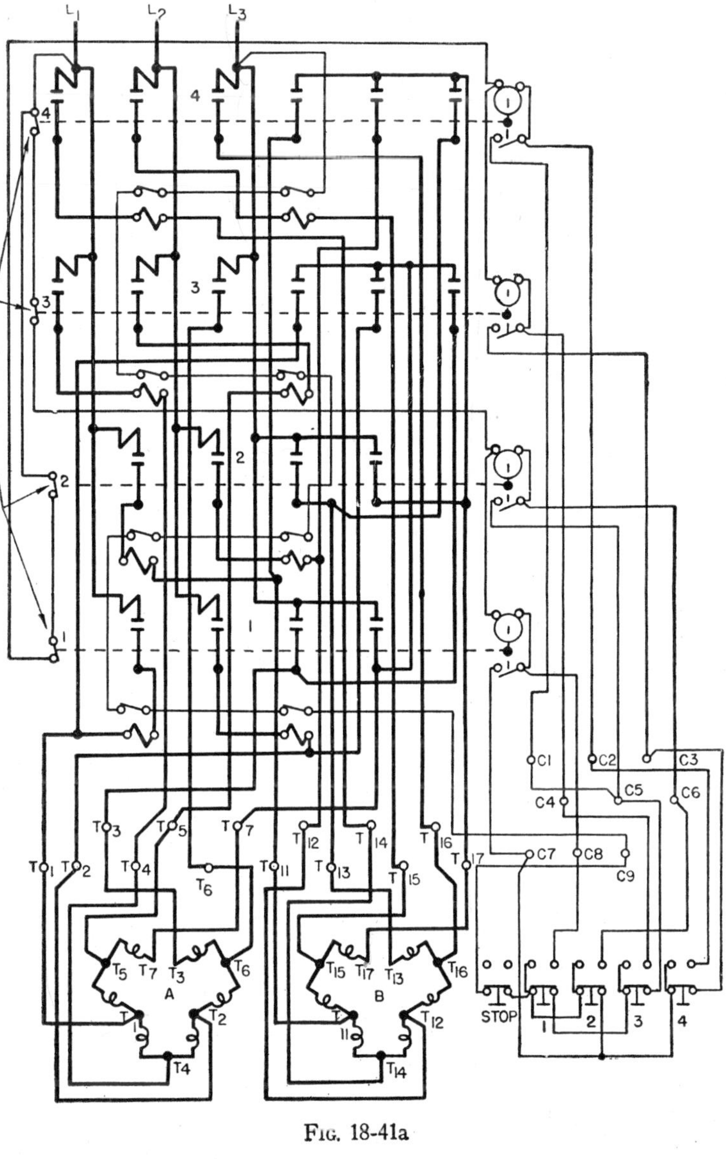
L1
L2
L3
INTERLOCKS
4
3
2
1
T1
T2
T3
T4
T5
T6
T7
T11
T12
T13
T14
T15
T16
T17
A
B
C1
C2
C3
C4
C5
C6
C7
C8
C9
STOP

Fig. 18-41a

Fig. 18-40b shows a method of connecting the push buttons to permit starting only at low speed, and to permit changing the speeds in steps (900-1200-1800, or in the reverse order). As in fig. 18-35 the hookup requires a single contact relay. The push buttons may also be connected as shown in fig. 18-40c.

Controller for a Four-Speed Motor—A wiring diagram of a four-speed controller and motor is shown in fig. 18-41a. The push buttons are connected so that the motor may be started at any speed, and the speeds may be changed in any order. The method of connecting the push buttons for sequence control is shown in fig. 18-41b.

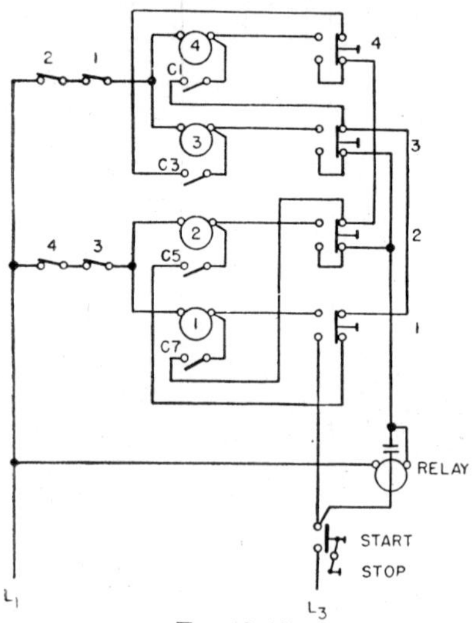

FIG. 18-41b

Reversing Multi-speed Motors—In some installations it is necessary to reverse multi-speed motors. This may be done with a multi-speed controller, together with a reversing controller (fig. 18-6a). In many cases drum controllers are used where reversing is desired, because they are simpler. A multi-speed reversing drum controller is shown in fig. 18-42a, and its developed view is shown in fig. 18-42b, in which the terminals are marked to correspond to the markings on the stator windings.

Suppose that a four-speed, constant torque motor (which has its leads brought out as shown in fig. 18-42c) is connected to the controller shown in figs. 18-42a and 18-42b. When the handle is in the off position, the fingers are not in contact with the segments, and consequently the motor is disconnected from the power line. When the handle is moved one notch in the forward direction, both sets of fingers are in position 1F. This connects winding A series delta (see fig. 18-42c), in which case the motor runs at the first speed. When the handle is at notch 2, both sets of fingers are at 2F, so that winding B is connected series delta to give the second speed. At 3F the two sets of fingers connect winding A parallel star, giving the third speed. At 4F the two sets of fingers connect winding B parallel star, giving the high speed.

If the handle is shifted from the off position in the reverse direction, the speeds will also change in steps, but the motor will run in the reverse direction because the connections of line wires L_1 and L_2 are interchanged at the motor.

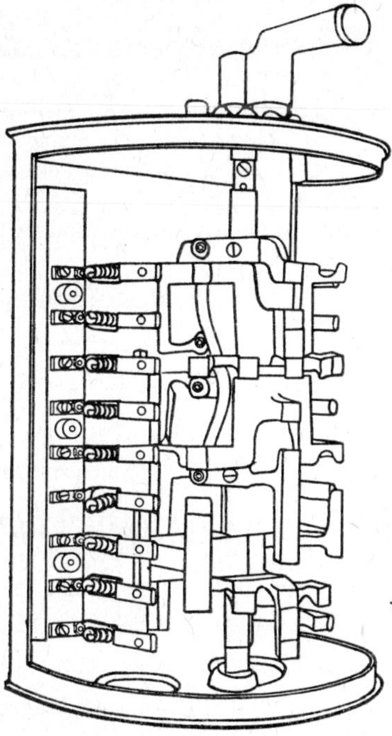

(a)

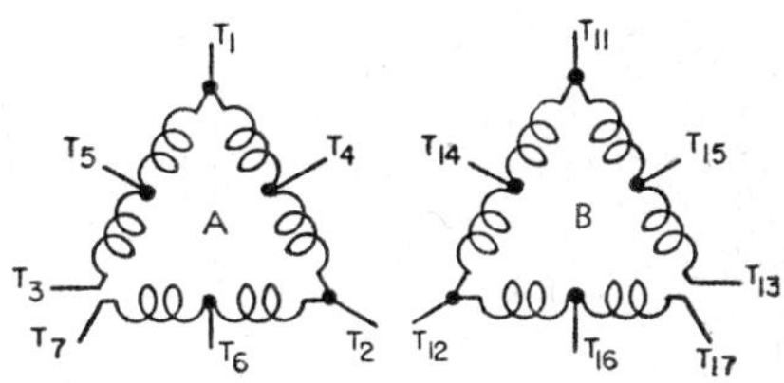

(c)

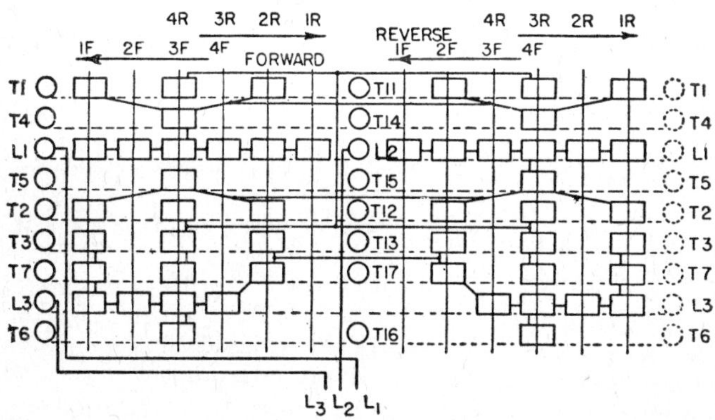

(b)

FIG. 18-42

Exercises

1. Give the advantages and disadvantages of starting squirrel cage motors directly across the line.
2. What factors limit the size of the motor that can be started directly across the line?
3. Why are thermal overloads used instead of fuses in across-the-line starters?
4. Show how a motor and a three-wire push button switch are connected to the starters in fig. 18-43a and fig. 18-43b.
5. Show how a motor and two push button switches are connected to the reversing controller shown in fig. 18-43c.

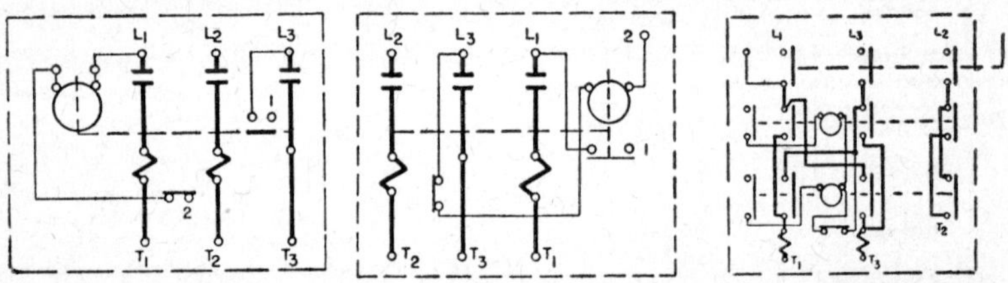

(a) (b) (c)

FIG. 18-43

6. Why are standard across-the-line starters (fig. 18-1) not satisfactory for motors that are so loaded that they come up to speed slowly?
7. Describe the different types of field relays used on controllers for synchronous motors.
8. What is the purpose of relay 7 in fig. 18-11?
9. How is dynamic braking accomplished with squirrel cage motors?
10. Fig. 18-44 illustrates a controller equipped for dynamic braking. Show how the line wires, push buttons, and motor are connected to it.
11. (a) What are the three methods of rapidly bringing a motor to a standstill? (b) Give the advantages and disadvantages of each method.
12. What is meant by plugging?
13. What advantage has the centrifugal type plugging relay over the other types?

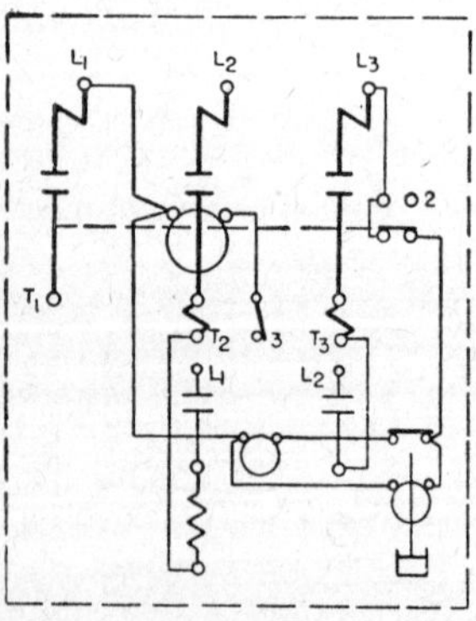

FIG. 18-44

14. Make a diagram showing the controller in fig. 18-43c connected for plugging a motor that is to run only in one direction (this controller, when so connected, requires the addition of an auxiliary contact).
15. When must a motor be started at a reduced voltage?
16. Make a diagram of the controller shown in fig. 18-19, replacing the dash pot by the definite time relay shown in fig. 18-20.
17. A standard squirrel cage motor, when started across the line, takes 600 per cent full load current and develops 150 per cent full load torque. If it is started by means of a star-delta controller, determine (a) the percentage of full load current and (b) the percentage of full load torque at starting.
18. Name the two types of resistance starters and explain how they differ.
19. When should a wound rotor induction motor be equipped with (a) a manual starter? (b) An automatic starter?
20. Make a diagram of the resistance starter shown in fig. 18-23, replacing the dash pot by the definite time relay shown in fig. 18-21.
21. In what ways is (a) a resistance starter for squirrel cage motors superior to a compensator? (b) A compensator superior to a resistance starter?
22. A three-phase, 440 volt, 10 horsepower squirrel cage motor takes 14 amperes and supplies 30.3 pound-feet at full load. On starting across the line, the motor takes 5.5 full load current and supplies 1.5 full load torque. (a) To what value must a resistance starter reduce the motor voltage so that the starting torque will equal the full load torque? (b) With this terminal voltage, what current will flow?
23. What line current must be supplied to the motor in problem 22 to produce a full load torque if the resistance starter is replaced by a compensator?
24. When starting multi-speed motors, why is it desirable to operate them at the lowest speed first?
25. By means of a diagram, show how a constant torque motor is connected to the controller in fig. 18-34 (connect the push buttons so that only the low speed button will start the motor).
26. Make a diagram showing how a two speed, variable torque motor is connected to the controller in fig. 18-34 (connect the push button so that only the low speed button will start the motor).
27. For what industrial application is the controller shown in fig. 18-39 best suited?
28. By means of a diagram, show how a four speed, constant horsepower induction motor is connected to the controller illustrated in fig. 18-39 so as to operate at speeds of 600 and 900 r p m.
29. For what kind of work is the controller shown in fig. 18-42a more suitable than the one shown in fig. 18-41a?

CHAPTER 19

ELECTRICAL INSTRUMENTS

Instruments—Some instruments are inherently suitable for direct current only, others for both direct and alternating current, and still others for alternating current only.

Permanent Magnet Instruments—For direct current measurements, the permanent magnet type of instrument is practically the only type used, because it is extremely sensitive. A permanent magnet type is shown in fig. 19-1a. Its movable coil, C, is generally wound

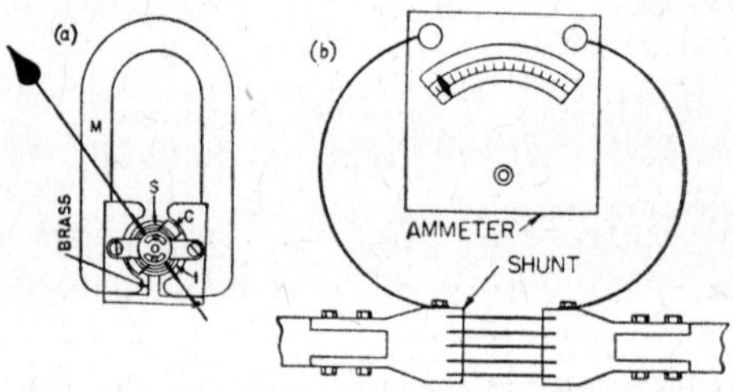

Fig. 19-1

on an aluminum bobbin, which is pivoted to turn between the poles of the permanent magnet, M. The bobbin, in addition to supporting coil C, highly dampens its movement. This coil is connected to the terminals of the instrument through the two spiral springs S, which also serve to supply a reacting force proportional to the deflection. When current flows in coil C, it rotates until the established force is just balanced by the springs.

The cylindrical soft iron core, I, is supported by a brass strip within the coil. This gives an annular air gap, in which a radial field is established. With such a field, the deflection of the coil, and therefore the indicating hand, is proportional to the current.

Permanent magnet instruments give full scale deflections with coil currents ranging from 5 to 20 milliamperes. When measuring higher currents than 20 milliamperes, a shunt is generally connected in parallel with the instrument as shown in fig. 19-1b.

EXAMPLE—An instrument, which has a resistance of 2.5 ohms, gives full scale deflection when carrying 20 milliamperes. To give full scale deflection when the current is 10 amperes, what resistance shunt is required?

The voltage across the instrument, and also that across the shunt, is 2.5×20 or 50 millivolts. The shunt resistance must then be $50 \div (10 - .02)$ or 5.01 milliohms.

In practice, the resistance is seldom marked on shunts, the millivolt drop and current rating being marked instead.

It is possible to use the same instrument for making current readings of any range provided the proper shunt is used to suit each case.

Voltmeter—Since the current through an instrument is proportional to the voltage impressed at its terminals, the instrument considered above may be used as a voltmeter. In this case, a resistor must be connected in series with the movable coil to restrict the current to 20 milliamperes or the voltage drop to 50 millivolts during full scale deflection.

EXAMPLE—What resistance must be connected in series with the instrument considered in the above example so that full scale deflection occurs at 150 volts?

The total resistance is $150 \div .02 = 7500$ ohms. Since the coil has a resistance of 2.5 ohms, the series resistor must be $7500 - 2.5$ or 7497.5 ohms.

For direct readings, the instrument should be equipped with a 150 volt range scale. Voltage readings of other ranges may be made with the same instrument, provided a suitable series resistor is used.

Electrodynamometer Type Instrument—The structure of an electrodynamometer type instrument is shown in fig. 19-2a. It consists of two stationary coils, S, and a movable coil, M, to which the indicating hand is attached. The three coils are connected in series through the two spiral springs, which also hold the movable coil in the position shown (fig. 19-2a) when the coils carry no current. When the coils carry current, coil M tends to turn in the clockwise direction, because its flux tends to line up with the flux of coils S. Apparently, this instrument is the same as the one shown in fig. 19-1a, except that coils S are substituted for the permanent magnet.

If the current through an instrument of the type shown in fig. 19-2a is reversed, the torque developed remains in the clockwise direction. Hence, such an instrument may be used for measuring alternating current as well as direct current.

In this instrument the scale is not divided uniformly, as in the permanent magnet type. As in the case of a series motor, the torque produced varies as the square of the current. Hence, the divisions near the start end of the scale are small (fig. 19-2a) and cannot be read accurately, and those near the finish end are large and can be read very accurately.

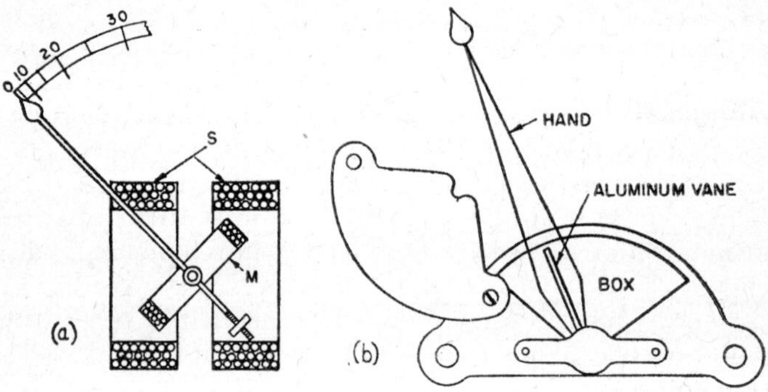

Fig. 19-2

Alternating current instruments are usually dampened. This is generally accomplished by an aluminum vane, which is attached to the same shaft as the hand and fits closely in an enclosed box (fig. 19-2b). When coil M (fig. 19-2a), and therefore the vane, moves, air is forced from one side of the vane to the other, thus producing a damping effect.

Wattmeter—The wattmeter is generally a dynamometer type instrument, constructed like the one shown in fig. 19-2a, but the coils are connected in a load circuit as shown in fig. 19-3a. The stationary coils are connected in series with the load, to carry the load current, and are thus known as current coils (C). The movable coil is connected across the line through a high resistance. As the current through this coil is proportional to the supplied voltage, it is known as a voltage coil (V). The circuit of the voltage coil has a negligible reactance and a high resistance. Its current and flux may therefore be considered to be in phase with the line voltage (E).

At unity power factor, the flux Φ_c of coils C and the flux Φ_v of coil V are in phase, and maximum deflection occurs for given values of I and E. When I lags E, only the component of Φ_c in phase with Φ_v is effective in producing torque (fig. 19-3b). Then the torque is proportional to $\Phi_c \cos \theta$ and Φ_v. Since Φ_c and Φ_v are respectively proportional to I and E, the torque is proportional to EI $\cos \theta$, the true power. Hence, the instrument may be calibrated directly in watts (or kilowatts) to indicate the true power supplied to the load.

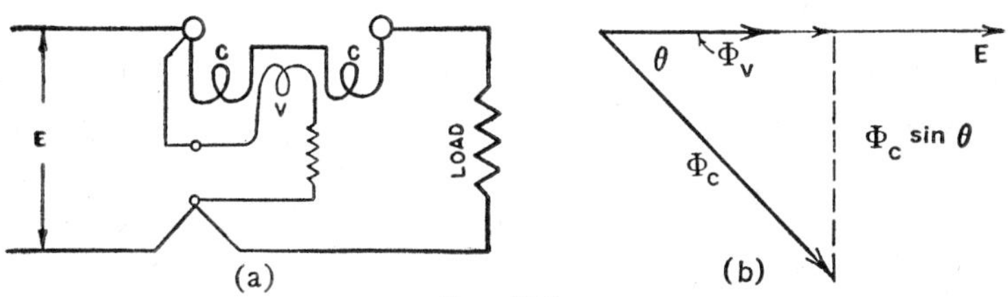

FIG. 19-3

Wattmeter Capacity—A wattmeter has two circuits, either of which may burn out on carrying too much current, without causing the hand to move beyond the upper limit of the scale. For this reason, current and voltage ratings are marked on the instrument, and care should be taken that neither of these limits are passed.

Polyphase Wattmeter—Two single phase wattmeters are required to measure the total power taken by a polyphase circuit. Two such instruments may be combined into one, in which case the two voltage coils are mounted on the same shaft and rotate in their respective current coils. The total power is read from a single scale.

Iron Vane Instrument—A common instrument used for measuring alternating current and alternating current voltages is shown in fig. 19-4a. A cylindrically bent soft iron vane (M) is mounted axially on a shaft to which the indicating hand is attached. A second vane (F), which is wedge-shaped and has a slightly larger radius, is fixed stationary alongside the first vane. The stationary coil, which surrounds the vanes, has only a few turns of heavy wire if the instrument is an ammeter, but has a large number of turns of fine wire connected in series with a resistor if it is a voltmeter.

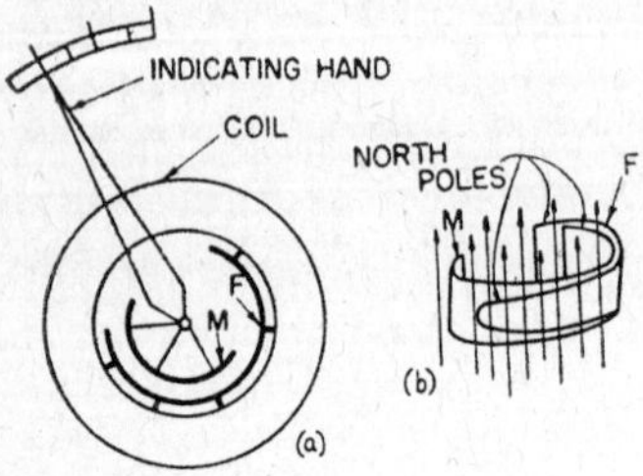

Fig. 19-4

When the instrument is connected in a circuit, the current through the coil establishes a field which magnetizes the two vanes. This establishes repulsion between the two, because their two upper edges (as well as their two lower edges) always have the same polarity, regardless of which way the current flows through the coil (see fig. 19-4b). Then the instrument may be used on direct or alternating current. The repulsion causes the movable strip M to slide away from the fixed strip F, and in so doing, strip M rotates the hand.

Inclined Coil Instrument—Another common type of instrument is shown in fig. 19-5a. It has a coil mounted at an angle to the shaft. On the shaft, to which the hand is fastened, two soft iron vanes

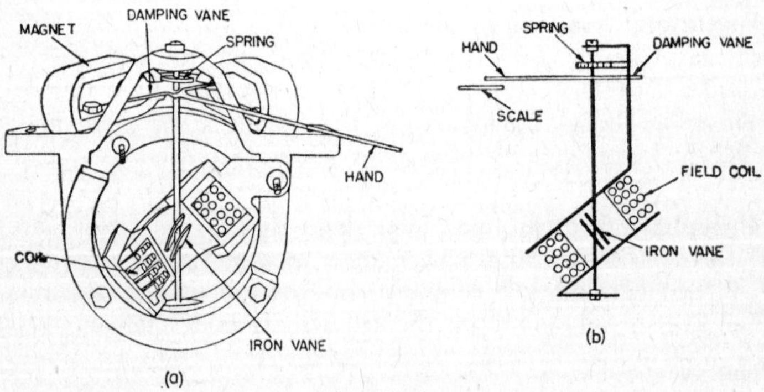

Fig. 19-5

are obliquely attached. When the coil carries no current, a spring holds the hand at zero, and the iron vanes lie in the plane of the coil. When the coil carries current, magnetic lines are established at its center, and the iron vanes tend to line up with them (fig. 19-5b).

This causes the vanes to rotate against the action of the spring and the hand moves up scale.

The iron vanes will tend to line up with the magnetic lines regardless whether the lines are coming out or going in the coil; hence, this instrument can be used to measure either direct or alternating current.

The iron vane and the inclined coil type instruments are used extensively because they are simple in construction and inexpensive.

Induction Type Instrument—The general structure of an induction type instrument is shown in fig. 19-6a. The principle of operation is the same as that of a split-phase motor. The aluminum disk D serves as the rotor, and to it the hand is attached. The restraining force is supplied by springs, as in the previously considered instruments. The winding P is connected in the circuit whose current is to

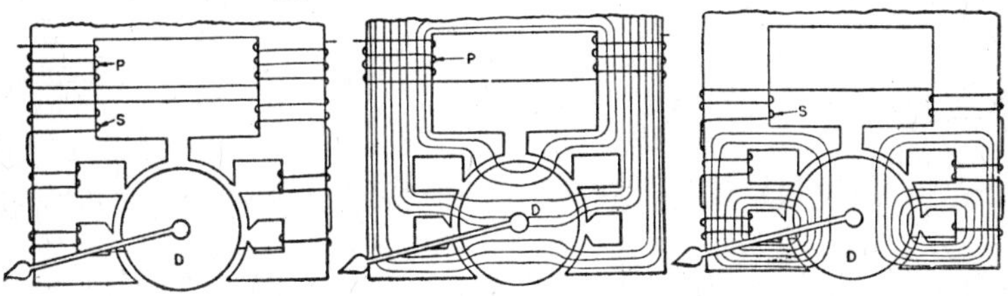

(a) (b) (c)

FIG. 19-6

be measured. The current through this winding sets up a flux, as shown in fig. 19-6b, and by transformer action establishes a current in winding S, which sets up a flux as shown in fig. 19-6c. The construction of the instrument is such that the fluxes shown in figs. 19-6b and 19-6c pulsate 90° out of phase, so that a rotating magnetic field is produced.

The interaction of the rotating flux and the eddy currents established in the rotor cause the rotor and the attached hand to turn to a point where the turning force is just balanced by the restraining spring.

This instrument can be used only for making alternating current measurements. It is quite accurate for frequency with very wide limits. It has a very long scale, extending for nearly 360 degrees, which makes it possible to take very accurate readings.

High Frequency Meter—The instruments previously mentioned are inaccurate for frequencies above about 100 cycles per second and would not indicate at all at frequencies as high as 100,000 cycles, due to the fact that the high reactances of the coils would permit practically no current to flow through them. Two types of instruments are suitable for making high frequency alternating current measurements. These are known as the hot wire and the thermocouple instruments.

Hot Wire Instrument—A simplified diagram of a hot wire instrument is shown in fig. 19-7. F represents a silk thread connected to spring S, wound around pulley P, and attached to a bead threaded on wire AB. Wire AB, which is made of a platinum alloy, is connected in the circuit whose current is to be measured. This wire heats and lengthens when current passes through it. The slack is taken up by spring S, causing pulley P to turn and move the hand over the scale.

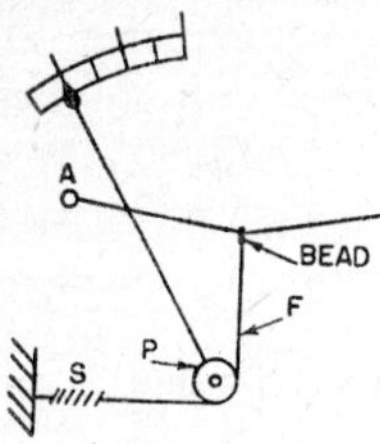

FIG. 19-7

This type of instrument is slow in indicating and the tension in wire AB changes with a change in room temperature. This means that the hand must be set to zero practically each time the instrument is to be used.

Thermocouple Instrument—The construction of a thermocouple instrument is shown in fig. 19-8. The direct or alternating current to be measured is sent through heater H, which heats the junction of two dissimilar metals. Two dissimilar metals, which are joined together and have their junction heated, develop a unidirectional e m f, which is proportional to the temperature difference between the junction end and the open end of the thermocouple. A sensitive milliammeter is connected to the open ends, and is generally calibrated to indicate the current through the heater. For very small readings, the heater and thermocouple are enclosed in an evacuated glass bulb, to eliminate oxidation.

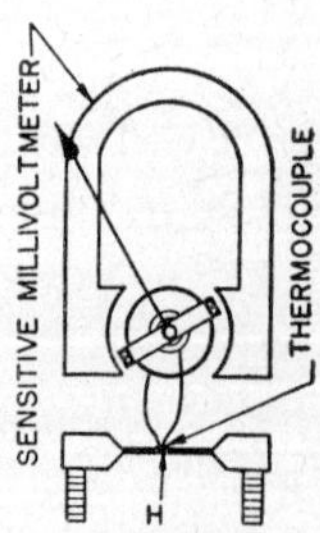

FIG. 19-8

Wheatstone Bridge—Resistances are measured most accurately by means of an instrument called a Wheatstone Bridge. A simplified diagram of this instrument is shown in fig. 19-9. It consists of three known resistances (R_1, R_2, and R_3) to which an unknown resistance (R_x) to be determined is connected. A low voltage battery and a galvanometer (G) are connected to the resistors as shown (a galvanometer is an instrument that measures extremely small currents). When the battery switch is closed, the current flows to A, then divides, one part through resistances R_x and R_1, and the other part flows through R_2 and R_3. The currents unite at C and return to the battery. After the galvanometer switch is closed, the variable resistances R_1, R_2, and R_3 are adjusted so that the galvanometer indicates zero. This adjustment is generally made by setting R_2 and R_3 at some fixed values such as 1, 10, 100, etc. Then R_1 is varied until the galvanometer hand does not deflect. Now the bridge is said to be balanced. Points B and D now are at the same electrical pressure; if any electrical pressure existed between these points, a current would flow through the galvanometer and its hand would deflect. Let I_1 represent the current through path ABC and I_2 represent the current through path ADC. The voltage drop from A to B equals I_1R_x and from A to D equals I_2R_2. But since points B and D have equal pressure, the drop from A to B must equal the drop from A to D, or

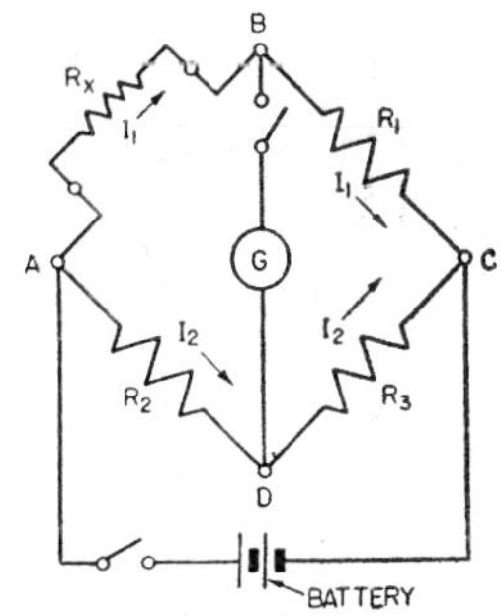

FIG. 19-9

$$I_1R_x = I_2R_2 \tag{19-1}$$

Similarly, the voltage drop from B to C must equal the drop from D to C, or

$$I_1R_1 = I_2R_3 \tag{19-2}$$

Dividing 19-1 by 19-2 gives

$$\frac{I_1R_x}{I_1R_1} = \frac{I_2R_2}{I_2R_3}$$

Cancelling I_1's and I_2's gives

$$\frac{R_x}{R_1} = \frac{R_2}{R_3} \text{ or } R_x = \frac{R_2}{R_3}R_1 \tag{19-3}$$

It is not necessary to know the values of R_2 and R_3, only the value of their ratio need be known. This multiplied by R_1 gives the unknown resistance (see equation 19-3).

EXAMPLE—After a bridge is balanced, $R_2/R_3 = 100$ and $R_1 = 34$. What is the value of R_x?

$$R_x = 100 \times 34 = 3400 \text{ ohms}$$

One common type of Wheatstone Bridge is shown in fig. 19-10a. In this type, each resistance is precision wound and connected between adjacent brass blocks as shown in fig. 19-10b. Resistance R_1 is

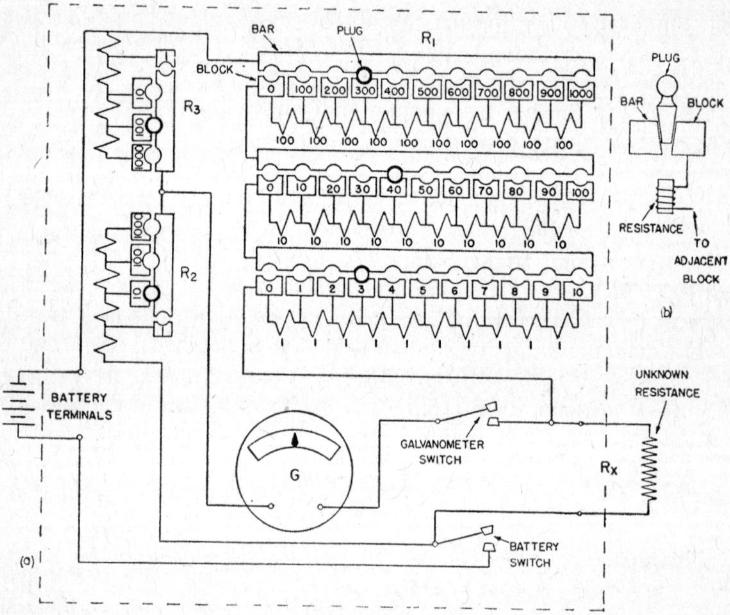

FIG. 19-10

divided into three parts. One part consists of ten 1 ohm coils; the second part consists of ten 10 ohm coils; and the third part consists of ten 100 ohm coils. One plug is used in each part. For the plug settings shown, $R_1 = 343$ ohms, $R_2 = 10$ ohms, and $R_3 = 100$ ohms. Then $R_x = 34.3$ ohms.

Slide Wire Bridge—Another common type of bridge is the slide wire bridge shown in fig. 19-11. It is the same as the one shown in fig. 19-9 except that R_2 and R_3 are replaced by a high resistance wire (ADC, fig. 19-11) stretched tightly alongside of a scale. After the switches are closed, sliding contact D is moved along the wire until the galvanometer indicates zero. At this point a balance is reached.

Since the resistance of any section of the high resistance wire is proportional to its length, then in fig. 19-11,

$$\frac{R_2}{R_3} = \frac{\text{length of AD}}{\text{length of DC}} \tag{19-4}$$

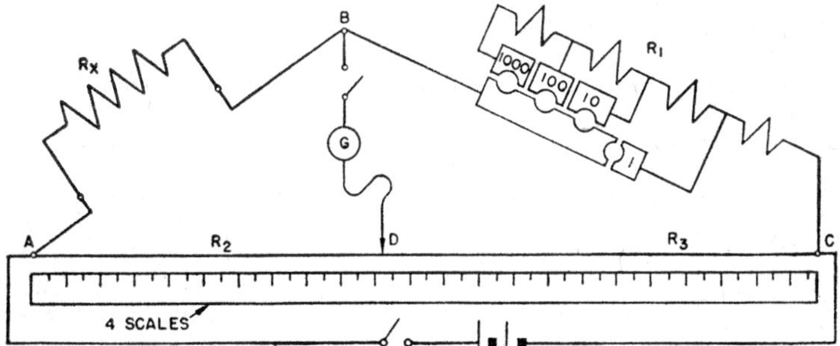

FIG. 19-11

Before equation 19-3 can be applied to the slide wire bridge, it must be expressed as follows:

$$R_x = \frac{\text{length of AD}}{\text{length of DC}} R_1$$

This instrument has four scales, one for each value of R_1, so that when the bridge is balanced it is merely necessary to read directly the value of the unknown resistance R_x.

Megger—In high voltage systems it is essential that the insulation on conductors be such as to permit a negligible current leakage. If the current becomes appreciable at some point, a high temperature is produced, causing the insulation to char and completely break down.

To determine whether or not an insulation is faulty, its resistance is measured by an instrument called a megger. This instrument is primarily suitable for measuring high resistances quickly, having a range of about 10^8 ohms. It is constructed as shown in fig. 19-12. The permanent magnets M supply the magnetic field for the hand driven direct current generator and the indicating instrument. Coils V1 and

V2 are in series and connect across the terminals of the generator. When a resistance to be measured is connected to terminals T, coil C is in series with it. Coils V1, V2, and C are rigidly fixed together, but are free to rotate about axis O. No spring is used, thus the hand may rest anywhere on the scale when the instrument is not being used.

Consider that no resistance is connected to terminals T and the generator is rotated. A voltage is induced, which sends a current through coils V1 and V2, causing these coils to take a position along the dotted line, in the gap of the iron ring. The hand then points to infinity on the scale.

Now consider that a resistance is connected to terminals T. A current will flow through coil C, establishing a force which moves the indicating hand toward zero. As the hand moves toward zero, coils V1 and V2 move into a field of increasing strength, and consequently, an increasing restraining torque is produced. When the torque produced by the current in coils V1 and V2 just balances that produced by the current through coil C, the hand comes to rest, indicating the value of the external resistance.

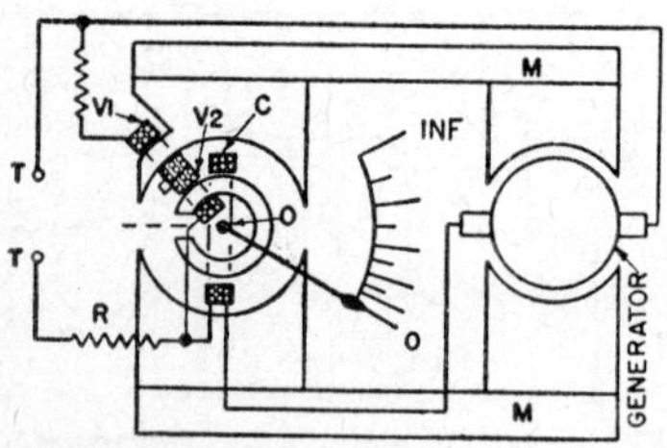

FIG. 19-12

If terminals T are short-circuited, the hand will come to rest at zero. This will not harm the instrument because the current is limited by resistance R.

The reading of the instrument does not depend upon the speed at which the generator is driven, because the currents through both circuits change to the same extent for a given change in the generated voltage.

Direct Current Watthour Meters—Since electrical energy is the product of power and time, it can be measured only with an instrument that takes into account both of these factors. Such an instrument is called a watthour meter, and in principle operates as a small motor. One type, known as the Thomson watthour meter, is constructed as shown in fig. 19-13. Its field coils F are wound with heavy wire and are connected in series with the load, thus producing a magnetic field in which armature A rotates. The small coil C, armature A, and resistance R are in series and connected across the line.

The magnetic field produced by coils F is proportional to the load current, because the magnetic circuit has no iron. The current in the armature is proportional to the line voltage, because it is connected across the line. Then the torque developed in the armature, which is proportional to the field flux and the armature current (see equation 7-10), is also proportional to the line current and the line voltage, or the power supplied to the load. This torque causes the armature to rotate and drive indicators through a train of gears, recording the number of kilowatt hours delivered.

Coil C produces a weak field, even when no current is supplied to the load. This field establishes enough torque to just overcome friction.

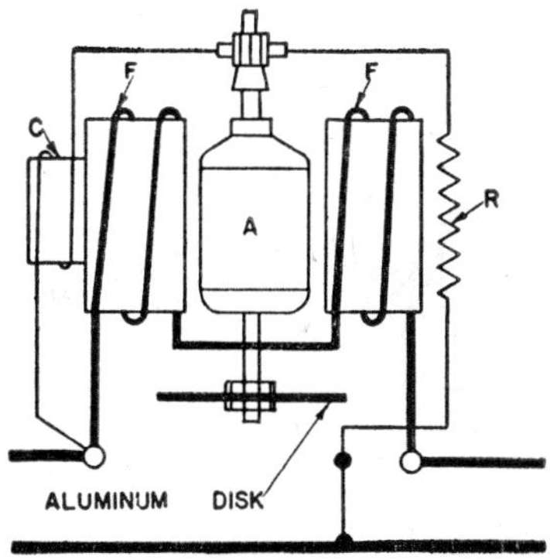

Fig. 19-13

For the meter to indicate correctly, its speed must be directly proportional to the power. Since the driving torque is directly proportional to the power, a resisting torque must be provided that is directly proportional to the speed. This resisting torque is obtained by mounting an aluminum disk on the shaft, which is arranged to rotate between the poles of two permanent magnets. The current induced in the disk is directly proportional to the speed, and therefore a resisting torque is developed directly proportional to the speed.

The construction of another type of watthour meter, known as the mercury motor type, is shown in fig. 19-14a. The rotating unit, which drives the dials through a train of gears, consists of a copper disk. Attached to the top of this disk is a cylindrical piece of wood, which causes the disk to float in the mercury, thus relieving the pressure on the lower bearing. The current through the voltage coil sets up a magnetic field as indicated in fig. 19-14b. The line current enters and leaves the rotating disk through the mercury. The mercury serves the same purpose as the brushes and commutator in the Thompson type watthour meter. The reaction between the current in the disk and the magnetic field (fig. 19-14b) establishes a torque, which varies directly with the line current and the field strength. The line current varies with the load and the field strength varies with the voltage across the coil. Then the torque is proportional to the power supplied to the load, which is the condition necessary in a watthour meter.

In fig. 19-14a it will be noted that the voltage coil is connected to a resistance that is shunted across the mercury. By moving the connection to this resistance, which is done by shifting the clamp, the torque at no load is adjusted to just compensate for friction. The friction established by the mercury increases with the speed of the

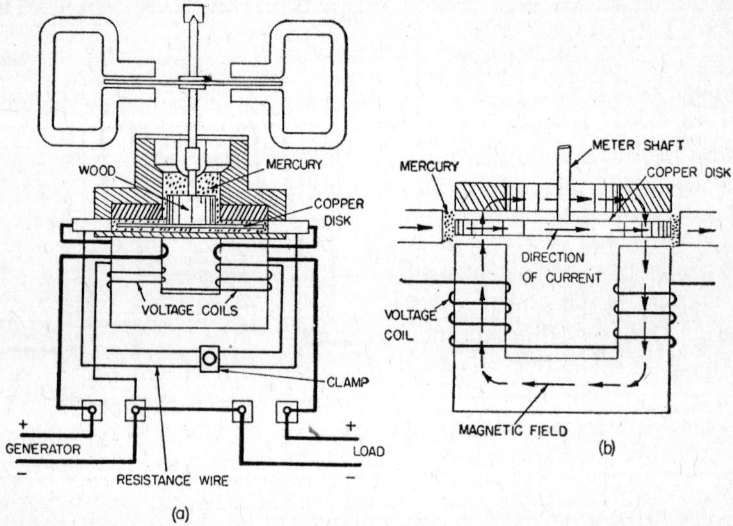

FIG. 19-14

copper disk. Compensation for this additional friction is secured by a winding connected in series with the ungrounded line.

As in the Thompson watthour meter, the necessary resisting torque is provided by an aluminum disk rotating between the poles of permanent magnets. This instrument is used for direct current only.

Alternating Current Watthour Meters—The watthour meter shown in fig. 19-13 will operate satisfactorily when connected in an alternating current circuit. However, in such a circuit an induction type watthour meter is generally used, because it is simpler and more reliable than the other types. An inductive type instrument is shown in fig. 19-15. V is the voltage coil and C the current coils. The aluminum disk D rotates between the poles, but is off center, and turns the indicating hands through a train of gears. In order that the instrument will indicate correctly at all power factors, the flux produced by coil V must lag the line voltage by 90 degrees (attained by varying resistance R). With a lagging current, if the instrument under-registers, the flux of coil V lags the line voltage less than 90 degrees, and therefore resistance R should be decreased. This is accomplished by twisting R further and soldering it.

Above the pole surrounded by coil V, a shading coil S is provided, and can be shifted by a lever. This shading coil causes the flux to

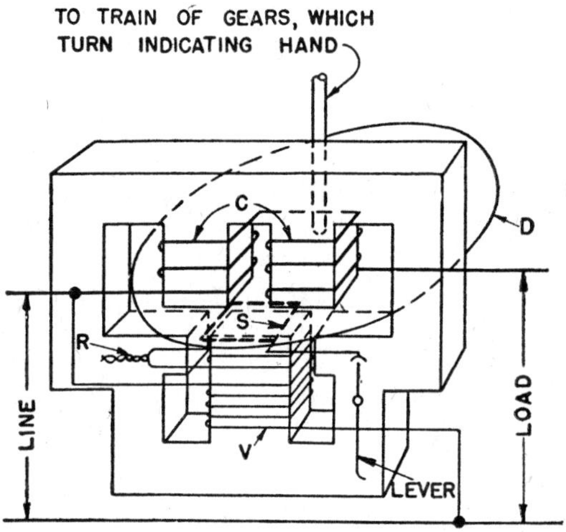

FIG. 19-15

sweep across the pole face as explained for single phase induction motors (chapter 15), and establishes a small torque. Shading coil S should be adjusted so that just enough torque is produced to overcome friction.

The driving torque is produced as follows: Assuming a unity power factor load, at the instant that the line current and the line voltage are zero (instant 1, fig. 19-16a), the flux in coil V is maximum and the flux in coils C is zero but changing at its maximum rate. The rapidly changing flux establishes maximum current in the aluminum disk, as indicated in fig. 19-16b. This current passes above the pole in coil V, which at this instant is carrying maximum flux, and therefore a torque is developed clockwise.

At instant 2 (fig. 19-16a), the flux in coils C is maximum and the flux in coil V is zero but is changing at its greatest rate, establishing a current in the aluminum disk as indicated in fig. 19-16c. This current passes under the poles of coils C, in which the flux is maximum, and again the torque developed is clockwise.

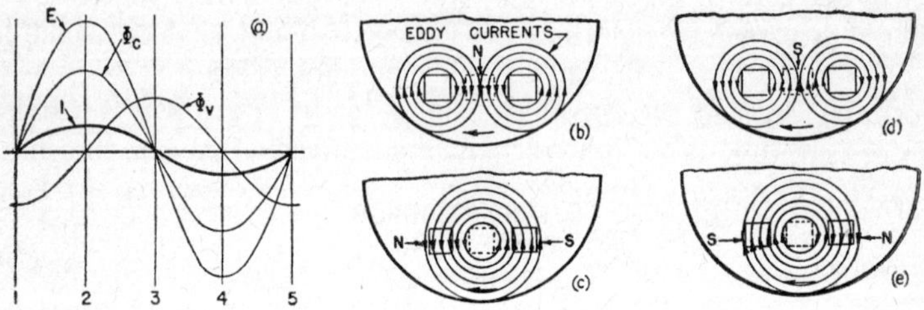

FIG. 19-16

At instant 3, the current and flux relations are as shown in fig. 19-16d; and at instant 4, the conditions are as indicated in fig. 19-16e. In both of these cases, the torque developed is also clockwise. Then a torque is produced continuously in one direction.

Three-Phase Watthour Meter—In a three-phase system, energy, like power, may be measured by means of two single-phase in-

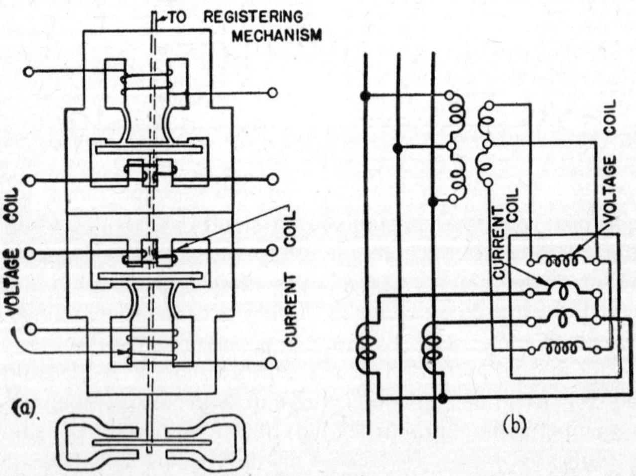

FIG. 19-17

struments. A three-phase instrument (fig. 19-17a) is merely a combination of two single-phase instruments, with their moving elements mounted on the same shaft, in which case the total driving torque is the sum of the torques exerted by both moving elements. Thus, only one registering mechanism is required. This meter is connected as shown in fig. 19-17b.

Power Factor Meter—One type of power factor meter is constructed as shown in fig. 19-18a. The two coils A and B, which are alike, are placed at right angles to each other. They are connected across the line and are therefore voltage coils. One path includes a high resistance and the other a high inductance. The currents in the two coils are nearly 90 degrees out of phase and therefore produce a

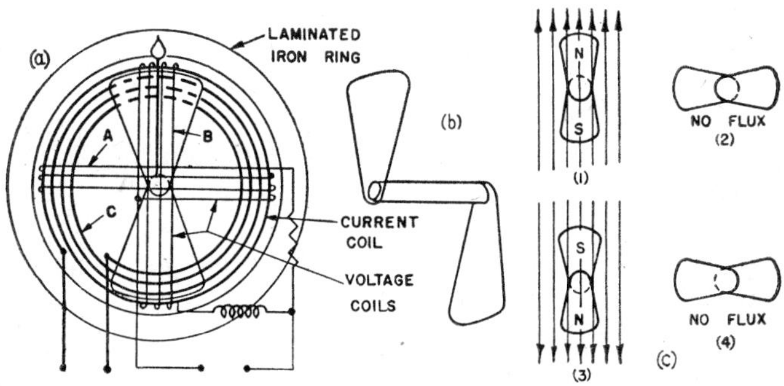

FIG. 19-18

rotating magnetic field, as explained in the discussion of split phase induction motors (chapter 15). Coil C, which is at right angles to both coils A and B, is connected in series with the load and is therefore a current coil.

A soft iron vane, constructed as shown in fig. 19-18b, is mounted so that its axis coincides with the axis of coil C, and is magnetized by the current through this coil. The laminated iron ring provides the return path for the flux. *The attraction between the iron vane and the rotating magnetic field will cause the vane to take a position parallel to the rotating magnetic field at the instant that the magnetization of the vane is maximum.* The direction of the rotating magnetic field and the magnetization of the iron vane for 90 degree intervals is shown in fig. 19-18c.

When the current leads the voltage, the magnetization of the iron vane will reach its maximum sooner than when the power factor is unity. As a result, the iron vane will line up with the rotating flux before it gets in the vertical position. Apparently, if the current lags the voltage, the vane will line up with the rotating flux after it has rotated beyond the vertical position.

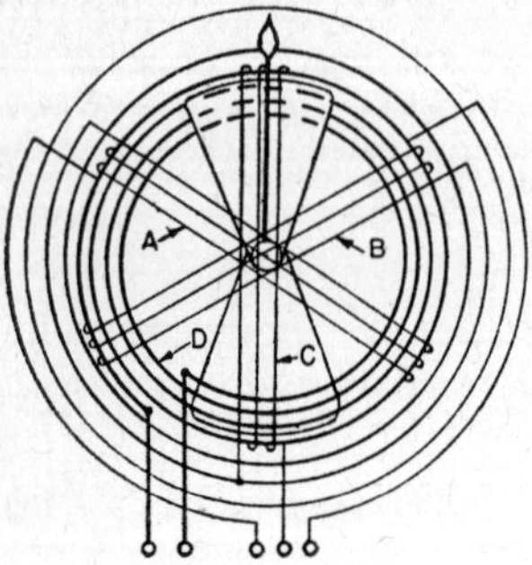

FIG. 19-19

In fig. 19-19 is shown a wiring diagram of a three phase power factor meter. In this case the rotating magnetic field is produced by a three phase winding, consisting of coils A, B, and C, instead of the split phase winding, which is used in single phase power factor meters. As in the case of single phase power factor meters, coil D produces an alternating field in the iron vane, which takes a position parallel to the rotating magnetic field when the magnetizing action in the iron vane is maximum.

Synchroscope—The power factor meter indicates the phase relation between the current and the voltage, but it will show the phase relation of two voltages if the current coil is replaced by a voltage coil and the leads brought out as in fig. 19-20. Such an instrument is called a synchroscope, because it is used to indicate whether or not two alternators are in synchronism (see chapter 14).

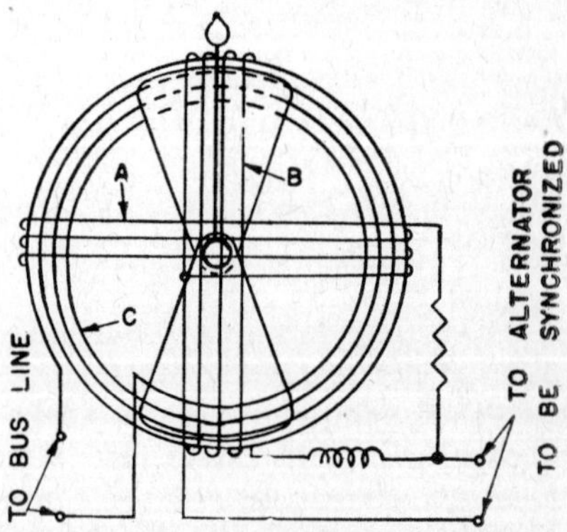

FIG. 19-20

Coils A and B (fig. 19-20) are connected to the alternator to be synchronized (through a voltage transformer), and coil C is connected to the bus line. If the frequency of the alternator is higher than that of the bus line, the hand of the synchroscope will turn in one direction; if the frequency is lower, the hand will turn in the opposite direction. If the two frequencies are nearly equal, the voltages will be in phase or nearly so for long periods. During the time the voltages are in phase, the current in coil C and the flux in the iron vane will reverse with the current in coil A, and the iron vane will remain stationary. The hand will be at standstill at a mark on the dial indicating synchronism, and the switch connecting the alternator to the power line may be closed.

Frequency Meter—Two types of instruments are manufactured for measuring alternating current frequencies. One type is shown in fig. 19-21a. The series of thin reeds and the soft piece of iron A are attached to bar B, which has a flexible support. When coil C is energized with a current whose frequency is to be measured, A is attracted twice each cycle, vibrating bar B. The reeds attached to bar B have natural frequencies that differ in sequence by one vibration per second, because the reeds have different lengths (in some types the reeds are all the same length but are loaded at their free ends with different amounts of solder). Then bar B, on vibrating, causes all the reeds to vibrate slightly. But the reed that has a natural frequency equal to that of the alternating current through coil C will vibrate violently (see fig. 19-21b).

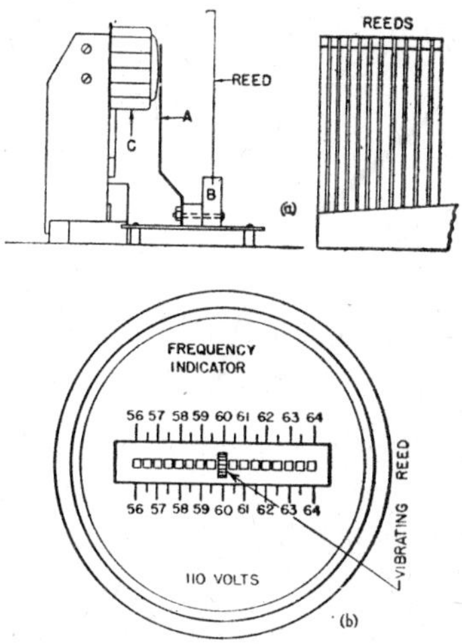

Fig. 19-21

Another common type of frequency meter is shown in fig. 19-22. In this instrument the shading coil principle is used. One shading coil tends to turn the disk in one direction and the other tends to turn the disk in the opposite direction. The two coils are connected in parallel, but one is in series with a high resistance and the other with a high inductance. For a given voltage, then, the current through A is practically constant, while the current through B depends on the frequency. If the frequency is constant and the voltage is changed, the currents in the two coils, and therefore the opposing torques, will change the same amount. Thus the indication of the instrument is not affected by voltage changes.

If the disk were circular it would tend to turn continuously. To prevent this, the left half of the disk is a semi-circle with its center on a straight line passing through the centers of coils A and B, and the right half is a semi-circle with its center slightly higher than that of the left half. Then the area of the disk in the air-gap of magnet A is constant, while that in the air-gap of magnet B varies.

When the frequency increases, the torque produced by magnet B becomes weaker and the disk turns clockwise. In so doing, the area

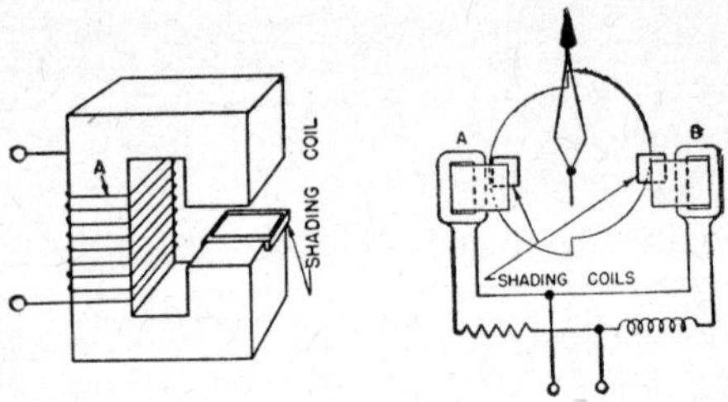

(a) (b)

FIG. 19-22

of the disk in the air-gap of magnet B increases until the two torques are equal. Then the disk stops.

When the frequency decreases, the torque produced my magnet B becomes stronger and the disk turns counter-clockwise. The area of the disk in the air-gap of B decreases until the two torques are equal.

Protective Relays—A protective relay is an electrical instrument which causes circuit breakers or oil switches to open when some abnormal condition occurs, thus protecting the circuit and attached apparatus. Such relays may be divided into three types: inverse time, definite time, and inverse-definite time.

It has become a common practice to indicate the function of relays in wiring diagrams by means of numbers prepared by the National Electric Manufacturers Association. In the discussion of relays that follows, these numbers and their meaning are given.

Inverse Time, Bellows Type Relay—Fig. 19-23a shows an overload relay in which the time limit is obtained by means of a bellows. When the current through the coil reaches a certain strength, plunger P is drawn upward. This plunger moves up slowly because it compresses the air in the bellows, which gradually escapes through opening 1 (fig. 19-23b). No air can escape through 2, because the air exerts a pressure at the top of cylinder C, holding it down and shutting off passage 3. Then, after a certain time, disk D (fig. 19-23a) moves against contacts C, completing the circuit of the trip coil and opening the oil switch (fig. 19-23c). Two such relays are connected to a three-phase line as shown in fig. 19-23d. This type of relay is

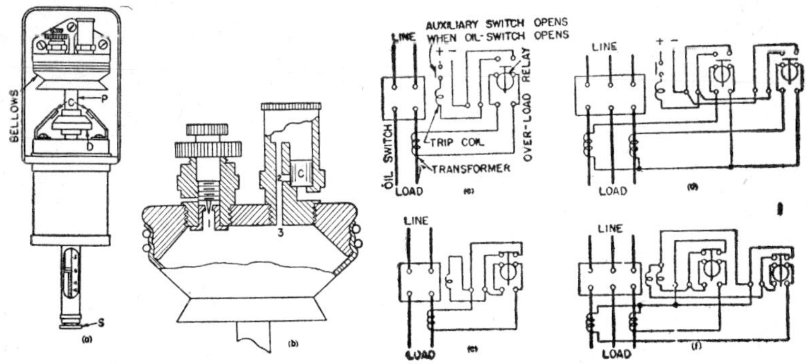

FIG. 19-23

known as a circuit opening type. It is used when the trip coil is energized with direct current. If the trip coil is energized with alternating current, a relay is required in which the movable contact is normally closed, and which opens when a high current flows through the coil (fig. 19-23e). In a three-phase system, the relays are connected as in fig. 19-23f.

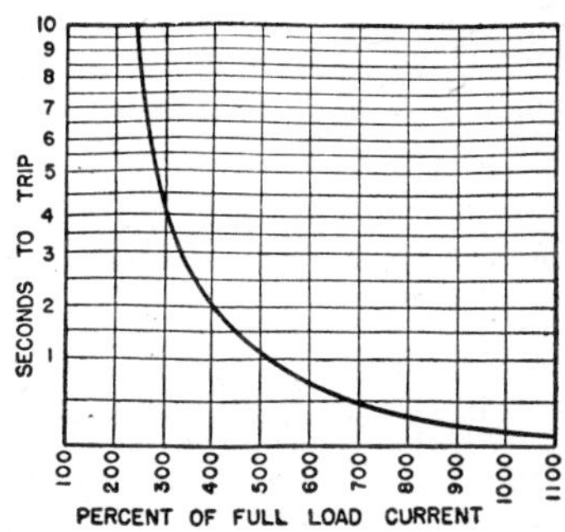

FIG. 19-24

When the coil of a plunger type relay is deenergized, the plunger starts to move downward and the air pressure in the bellows (fig. 19-23b) decreases below that of the outer air. Then the outer air lifts cylinder C and passes through 2 and 3 and into the bellows, thus allowing the plunger to return to its normal position quickly.

The relay shown in fig. 19-23a may be adjusted to operate at different current values by adjusting screw S. The time required for it to operate varies inversely with the current. This is indicated by the curve in fig. 19-24.

Definite Time Relays—A definite time relay (fig. 19-25) is constructed the same as the relay shown in fig. 19-23, except that the

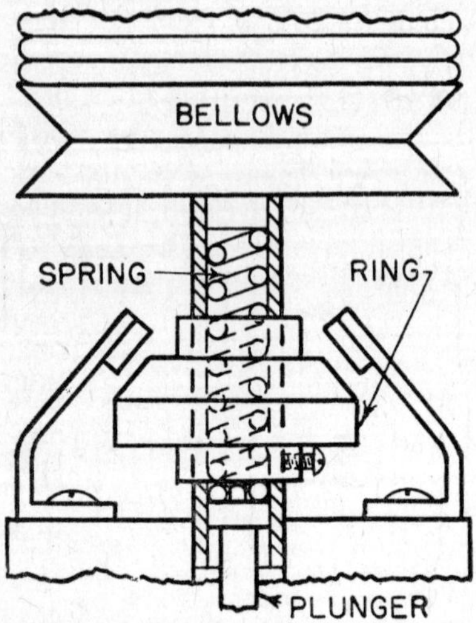

FIG. 19-25

plunger is not rigidly attached to the bellows and the contact ring. When its coil is energized, the plunger instantly rises and compresses a spring, which in turn raises the shaft attached to the bellows and contact ring.

When this type of relay is used, the spring is compressed instantly when an overload occurs. Then the bellows allows the contacting ring to move slowly upward and make contact. This takes place in a definite time for an overload of any value.

Inverse time and definite time relays are denoted by 62, which means time delay relays.

Inductive Type Current Relays—A relay is made that gives a more accurate and more reliable protection than the plunger type. Its basic principle of operation is the same as that of the inductive

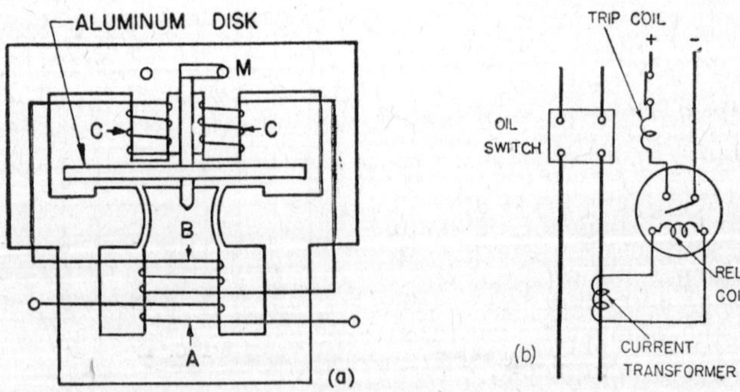

FIG. 19-26

type watthour meter, and it is therefore called an inductive type relay. This relay is constructed as shown in fig. 19-26a. It is connected to a single phase line as shown in fig. 19-26b. When current flows through coil A (fig. 19-26a), a high voltage is induced in coil B, energizing coils C. If this current is high, sufficient torque will be developed to cause the aluminum disk to rotate and close the main contact M.

To make sure that the disk will remain at the starting point until overload occurs, holes are made in the disk between the poles of the electromagnet. Due to these holes, the torque is low when the disk is in the normal position. As soon as the current is high enough to cause the disk to move slightly, the holes move out from between the poles, after which a high torque is developed. The disk then rotates and closes the main contact.

This relay has inverse time limit characteristics, as indicated by the curve shown in fig. 19-24. At 300 per cent overload the relay trips at the end of four seconds (fig. 19-24), and at 400 per cent overload it trips at the end of two seconds. For loads above this the relay is nearly instantaneous, which is generally undesirable.

Inverse-Definite Time Relay—An inverse-definite time relay is shown in fig. 19-27a. It is the same as the one shown in fig. 19-26a, except that it has a small transformer in the secondary circuit as indicated in fig. 19-27b. At low overloads the current in S increases in the same proportion as the current in P, thus providing inverse time. When the currents in windings P and S reach certain values, the transformer core becomes saturated. The current in S, and therefore the flux in poles C, cannot increase further, regardless of what the overload may become. The flux entering the disk from B increases slightly because most of the additional flux set up in this pole passes through air gaps G. Thus the relay has inverse time limit on moderate overloads and definite time limit on heavy overloads. This characteristic is indicated by the curve in fig. 19-28. This relay is denoted by 51 which means A C overcurrent relay.

Auxiliary Contact—The auxiliary contact (fig. 19-27b) is near the bottom of the relay (not visible in fig. 19-27a) and short-circuits the main contact, so that no arc is produced at this contact when it opens. This contact is very small and would burn easily. The circuit of the trip coil is kept closed by the auxiliary contact until the oil switch opens, whether or not the main contact remains closed.

Taps—The current coil of the relay (fig. 19-27b) is provided with a number of taps, so that it may be adjusted to operate at selected

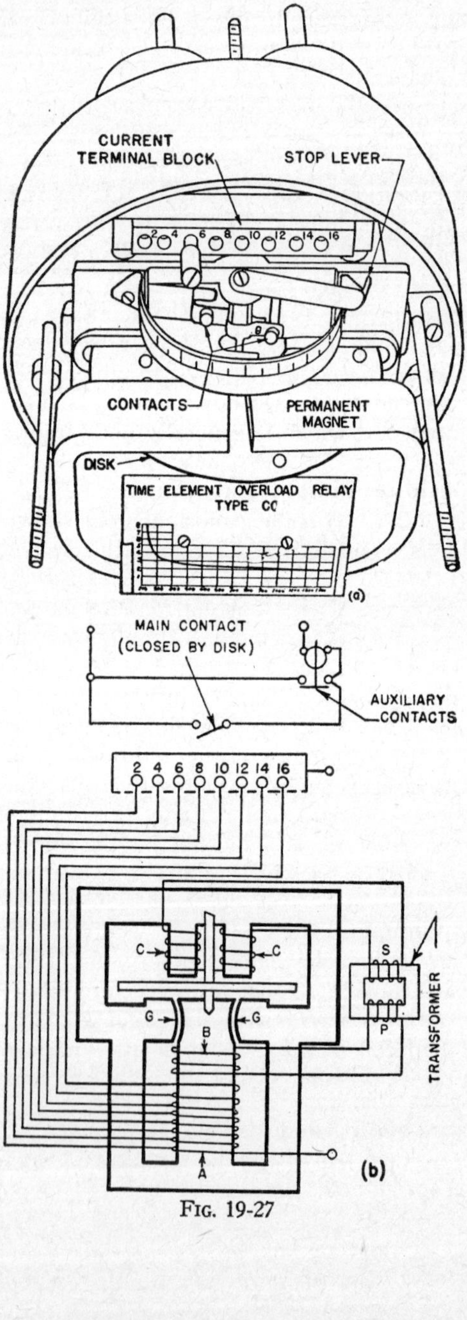
CURRENT
TERMINAL BLOCK
STOP LEVER
2 4 6 8 10 12 14 16
CONTACTS
PERMANENT
MAGNET
DISK
TIME ELEMENT OVERLOAD RELAY
TYPE CO
(a)
MAIN CONTACT
(CLOSED BY DISK)
AUXILIARY
CONTACTS
2 4 6 8 10 12 14 16
C
C
S
G
G
B
P
TRANSFORMER
A
(b)

FIG. 19-27

values of current. Referring to figs. 19-27a and 19-27b, a tap connection for a current of 10 amperes is made by inserting a screw in the hole marked 10.

Contact Stop—It is necessary to have a means of adjusting the time limit of a relay for a given load. This is accomplished by adjusting the angle through which the main contact rotates before closing. For a given load, suppose that the contact travels half a revolution in six seconds to close. If the stop lever (see fig. 19-27a) is set so that the main contact needs travel only one-quarter of a revolution to close, this will take place in three seconds. By means of the stop lever, then, the time limit may be set as desired.

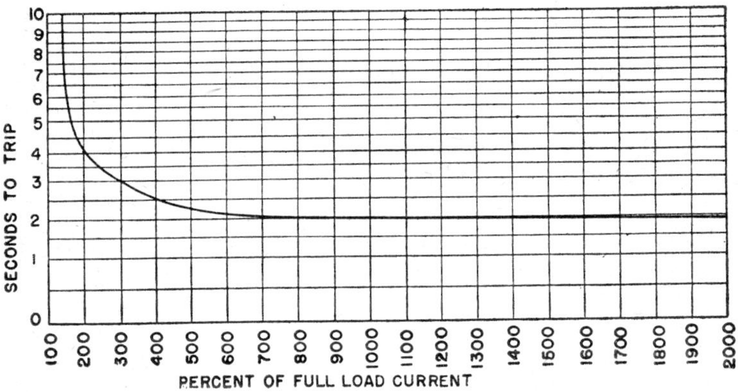

Fig. 19-28

Voltage Relay—To protect lines against abnormally high or abnormally low voltages, voltage regulating relays are used. Over-voltage causes lamps to burn out, while under-voltage causes them to burn dimly and less efficiently. Low voltage impressed on motors causes them to run slowly and draw excessive current. Relays used to guard against these conditions are known as over-voltage and under-voltage relays. Both are constructed the same as the 51 relay (fig. 19-27), except that the main coil in this type is wound with many turns of fine wire and is connected across the line.

In the over-voltage relay, denoted by 59 and connected as shown in fig. 19-29a, a spring holds the contact open. If the voltage becomes abnormally high, sufficient torque is developed to overcome the spring and close the contact, energizing the trip coil and thus opening the oil switch.

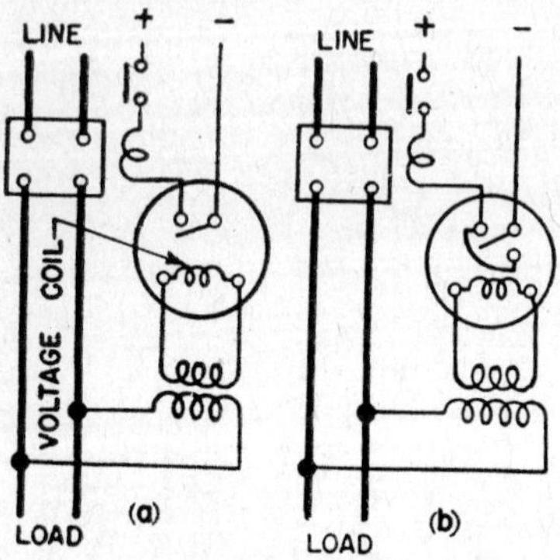

Fig. 19-29

In the under-voltage relay, denoted by 27, a spring holds a contact closed when the winding is deenergized. The contact is held open when the voltage is normal. If the voltage drops below whatever value the relay is set the developed torque is not sufficient to hold the contact open, so it closes, opening the oil switch.

A voltage regulating relay, denoted by 58, is manufactured with two contacts (see fig. 19-29b). One contact closes when the voltage drops below the normal value, and the other closes when the voltage rises above the normal value. In closing either way, the oil switch is tripped.

Power Relay—A power relay is the same as the inductive type watthour meter except that the shaft attached to the revolving disk operates a contact. In this relay, the torque developed holds the contacts open when the power is in the right direction and closes the contacts when the power is in the wrong direction. While this type of relay is generally used for protecting against a reversal of power, it is sometimes used for protection against an excess of power in the right direction. When so used, either its current or its voltage coil must be connected in reverse.

Power Directional Relays—Consider that power is supplied from a power house bus to a substation bus over two feeders, as shown in fig. 19-30. Suppose that a short circuit occurs at S in feeder A. Power will feed into the short circuit, not only through feeder A, but also through feeder B, the substation bus, and back through feeder A to S. If overload relays only are used on each feeder, they will operate, opening the oil switches in both feeders. But in this case it is desirable that only the oil switches in feeder A open.

Suppose that in this case power relays are used at oil switches 2 and 4 as indicated in fig. 19-30. The power through 4 is in the right direction; then its relay contacts are held open. The power through 2 is in the wrong direction; then its relay contacts close, opening it. After 2 opens, a high current will continue to flow from 1 to the short circuit S. Oil switch 1 will then be opened by its overload relays.

In fig. 19-30, if the short circuit is near the power relays of oil switch 2, a low voltage will be impressed on the voltage coils of these relays and they will not operate. Moreover, in a three-phase system,

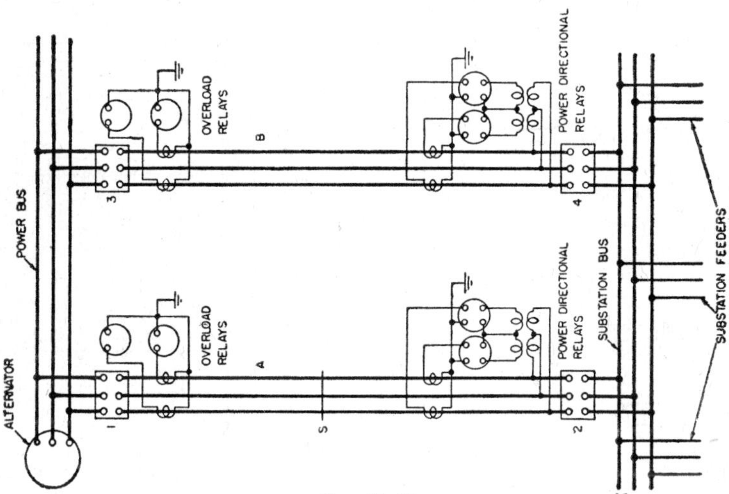

Fig. 19-30

when a short circuit occurs between two line wires, the current and voltage actuating relays are frequently nearly 90 degrees out of phase. Under this condition, almost no torque is developed to close the relays.

Relays are manufactured especially for feeders connected as shown in fig. 19-30. A wiring diagram of one type, known as a power directional relay (denoted by 67), is shown in fig. 19-31a. It is connected to a feeder as shown in fig. 19-31b. The lower unit of this relay is the same as the type 51 relay (see fig. 19-27) and the upper unit is the same as the power relay just discussed, except that it has a weak spring so that the least amount of power in the opposite direction will cause its contacts to close. The relay has two contacts in series, so that both must close to energize the trip coil.

If a short occurs near a power directional relay, the direction of power through this relay reverses, and the upper unit closes its contact instantly (see fig. 19-31a). But the lower unit, being an overcurrent relay, closes after a certain time, depending upon the nature of the overload. After its contact closes the trip coil circuit is complete, and the oil switch disconnects the faulty feeder.

For a three-phase system, three such instruments as that shown in fig. 19-31a are required for each feeder. To serve the same pur-

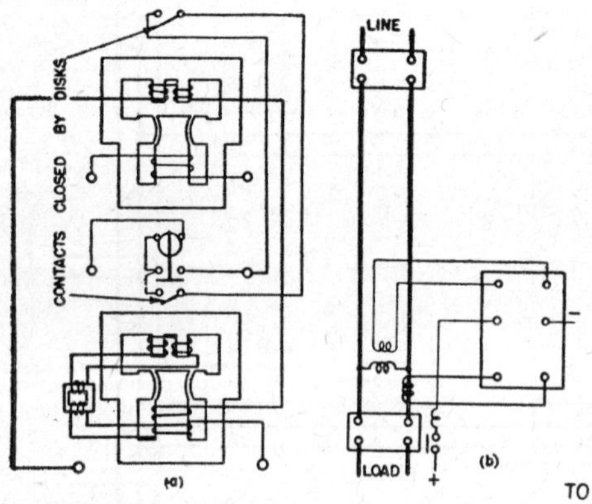

FIG. 19-31

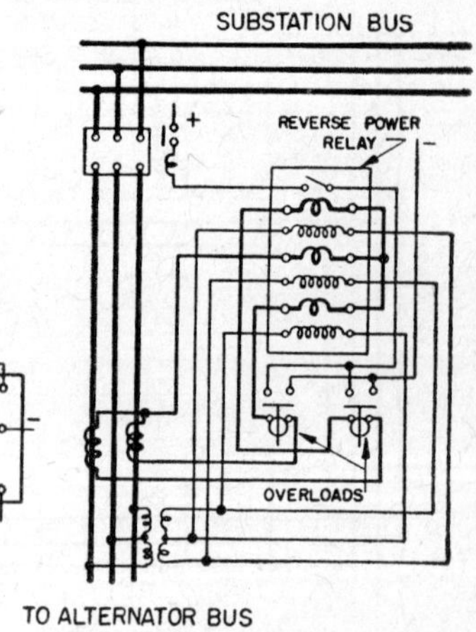

FIG. 19-32

pose, a three-phase instrument is manufactured. This instrument contains three power units, arranged to drive two aluminum disks on the same shaft, to which a single contact is attached. The lower disk is driven by two units and the upper by one. This relay is connected with two or three overload relays (see fig. 19-32).

Phase Balance Current Relay—The phase balance current relay (denoted by 46) is used to protect a motor against an open phase, against a reversal of phase, and against a short circuit existing in one of the phases.

The internal connection of the relay is shown in fig. 19-33a. The lower part is similar in construction to the 51 relay and the upper part consists of stationary resistors and reactors. The relay is connected to the current transformers as shown in figs. 19-33b and 19-33c. The resistors and reactors have such values that no current flows through coil R (fig. 19-33c) when a normal condition exists. If an abnormal condition occurs, such as an open phase, current will flow through coil R and the relay will operate, closing the main contact. This energizes the trip coil, opening the oil switch.

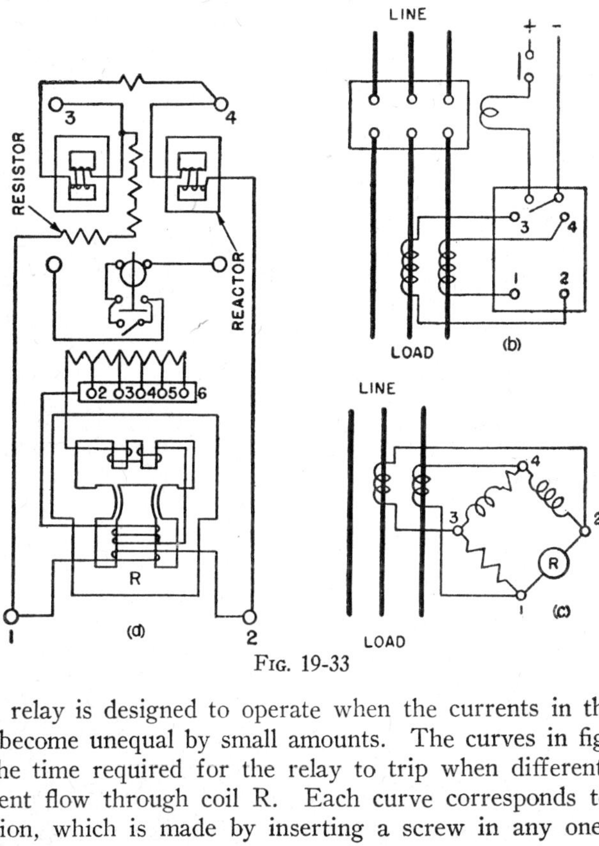

FIG. 19-33

The relay is designed to operate when the currents in the three phases become unequal by small amounts. The curves in fig. 19-34 show the time required for the relay to trip when different values of current flow through coil R. Each curve corresponds to a tap connection, which is made by inserting a screw in any one of the numbered holes.

Differential Protection—Relays are connected differentially to protect alternators, A C motors, buses, circuit breakers, and transformers against faults. A diagram to illustrate the principle of differential protection in one phase is shown in fig. 19-35. In this case a standard over-current relay is used, which is shown in fig. 19-27.

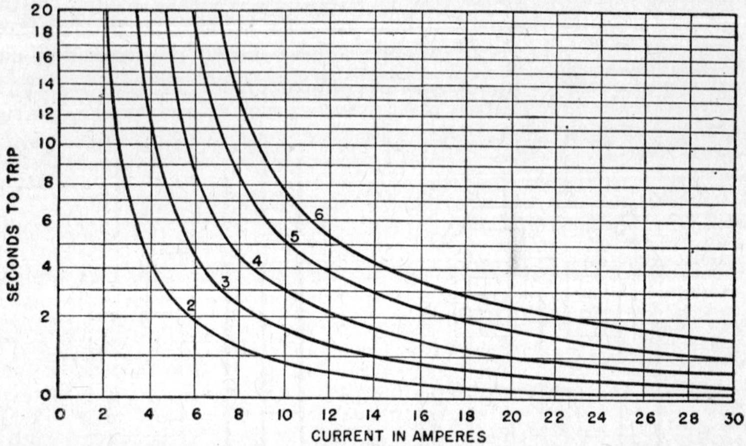

Fig. 19-34

Each of the other phases is protected in the same way. The current transformers have the same turn ratio and consequently give the same secondary currents if no fault exists. The secondaries are connected in series so that current will circulate in them, but none will pass through the relay, since there is as much tendency for the current to flow through it in one direction as in the other.

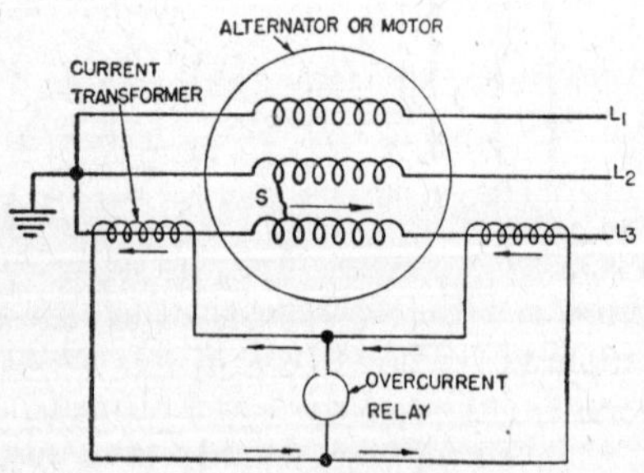

Fig. 19-35

Suppose that a short-circuit occurs in the winding (fig. 19-35) at S. This will cause the currents in the two secondaries to be different; consequently, current will flow through the relay. The relay will close its contacts and thus trip the circuit breaker.

Bus systems are frequently protected by over current relays connected differentially. This is illustrated in fig. 19-36. All the current transformers have the same turn ratio. The secondaries of the current transformers in the incoming lines are connected so as to send current in the same direction as those in the outgoing lives. Under normal conditions the vector sum of the currents flowing in outgoing lines connected to a bus equals the current flowing in the incoming line connected to the same bus. Then, in this case, the currents in current transformers' secondaries of outgoing lines just equal the currents in the current transformers' secondaries of in-

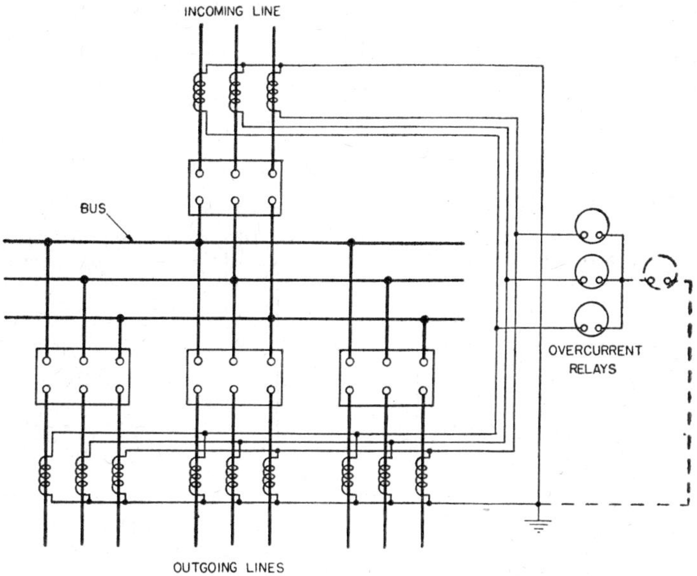

FIG. 19-36

coming lines and no current will flow through the relays. When a fault occurs in the buses or in the circuit breakers, these currents will not be equal. Consequently, current will flow in the relays proportional to the fault current, and the relay will then trip all the circuit breakers connecting to the faulty bus.

In most cases a fourth relay (shown dotted) is added to each group. Current flows through it only when a ground occurs. Because it is very sensitive, it gives sensitive protection against ground faults. This relay (denoted by 64) is known as a ground protective relay.

In fig. 19-35 the secondaries will generally not supply equal currents due to unequal lengths of leads to the relay and also due to the connection of other relays and meters to one secondary and not to the other. When either of these conditions exists, a small current will flow in the relay when the line current is small and a high current will flow when the line current is high. Then, the relays must be set to trip the circuit breakers only when the relay currents are abnormally high so that they will not trip under normal conditions. Because of

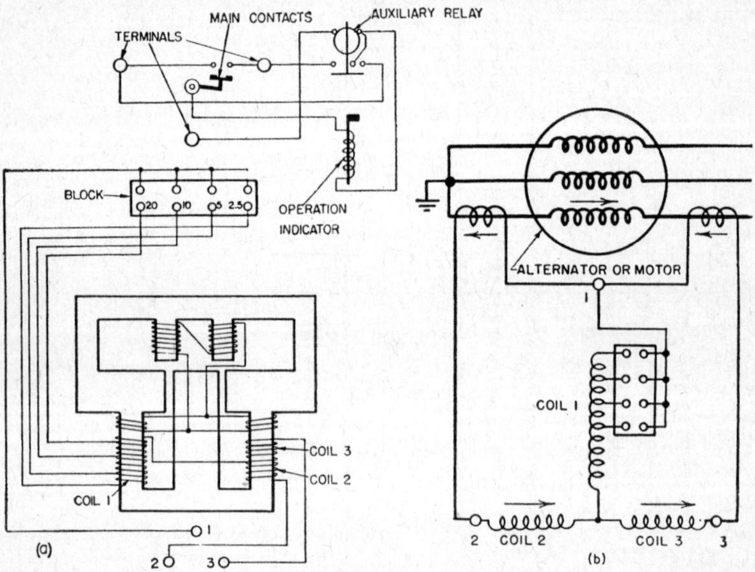

Fig. 19-37

this, the over current relays cannot be set to operate when a high resistance short or ground occurs. Hence, with current overload relays, a high degree of sensitivity cannot be obtained, but this may be secured with differential relays (denoted by 87). These relays are available in two types—one type for alternators and motors (fig. 19-37a), and the other type for transformers (fig. 19-40). They are similar in construction to over current relays.

Alternator and Motor Differential Relay—The connection for generator and motor type relay is shown in fig. 19-37b. Only one is shown for simplicity even though three are used as shown in fig. 19-38. Current through coils 2 and 3 (fig. 19-37b) produce a magnetizing action that is additive and develops a torque that, together with a spiral spring, tends to hold the contacts open. Current through coil 1 produces a magnetizing action which tends to close the contacts. Under normal conditions, the strength of coils 2 and 3 slightly exceeds that of coil 1 by the same amount at all loads. But when the secondary currents become unequal due to a fault, the magnetizing action of coil 1 exceeds that of coils 2 and 3. Then the relay contacts will close and trip the circuit breaker. Coil 1 is provided with taps

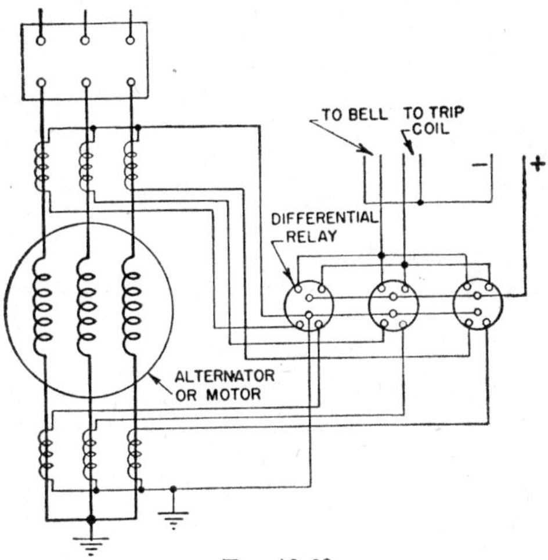

FIG. 19-38

so that the relay may be set to operate on a current unbalance equal to 2.5, 5, 10, or 20 per cent of the smaller secondary current. The tap setting used depends upon the degree of protection desired. In order to make a tap setting, two tap screws are inserted into the block, one directly above the other.

The relay contacts are open only a short distance, and because of this, the contacts close almost instantly, clearing a fault before it does excessive damage to the equipment.

Each relay is provided with a trip circuit suitable for tripping two circuit breakers at the same time. The solenoid operated auxiliary switch in the trip circuit serves the same purpose as the one in the over current relay (fig. 19-27b). It shunts the main contacts in the relay and maintains the circuit until the circuit breaker opens, whether or not the main contacts remain closed. The method of connecting the trip circuits in a three phase system is shown in fig. 19-38.

Differentially connected relays provide protection against a phase to phase or phase to ground fault. They will not protect against an open circuit or a short circuit occurring in the coils in any one phase. Protection against these faults is provided by means of a phase balance current relay (fig. 19-33).

Transformer Differential Relay—Differential relays of the type shown in fig. 19-37a can be used to protect transformers against shorts and grounds. In this case the relay is connected as shown in fig. 19-39.

Sometimes it is not possible to obtain current transformers with turn ratios that will produce equal secondary currents. Also, at times,

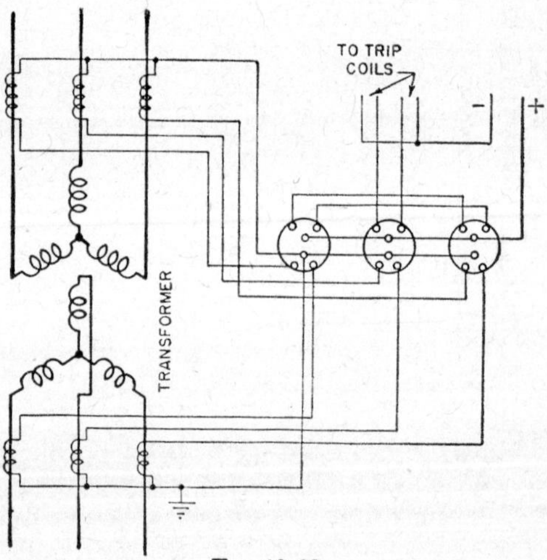

Fig. 19-39

power transformers are connected star-delta, and in this case the currents in the secondaries will be out of phase. For either case, relays of the type shown in fig. 19-37 cannot be used since during normal conditions sufficient current would flow through coil 1 (fig. 19-37) to cause the relay to operate and trip the circuit breaker. For these two cases, the type of relay shown in fig. 19-40a is used. A

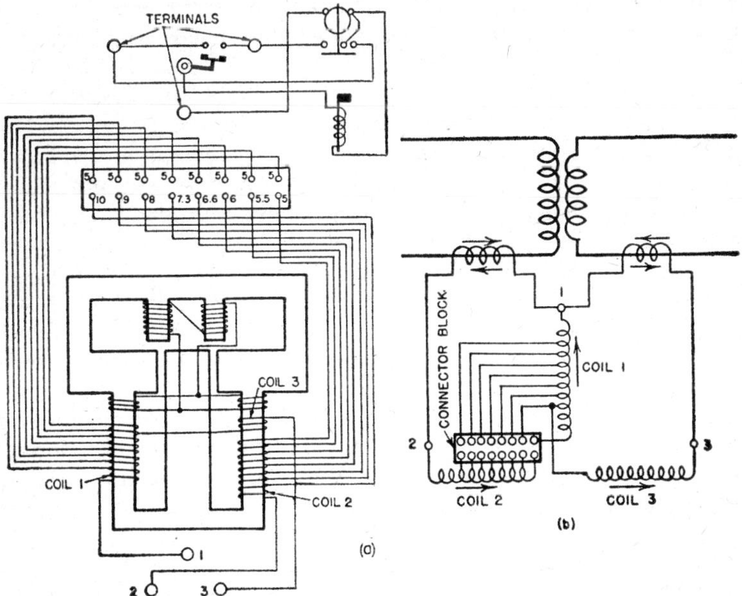

FIG. 19-40

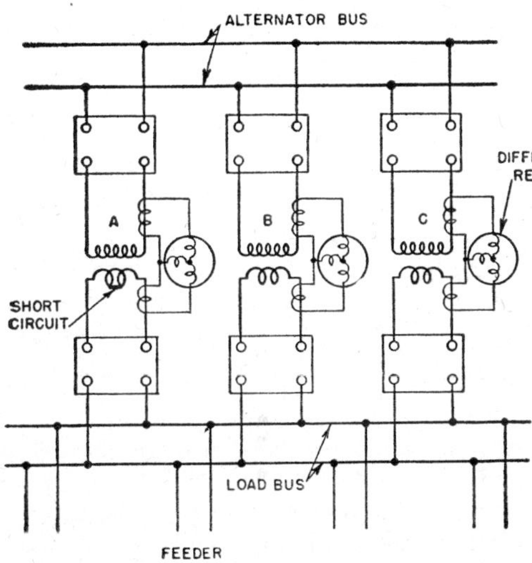

FIG. 19-41

simplified diagram of this relay is shown in fig. 19-40b. Taps are brought out which make possible adjusting the magnetizing action of coil 1 to a value slightly less than that of coils 2 and 3. As in the alternator and motor type, a tap setting is made by inserting two tap screws into the block, one directly above the other.

A very common application of differential relays is to disconnect faulty transformers in a system in which several transformers are connected in parallel. To make clear how this is accomplished, three single phase transformers in parallel will be considered as shown in fig. 19-41, in which case it is assumed that a short circuit occurs in transformer A. This short circuit will cause a reversal of power in transformer A and as a result transformers B and C become heavily loaded. If the three transformers were protected by current overload relays only, the relays would disconnect the three transformers at the same time, interrupting the total power supplied. With the use of differential relays, however, only the faulty transformer is disconnected, and the load is carried by the other two.

Exercises

1. (a) Show by a diagram how the resistors are connected in a three scale voltmeter. (b) How the shunts are connected in a three scale ammeter.
2. A D C ammeter has a 1 ampere range and .05 ohm resistance. Determine the resistance of the shunt required to extend the range to 200 amperes.
3. An ammeter has a resistance of 10 ohms and its shunt has a resistance of .001 ohm. If 100 amperes flow through the line, determine the current flowing through the instrument and the current through the shunt.
4. A voltmeter has a range of 3 volts and a resistance of 300 ohms. How much resistance must be connected in series with it to increase its range to 150 volts?
5. A 30,000 ohm voltmeter indicates 250 volts when connected across a D C power line. It indicates 20 volts when it is in series with a coil and the two are connected across the power line. Determine the resistance of the coil. (This problem illustrates how resistance is determined with the use of a voltmeter.)
6. Can the same wattmeter be used for measuring alternating current and direct current power?
7. Under what conditions will a wattmeter tend to indicate in reverse?
8. Is it possible to burn out one of the coils in a wattmeter without the indicating hand moving off scale? Explain.

9. What is the torque proportional to in a wattmeter?
10. Name the types of ammeters used to measure both direct and alternating current.
11. How does an iron vane voltmeter differ from an iron vane ammeter?
12. An iron vane type instrument, having a resistance of 700 ohms and a reactance of 100 ohms, has a 30 volt range. If the range of this instrument is to be 150 volts, how much resistance must be connected in series with it?
13. The coil of the instrument considered in problem 12 contains 3182 turns. If this instrument is to be converted into a 5 ampere range ammeter, how many turns must the coil have?
14. What type of instruments are suitable for measuring high frequency currents?
15. In A C instruments the scale divisions are not uniform, but increase as the upper end is approached. Explain why such a scale is required for the hot wire type of instrument.
16. The instrument shown in fig. 19-6 can be made into a wattmeter. By means of a diagram, show how this is done.
17. In the bridge shown in fig. 19-10, a balance is obtained when $R_1 = 682$, $R_2 = 1000$, and $R_3 = 10$ ohms. What is the value of R_x?
18. What maximum value of resistance can be measured by the Wheatstone bridge shown in fig. 19-10? What minimum value?
19. When the bridge shown in fig. 19-10 is balanced, $R_x = .429$ ohm. What is the value of R_1, R_2, and R_3?
20. Suppose that in fig. 19-11, AC is 30 inches and R_1 is adjusted for 100 ohms. If the bridge balances when AD = 14 inches, what is the value of R_x?
21. For what purpose is a megger used?
22. Why does a megger indicate correctly at all speeds?
23. Give the advantages and disadvantages of the following instruments: (a) iron vane type; (b) inductive type; (c) hot wire type; (d) thermocouple type.
24. Why is the inductive type of watthour meter in more general use than the Thompson watthour meter for A C measurements?
25. (a) How must the retarding torque vary in a watthour meter? (b) How is this retarding torque produced?
26. What is the purpose of the permanent magnets and the aluminum disk in a watthour meter?
27. Classify all watthour meters considered according to the circuits (direct or alternating current) in which they may be used.
28. How is friction compensated in a (a) Thompson watthour meter? (b) In an inductive type watthour meter?
29. How is a watthour meter adjusted so that it will read correctly?

30. Describe briefly the construction of a three-phase watthour meter.
31. Describe the principle of operation of the power factor meter.
32. How does a synchroscope differ from a power factor meter?
33. What does the synchroscope indicate?
34. Describe the vibrating reed type of frequency meter.
35. Why cannot a circular disk be used in the inductive type of frequency meter?
36. Why does not the reading of an inductive type of frequency meter change when the voltage is changed?
37. When are normally open over current relays used? Normally closed over current relays?
38. Describe the characteristics of an (a) inverse time relay, (b) a definite time relay, and (c) an inverse-definite time relay.
39. Why is the current coil of the relay shown in fig. 19-27 provided with taps?
40. What is the purpose of the stop lever in fig. 19-27a?
41. What is the detrimental effect of (a) over voltage? (b) Undervoltage?
42. Describe the three types of voltage relays.
43. What two adjustments must be made on an over current relay of the type shown in fig. 19-27?
44. For what purpose is a power directional relay used?
45. Why cannot power relays be used instead of power directional relays to disconnect a feeder when a short circuit occurs?
46. By means of a diagram, show how a three-phase power directional relay (fig. 19-32) is connected, using three overloads.
47. How must the phase balance current relay shown in fig. 19-33 be adjusted so that it will operate in 5 seconds when a current of 10 amperes flows through its current coil?
48. For what purpose is a phase balance current relay used?
49. Give three applications of differential protection.
50. When are motor and generator differential relays not suitable for protecting transformers?

CHAPTER 20

ELECTRON TUBES

Vacuum Tubes—An electron tube of the simplest type consists of a filament and a plate mounted in an evacuated tube (fig. 20-1a). This tube, as well as all others that contain two elements, is called a diode. When current is sent through the filament of a diode, the filament becomes very hot, causing electrons to have enough speed to pass through the surface of the filament, just as molecules pass through the surface of water being heated. The temperature required for the filament to emit electrons depends upon the kind of material from which it is made. Pure tungsten filaments emit few electrons. They are made to emit an abundance of electrons by either impregnating them with thorium or coating them with calcium or barium oxide. For the same amount of power consumed by each filament, the thoriated tungsten type emits about 20 times as many electrons and the oxide coated type emits about 100 times as many electrons as the pure tungsten type. The thoriated tungsten filaments are generally operated at a temperature of about 1600° C to give sufficient emission, while oxide coated filaments are generally operated at a temperature of about 750° C because at this low temperature they emit sufficient electrons and have a longer life.

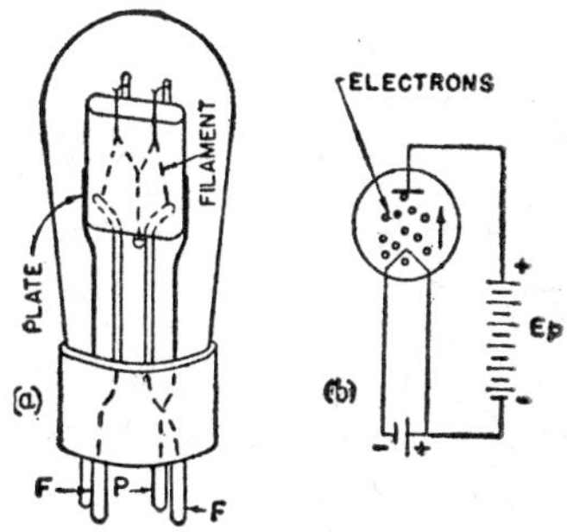

Fig. 20-1

The filament is left positive when the electrons are emitted from it. If no connection is made to the plate, the electrons boiled off will finally be attracted back into the filament. However, if the plate is made highly positive with respect to the filament, as indicated in fig. 20-1b, the plate will attract some electrons with greater force than the filament and these will travel to it. The electrons emitted from the filament form a cloud around it, which is known as a "space charge." These electrons, being negative, tend to repel other electrons leaving the filament. Then the space charge practically neutralizes the attractive effect of the plate, unless the plate voltage is high enough to give a force of attraction that is higher than the repulsive action of the space charge.

The amount of current that flows between the filament and the plate, called the plate current, is limited by the number of electrons emitted by the filament and the ability of the plate to attract the electrons emitted. In a given tube, the number of electrons emitted from the filament depends upon its temperature, which in turn depends upon the filament current. The ability of the plate to attract the electrons depends upon the plate voltage (fig. 20-1b). Then the plate current of a two-electrode vacuum tube may be varied by varying the filament current or the plate voltage.

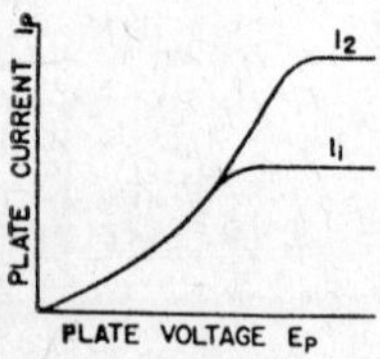

FIG. 20-2

Suppose the filament current is kept constant and the plate voltage (E_p) is varied. The plate current will vary as indicated by curve I_1 in fig. 20-2. This curve shows that the plate current rises rapidly at first, but the curve becomes horizontal after a certain point is reached, indicating that above that point the plate current no longer rises with an increase in plate voltage. Apparently, for a constant filament current, the number of electrons attracted to the plate will increase with the plate voltage until a condition is reached where the plate attracts electrons as fast as they are emitted from the filament. After this, any increase in plate voltage cannot cause an increase in plate current, unless the number of electrons given off by the filament is increased by increasing the filament current. If the filament current is changed to some higher value and the plate voltage is again varied, the plate current will vary as indicated by I_2.

Rectifiers—Direct current is sometimes needed for energizing auxiliary circuits and for charging batteries. If only alternating current is available, it may be converted into direct current by using devices which offer a high resistance to the flow of current in one direction and a low resistance to the flow in the opposite direction. These devices are called rectifiers. There are several types of rectifiers, but the most common for low currents are electron tube and copper oxide rectifiers.

Diode Tube Rectifiers—If an A C voltage is supplied to the plate of a diode, current flows during the time the plate is positive with respect to the filament and no current flows during the time the plate is negative, because during this time electrons are repelled from the plate and driven back into the filament. Thus, current flows only during the positive half cycle of voltage, as indicated in fig. 20-3a,

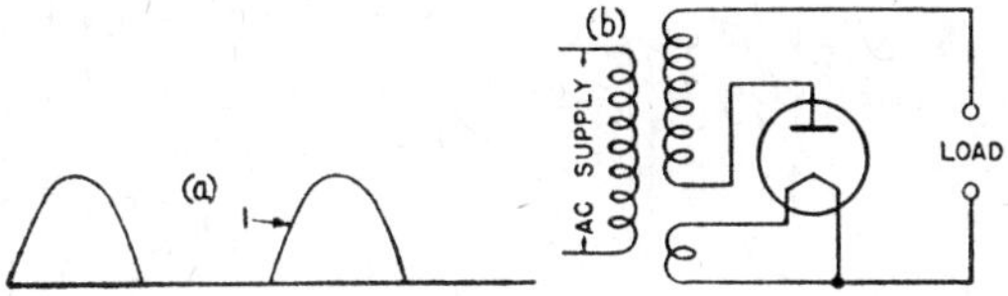

FIG. 20-3

and therefore the current is in only one direction. This is known as half wave rectification. A typical connection of a half wave rectifier is shown in fig. 20-3b. The filament is energized from a low voltage winding on the same transformer that supplies the high voltage to be rectified.

Both halves of each cycle may be rectified by using two rectifiers connected as shown in fig. 20-4a. The current output will vary as

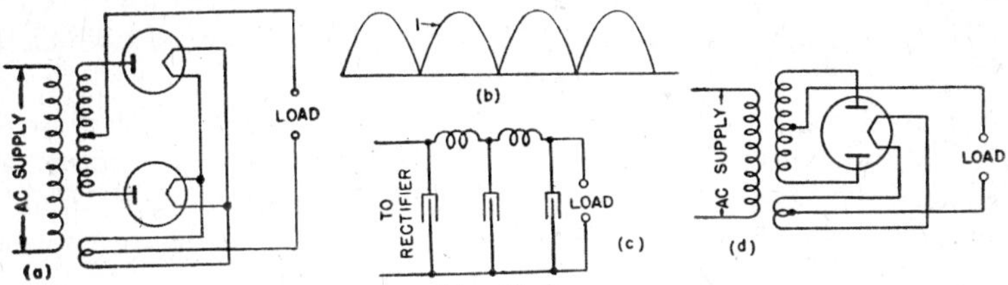

FIG. 20-4

indicated in fig. 20-4b, which shows a unidirectional pulsating current. A practically constant output current may be obtained with the use of three condensers and two inductances connected as shown in fig. 20-4c.

A single tube with two separate plates may be used instead of the two tubes shown in fig. 20-4a. Such a tube is known as a full wave rectifier and is connected in the circuit as shown in fig. 20-4d.

Copper Oxide Rectifier—One element of a copper oxide rectifier consists of a copper disk with a copper oxide layer formed on one side, and a soft metal, such as lead, pressed on this layer (fig. 20-5a). A current can readily pass from the copper oxide to the copper disk, but a very high resistance is offered to the flow of current in the reverse direction. One such element can stand an alternating e m f

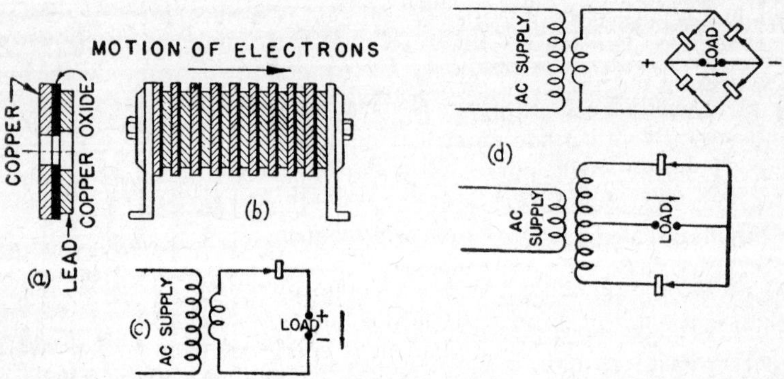

FIG. 20-5

of about 11 volts. For higher voltages, a number of elements must be stacked together as shown in fig. 20-5b. Fig. 20-5c shows a half wave rectifier in which current will flow only when the voltage is in the direction indicated by the arrow. Both halves of each cycle may be rectified by connecting rectifiers as shown in fig. 20-5d.

Three-Electrode Tubes—The three-electrode tube, called the triode, is the same as the diode except that the former has an additional electrode, called the grid. This grid consists of a wire screen

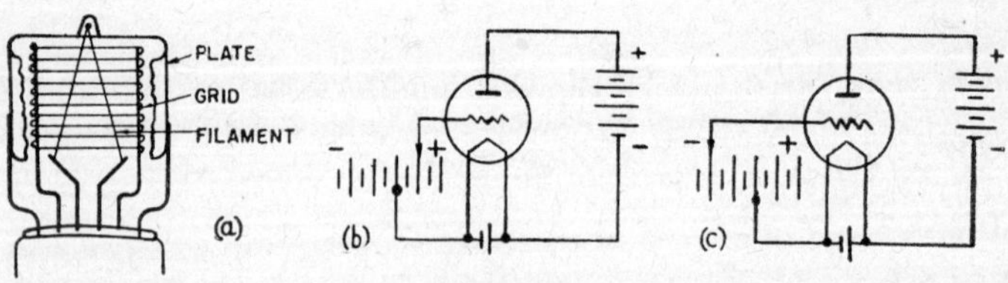

FIG. 20-6

that completely surrounds the filament and connects to a third prong (fig. 20-6a). In passing from the filament to the plate, the electrons must pass through the openings in the grid.

If the grid is positive with reference to the filament, as shown in fig. 20-6b, it will assist the plate in attracting the cloud of electrons which surrounds the filament. Some of the electrons strike the screen grid and therefore establish a small grid current. But most of the electrons are traveling at a high speed when they reach the screen grid, and therefore pass through it and are attracted to the plate.

If the grid is negative with reference to the filament, as shown in fig. 20-6c, it will repel electrons back into the filament, decreasing the plate current. If the grid is made sufficiently negative, the plate current will become zero. Thus the plate current can be varied by varying the grid voltage.

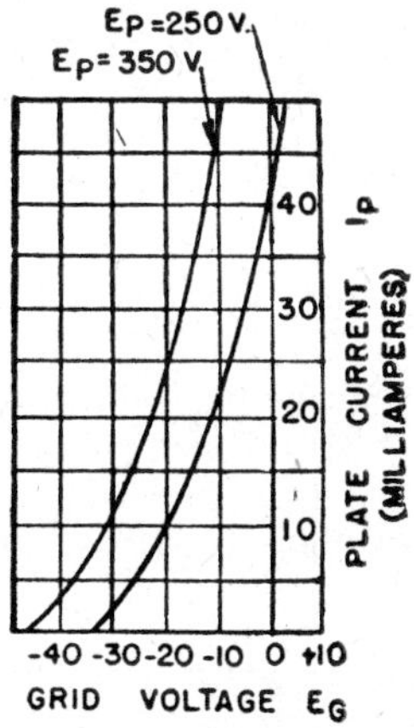

FIG. 20-7

Suppose that the plate voltage and filament current are kept constant and that the grid voltage is varied (fig. 20-6). When the grid voltage is made more positive, the plate current increases; as it is made more negative, the plate current decreases. This is indicated by the characteristic curves shown in fig. 20-7. The grid is generally placed much nearer the filament than the plate, and because of this a small change in grid voltage causes a large change in plate current. Then a small change in grid voltage has the same effect as a large change in plate voltage in varying the plate current. Such operating characteristics can be expressed in terms of tube constants known as amplification factor and plate resistance.

Amplification Factor—*The amplification factor of a tube is defined as the ratio of a change in plate voltage necessary to produce a change in plate current to the change in grid voltage necessary to produce the same change in plate current.* The amplification factor (denoted by μ) may then be expressed

$$\mu = \frac{\Delta E_p}{\Delta E_g} \qquad (20\text{-}1)$$

where Δ means a small change. The greater the distance that the grid is from the plate, the closer the spacing of the grid turns, the larger the grid wires, the greater will be the amplification factor of a tube.

EXAMPLE—Determine the amplification factor of a type 10 tube which has the characteristics shown in fig. 20-7.

At a grid voltage of −10, a change from 250 to 350 volts in plate voltage produces a change from 22.5 to 47.5 milliamperes in plate current. The plate current will be changed the same by changing the grid from −21 to −10 volts. Then, in this case

$$\mu = \frac{350 - 250}{21 - 10} = 9.1$$

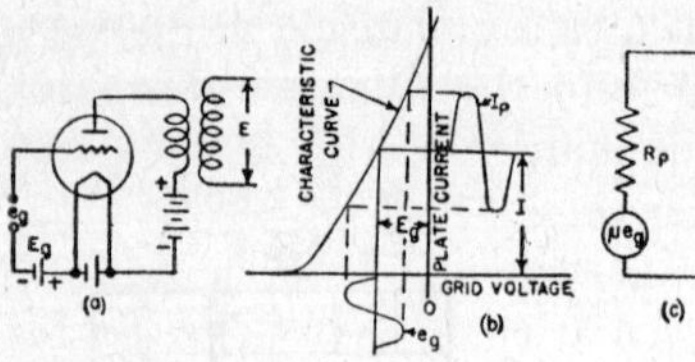

FIG. 20-8.

Since μ is always much greater than unity, a triode is used to amplify small voltages. This tube is normally operated with a constant voltage impressed on the filament and on the plate circuit, but the voltage impressed on the grid varies due to the input alternating voltage, e_g.

If the grid circuit in fig. 20-8a is closed, the grid voltage E_g permits the flow of a steady current I in the plate circuit, which is indicated in fig. 20-8b. If a small alternating voltage e_g is impressed on the grid

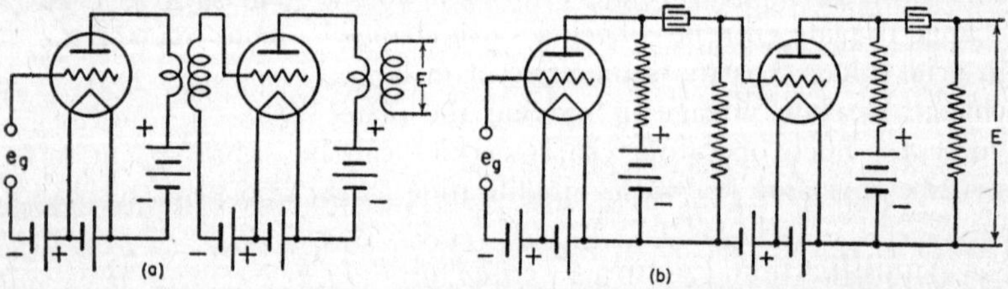

FIG. 20-9.

circuit, the plate current will fluctuate about line I, as indicated by line I_p (fig. 20-8b). Suppose the grid voltage varies as a sine curve and it is desired that the plate current vary in the same manner. Then the grid voltage E_g and the plate voltage E_p must be so-selected that the alternating curves e_g and I_p vary along the straight portion of the characteristic curve. The grid voltage E_g is always made negative so that the grid current will be zero. The constant value of plate current (indicated by I in fig. 20-8b) in flowing through the primary of the transformer shown in fig. 20-8a, induces no voltage in the secondary and therefore has no effect on the secondary turns. The alternating component of the plate current (I_p) causes an e m f to be induced in the secondary turns. With a transformer that has a 1 to 3 ratio, E is generally about 25 times as great as e_g. Thus a high amplification is produced. If a greater amplification is desired, the output voltage E must be impressed on a second tube connected in the same way as the first. Two stages of amplification are shown in fig. 20-9. Fig. 20-9a shows a transformer coupled amplifier and fig. 20-9b shows a resistance coupled amplifier.

A grid voltage e_g will cause the plate current to vary exactly in the same way as if the plate voltage were varied by μe_g. Then as far as a variation in plate current is concerned, the plate circuit of an amplifying tube is equivalent to a series circuit having an alternating voltage μe_g acting on the plate resistance, R_p, in series with the load impedance Z. Such a circuit, which is known as the equivalent circuit of an amplifying tube, is shown in fig. 20-8c.

Plate Resistance—Since in an amplifier we are concerned with the extent the plate current is varying, it is important to express a relation between the change in plate voltage, ΔE_p, corresponding to a change in plate current, ΔI_p. The ratio of these two changes is called the plate resistance, R_p, that is,

$$R_p = \frac{\Delta E_p}{\Delta I_p}$$

Example—Determine the plate resistance of a type 10 tube, which has the characteristics shown in fig. 20-7.

At a grid voltage of -10 and a plate voltage of 250, the plate current will be 22.5 milliamperes. With the same grid voltage, if the plate voltage is changed from 250 to 350, the plate current changes from 22.5 to 47.5 milliamperes. Then the plate resistance is

$$R_p = \frac{350 - 250}{.0475 - .0225} = 4000 \text{ ohms}$$

Cathodes—In a tube the electron-emitting element is known as the cathode and the plate is known as the anode. Two types of cathodes are common, the usual filament type (fig. 20-1) and the indirect-heated cathode type (fig. 20-10). The latter type consists of an oxide-coated metal cylinder which surrounds a filament called the heater. The heater, which is made of tungsten, serves only for heating the cylinder, and is generally energized with alternating current. It is undesirable to use batteries for supplying the various D C voltages required in amplifiers. Amplifiers without batteries are shown in fig. 20-11. The grid bias voltages are produced by the plate currents flowing through resistances R. The voltage drops in these resistances are kept practically constant by the condensers shunted across them. The plate voltages are supplied by rectifiers with filters (fig. 20-4).

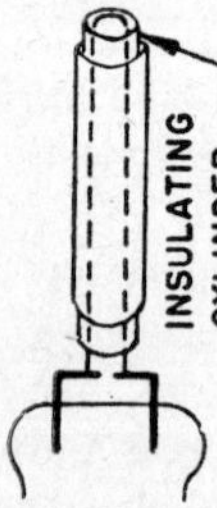

FIG. 20-10.

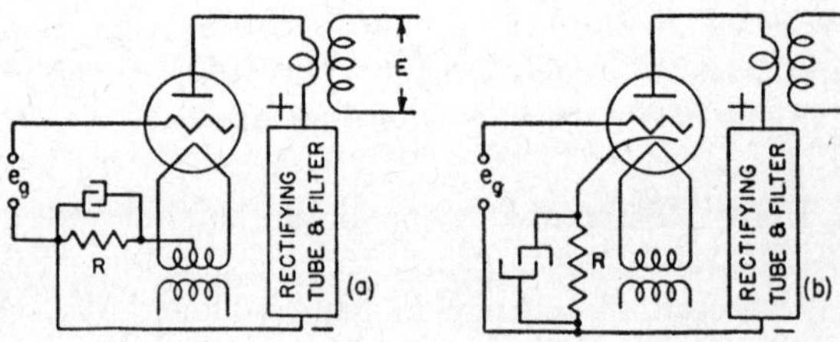

FIG. 20-11.

The filament (fig. 20-11a), which is designed for a low voltage, and the cathode (fig. 20-11b) have large cross-sections and therefore their temperatures remain constant. Then they emit electrons at a constant rate.

Gas Filled Tubes—A vacuum tube has very desirable characteristics for certain industrial applications, such as for rectification and amplification. However, the vacuum tube has an outstanding undesirable feature, which is high power loss within the tube, and therefore has a low efficiency. Part of the power loss is that required to heat the filament, and the remainder is due to the high internal resistance of the tube. Moreover, in a vacuum tube a high plate voltage is required to attract electrons to the plate (see Table 20-I). *The high filament and plate losses and the high voltage drop in a vacuum* *tube can be reduced considerably if an inert gas or mercury is placed in the tube, but when this is done the operating characteristics are entirely changed.*

TABLE 20-I

Tube Number	Filament		Plate Voltage	Grid Voltage	A C Plate Resistance	Amplification Factor
	Voltage	Current				
10	7.5	1.25 (A C or D C)	350	—31	5150	8
12A	5	.25 (D C)	135	— 9	5100	8.5
26	1.5	1.05 (A C or D C)	135	—10	7600	8.3

To consider how a gas filled electron tube operates, suppose that an electron leaves the filament. It will soon collide with a gas molecule and knock one or more electrons out of it, and these free electrons will move toward the plate. The molecule that lost electrons is now positively charged and will attract electrons in the space charge. In the same way many other molecules become positively charged and together render a valuable service by neutralizing the space charge (molecules carrying a charge are called ions, and a gas with many ions is said to be ionized). Neutralizing the space charge permits more electrons to get to the plate.

In a gas filled tube, the plate resistance is much lower than in a vacuum tube and the drop is only 10 to 15 volts, which remains practically constant. Thus, such a tube has a much higher current carrying capacity and a higher efficiency than a vacuum tube of the same size.

Diode Gas Tubes—A two-electrode gas filled tube can be used as a rectifier. One type, known as a tungar tube, is shown in fig. 20-12. It was designed especially for battery charging.

Triode Gas Tubes—Three-electrode gas filled tubes, generally called grid glow tubes, are used in alternating current circuits, and are classified according to the cathode they employ. That is, they are shown as (1) hot cathode, (2) cold cathode, and (3) mercury pool cathode types.

Hot Cathode Tube—The hot cathode tube (fig. 20-13) resembles the conventional three-electrode vacuum tube in construction, but its operating characteristics are much different.

For each value of plate voltage impressed on this tube, there is a definite value of grid voltage at which ionization will begin and allow the tube to carry current. This value of grid voltage is known as the critical voltage. If the grid is more negative than this critical voltage, the plate current cannot start to flow. If the grid becomes more positive than the critical value, the plate current starts to flow. Once the plate current starts to flow, the gas becomes ionized and the grid no longer has any control. Then the tube functions the same

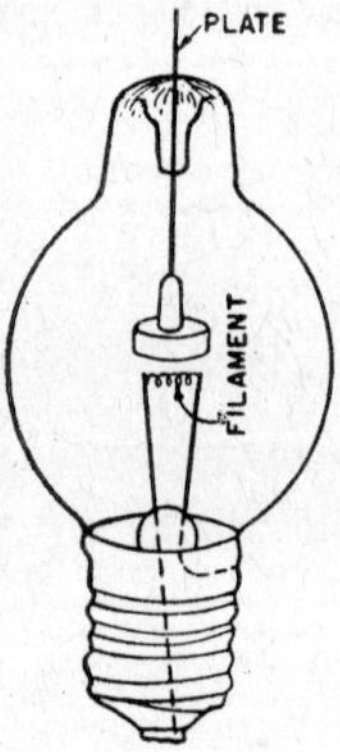

FIG. 20-12

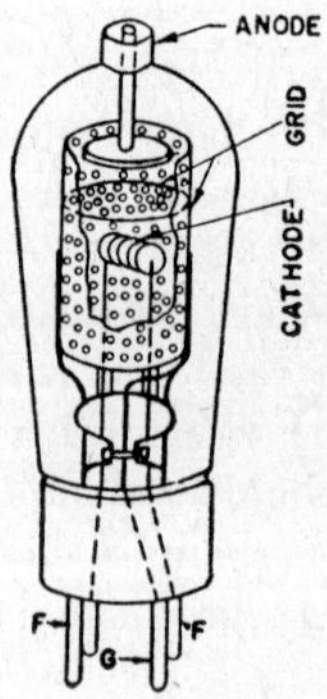

FIG. 20-13

as a diode gas rectifier until the half-cycle of voltage is complete, at which point the grid again gains control. With both the grid and plate circuit supplied with alternating current, the phase relation between the grid and plate voltages governs the point where the plate current begins to flow in each cycle. Then by shifting the grid voltage it is possible to govern the amount of current passing through the tube.

If the grid voltage is in phase with the plate voltage (fig. 20-14a), conduction will begin as soon as the plate voltage becomes positive, because in this case the grid is always less negative than the critical voltage.

If the grid voltage lags the plate voltage (fig. 20-14b), the grid will be more negative than the critical voltage when the plate becomes positive, and a flow of current will not occur until the grid becomes less negative than the critical voltage.

If the grid voltage is 180 degrees out of phase with the plate voltage (fig. 20-14c), no current flows because the grid never becomes less negative than the critical voltage.

A grid shift control connected to a full wave rectifier is shown in fig. 20-15a; its vector diagram is shown in fig. 20-15b. The voltage E_g, which is the output voltage of the control, lags the secondary voltage E by θ degrees. E_g may be shifted relative to E without changing its maximum value merely by changing R.

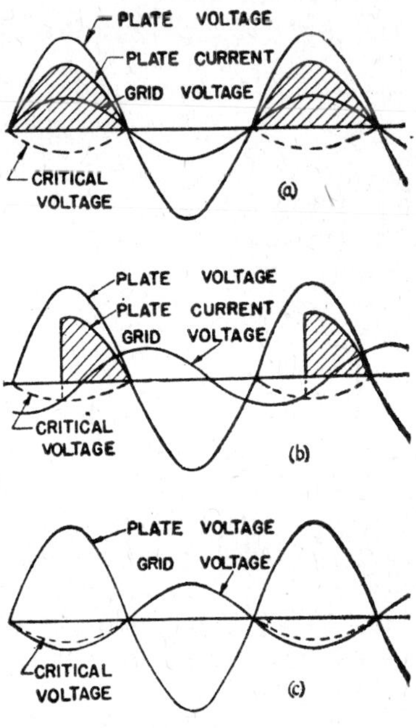

FIG. 20-14.

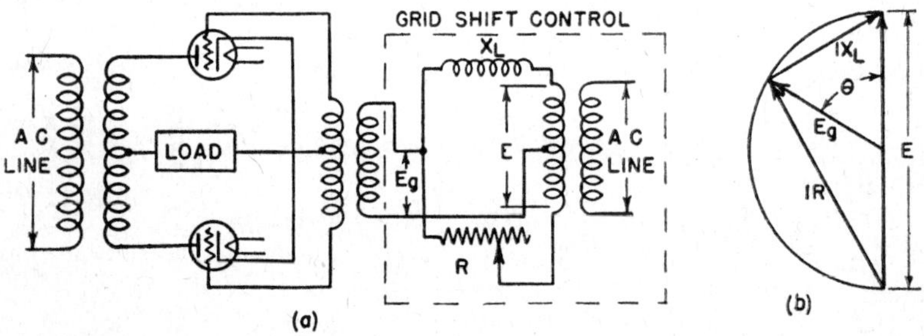

FIG. 20-15.

Cold Cathode Tube—It is possible to produce ionization in a tube that does not have a hot cathode. Such a tube is shown in fig. 20-16. On applying a voltage to the tube, the electrons will be attracted by the anode and will move toward it. In so doing, they will strike molecules of gas, therefore producing ionization. Electrons will be emitted from the cathode, due to the positive ion bombardment, and electrons will therefore continue to pass from the cathode to the anode.

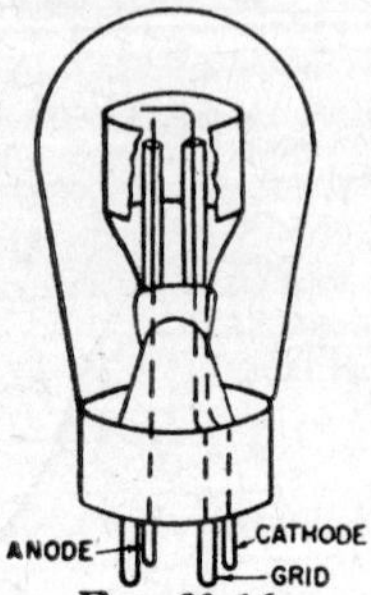

FIG. 20-16

This type of tube has a higher voltage drop and consequently a greater power loss than the gas filled, hot cathode type.

Mercury Pool Cathode Tube—The mercury pool cathode tube consists of a glass tube with a mercury pool at the bottom, which is connected to one prong, and a plate at the top. The tube also has a third electrode called an igniter, that is generally made of silicon carbide, and connects to a prong that sticks down into the mercury pool (fig. 20-17). This prong is always connected to an auxiliary tube that carries current to start the arc at the igniter (fig. 20-18a).

Fig. 20-18a shows a mercury pool tube connected for half wave rectification. As the tube voltage increases from zero, as indicated by curve E in fig. 20-18b, the igniter current increases as indicated by i. When i reaches a certain value, an arc forms between the end of the igniter and the mercury pool. The arc instantly transfers to the anode and the voltage between the cathode and the anode reduces to about 15 volts, while the load current rises instantly and then varies as a sine wave for the remainder of the half cycle.

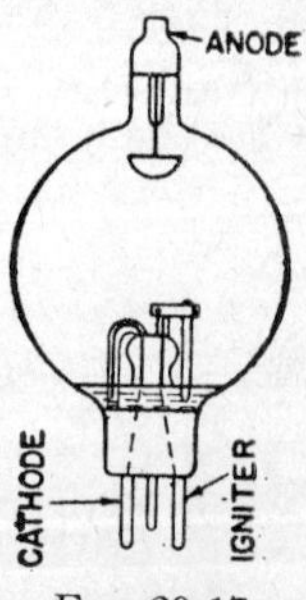

FIG. 20-17

The igniter carries the current from the beginning of the cycle until the arc jumps to the anode, at which time the igniter current drops to nearly zero because of the reduction in tube voltage. During the second half of the cycle the tube is non-conductive.

The arrangement required for full wave rectification is shown in fig. 20-19.

Phototube—The construction of a typical phototube is shown in fig. 20-20a. The cathode consists of a plate bent in a semicircle and is connected to one prong. It is made of a material that emits electrons when light strikes it.

The anode consists of a wire that extends vertically in the center of the tube and is connected to a second prong. These electrodes are sealed in either an evacuated or gas filled tube.

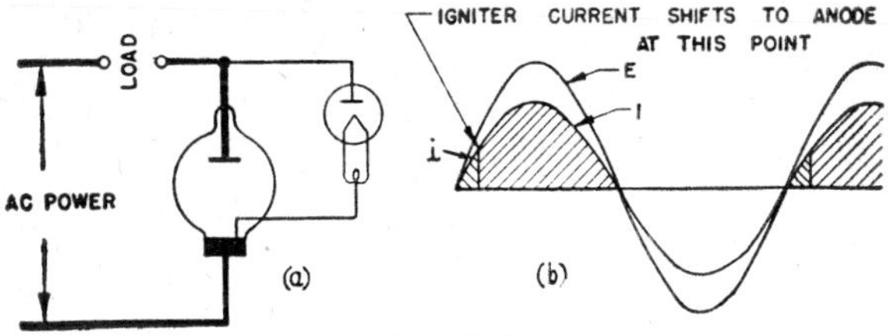

FIG. 20-18

When a phototube is connected as shown in fig. 20-20b and light is directed on the cathode, electrons are emitted from it, pass to the anode, and then through the external circuit. When no light strikes the coating, no current flows.

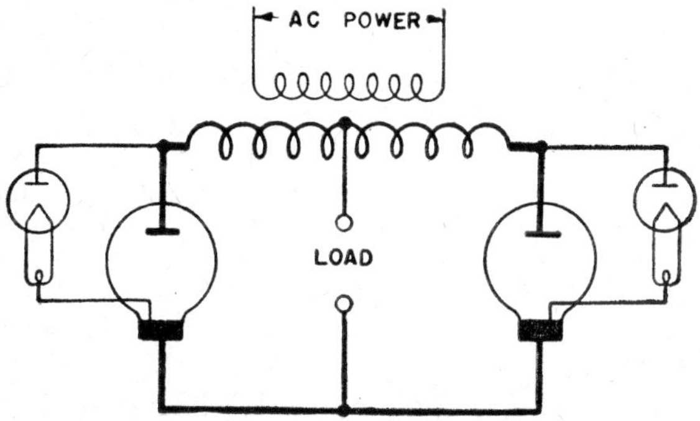

FIG. 20-19

The manner in which the current varies with the illumination in both the vacuum tube and the gas filled tube, for the same voltage, is shown in fig. 20-21. These curves indicate that in both cases the current is approximately proportional to the illumination.

The current that flows in a phototube with normal light intensity is too small to operate even the most sensitive relays. This output current may be amplified, however, after which it can operate a relay or any other device. A phototube can be connected to an amplifier so that the output current either increases or decreases when the tube is illuminated. Both connections are shown in fig. 20-22.

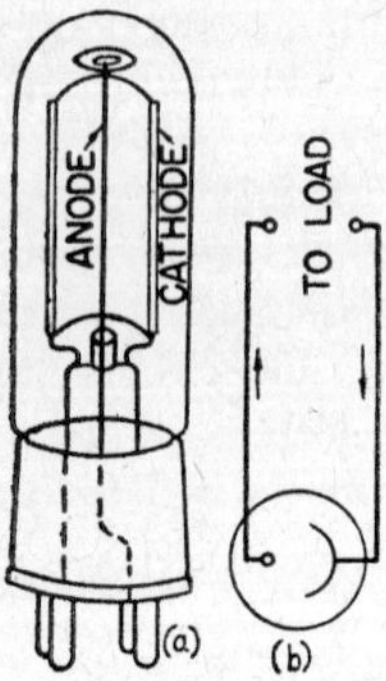

FIG. 20-20

In fig. 20-22a, when the tube is illuminated a current is established in resistance R. This current produces a voltage drop in such a direction as to make the grid more positive with respect to the filament. This causes the plate current to increase and operate a relay or other device. When the tube is not illuminated, no current, and therefore no voltage, occurs in resistance R. The voltage E_g makes the grid sufficiently negative with respect to the filament to cause the amplifying tube to pass no current.

The circuit shown in fig. 20-22b is the same as that of fig. 20-22a except that the phototube and its battery are reversed and the grid voltage E_g is also reduced. When the phototube is illuminated in this case, establishing a current and voltage drop in resistance R the voltage drop is in such a direction as to make the grid more negative, and therefore the tube passes no current. When the phototube is not illuminated, there is no voltage across resistance R, the grid is less negative, and the tube passes current.

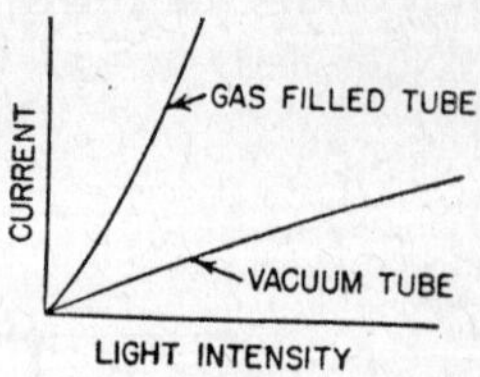

FIG. 20-21

If alternating current only is available, it is unnecessary to rectify it for the phototube and amplifier because they will operate on alternating current. Typical alternating current circuits are shown in fig. 20-23.

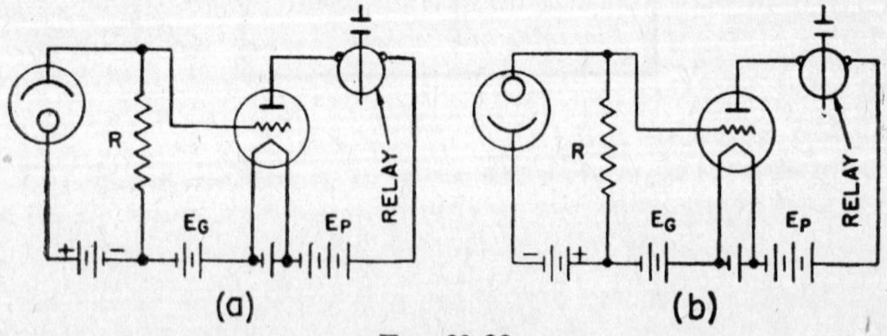

FIG. 20-22

In fig. 20-23a, suppose the phototube is not illuminated. During the half cycle that the plate is positive, the grid is negative with respect to the filament and the amplifying tube does not conduct. During the next half of the cycle, the grid is positive, but the plate is negative and again the tube does not conduct. Thus the amplifying tube does not conduct.

If the phototube is illuminated, it establishes a current in resistance R. This current produces a voltage drop in this resistance in such a direction as to make the grid positive during the half cycle that the plate is positive and the amplifying tube conducts. Since conduction cannot take place during the half cycles that the plate is negative, the output is pulsating direct current. Such output is satisfactory for operating slow speed relays.

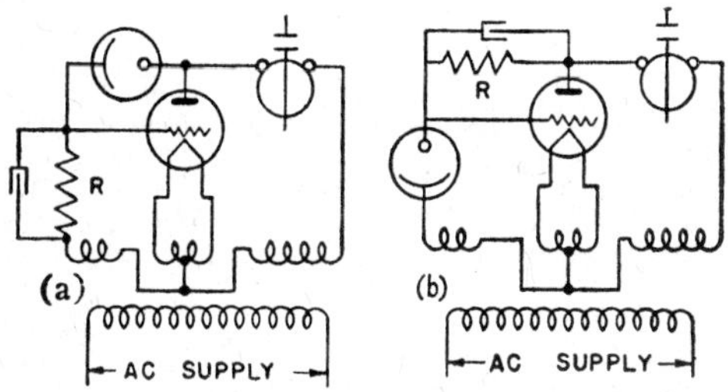

FIG. 20-23

When R is high (fig. 20-23a), a condenser must be connected across this resistance. Without this condenser, the negative grid voltage, which occurs when the phototube is not illuminated, is thrown out of phase with the plate voltage and the grid is ineffective.

If it is desired that the amplifying tube conduct only when the phototube is not illuminated, the circuit shown in fig. 20-23b is generally used.

Industrial Uses of Phototube—Fig. 20-24 shows a counter operated by a phototube. The circuit may be as shown in fig. 20-22b or 20-23b except that a plunger type magnet is used instead of a relay. A light beam is directed across the conveyor to strike the phototube. Each time an article carried on the conveyor interrupts the beam, the amplifier sends a current surge through coil C. The plunger is drawn downward, causing the counter to register.

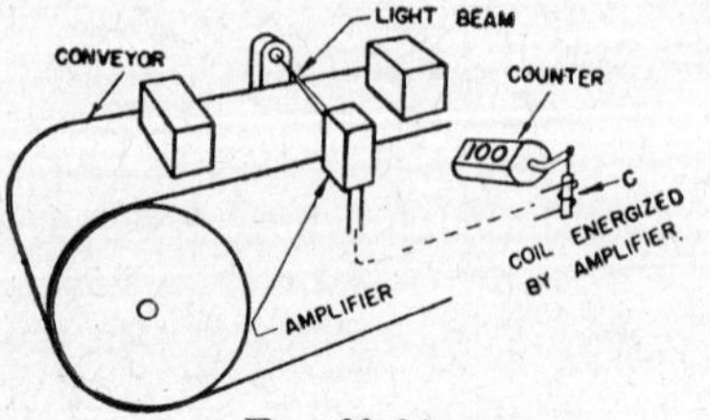

FIG. 20-24

Injury to press operators may be eliminated by the use of a safety device including a photocell and amplifier. A light beam is placed in such a position that the operator cannot place a hand under the press without interrupting it. This has no effect if the press is opening, but the interruption of the ray causes the machine to stop instantly if the press is closing.

Color Matching and Sorting—An important application of a phototube is its use in a device for matching colors or for sorting materials. Two materials that appear alike in color under one source of light may appear different in color under a light of another source. The two materials will appear alike under any light, however, if their colors are actually alike. In a commercial color matcher, samples are compared under three colors—red, green, and blue.

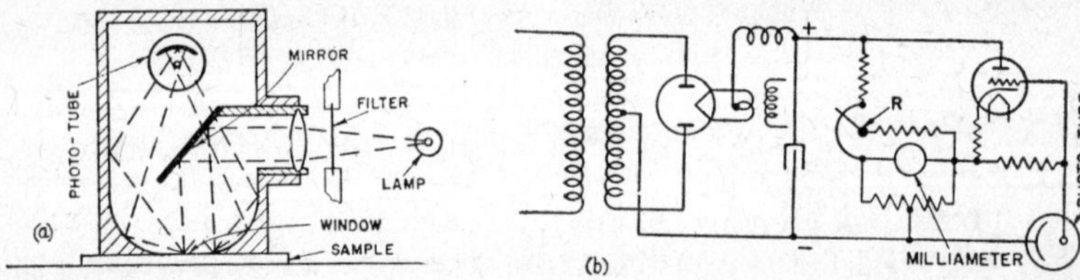

FIG. 20-25

A cross section of a color matcher is shown in fig. 20-25a. Light which passes from the lamp through one of the three color filters is made to travel in parallel rays by the lens, and is reflected on the sample by the mirror. The sample in turn reflects the light, which reaches the phototube. The phototube actuates an amplifier (fig. 20-25b) in which its output current decreases with an increase in illumination.

To determine if a sample matches with a standard, the standard is placed at the window and one of the three screens, say the red one, is placed in position. Then the resistor R (fig. 20-25b) is adjusted so that the milliammeter indicates zero on a zero center scale.

The standard is then replaced by the sample to be tested. If the meter reading does not change, this procedure is repeated with the green and then the blue filter. If the meter reading is the same in all the cases, the sample and the standard may be considered a perfect match.

An appliance similar to that shown in fig. 20-25a may be used for sorting by color, in which case an ejecting mechanism is used instead of a milliammeter.

Fig. 20-26 shows the method of sorting stock according to size. When the diameter of the stock is below normal, the light beam strikes mirror A and is reflected to phototube A. This phototube

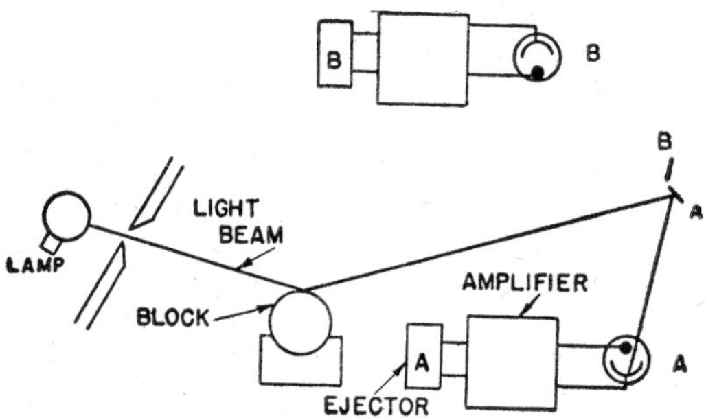

FIG. 20-26

actuates the magnetic ejector A, which ejects the stock in one container. If the diameter of the stock is above normal, the light beam strikes mirror B and is reflected to phototube B, actuating ejector B, which ejects the stock in another container. When the diameter is normal, the light strikes neither mirror and the stock moves onward.

Temperature Control—A very simple but important industrial application of a phototube is illustrated in fig. 20-27a. The light beam passes through a hole in the scale of an indicating instrument and strikes a phototube. The instrument may indicate temperature, pressure, or a number of other conditions. When a limiting condition is reached, the indicating hand covers the hole. Then no light strikes the phototube, causing the amplifier to operate a relay.

The same effect may be obtained by employing a mirror instead of making use of a hole in the instrument. The method of doing this is illustrated in fig. 20-27b. In this case, when the limited condition is reached, the indicating hand eclipses the light beam to the mirror. Then no light strikes the tube, and this causes the relay to operate.

Phototubes controlled by light beams have numerous other industrial applications. They are used to automatically control spray guns. In this case the article to be sprayed is carried by a conveyor. When it starts passing before a spray gun, a light beam is eclipsed, causing the spray gun to be turned "on." Phototubes actuated by light are also used for automatic testing, opening doors, etc.

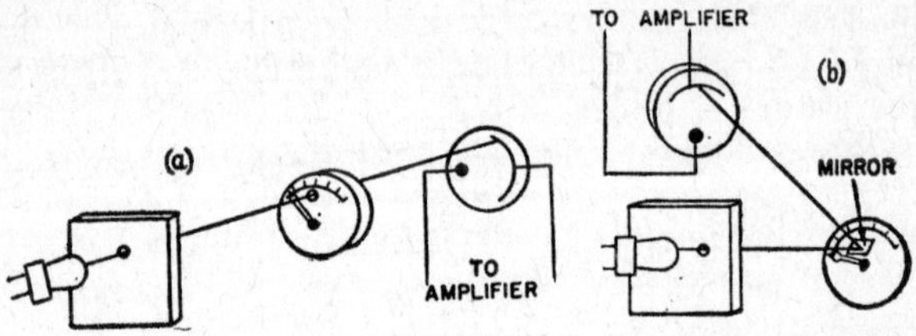

FIG. 20-27

Photocells—It was shown above that copper coated with copper oxide can be used as a rectifier. This element is photo-sensitive and has industrial uses similar to those of the phototube. It is mounted in a convenient case (fig. 20-28a) and is known as a photocell.

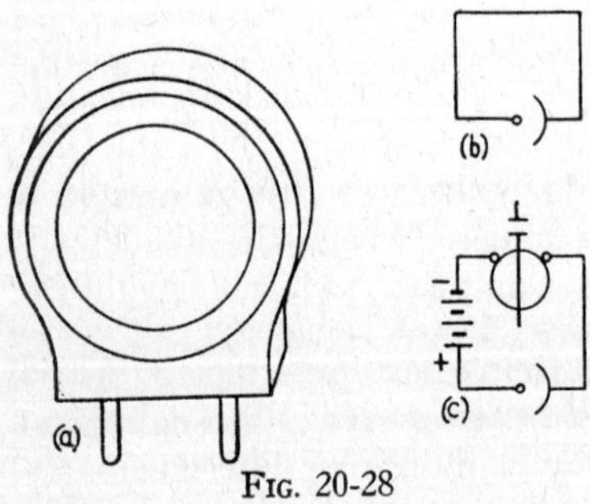

FIG. 20-28

The oxide is coated with a metallic film, through which light passes.

In the circuit of fig. 20-28b, a current will flow when light strikes the copper oxide.

An outstanding characteristic of the photocell is the amount of current it will supply. When subjected to the same illumination as a phototube, it will deliver considerably more current, and the cell has a low internal resistance.

The current output of a photocell can be easily amplified, not as is done with a phototube, but by connecting a battery in the circuit of the cell as indicated in fig. 20-28c. When the cell is illuminated, its resistance decreases and more current passes. This current is sufficient to operate a small relay.

Exercises

1. (a) What is the most common use of a two element vacuum tube? (b) A three element vacuum tube?
2. What is the purpose of a rectifier?
3. By means of diagrams show how the voltages required in fig. 20-9 may be supplied without batteries.
4. How and why is the grid of a vacuum tube kept negative?
5. By means of arrows, show the direction of the electrons in fig. 20-9b.
6. Give the different methods by which the plate current of a triode can be increased.
7. Why is the grid of a vacuum tube placed near the filament instead of near the plate?
8. Describe the two types of cathodes used in a vacuum tube.
9. In a type 27 tube, a change of 4 milliamperes in the plate current is produced by changing the plate voltage by 40 volts or by changing the grid bias by 4.5 volts. Determine (a) the amplification factor and (b) the plate resistance.
10. A type 26 tube, which has an amplification factor of 8 and a plate resistance of 7300 ohms, has a load resistance of 8000 ohms and an alternating e m f of 10 volts on the grid. Determine (a) the effective value of the plate current and (b) the effective voltage across the load.
11. A type 40 tube requires a grid bias of —3 volts for a plate current of .2 milliampere. What is the value of bias resistor when the tube is connected (a) as shown in fig. 20-9a? (b) As shown in fig. 20-9b?
12. Determine the effective voltage supplied by a type 26 tube to a load that has 400 ohms resistance and 2000 ohms reactance. (Consider the grid voltage, amplification factor and internal resistance the same as in problem 10.)
13. (a) What is meant by ionization? (b) Explain how ionization is produced.
14. How do the characteristics of triode vacuum and gas filled tubes differ?
15. Why is the efficiency of a gas filled tube considerably higher than that of a vacuum tube?
16. Can a triode gas filled tube be used as an amplifier? Why?
17. What are the three types of cathodes used in gas filled tubes?
18. What advantages do a cold cathode tube and a mercury pool cathode tube have over a hot cathode tube?
19. For what class of work are mercury pool cathode tubes suited?
20. (a) Explain how the output of a phototube is amplified; (b) how the output of a photocell is amplified.

CHAPTER 21

WELDING SYSTEMS

Two pieces of metal may be united by a process known as electric welding, either by arc welding, in which an electric arc is used to produce the heat, or by resistance welding, in which the pieces are placed in contact and an electric current is sent through the joint to heat it.

Arc Welding—A wiring diagram of an arc welding outfit is shown in fig. 21-1. The generator, which is directly coupled to a

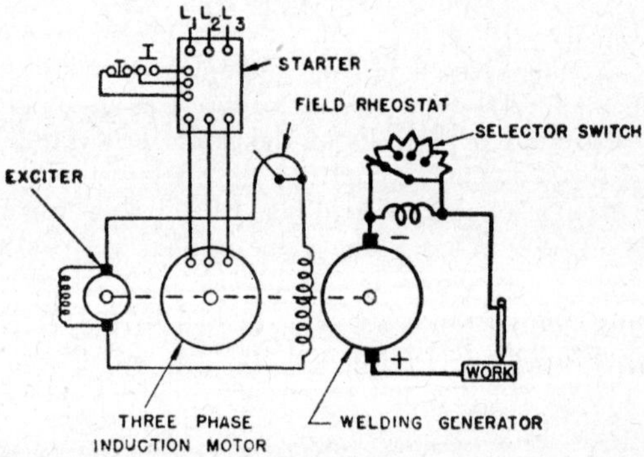

FIG. 21-1

motor, is differentially connected and generates about 60 volts at open circuit. The terminal voltage and welding current are varied by a divertor, called a selector switch, across the series field, and a rheostat in the shunt field circuit (the large current variations are obtained by the selector switch and the finer variations are obtained by the rheostat). To start the welding current, the electrode is touched to the work. This short-circuits the generator and a high current flows. Next the electrode is withdrawn slightly. As it breaks contact with the work, electrons shoot off at extremely high velocities, striking air molecules in their paths. Electrons are knocked out of the strucked molecules, and these are strongly attracted toward the positive (the material being welded). The large number of electrons set free causes the air gap to be a good conductor and it carries current readily. The electrons bombard the work, causing it to heat and ultimately forming a pool of molten metal. The molecules in the air gap, having lost electrons, become positively charged. Being much larger than electrons, these move toward the negative (electrode) at a much slower rate. At this terminal the impact is low and less heat is developed. About 75 per cent of the heat produced is liberated at the work and the remaining 25 per cent is liberated at the electrode.

A standard differentially connected compound generator has two undesirable features when used for arc welding:

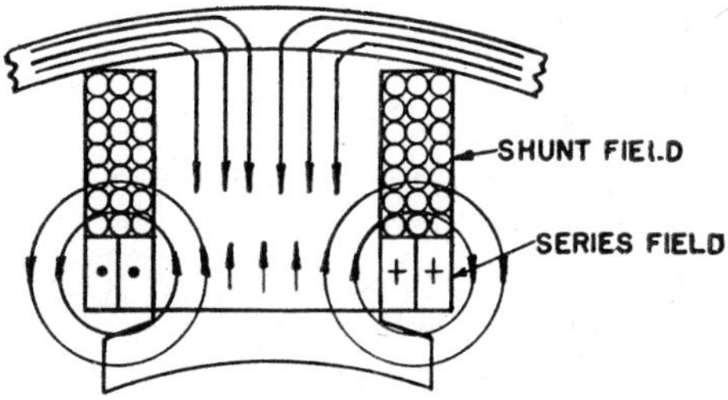

FIG. 21-2

(1) On touching the work with the electrode, the series field current suddenly increases, establishing an increasing flux, which induces a voltage in the shunt field (see fig. 21-2). This induced voltage raises the field current and consequently causes the welding current to increase to a high value, as suggested by curve A in fig. 21-3a. This high welding current produces sufficient heat to cause the electrode and work to melt, and the two fuse together, forming a low resistance joint. The heat being produced immediately decreases and the molten metal freezes, causing the electrode to stick to the work. For an ideal condition, the electrode and work should not melt until after the electrode is withdrawn from the work. This occurs only if there is no current surge.

(2) Suppose that an existing drawn arc is lengthened slightly. The magnetism established by the series field current decreases, inducing a voltage in the shunt field, which reduces the shunt field current. This reduces the welding current, as indicated by curve A in fig. 21-3b, and in many cases the arc is extinguished.

Methods of Reducing Current Fluctuations—An arc is easily struck and easily handled if the generator has short-circuiting characteristics (as shown by curve B in fig. 21-3a) and arc lengthening characteristics (as shown by curve B in fig. 21-3b).

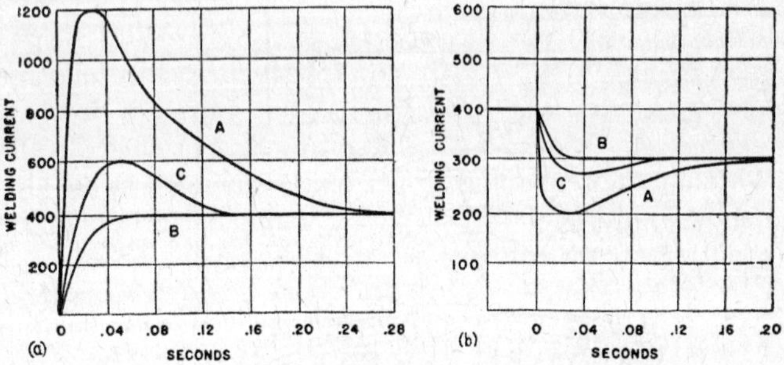

Fig. 21-3

One method of obtaining these characteristics is to neutralize the voltage induced in the shunt field. This may be done with a transformer whose primary is connected in the welding circuit and whose secondary is connected in the field circuit (see fig. 21-4). This scheme may be used with either a separately excited or self-excited shunt field.

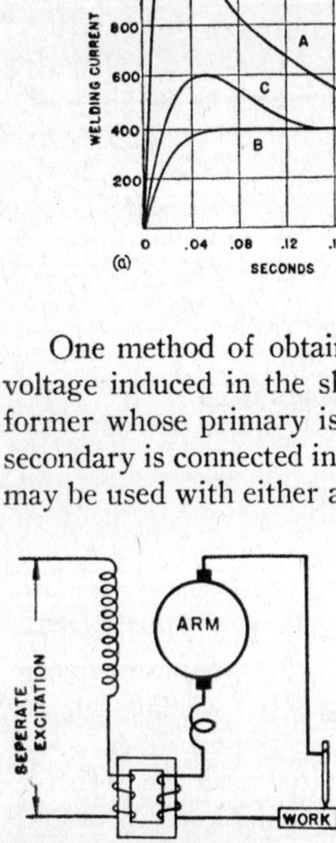

Fig. 21-4

A change in the welding current causes a change in the series field and primary current. Then a voltage is induced in the shunt field and in the secondary, and these neutralize each other.

Instead of using a transformer to reduce current surging, a reactor may be used, and is connected in the welding circuit. According to Lenz's law, when the welding current is increasing, the e m f induced in the reactor opposes this increase; and when the current is decreasing, the induced e m f reverses and acts to keep the current flowing (see chapter 10). This tends to keep the welding current constant. The most undesirable feature of this system is that the reactor must be undesirably large and consumes considerable power.

The field structure of a generator that has good operating characteristics for arc welding is shown in fig. 21-5a. It has four main poles and four interpoles. The two main poles (N) have a shunt field wind-

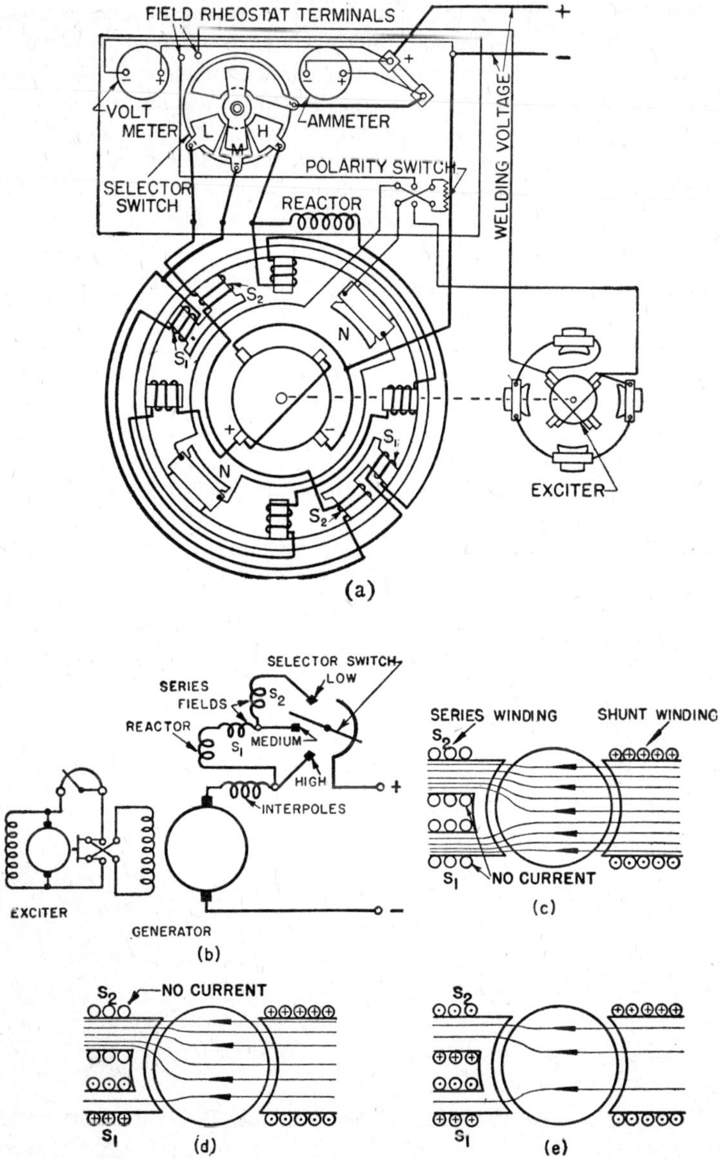
FIELD RHEOSTAT TERMINALS
+
−
VOLT METER
AMMETER
L
H
M
POLARITY SWITCH
SELECTOR SWITCH
REACTOR
WELDING VOLTAGE
S_2
S_1
N
N
EXCITER
(a)
SELECTOR SWITCH
LOW
SERIES FIELDS
REACTOR
MEDIUM
HIGH
INTERPOLES
EXCITER
GENERATOR
(b)
SERIES WINDING
SHUNT WINDING
NO CURRENT
(c)
NO CURRENT
(d)
(e)

Fig. 21 5

ing, and the two S poles are each split with series windings on each part. The method by which the machine is connected is shown schematically in fig. 21-5b. When the selector switch is in the "high" position, the series field windings and the reactor are not in the circuit, and the generator operates as a shunt machine. The flux distribution for this condition is shown in fig. 21-5c, where a two pole machine is considered for simplicity. The welding current is adjusted by the rheostat. The reactor is not included in the circuit because it is not necessary for high current welding.

When the selector switch is in the "medium" position, the reactor and series winding (S_1) are in the circuit. Winding S_1 reduces the flux and therefore the supplied voltage (see fig. 21-5d).

When the selector switch is in the "low" position, series winding S_2 is also inserted in the circuit. This winding further reduces the flux, and hence the voltage (see fig. 21-5e).

For medium and low current welding, the reactor is used to stabilize the arc.

Constant Voltage Generator—Several welding units cannot be connected to a generator with drooping characteristics, such as those previously mentioned, because when an electrode is touched to the work, the established short circuit causes the generator terminal voltage to drop to nearly zero. As a result, it becomes impossible for the other welders to maintain their arcs. Thus, if generators with drooping characteristics are used, each welder must be supplied with a separate machine.

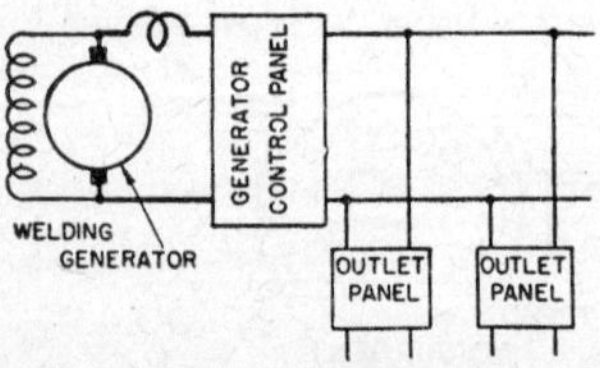

FIG. 21-6

When it is desired to have several welders work from a single generator, a flat compound machine is used. In this case the current is supplied through a generator control panel and through outlet panels, one of which is required for each welder (see fig. 21-6). The resistance in the outlet panels, which is varied by the knife switches, provides the drooping characteristics and limits the current to a safe value during the short-circuiting period.

Automatic Arc Welding—The main units of an arc welding outfit, excluding the motor-generator set, are the control panel and the welding head. Such units, as manufactured by the General Electric Company, are shown in figs. 21-7a and 21-7b. These will be con-

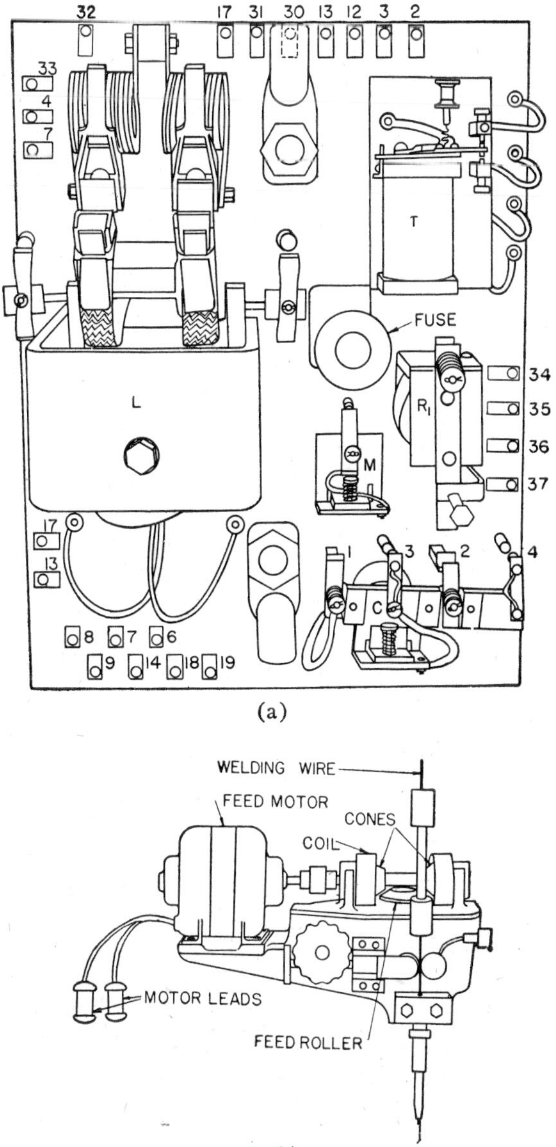
32
17 31 30 13 12 3 2
33
4
7
T
FUSE
L
R1
M
34
35
36
37
17
13
1 3 2 4
C
8 7 6
9 14 18 19
(a)
WELDING WIRE
FEED MOTOR
CONES
COIL
MOTOR LEADS
FEED ROLLER

(b)

Fig. 21-7

sidered first by means of an elementary diagram, which is shown in fig. 21-8a.

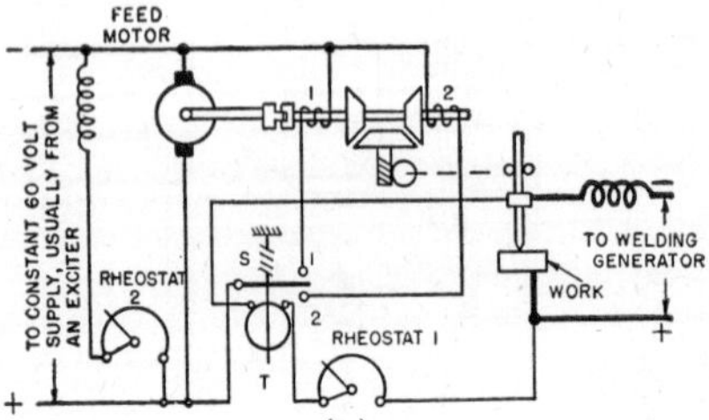

(a)

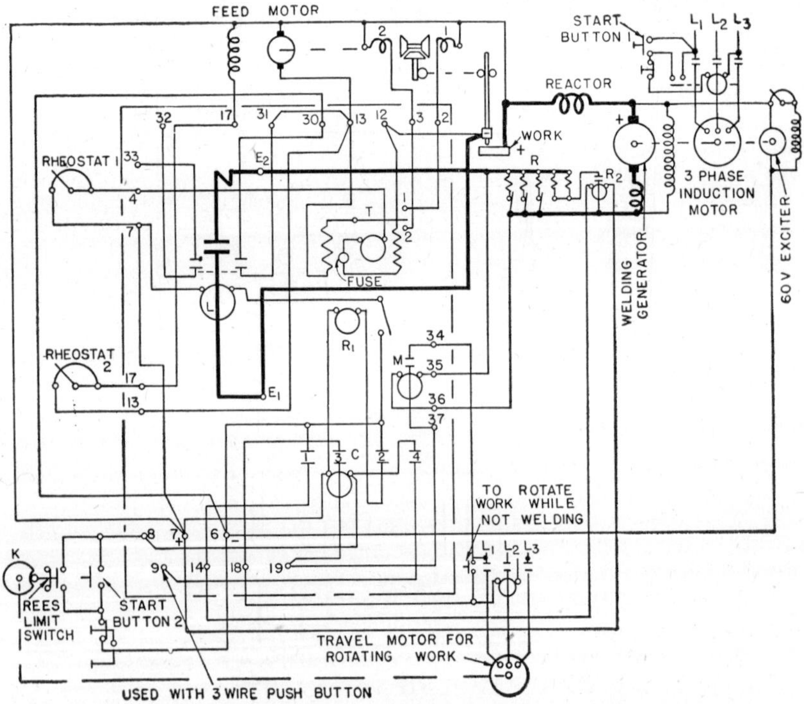

(b)

Fig. 21-8

The coil of relay T is connected across the arc, through rheostat 1, and is therefore energized by the arc voltage. When starting to weld, the coil of relay T is across the open circuit generator voltage. Hence, current flows through the relay coil and the vibrating arm is pulled down. This closes contact 2, energizing coil 2 in the welding head. The shaft with cones is connected to the motor shaft in a way that allows it to be free to slide back and forth while rotating. The cone that feeds the electrode downward by means of rollers is forced into contact with the disk by coil 2, as this coil is energized (see fig. 21-7b). At the instant that the electrode touches the work, relay coil T is short-circuited and deenergized. Spring S then pulls the vibrating arm upward, closing contact 1. This deenergizes coil 2 and energizes coil 1. The cones then pull over and the disk rotates in the opposite direction, pulling back the electrode and forming the arc, which reaches the voltage for which rheostat 1 is set. As soon as this voltage is reached, the arm is again pulled down, again closing contact 2. The relay arm will continue to vibrate between contacts 1 and 2, controlling the rate at which the wire is fed to the arc.

A complete wiring diagram of a General Electric arc welding system is shown in fig. 21-8b. The generator is started by pressing the start button at 1. Then start button 2 is pressed, closing the four finger contactor C. Contact 1 completes the coil circuit of contactor R_2, which parallels the resistors in the welding circuit. Contact 2 completes the coil circuit of relay R_1. This relay, on closing, energizes the coil of contactor L, which also closes. An auxiliary contact of contactor L then energizes the feed motor and connects the arm of relay T to the negative live wire. This connects relay T as shown in fig. 21-8a. The relay then controls the operation of the welding head as previously explained. The instant the electrode touches the work, a voltage is established across resistors R, which energizes contactor M. This contactor, on closing, starts the travel motor, which rotates the work and cam K. Cam K closes the limit switch, which maintains the coil circuit of contactor C until the weld is complete. If the arc is broken, no voltage is established across the coil of contactor M and it opens, stopping the travel motor. As soon as the arc is reestablished, contactor M again closes, restarting the travel motor. Thus the bead formed will be continuous and will require no touching up.

When cam K returns to the start position, the limit switch is released, deenergizing the coil of contactor C, which opens. Contact 2 deenergizes contactor R_1, which stays closed about two seconds because it incorporates the definite time element principle (see chapter

8). Then the arc exists two seconds after the feed motor stops, to eliminate the formation of a crater at the end of the weld. When the crater is being filled, some resistance in R is disconnected by

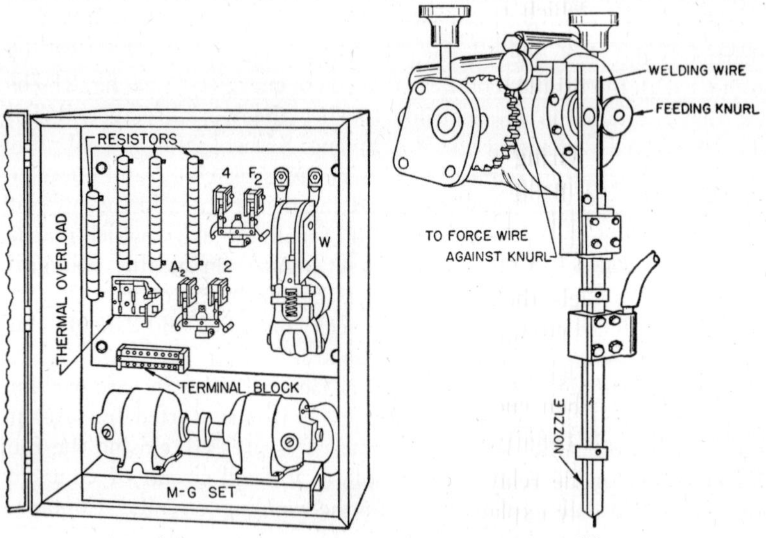

(a) (b)

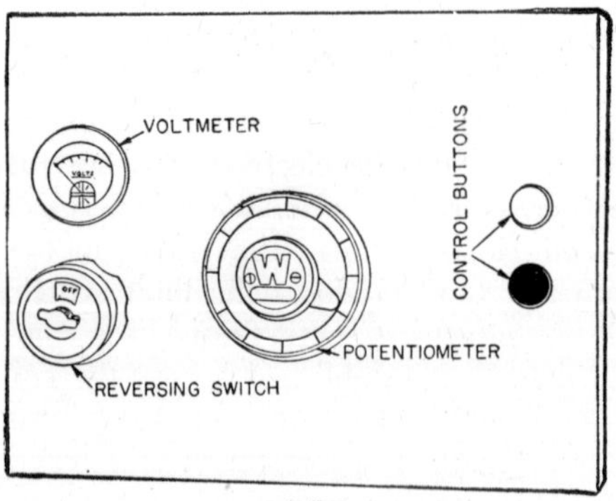

(c)

Fig. 21-9

contactor R_2. If resistors were not removed, the welding current would burn a hole in the work.

If the generator has drooping characteristics, in which case no resistance R is required in the welding circuit, contactor M must have a series coil. This is connected in series with the main contactor L.

Another type of automatic arc welding outfit is shown in fig. 21-9. It consists of three principle units, known as the main control panel (fig. 21-9a), the feed motor (fig. 21-9b), and the operator's control panel (fig. 21-9c). A simplified wiring diagram of this outfit is shown in fig. 21-10a. The feed motor is automatically controlled, both for a change in speed and direction of rotation, by generator G driven at a constant speed by motor M. Generator G receives its field excitation from a source governed by the arc voltage.

When button S is pressed, contacts 2 and A_2 close. Contact 2 connects the feed motor to the generator, and A_2 energizes the coil

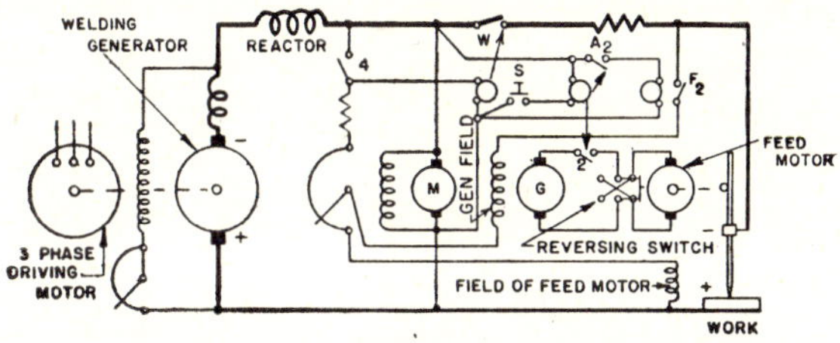

(a)

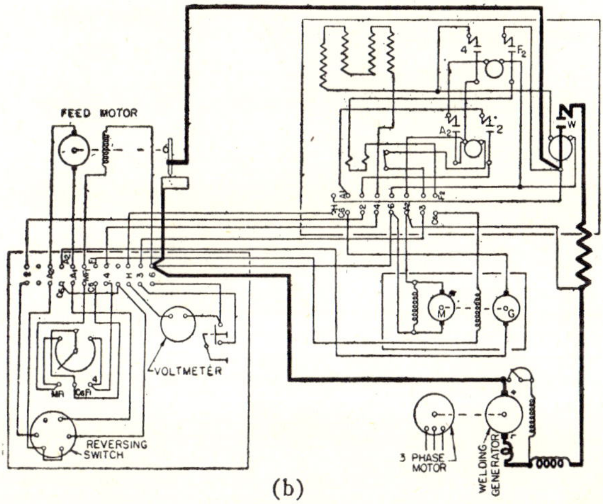

(b)

FIG. 21-10

that closes contacts F_2 and 4. Contact 4 completes the field circuit of the feed motor and energizes the coil of the main contactor W, which closes. F_2 completes the field circuit of generator G. Now generator G supplies power to the welding head, which feeds the electrode toward the work. At the instant that the electrode strikes the work, the field current of generator G reverses, reversing the generator polarity and therefore the feed motor. Thus the electrode draws an arc. As the electrode becomes more negative, the field voltage of generator G decreases, and the feed motor speed consequently decreases. When the field voltage of generator G becomes zero, the generator supplies zero voltage and the feed motor stops.

As soon as the arc becomes slightly more lengthened, due to the fusing of the welding wire, the electrode becomes slightly more negative, and current will flow in the generator field in a direction that will cause the feed motor to move the wire toward the work. When the arc is again correct in length, the generator field voltage is zero and the head stops.

The complete wiring diagram of the welder is shown in fig. 21-10b. The reversing switch is used to run the welding wire back.

A C Arc Welding—Arc welding with 60 cycle alternating current is not possible with bare wire because the arc is extinguished when the current passes through the zero value. However, an A C arc can be maintained with coated wire.

Transformers for 60 cycle arc welding are generally made as suggested in fig. 21-11. The gap establishes a large leakage reactance,

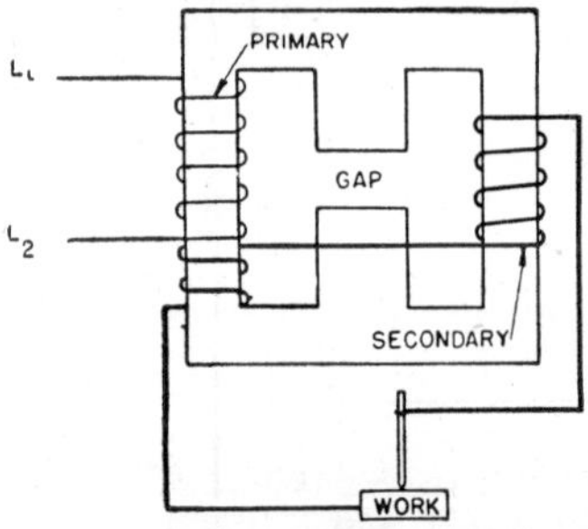

FIG. 21-11

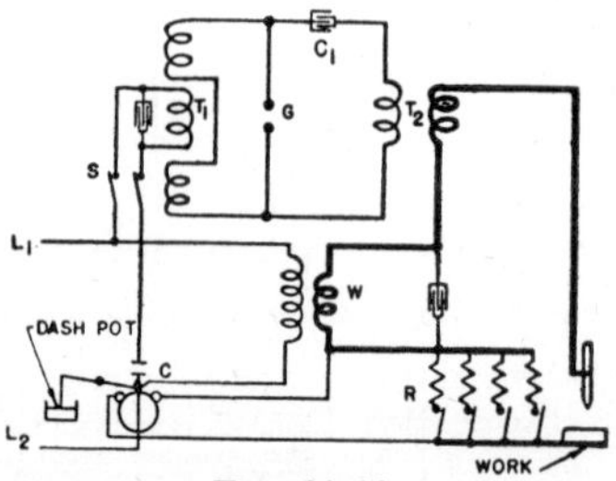

FIG. 21-12

causing the transformer to have drooping characteristics. A part of the secondary winding is placed on the same leg as the primary. This causes the generated secondary voltage to drop considerably when the secondary is short-circuited.

Bare electrodes can be used with the type of A C arc welder shown in fig. 21-12. The primary of transformer W is connected to a source of power, such as a 440 volt, 60 cycle line. The secondary supplies the welding current at about 70 volts. This unit includes a high frequency outfit, which functions as follows: With switch S closed, transformer T_1 impresses several thousand volts on condenser C_1, a voltage sufficient to cause a spark to jump gap G. This spark short-circuits transformer T_1, and condenser C_1 through the primary of transformer T_2 (the current in T_1 does not increase much because this transformer has a high flux leakage). Condenser C_1 discharges, sending a rush of current through the gap and primary of T_2. After the condenser is fully discharged, the current is kept flowing by the primary inductance of transformer T_2. This charges the condenser in the opposite way, but not quite as strongly as in the first case, because energy is lost due to the resistance in the circuit. The condenser again discharges and charges up in the original direction, this action continuing until all the energy is used up (see fig. 21-13). The frequency is about 100,000 cycles per second.

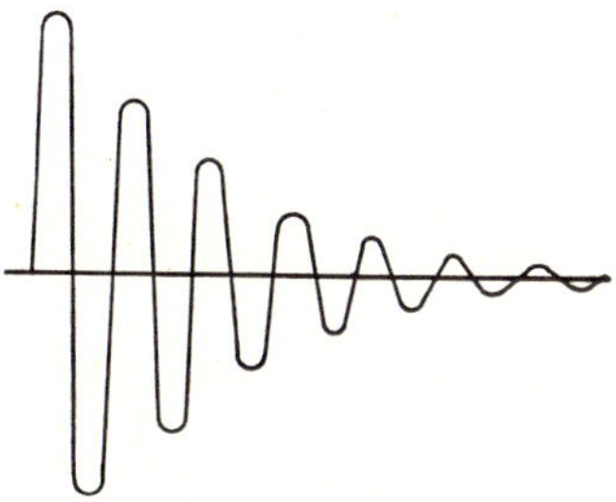

FIG. 21-13

The high frequency current in the primary establishes a high frequency high voltage in the secondary, which is superimposed on the welding current and keeps the air between the electrode and the work ionized. In such a case, the arc will not go out when the welding current is passing through zero, and the arc resistance will remain practically constant, eliminating extreme fluctuations in the welding current.

The high voltage supplied by transformer T_2 is not dangerous, because it supplies a current at a high frequency, which will travel only near the surface of the skin, where it does no harm.

Contactor C assures that the high frequency voltage is off when the welding current is off. This is desirable when changing the electrode. The dash pot prevents contact C from chattering, which it tends to do when the welding arc breaks.

Resistance Welding—Resistance welding may be divided into spot, butt, and seam or line welding. In these methods a single-phase transformer is used, which changes the relatively high shop voltage, capable of supplying a low current, to a low voltage (5 to 8 volts) and an extremely high current. The actual secondary voltage required depends upon the kind of material being welded and its cross-sectional area. A tap transfer switch is generally included in the primary circuit, enabling the secondary voltage to be adjusted as desired.

Fig. 21-14a shows the top section of a flat top butt welder, equipped with dies for butt welding the tube and bell of a rear axle housing

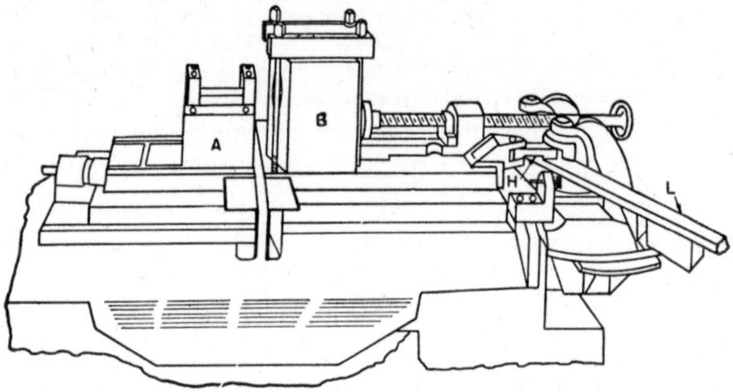

(a)

for an automobile. The tubing is rigidly bolted in head A, which is stationary, and the casting is set in the face of head B, which is joined to toggle lever L through link H. The wiring diagram of the welder is shown in fig. 21-14b. The welding is accomplished as follows: The surfaces to be welded are brought together by lever L, and the control switch in the handle is then pressed. This closes the magnetic switch in the primary circuit, and a high current is sent through the butted ends. These ends are actually in contact at only a few points, and these contacts offer a high resistance to the flow of the current. Hence the contacting points are highly heated and become molten in a short time. The control switch is released when sufficient heat has been obtained, and the ends in contact are brought together with a blow, completing the weld. An upset butt is then formed at the welded section, which is removed by machining.

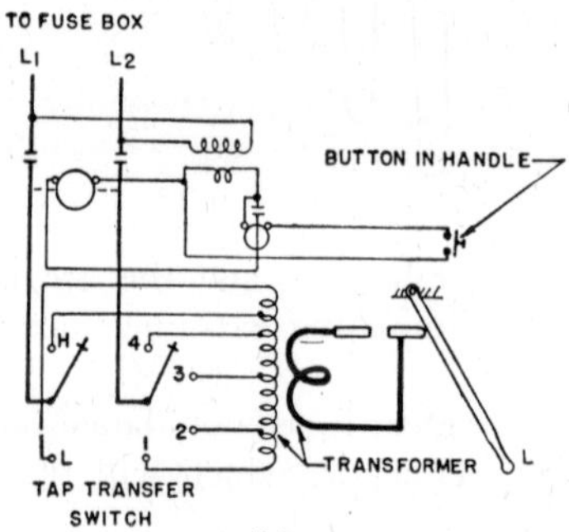

(b)

FIG. 21-14

Butt welding is frequently done automatically. A wiring diagram of a machine that automatically butt welds sheet stock is shown in fig. 21-15. The stock is placed in position and clamped with air operated jaws, and the start button is then pressed. This starts motor M, which turns cams A and B. Cam A operates limit switch L, completing the primary circuit and sending current through the stock, which begins to heat. The high current soon burns away the material in contact, and an arc is formed. Cam B slowly brings one piece toward the other, maintaining the arc. The instant the ends reach the welding temperature, cam B forces the moving sheet against the fixed sheet with a high pressure, forming the weld. At the same instant, cam A returns to the starting point, turning off the welding

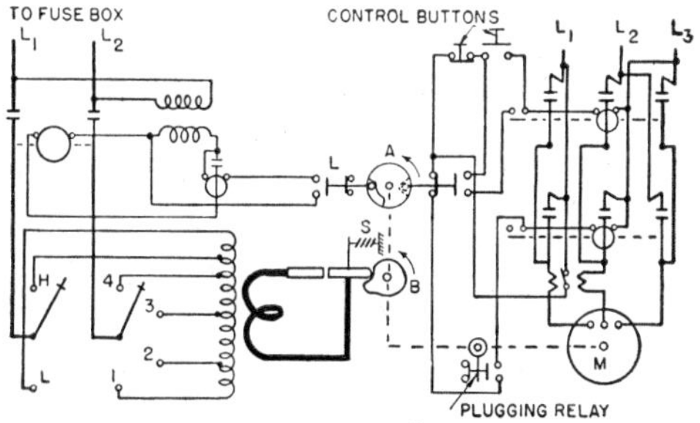

Fig. 21-15

power and plugging the motor. This causes cam B to come to a standstill at the starting position. Next, the air pressure on the clamping jaws is removed, and the jaws release the work. The jaws are pulled back to their normal position by spring S.

Spot Welding—Thin sheets of metal are generally joined by welding them together in spots by means of a machine called a spot welder. One type, manufactured by Taylor-Winfield, Inc., is shown in fig. 21-16. The metal sheets are placed between the electrodes and the pedal is pressed down. This first moves the electrodes together, clamping the sheets under spring pressure, and then closes a switch in the primary circuit, establishing a high current in the secondary. This current passes through the spot of metal between the electrodes. As soon as this spot reaches the welding temperature, the spring pressure between the electrodes completes the welding process. On releasing the pedal, the primary switch opens, interrupting the power, and the material is then released.

Buck Welding—Fig. 21-17 shows a wiring diagram of a buck welder, which in this case is used to make spot welds on an automobile body that is being assembled. To make a weld, the electrode at

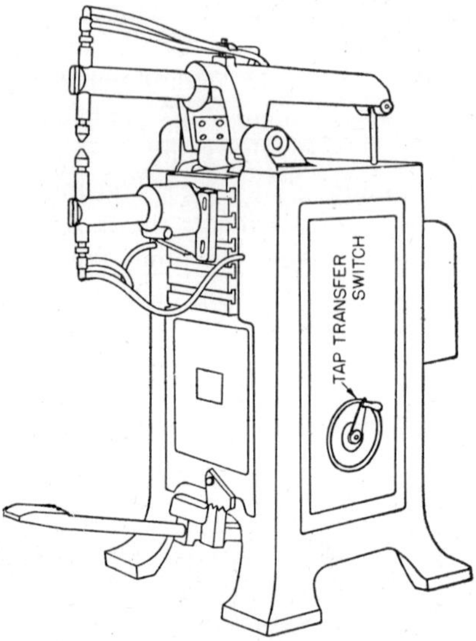

Fig. 21-16

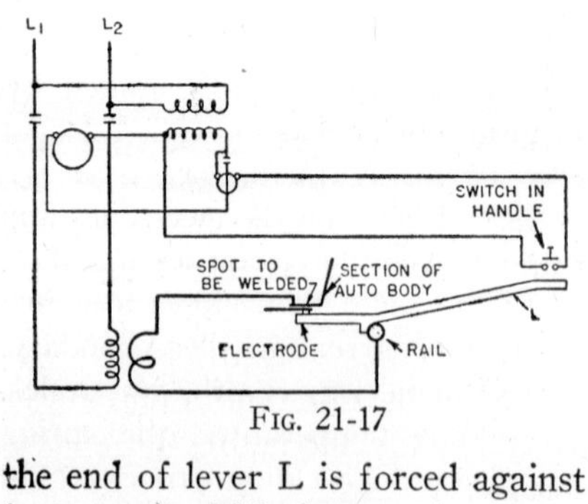

Fig. 21-17

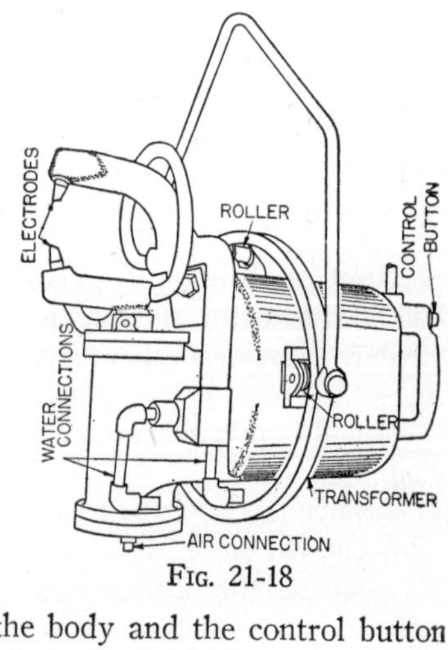

Fig. 21-18

the end of lever L is forced against the body and the control button is pressed. This button should be released before the pressure on the electrode is removed, or an arc will burn the metal.

Control for Portable Spot Welders—A type of control suited for portable spot welders of the air clamp type (shown in fig. 21-18)

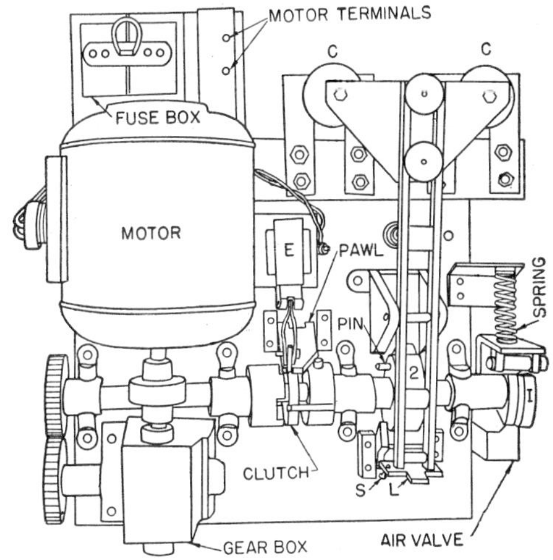

is shown in fig. 21-19. It operates as follows: With the knife switch closed, the motor runs continuously. When the control button is pressed, electromagnet E is energized, operating a pawl which allows the clutch to engage. The motor then drives the cam shaft through a train of gears. Cam 1 opens the air valve, causing the electrodes to clamp the work. A moment later, cam 2 closes contacts C, which complete the primary circuit, and these contacts are held closed by engaging latch L. Welding current is thus supplied. After a definite time, a pin on cam 2 trips latch L and a spring opens contacts C, interrupting the welding current. Cam 1 next allows a spring to close the air valve and the

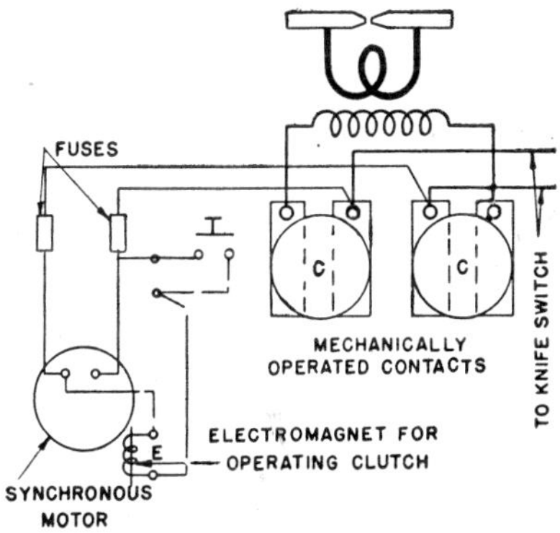

Fig. 21-19

work is released. It is impossible for the operator to release the electrodes while the welding is taking place. If this were done, the material would be severely burned.

The cam shaft is driven by a synchronous motor at 1800 r p m. Such a drive makes it possible to have an adjusting screw S on latch L, so that contacts C open at the instant that the current is at or near zero. When so adjusted, the contacts have minimum arcing and therefore maximum life.

Each cam is provided with a number of holes so that it may be shifted on the shaft to give changes in the welding period.

Automatic Spot Welder—To operate the machine shown in fig. 21-20, a switch must first be closed to start the motor. To make a series of welds joining two strips placed between the electrodes, the foot pedal is pressed and held down. This causes a clutch to

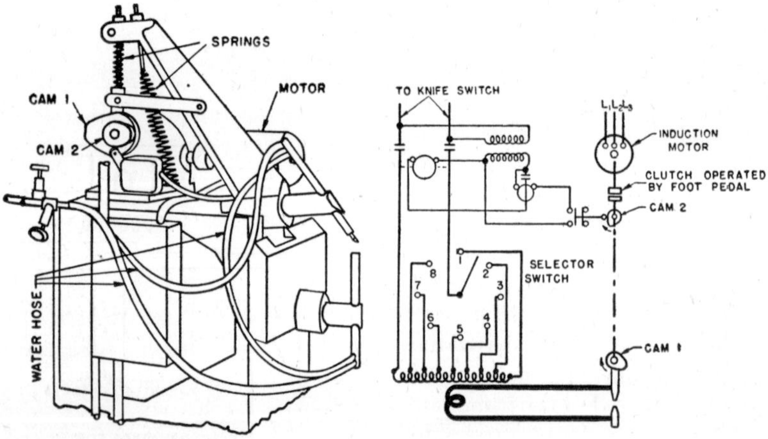

Fig. 21-20

engage and turn cams 1 and 2. Cam 1 moves the top electrode down, clamping the metal, and cam 2 then turns on the power for about a second to form the weld. After this, cam 1 releases the stock. While the electrodes are apart, the material is shifted and the welding cycle is repeated. Welding cycles occur repeatedly as long as the pedal is held down.

Constant Heat Welding—The amount of heat produced by current flowing through two pieces which are being welded may be expressed by Equation 2-22, which is

$$H = .000948I^2Rt \qquad (21\text{-}1)$$

where I = welding current, R = resistance of material between electrodes, t = time in seconds, and H = heat in B T U.

If the resistance R of the pieces between the electrodes is always the same, all welds will be the same for a fixed electrode voltage and a fixed time. But unfortunately, many factors affect the resistance and the welding current, and consequently, the nature of the weld produced. The contact resistance of the pieces may vary over a wide range, due to the presence of oil or scale, or to the condition of the electrode points, which become rough and oxidized after being used for some time. Variation in electrode pressure, thickness of the pieces being welded, and the supplied line voltage also affect the resistance and the welding current.

To decide how the heat produced during each weld may be kept constant, suppose we refer to Equation 21-1, and express it in the following way:

$$H = (.000948IR)It \qquad (21\text{-}2)$$

The quantity within the parentheses remains practically constant, since I decreases when R increases. To keep the heat produced during each weld constant, then, the product of the current and the time must be kept constant. That is, during welds in which the current decreases, the time must be correspondingly increased.

Fig. 21-21a shows a welding timer which has such operating features. Its wiring diagram is shown in figs. 21-21b and 21-21c. With reference to these figures, contactor M closes when switch B is pressed (a hand or foot switch). This energizes the magnetic valve V, which opens and allows air (or liquid) to enter the cylinder and cause the electrodes to clamp the stock. As soon as the required electrode pressure is reached, the pressure switch P closes and completes the circuit of the coil of contactor W. The contactor then closes and the welding starts. At the same time, the series transformer begins supplying a voltage, which is rectified by the rectifying tube R. Hence, unidirectional current is supplied to condenser C through rheostat R_1. As the condenser becomes charged, its terminal voltage

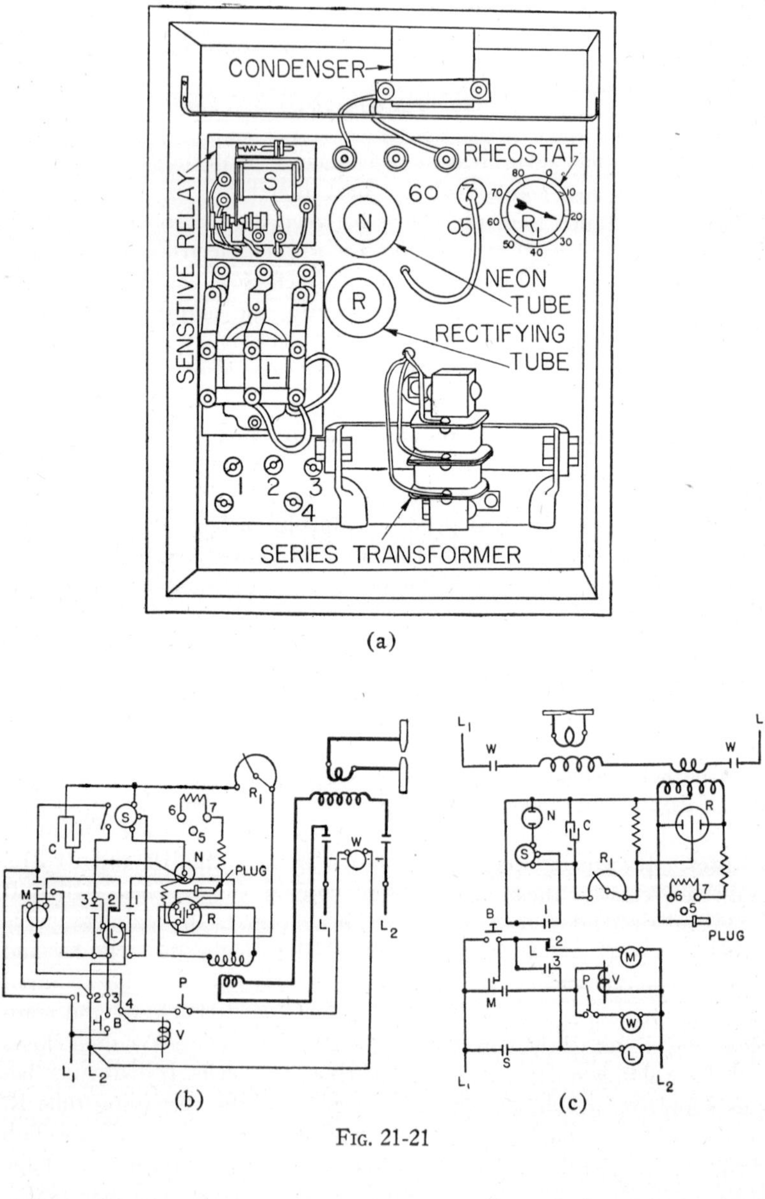
CONDENSER
RHEOSTAT
SENSITIVE RELAY
S
N
R
L
NEON TUBE
RECTIFYING TUBE
SERIES TRANSFORMER
(a)
PLUG
(b)
PLUG
(c)

FIG. 21-21

increases. When its voltage reaches a predetermined value, the neon tube N becomes conductive. The condenser then discharges through the coil of the sensitive relay S, causing the contact of this relay to close, energizing the coil of contactor L. Contact 2 opens and contacts 1 and 3 close. Contactor 2 deenergizes the coil of contactor M, which in turn deenergizes contactor W and magnetic valve V. Contactor W stops the welding and valve V releases the air, allowing the electrodes to open.

Contact 1 discharges the condenser, and contact 3 maintains L energized as long as the switch is pressed. This maintains contacts W deenergized, in which case no welding current can be supplied. When it is desired to form the next weld, the switch must be released and then pressed.

For a low current range, the plug is placed in receptacle 5. For higher current ranges, the plug must be placed in receptacles 6 or 7. This by-passes current from the rectifying tube, preventing it from becoming overloaded.

The welding contactor W remains energized while the condenser is being charged. The time required for charging is inversely proportional to the voltage impressed on the condenser, and this voltage varies directly with the welding current. When the welding current is low, the time is long, and when the current is high, the time is short. According to Equation 21-2, then, the same amount of heat is produced for all welds.

Seam Welding—Two sheets of metal may be joined by a seam or line weld. This is accomplished by butt welding the sheets as stated in the discussion pertaining to fig. 21-14, or by lap welding them. A machine that forms a lap type seam weld is shown in fig. 21-22. It has two electrodes in the form of thin rollers or disks, and the material being welded is forced through these by a small motor turning one of the rollers. The current passes from roller to roller, the rollers being connected to the terminals of the welding transformer secondary.

If the current were continuous, the welding would not be satisfactory, for the following reasons:

(1) The material being welded would require a large amount of heat at the start. As the welding progressed, the amount of heat supplied would have to be reduced, otherwise the heat supplied at the

start would travel ahead of the point of contact of the rollers and cause the temperature to continually build up and burn the material.

(2) In the material being welded, high currents would flow through paths of low resistance and little or no current would flow through paths of high resistance, making imperfect welds.

(3) The weld would be wide and deeply penetrate into the material, severely warping the surface.

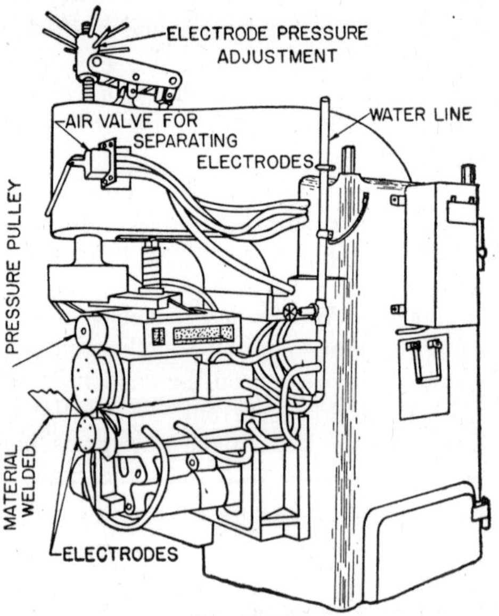

FIG. 21-22

(4) The material being welded would not cool fast enough, and a hot spot would exist just behind the point where the rollers were in contact with the material. At this spot the two sheets being welded would be nearly at the welding temperature, and would sometimes stick to the rollers and tear apart.

These difficulties are overcome by supplying a current interrupted at frequent intervals. One method of doing this is to use an interrupter (a device for making and breaking the circuit) in the primary circuit (see fig. 21-23). This produces a seam weld which is formed by a series of overlapping spot welds. With this method, a voltage high enough to break down the contact resistance of the material being welded is applied intermittently. The metal between the spots chills fast enough to prevent the material from being torn apart, and also prevents the building up of heat ahead of the electrode as progress is made.

In practice, seam welding is done at high speed, in which case it is necessary to interrupt the current about twice each second so that the spots overlap. The interval during which the current is "on" and during which it is "off" are of utmost importance and must be controlled very accurately. To do this with a mechanical interrupter is difficult, because the operation of the interrupter changes, due to wear, change in friction, change in temperature, and pitting of the contacts. Pitting of the contacts is the most serious hindrance, and can be eliminated only if the contacts are opened when the primary current passes through its zero value. This is very difficult to do.

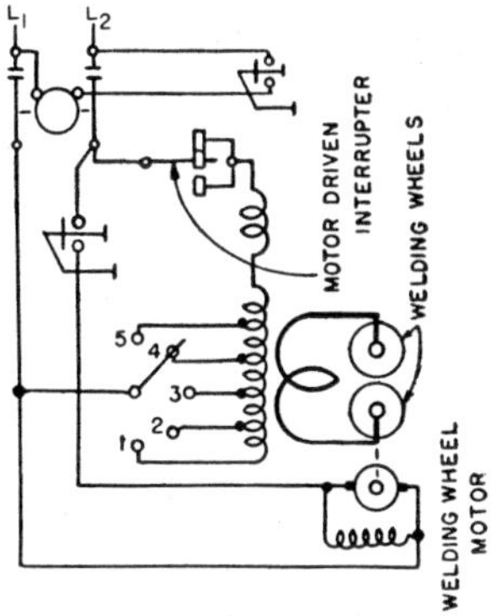

Fig. 21-23

Another system of interrupting the current is to use an electron tube control. With this control, more interruptions per second can be obtained than it is possible to obtain with mechanical interrupting devices. Such a control can be more easily and more accurately adjusted, and has no make and break contacts or moving parts. When set to suit a particular job, its operating characteristics do not change.

A simplified diagram of an electron tube welding control is illustrated in fig. 21-24a. The primary of a second transformer, whose secondary is alternately short-circuited and open-circuited at a rapid rate by the two hot cathode tubes 1 and 2, is connected in series with the primary of the welding transformer. When the secondary of the series transformer is short-circuited, the impedance of its primary is very low and full voltage is impressed on the primary of the welding transformer. When the secondary is open-circuited, considerable reactance is imposed in series with the primary of the welding transformer. As a result, practically no power is supplied to the welding transformer and no welding occurs at this instant.

The problem that concerns us now is how tubes 1 and 2 are controlled. As explained in chapter 20, for each value of plate

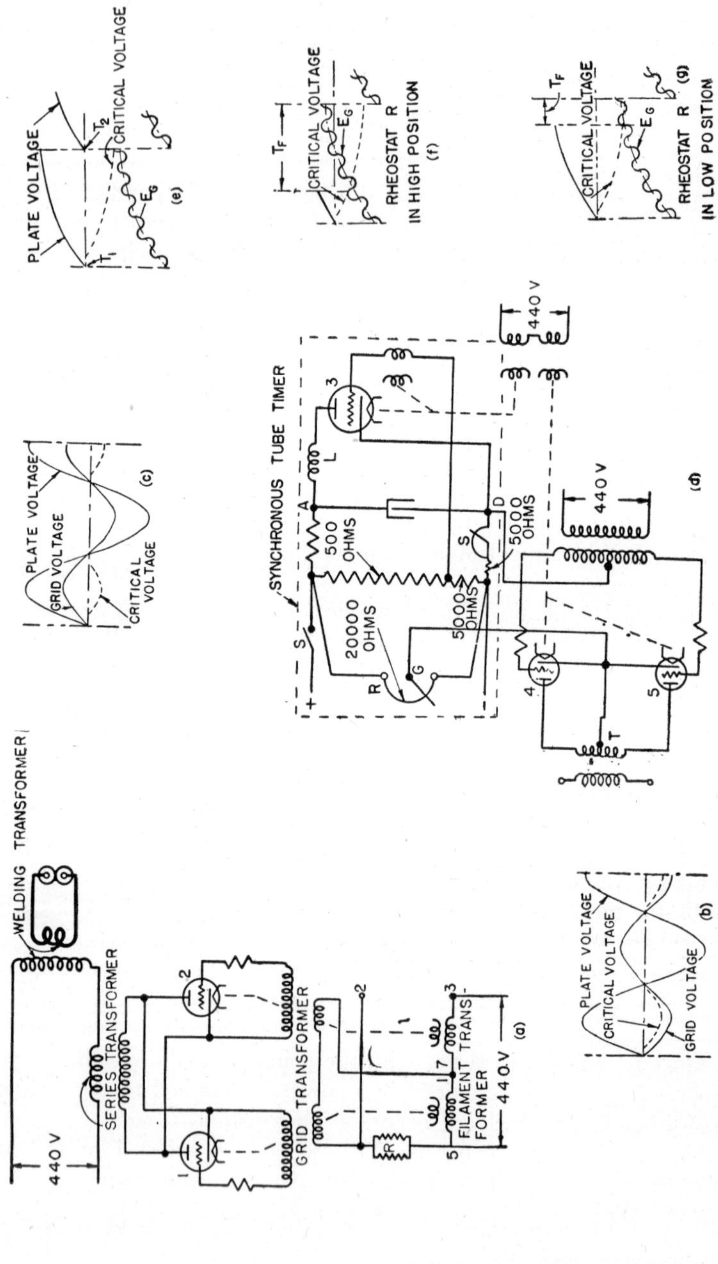
WELDING TRANSFORMER
SERIES TRANSFORMER
440 V
GRID TRANSFORMER
FILAMENT TRANSFORMER
440 V
(a)
PLATE VOLTAGE
CRITICAL VOLTAGE
GRID VOLTAGE
(b)
PLATE VOLTAGE
GRID VOLTAGE
CRITICAL VOLTAGE
(c)
SYNCHRONOUS TUBE TIMER
500 OHMS
20000 OHMS
5000 OHMS
5000 OHMS
440 V
440 V
(d)
PLATE VOLTAGE
CRITICAL VOLTAGE
(e)
CRITICAL VOLTAGE
RHEOSTAT R IN HIGH POSITION
(f)
CRITICAL VOLTAGE
RHEOSTAT R IN LOW POSITION
(g)

FIG. 21-24

voltage there is a given value of grid voltage, called the critical value, below which no ionization occurs and no plate current flows (fig. 21-24b). If the grid voltage is more positive than this critical value (fig. 21-24c), ionization occurs and a plate current flows. Once the plate current starts, the grid cannot stop or limit its flow. If the grid is made sufficiently negative, however, it will regain control when the plate current passes through zero, because ionization stops at that instant.

When terminals 2 and 3 are not connected, the primary of the grid transformer is connected between 17 and 5 of the filament transformer. In this case the plate and grid voltages are 180 degrees displaced, as shown in fig. 21-24b, and with this relationship no plate current flows.

Now suppose that terminals 2 and 3 are connected. This connects the primary of the grid transformer between terminals 3 and 17 of the filament transformer, which shifts the grid voltage 180 degrees. The plate and grid voltages are now in phase, as shown in fig. 21-24c, and plate current flows, short-circuiting the secondary of the series transformer. Resistance R is now across the line and prevents a short circuit.

For seam welding, terminals 2 and 3 must be brought in contact and then separated, at a rate that will cause the series of spot welds to just overlap. The connecting and disconnecting of terminals 2 and 3 cannot be done satisfactorily with a mechanical switch because its operating characteristics change. Accurate results can be had only with a synchronous tube timer, the wiring diagram of which is shown in fig. 21-24d.

At the instant that switch S is closed, the voltage across the condenser is zero, making the potential of A and D the same, and making the grid of tube 3 negative with respect to the cathode. This instant is shown at T_1 in fig. 21-24e. As the condenser becomes charged, the plate becomes more positive and the grid becomes less negative with respect to the cathode (see fig. 21-24e). The grid voltage varies as indicated by curve E_G, because it has a low superimposed voltage impressed on it. As soon as the grid voltage becomes less negative than the critical voltage, a high current will surge through the tube, instantly discharging the condenser, after which the grid again gains control and the cycle repeats, starting at T_2 (fig. 21-24e). This continues repeatedly.

The charging and discharging of the condenser causes the voltage between G and D to vary between limits which can be adjusted by rheostat R. The value of the voltage between G and D depends upon the setting of rheostat R and upon the amount of current through the 5000 ohm resistor and rheostat S. This voltage controls the grid bias and therefore the time during which tubes 4 and 5 carry current. A flow of current occurs in these tubes only when the grid voltage becomes less than the critical voltage. If rheostat R is adjusted so that the fixed voltage between G and D is low, the current in tubes 4 and 5 flows for a long period during each cycle, as indicated by T_F in fig. 21-24f. If adjusted so that the fixed voltage is high, the current flows for a short period (see T_F in fig. 21-24g). *Then rheostat S controls the period of one welding cycle and rheostat R governs the fraction of the cycle during which welding current flows.*

If the secondary of transformer T is now connected to terminals 2 and 3 (fig. 21-24a), these terminals will, in effect, be short-circuited when tubes 4 and 5 carry current and will be open-circuited when these tubes carry no current. This action, as previously explained, controls the operation of tubes 1 and 2, which intermittently short-circuit the secondary of the series transformer.

In considering tube 3 it was found that an A C voltage was superimposed on the grid bias. This was done to provide more accurate timing. If the grid voltage varied as indicated by the dotted line (fig. 21-24e), the intersection of this line with the critical voltage line would not be as definite as that obtained with the superimposed voltage.

Mercury Pool Electron Tube Control—A General Electric welding timer has been developed that is unlike the one shown in fig. 21-24a in having no series transformer. Instead, two mercury pool electron tubes are connected directly in the primary circuit (fig. 21-25). These are placed in parallel in reverse relation to each other, so as to pass full wave alternating current. Tubes L and T control the number of cycles for each weld, and start the conduction of tubes 1 and 2, as explained in chapter 20.

The control (fig. 21-25) contains a conventional full wave rectifier for supplying direct current voltage. The filter inductance and filter capacity make this voltage steady. Connected across the output of the rectifier is a voltage divider, which consists of resistors R1, R3, and rheostat R2. By means of R2 it is possible to adjust the grid bias of tube L. Resistance R3 furnishes the negative bias on the keying tube, K.

When the control switch is in the "off" position, as shown in fig. 21-25, no current flows through R5, and therefore the cathode of tube L is positive. The grid is negative relative to the cathode by an amount equal to the voltage drop in R1 and in a portion of rheostat R2. Then tube L is non-conducting.

When a spot weld is to be made, the control switch is turned on. The grid of tube K holds this tube non-conducting except for an interval of five degrees during each cycle. This interval can be changed by adjusting resistance R4. After the first positive peak occurs, the

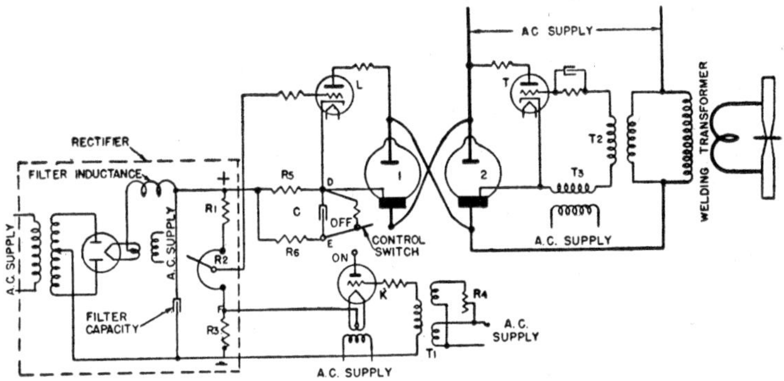

Fig. 21-25

tube becomes conductive and remains so until the control switch is turned off. When tube K becomes conductive, points E and F are brought to the same potential (not considering the tube drop, which is about 15 volts). At the instant that conduction starts, no voltage exists across condenser C, because it takes time to build up a voltage across a condenser. Then point D and therefore the cathode of tube L are at the same potential as point F. Now the grid bias is positive relative to the cathode of the timing tube by an amount equal to the voltage drop in a portion of rheostat R2. Hence, tube L becomes conductive as soon as its plate becomes positive, causing the mercury pool tube 1 to conduct as explained in chapter 20.

Condenser C gradually becomes charged, causing the potential at point D and the cathode of tube L to become more positive than the grid, or, in other words, when the grid of tube L has become more negative than the critical voltage, this tube and tube 1 will stop conducting. While condenser C is becoming charged, resistor R6 carries enough current to keep tube K ionized and therefore conductive.

When the current in tube 1 becomes zero at the end of the first half of the current cycle, the line voltage is high because the power factor of the welding transformer is poor, and as this voltage is across the primary of transformer T2, it causes the grid of tube T to be positive. Thus tube T and therefore tube 2 conduct. Then when tube 1 conducts a half cycle of current, tube 2 completes the cycle. When tube L stops conducting (which occurs at the instant that condenser C becomes fully charged), no voltage is impressed on transformer T2; and transformer T3, being connected to oppose T2, makes the grid of tube T negative. Thus tube T and 2 also become nonconductive.

To make a second spot weld, the control switch must be turned off and then turned on again, in which case the cycle of operation repeats.

Exercises

1. Resistance welding is divided into what classes? Briefly explain each.
2. In the automatic butt welder shown in fig. 21-15, what will take place if motor M is not plugged after a weld is completed?
3. What are the undesirable features of the welder shown in fig. 21-16?
4. Explain how the controller shown in fig. 21-19 operates.
5. What are the purposes of cams 1 and 2 in fig. 21-20?
6. What conditions affect the resistance between the electrodes when welding?
7. How must a welder operate to supply the same amount of heat to all welds?
8. (a) For what operating condition should the plug in fig. 21-21 be inserted in jack 7? Why? (b) What is the purpose of rheostat R_1?
9. Why is uninterrupted alternating current not suitable for seam welding?
10. Why is an ordinary vibrating contactor, such as the one shown in fig. 21-23, not satisfactory for interrupting the primary current?
11. In fig. 21-24, what is the purpose of rheostat S? Of rheostat R?
12. (a) In fig. 21-25, why are tubes L and T known as leading and trailing tubes? (b) In this controller, how is the welding period increased?
13. The fundamental principle of operation of the controllers shown in figs. 21-24 and 21-25 is the same. Explain this fundamental principle.

CHAPTER 22

ELECTRICAL DRAWING

Symbols—For electrical installations, such as those consisting of a controller, motor, and auxiliary equipment, two types of wiring diagrams are used, known as main diagrams and simplified diagrams. In both, the contactors, relays, resistors, etc., are represented by symbols, which are shown in fig. 22-1. Symbols are made as simple as possible so that the least amount of time is required in making a wiring diagram, and they are drawn in such a manner as to suggest the device they represent. Meters and relays are represented by a circle in which is shown the terminals indicated in their actual position. Terminals of voltage coils are indicated by dots and those of current coils are indicated by small circles.

The main circuits are shown by heavy lines and the control circuit by light ones. The heavy lines should be considerably heavier than the light ones to establish a contrast. The light lines should not be too fine, otherwise they will not appear on a blueprint. All lines representing connecting wires should be made either horizontally or vertically, except very short ones. Lines should not be less than one-eighth inch apart and should be uniformly spaced. They should be made with as few bends as possible and should cross as little as possible.

Main Diagram—A main diagram (see fig. 22-2a) shows the switches, contactors, relays, etc., as they are actually arranged on the panel and the electrical connections where they actually are. Connections of devices on a panel are generally made in the rear of the panel and wiring diagrams are generally drawn to show this view. All the devices on a panel should be shown enclosed in a rectangle made up of broken lines to make clear what the panel includes. The parts of the motor, such as the armature, shunt field, and series field, should be shown together. Machines coupled together (such as a motor, generator, and exciter) should be so indicated. Main diagrams are considerably useful when connecting the equipment, making changes in the hookup, and locating faults.

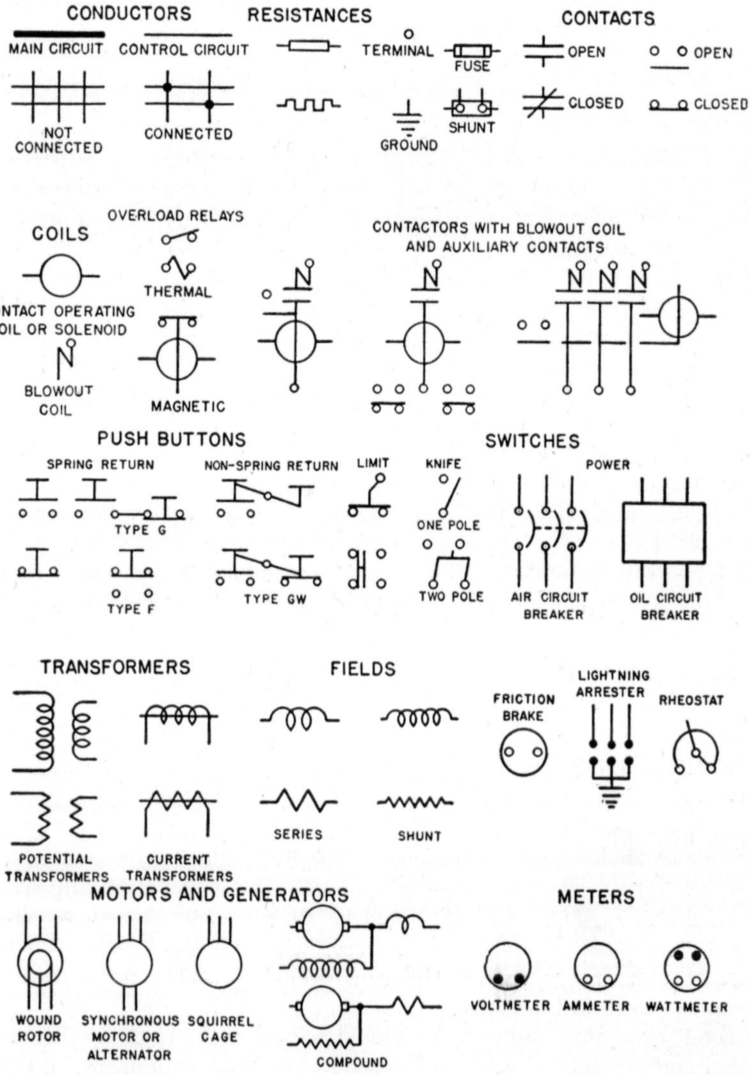
CONDUCTORS
MAIN CIRCUIT
CONTROL CIRCUIT
NOT CONNECTED
CONNECTED
RESISTANCES
TERMINAL
FUSE
SHUNT
GROUND
CONTACTS
OPEN
CLOSED
OPEN
CLOSED
COILS
CONTACT OPERATING COIL OR SOLENOID
BLOWOUT COIL
OVERLOAD RELAYS
THERMAL
MAGNETIC
CONTACTORS WITH BLOWOUT COIL AND AUXILIARY CONTACTS
PUSH BUTTONS
SPRING RETURN
TYPE G
TYPE F
NON-SPRING RETURN
TYPE GW
LIMIT
SWITCHES
KNIFE
ONE POLE
TWO POLE
POWER
AIR CIRCUIT BREAKER
OIL CIRCUIT BREAKER
TRANSFORMERS
POTENTIAL TRANSFORMERS
CURRENT TRANSFORMERS
FIELDS
SERIES
SHUNT
FRICTION BRAKE
LIGHTNING ARRESTER
RHEOSTAT
MOTORS AND GENERATORS
WOUND ROTOR
SYNCHRONOUS MOTOR OR ALTERNATOR
SQUIRREL CAGE
COMPOUND
METERS
VOLTMETER
AMMETER
WATTMETER

Fig. 22-1

If a wiring diagram is to include two or more panels that are alike, it is necessary to show a complete wiring diagram of one panel

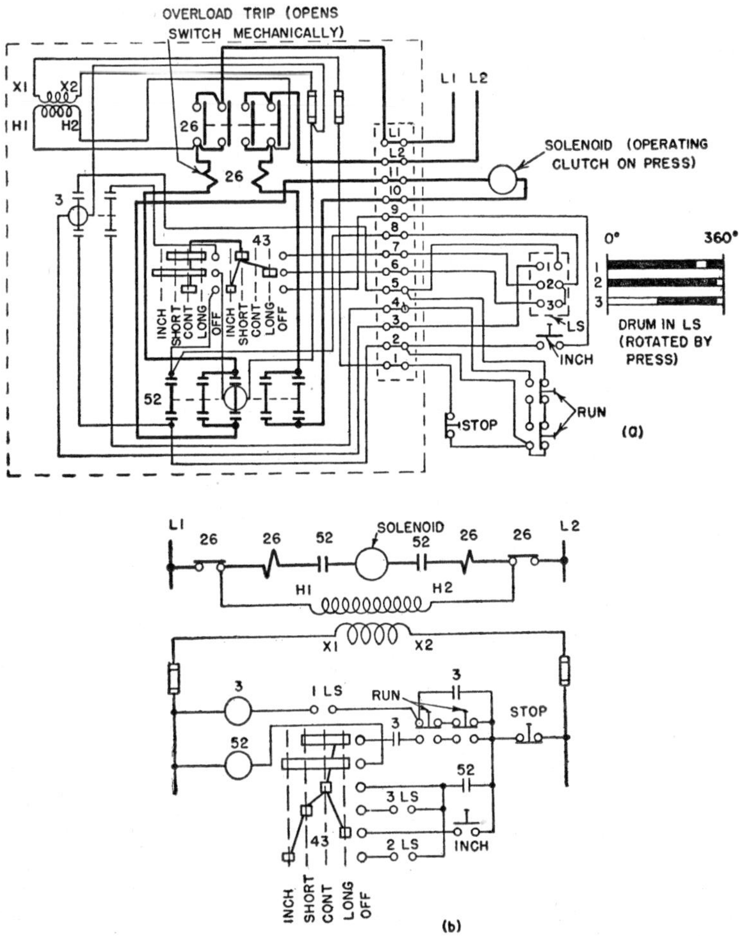

Fig. 22-2

only. The others may be represented by rectangles drawn with broken lines, and should include terminals to which the outgoing wires are connected.

External Wiring Diagrams—For some systems external wiring diagrams are drawn. Such diagrams indicate only how the external connections are made to the controller, motor, and auxiliary equipment. This type of diagram is useful principally when the equipment is being installed.

Simplified Wiring Diagram—It is sometimes difficult to figure out the operation of a controller from the main diagram. This can be done much more easily from a simplified diagram. This diagram generally consists of two vertical lines between which all the circuits are shown in a straight line, usually arranged in the order in which

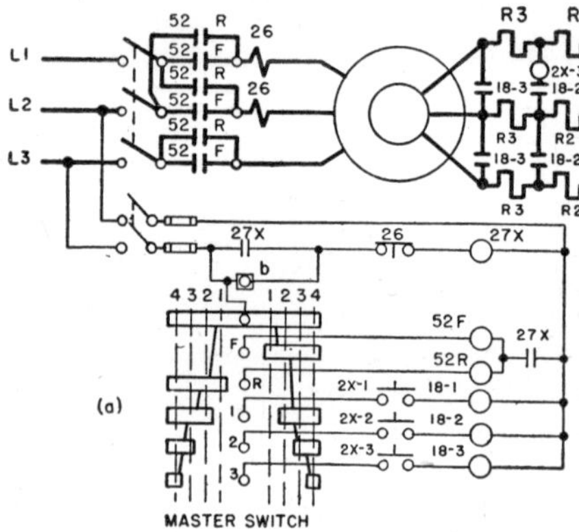

CONTROL SEQUENCE TABLE

CON-TACTS	REVERSE 4	3	2	1	OFF	FORWARD 1	2	3	4
27X					X				
52F						X	X	X	X
52R	X	X	X	X					
18-1	X	X	X				X	X	X
18-2	X	X						X	X
18-3	X								X

(b)

FIG. 22-3

they are energized so that they may be easily traced (see figs. 22-2b and 22-3a). No consideration is given to the actual position of the appliances; their parts are shown in the most convenient place. In simplified drawings the part of an appliance are given the number that appears on this appliance in the main diagram.

Control Sequence Table—A control sequence table is generally included with the more complicated diagrams to make clear the sequence in which the contactors close. A sequence table for the diagram shown in fig. 22-3a is drawn in fig. 22-3b. The first column contains the contactor numbers. The remaining columns indicate the order in which the contactors close for both directions of rotation of the motor. The letter X indicates that a contactor is closed. The vertical columns should be read in each direction from the one marked "off." For instance, this table shows that on moving the master switch to the fourth position in the reverse direction, contactors 52R, 18-1, 18-2, and 18-3 close in sequence.

Symbol Numbers—Due to the fact that all draftsmen do not use the same symbols to designate a certain device, it has become the common practice to place near each symbol a number, a letter, or a number and a letter which have been given a meaning by the National Electrical Manufacturers Association. The meaning of the numbers and letters used in figs. 22-2 and 22-3a are given below.

No.	Meaning	Letters	Meaning
2	Time delay starting or closing relay	F	Forward
18	Accelerating contactor or relay	R	Reverse
26	Thermal relay	S	Solenoid
27	A C under voltage relay	X	Auxiliary device
43	Transfer device		
52	A C circuit breaker or contactor		

Terminal Markings.—Terminals of equipment should be marked, and the markings generally used are given in the following:

Equipment	Abbreviations	Marking at Terminals
Armature	Arm	A1–A2
Brake	Br	B1–B2
Field (series)	Ser fld	S1–S2
(shunt)	Sh fld	F1–F2
Line wires	——	L1–L2–L3
Motor terminals	——	T1–T2–T3
Transformer (high voltage)	Trans	H1–H2–H3, etc.
(low voltage)	——	X1–X2–X3, etc.
(potential)	P. T.	——
(current)	C. T.	——

Manual Controllers—In fig. 22-4a is shown a wiring diagram of a manual controller for a hoisting motor on a crane. The dotted lines indicate additions that may be made. Even though in this controller, the fingers are stationary and the drum rotates, the diagram can be traced more easily by imagining that the drum is stationary and that the fingers move to position 1, then 2, etc.

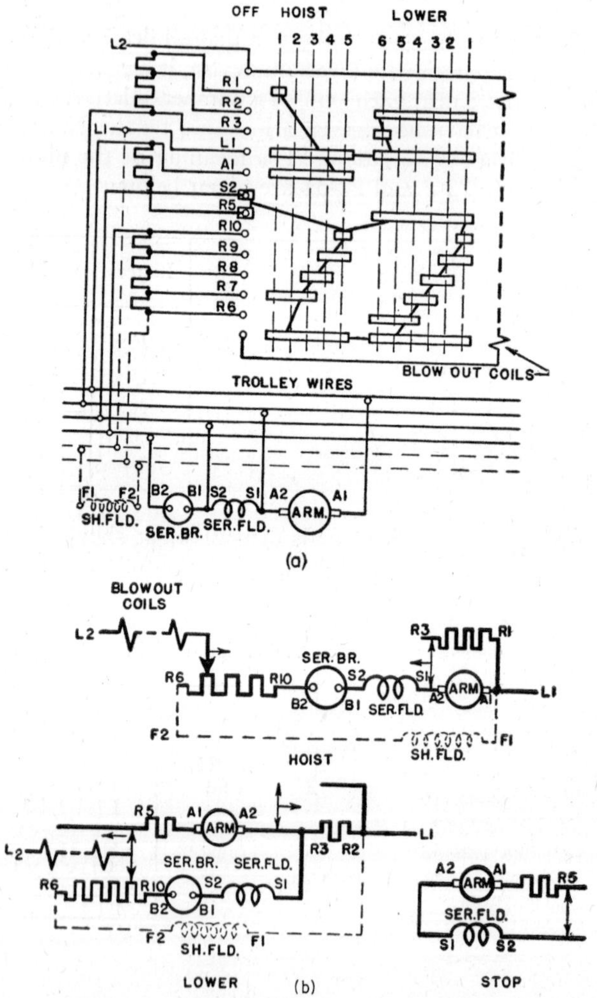

Fig. 22-4

The most convenient method of making simplified diagrams of manual controllers is to draw separate diagrams for each function of the controller. Diagrams illustrating this, for the controller in fig. 22-4a are shown in fig. 22-4b. These diagrams show that the machine operates as a series motor when hoisting the load and as a shunt motor when lowering it.

FIG. 22-5

Symbols for Wiring Plans—In fig. 22-5 are shown symbols for wiring plans, used in making single line diagrams. Ceiling outlets are designated by the same symbols as wall outlets, except that the short lines drawn to the circles are omitted. A standard symbol with a lower case subscript denotes special equipment. A description of such equipment should be given in the specifications on the drawing.

Wiring Plans—To clarify the method of making wiring plans, a typical factory building will be considered. The symbols used are identified in fig. 22-5. A section through this building is shown in

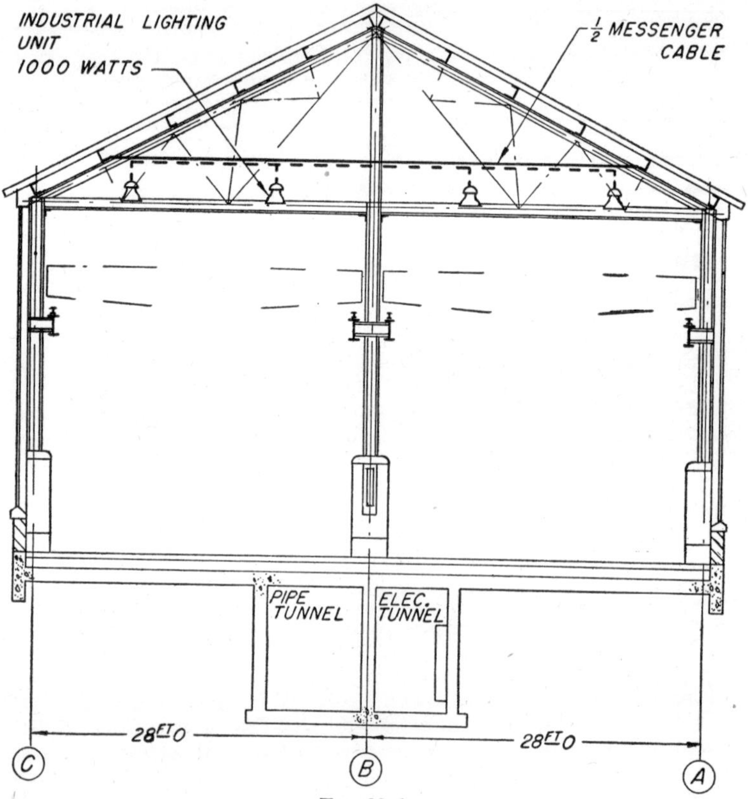

FIG. 22-6

fig. 22-6. The master cabinets for both lighting and power are located in the electric tunnel. This figure shows that the lighting circuits are installed in conduit, which, with the luminaires, are fastened to messenger cables.

The lighting and power systems are generally installed separately, and due to the large number of conduit runs, they are shown in different drawings. Fig. 22-7 shows the location of the lighting outlets and how the conduit runs are made. The arrows point to the cabinets to which the conduit runs connect.

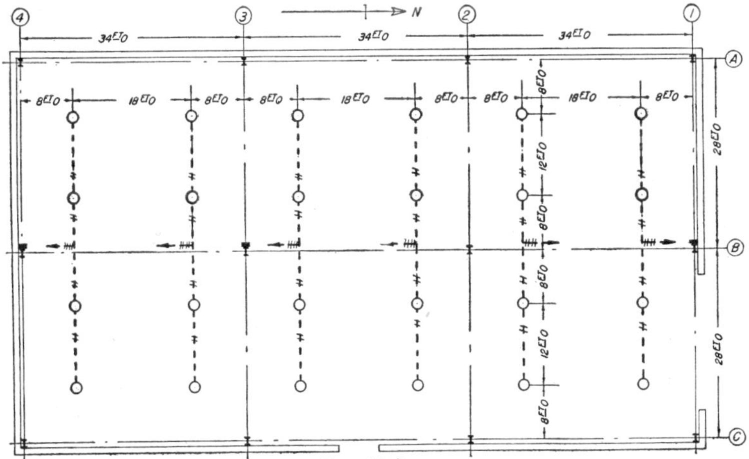

Fig. 22-7

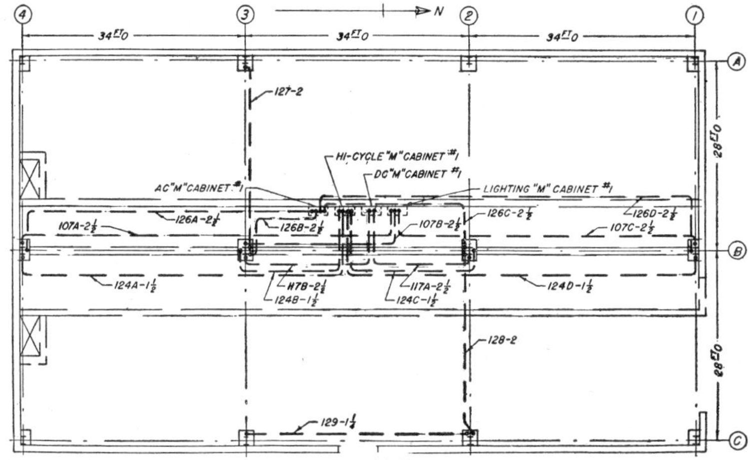

Fig. 22-8

Fig. 22-8 shows the conduit runs between master and sub-cabinets. The cabinet at 3B is built around the column and is divided into four compartments, which make available 110-220 volts and 440 volts at 60 cycles, 220 volts at 180 cycles, and 220 volts D C. In some cases the conduit is shown curved, where actually it is straight. It is drawn

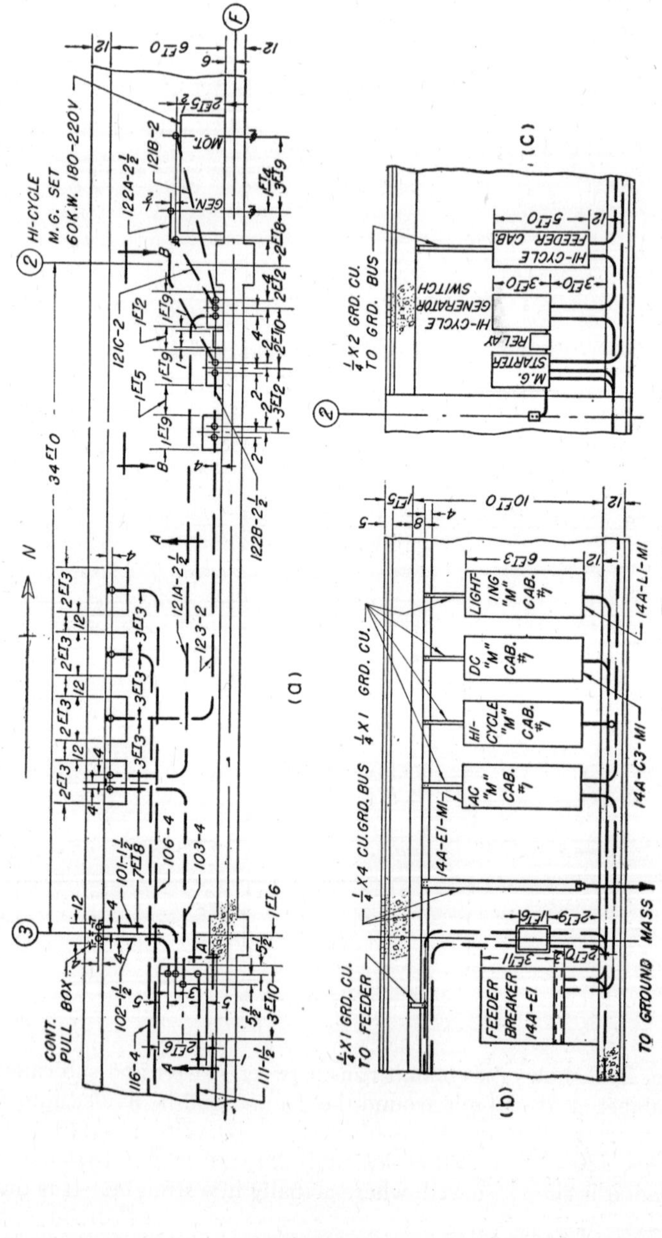

HI-CYCLE
M.G. SET
60K.W. 180-220V
CONT.
PULL BOX
34 FT 0
N
(a)
1/4 X 4 CU.GRD.BUS
1/4 X 1 GRD. CU.
1/4 X 1 GRD. CU.
TO FEEDER
FEEDER
BREAKER
14A-E1
AC "M" CAB. #1
HI-CYCLE "M" CAB. #1
DC "M" CAB. #1
LIGHTING "M" CAB. #1
TO GROUND MASS
14A-E1-M1
14A-C3-M1
14A-L1-M1
(b)
1/4 X 2 GRD. CU.
TO GRD. BUS
M.G. STARTER
RELAY
HI-CYCLE GENERATOR SWITCH
HI-CYCLE FEEDER CAB.
(C)

FIG. 22-9

this way to miss architectural details and to avoid confusing the wires with the architecture.

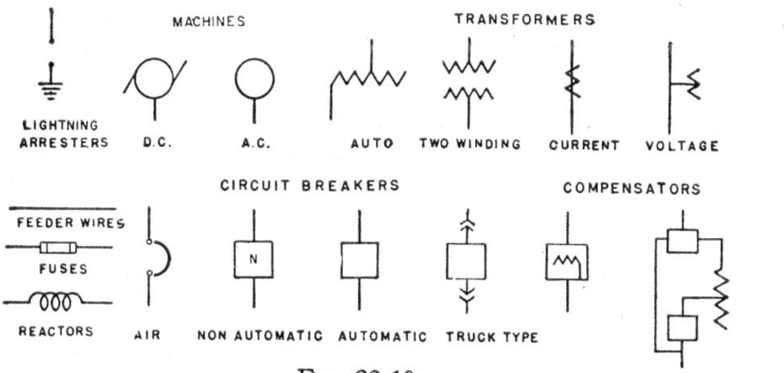

Fig. 22-10

The section of the tunnel that includes the master cabinets is shown in fig. 22-9a. This drawing shows the conduit runs concealed in the

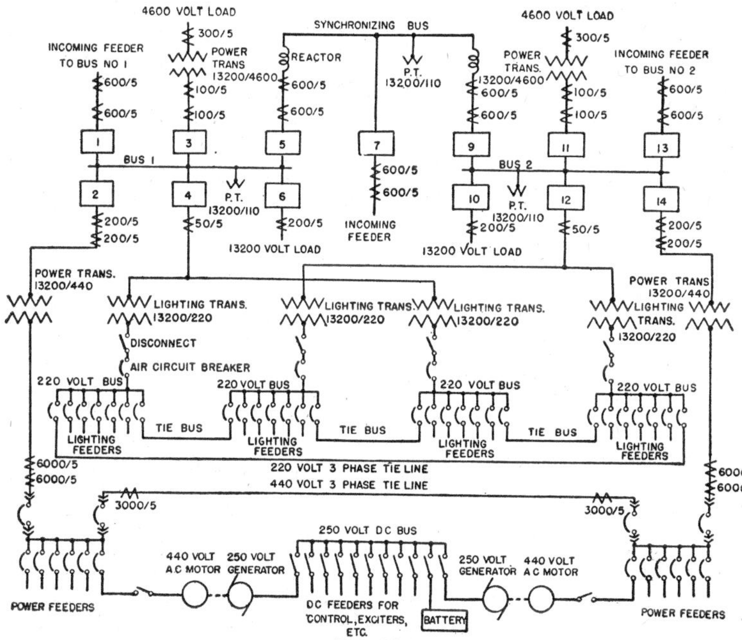

Fig. 22-11

floor of the tunnel. It also includes dimensions that show the exact location of the cabinets and additional equipment. To show more clearly the layout of the equipment in the tunnel, elevation views are drawn in figs. 22-9b and 22-9c for the planes A-A-A and B-B (fig. 22-9a). The arrows at the ends of these planes indicate the direction in which the views are taken. The portion in front of these planes is considered removed. Because it is the standard practice to show the floor at the bottom in a drawing, the sectional view for plane B-B is shown turned 180 degrees.

The tunnel passes under several buildings and contains several groups of cabinets such as the one shown in fig. 22-9a. A feeder (not shown) extends from substation 14A to circuit breaker 14A-E1 (fig. 22-9b). Sub-feeders from this breaker connect to several 440 volt, three phase, master cabinets. One of these cabinets, which is shown in fig. 22-9b, is denoted by 14A-E1-M1. Marking 14A designates the substation from which the power is supplied and E1 indicates the circuit breaker through which this cabinet receives its power (E signifies 440 volts, 3 phase). Marking M1 denotes master cabinet number 1.

If the exact locations of the motors are known, a drawing should be made indicating them. This drawing should also show the conduit runs from the sub-cabinets to the motors.

Conduit Chart—Since it is impossible to give all the required information in the drawings, it is necessary to include a conduit chart. Such a chart, partly filled out for the wiring plans given above, is shown in table 22-I. It gives the conduit sizes, number and size of wire in each, etc.

TABLE 22-I.—CONDUIT CHART

Conduit		Wire				
No.	Size	No. of	Size	Function	From	To
106	4″	3/C	750 MCM	Lighting "M" Cabinet feeder	14A Lighting Feeder cubicle	14A–L1–M1 Lighting "M" cabinet
116	4″	4	500 MCM	DC "M" Cabinet feeder	14A D C feeder Cubicle	14A–C3–M1 D C "M" cabinet

Single Line Diagrams—The layout of power generating and distributing systems is generally shown by a single line diagram. The wires to an instrument or the wires in a power line or feeder are shown by a single line. Symbols used in single line diagrams are shown in fig. 22-10. The method of making a single line diagram of a substation is illustrated in fig. 22-11.

INDEX

Across- the-line starter, 465
Algebra, 1
  addition, 1
  axioms, 9
  cancellation, 6
  curves, 17, 18
  division, 6
  equations, 8, 11, 16
  exponents, 5
  expressions, 4
  factoring, 15
  factors, 3
  members, 8
  multiplication, 2, 3
  parentheses, 14
  powers, 5
  root, 7
  subscripts, 7
  subtraction, 2
Alternating current, 283
Alternating current controllers, 465
Alternators, 375
Alternators, characteristics, 386
  field excitation, 394
  frequency, 377
  hunting, 395
  loading, 393
  panels for, 392
  parallel operation of, 393
  paralleling, 388
  power factor of, 394
  salient pole type, 375
  synchronizing, 390
  three-phase, 361
  turbo type, 376
  two-phase, 359
Alternator windings, 378
  basket type, 385
Alternator windings
  chain type, 384
  concentrated, 380
  distributed, 380
  lap type, 378
  pitch of, 382
  single-phase, 379
  two-phase, 379
Ampere, 24
Ampere-turns, 110
Amplification factor, 545
Amplifiers, 546
Arc welding, 560
  alternating current, 570
  automatic, 564
Armature (see also windings), 170
  reaction in, 173, 176, 203
  testing of, 165
Armature winding, 130
Automatic controllers, 221

Battery (see cells), 43
  care of, 44
  charging rate, 43
  freezing, 45
  resistance, 43
  sulphation, 45
Belts, 53
Benders for conduit, 76
Boxes, 81
Brake, dynamic, 228, 239, 257, 258, 473
  friction or magnetic, 124, 250, 473
  prony, 51
Breaker, 221
British thermal unit, 49
Brushes, 140, 146, 153, 159, 171

Brushes, shifting of, 175
Buck welder, 574
Bus protection, 532
Butt welder, 571
B X (see cable, armored)

Cable, armored, 91
  fittings for, 91
  non-metallic sheathed, 91
Cabinets, 79
Cathode, 548
Cells, 39, 40, 41
Chording, 458
Circuits, parallel, 31, 323
  series, 28, 318
  series parallel, 34, 325
  three-phase, 361
  three wire, 69
  two-phase, 359
Circular mil, 61
Coil groups, 381, 435, 444
Color matching, 556
Commutating poles, 182
Commutation, 179
Compass, 105
Compensating winding, 179, 426. 428
Compensator, 484
  automatic, 486
  for starting several motors, 492
  high voltage, 488
  manual, 484
Compound motor, 211
Condenser, 299
  capacity of, 301, 306
  circuits, 302
  motor, 421
Conductors, 27
  approximate determination of size, 65
  current capacity, 67
  tables, 64, 66, 67
Conduit, 75
  bending, 77
  chart, 598

Conduit
  fittings for, 90
  flexible, 89
  offset in, 78
  reaming, 78
  size of, 87
Constant heat welding, 577
Constant voltage generator, 189, 564
Contact, auxiliary, 525
Contactors, 231
Controller, 216, 465
  across-the-line, 465
  automatic, 221
  counter e m f type, 229
  current limit type, 231
  drum, 219, 221, 470, 496, 500
  face plate type, 217
  for high torque starting, 470
  for synchronous motor, 472
  for wound rotor motor, 482
  four speed, 500
  hoisting, 250
  resistance type (A C), 481
  reversing, 218, 242, 469, 500
  three speed, 497
  time element, 242
  two speed, 492
Controls, 222, 494
Cord, 94
Coulomb, 24, 302
Counter e m f, 201
Counter e m f controller, 229
Current, 24, 283
Current limit controllers, 231
Current overload relay, 226, 466, 522

Damping, 506
Dash pot timing, 243
Definite time relay, 524
Delta connection, 364
Dielectric strength constants, 307
Differential protection for
  alternator and motor, 531

Differential protection for
transformer, 536
Drawing, 587
control sequence table, 590
main diagram, 587
single line diagrams, 599
symbols for wiring diagrams, 587, 591
symbols for wiring plans, 593
symbol number, 591
Drum type controller, 219, 221, 470, 496, 500
Dummy coils, 153, 159, 442
Dynamic braking, 228, 239, 257, 258, 473, 478

Eddy currents, 346, 518
Efficiency, 48, 209, 210, 403, 410
Electrons, 21
direction of flow, 23
Electroplating, 38
Equalizer, 146, 193
Equations, 1, 8, 11, 16

Field relay, 227
Fishing, 85
Fish tape pullers, 87
Fittings, 81, 84
Fixtures, 92, 93
Fleming's left-hand rule, 198
right hand rule, 130
Flux, 105, 110, 173
Flux density, 113
Frequency, 283, 377
Frequency converter, 411

Gage, wire, 65, 67
Galvanometer, 511
Gas filled tubes, 548
Generator (see also alternator), 130, 170
building up of, 186, 190
characteristics, 189
compound, 188
constant voltage, 189, 564
Generator
differential relays for, 534
excitation of, 172, 184
frame, core, and shoes in, 172
paralleling, 192
power to drive, 200
self-excited or shunt, 185
series, 187
welding, 560
Geometry, 260
angle, 260
axioms, 261
corollary, 262
definitions, 261
oblique triangles, 261, 265
parallelogram, 261, 265
right triangle, 261
theorems, 261

Heat, 49
Hickey, 76
Hoisting, 250
Horsepower, 47, 52
of motor, 202
Hunting of alternators, 395
Hydrometer, 45
Hysteresis, 119

Impedance, 319
Induction motor, 397, 418, 453
characteristics, 405
direction of rotation, 401
efficiency, 403
load, 404
rotating magnetic field, 398
rotor current, 402
rotor resistance, 406
slip, 403
speed, 400
starting, 402
stator current, 404
torque, 401, 406
Induction motor types,
adjustable speed, 409
double squirrel cage, 408

Induction motor types
high frequency, 411
multispeed types, 453, 497
constant H P, 456
constant torque, 459
variable torque, 461
squirrel cage, 397, 408
wound rotor, 407
single-phase motors, 418
Inductance, 292, 294, 295, 298
Inductive type relay, 524
Instruments, 504
electrodynamometer type, 505
frequency meter, 521
high frequency, 509
hot wire type, 510
inclined coil type, 508
induction type, 509
iron vane type, 507
megger, 513
permanent magnet type, 504
power factor meter, 519
slide wire bridge, 512
synchroscope, 520
thermocouple type, 510
voltmeter, 505
watthour meters, 514
wattmeter, 506
Wheatstone bridge, 510
Insulation, 94, 513
Inverse time relay, 522
Ions, 38

Lap winding, 132
Law of cosines, 276
Law of sines, 274
Left-hand rule, 198
Lenz's law, 293
Limit switches, 248, 470
Low voltage protection, 225
Low voltage release, 217

Magnetic blowout, 220
brakes, 124, 250, 473
chucks. 123
Magnetic
circuits, 115, 117
clutches, 123
fields, 100, 105, 110
pull, 125
screens, 105
switch, 221
Magnetism, 100
Magnetomotive force, 110
Magnets, 100, 121
Magnet wire, 71
Main wiring diagram, 587
Manual controller, 217, 484, 500, 591
Matching colors, 556
Matter, 21
Megger, 513
Mercury cathode tube, 552
Metallic tubing (for wiring), 89
Meters (see also instruments), 504
ammeter, 24, 504
frequency, 521
power factor, 519
voltmeter, 26, 505
watthour meter, 514
wattmeter, 506
Mil-foot, 62
Millivolt, 505
Mils, 61
Molecules, 21, 41, 103
Motor, 197, 397, 453
adjustable speed, 409
compound, 211
horsepower of, 52, 202
induction, 397
multi-speed induction, 453
series, 209, 425
shunt, 204, 425
single-phase, 418
synchronous, 412
Motor differential relay, 531, 534
Motor rule, 198
Multispeed controllers, 492

N.E.M.A. standards, 591

Ohm's law, 27
Ohm's law for magnetic circuits, 110
Oil switch, 487
Operator switch, 239, 242, 248
Overcurrent relay for differential protection, 532
Overload relays, 226, 466, 522

Paths, 132, 137, 142, 151, 156
Permanent magnet instruments, 504
Permeability, 112
Phase balance relay, 530
Photo cell, 225, 558
Photo tube, 552
Pitch, 134, 135, 149, 382
Plugging, 475
Poles, commutation, 204
Power, 46, 308, 311, 367
  cost of, 332
  reversal of, 528
  to drive generator, 200
Power factor, 314, 319, 328, 394
Power factor meter, 519
Power relay, 528
Pressure, 25
Prony brake, 51
Protection, bus, 532
  low voltage, 225
  overload, 226
Pulley, 53, 124

Raceway, 93
Reactance
  capacitive, 305
  inductive, 296
Reaming, 78
Rectifier, 542
Rel, 110
Relays, 522
  alternator differential, 534
  auxiliary contacts, 525

Relays
  current overload, 226, 466, 522
  definite time, 479, 524
  field, 227
  inductive type, 524
  inverse definite time, 525
  motor differential, 534
  phase balance, 530
  plugging, 475
  power, 528
  tap setting of, 525, 535, 538
  time adjustment of, 527
  transformer differential, 536
  trip circuit of, 523, 525, 528
  voltage, 527
Reluctance, 110
Resistance, 26, 29, 74, 547
  welding, 571
Resistivity, 65
Reversing controller, 218, 242, 469, 500
Rewinding stator, 446
Rotating field, 398, 413
Rotor, 397, 407

Saturation, 113, 183
Scales, Centigrade, and Fahrenheit, 73
Seam welder, 579
Separator, magnetic, 124
Sequence control table, 590
Series motor, 209, 425
Series relay, 238
Shading coil motor, 422
Short circuit in power systems, 529
Shunt motor, 204, 425
Single-phase motors, 418
  repulsion induction type, 423
  series type, 425
  shading coil type, 422
  shunt type, 425
  split phase type, 420
  unity power factor type, 427

Slide wire bridge, 512
Slip, 403
Speed ratio, 53
Split-phase motor, 420
Spot welding, 573
Square mils, 63
Star connections, 362
Starter (see controller)
Stator rewinding, 446
Stator winding, form, hand, and skein, 428
  single-phase, 428
  three-phase, 438
Switch, knife, snap, toggle, 96
  centrifugal, 421, 428
  limit, 248, 470
  magnetic, 221
  operator, 239, 242, 248
Symbol numbers, 591
Symbols, 588, 593, 597
Synchronous motor, 412
  field excitation, 415
  multispeed, 463
  power factor, 418
  starting, 414
  torque, 415
  uses of, 418
Synchroscope, 391, 520

Tap setting in relays, 525, 535, 538,
Temperature control by phototube, 557
Temperature scale, Centigrade and Fahrenheit, 73
Thermionic tubes, 541
Three-phase circuits, 361
Thumb rule, 107, 109
Time element controllers, 242
  dash pot type, 243
  inductive type, 247
  pendulum type, 244
  pilot motor type, 245
Torque, 52, 401, 406, 415, 455
Torque
  in instruments, 506, 507, 517, 525
  of motors, 199, 202, 206
Transformer, 337
  auto, 350
  construction, 340
  cooling, 346
  current, 351
  design, 342
  differential relay, 536
  losses, 346
  polarity, 369
  potential (or voltage), 351
  tap changer, 348
  three-phase, 370
Trigonometric functions, 269
Trigonometry, 269
  component of vectors, 280, 281
  law of cosines, 276, 277, 278
  law of sines, 274
  oblique triangles, 261, 265
  resultant of vectors, 279
  right triangles, 267
  table, 270
Tubes, cold cathode, 552
  gas filled, 548
  hot cathode, 550
  mercury cathode, 552
  photo electric, 552
  rectifier, 542
  vacuum, 541
Two-phase circuits, 359

Underwriter's knot, 95
Unity power factor motor, 427

Vacuum tubes, 541
Voltage, 25, 68
  generation of, 160
  in rotating coil, 131
Voltage relay, 527
Voltmeter, 505

Ward-Leonard system, 257
Watt, 47
Watthour meter, 514
Wattmeter, 506
Wave winding, 147
Weber's theory, 103
Welder, arc, 560, 564
  buck, 574
  butt, 571
  constant heat, 577
  generator, 560
  resistance, 571
  seam, 579
  spot, 573, 575
Welding control, electron tube, 584
Wheatstone bridge, 510
Winding, basket, 385
  chain, 384
  changing of, 446
  concentrated and distributed, 380
  double layer, 134, 435
  double voltage, 441
Winding, form, 134, 429
  frog leg, 159
  hand, 428
  lap, 132, 378
  multiplicity, 134, 141, 155
  parallel connections, 437, 441
  reentrancy, 143, 157
  selection of, 160
  single-phase, 379
  skein, 430
  split, 162
  three-phase, 438
  two-phase, 379, 436
  unequal coil grouping in, 442
  V-type, 164
  wave, 147
Wire gage, 66
Wire, magnet, 71
Wire size, 67, 118, 126
Wire tables, 64, 66, 67
Wiring methods, 75
Wiring plan, 594
Work, 46, 49